Infiltration Measurements for Soil Hydraulic Characterization

Rafael Angulo-Jaramillo • Vincenzo Bagarello
Massimo Iovino • Laurent Lassabatere

Infiltration Measurements for Soil Hydraulic Characterization

Springer

Rafael Angulo-Jaramillo
Université Claude Bernard Lyon 1,
CNRS, ENTPE, UMR5023
Ecologie des Hydrosystèmes
Naturels et Anthropisés
University of Lyon
Vaulx-en-Velin, France

Vincenzo Bagarello
Scienze Agrarie e Forestali
University of Palermo
Palermo, Italy

Massimo Iovino
Scienze Agrarie e Forestali
University of Palermo
Palermo, Italy

Laurent Lassabatere
Université Claude Bernard Lyon 1,
CNRS, ENTPE, UMR5023
Ecologie des Hydrosystèmes
Naturels et Anthropisés
University of Lyon
Vaulx-en-Velin, France

ISBN 978-3-319-81118-5 ISBN 978-3-319-31788-5 (eBook)
DOI 10.1007/978-3-319-31788-5

Softcover reprint of the hardcover 1st edition 2016

Printed on acid-free paper

This Springer imprint is published by Springer Nature
The registered company is Springer International Publishing AG Switzerland

We listed the authors of this book in alphabetic order since each of us gave a similar contribution to the outcome of the project. Speaking, listening and understanding was easy thanks to the team spirit that moved our work, and our cooperation was pleasant and fruitful at any moment.

To my beloved wife Anne and our children Adrien, Coralie, Victor, Simon and Jeanne whose affection gives me support and energy.
R.A.J.

To Rosalba, Sara and Sergio. I couldn't wish for anything better from my family.
V.B.

To my wife, Maria, and my children, Marta and Paolo, for their endless support.
M.I.

To my family my brother, Thierry, my parents, Maité and Louis, my uncles, Albert and Jean-Baptiste, and to Lea for accompanying me. To water, without which none of this could have existed.
L.L.

To all readers of this book. We hope they will enjoy their reading and find useful information to address their environmental studies and solve practical problems.
R.A.J., V.B., M.I., L.L.

Foreword

The soil and plant-based ecosystems which cloak the lands of our earth are the planet's critical zones. They provide valuable ecosystem services. Through these soil and plant systems on the earth's surface, there are massive fluxes and storages of mass and energy. These flows and storages provide valuable ecosystem goods and services. Water is the prime natural capital stock. We are vitally dependent on the myriad of ecosystem services that water delivers to us via our soils.

Over the last 50 years, we have rapidly developed theories and models of the biophysical functioning of our coupled soil, water and plant systems. This quantitative modelling is a valuable means by which we can organise our hard-won biophysical knowledge so that it can be applied to manage better our productive ecosystems. Our scientific understanding can be used for developing policies, implementing actions and monitoring outcomes to protect our natural capital stocks and maintain the ecosystem services that flow from them. There is a strong end-user pull for scientists and engineers to develop better models and sophisticated decision-support tools to enable sustainable management of the soil, plant and water systems that we are critically dependent upon.

Despite this comprehensive theoretical knowledge of the flows and storages of water in our diverse soil-plant-atmosphere system, we are still somewhat hampered by our inability to measure those parameters which our modelling schema is reliant upon for their operation. That is because the flows of water and energy in these systems are governed by a complex of linked and highly non-linear processes. Simply put, 'it is difficult' to obtain measurements of those parameters that our models rely on. Or as the French physicist Joseph Fourier (1768–1830) said more elegantly '... nature is indifferent to the difficulties it causes mathematicians'.

But thankfully over the last couple of decades, we have seen the development of many new measurement techniques and the creation of innovative devices that have improved the acuity of the vision by which we can realise useful hydraulic characterisations of our soils. These characterisations are resulting in better modelling and enhanced predictions of the flows of water through the soils of the root zones of our plant ecosystems and of the fluxes through the vadose zone towards our groundwater reserves.

Given the rapidity of these empirical developments, and the diversity of these new measurement methods and infiltration devices, it is timely that there now be a compendium of the techniques that are now available for soil hydraulic characterisation. Hitherto, no one book has provided this.

In this volume, Drs Angulo-Jaramillo, Bagarello, Iovino and Lassabatere have done exactly that. They have written a comprehensive and detailed book on infiltration measurements for soil hydraulic characterisation. The book is a tour de force. It will appeal to a broad ambit of soil scientists and engineers: from those who revel in the equations that describe water flow and storage in porous media through to those who simply wish to know how to use measurement devices in the field and to read about learnt tricks for successfully operating them.

The book begins with a full and detailed description of the theories that govern water flow through and storage in soil. This is mathematically detailed, but with descriptions in the text to explain what is being modelled.

The second chapter focuses on methods to measure the critical parameter of the soil's saturated hydraulic conductivity, K_s. The techniques described include the use of wells and bores, plus rings and drippers. Not only is the underpinning theory described, practical tips and worked examples are also provided. Throughout the book, field data are used to show how the parameter values are computed, with actual results from such diverse places as Burundi and Sicily.

The third chapter describes methods for measuring the hydraulic properties of unsaturated soil using either the tension (disk) infiltrometer – in all of its guises; or BEST – the Beerkan Estimation of Soil Transfer parameters. There is a comprehensive bibliography to an exceptionally wide range of published works of field measurement of unsaturated properties, along with 11 tables, and some 46 figures or 'how to' photographs. The chapter concludes with a practical section on how to use the properties determined with the BEST technique.

Of course, as Fourier alluded to, not all soils comply with the isotropy and wettability that our theories demand. So the fourth chapter is dedicated to measurement techniques for soils that provide us with additional challenges such as water repellency, surface crusts, exceptionally low permeabilities, and macropores. Fourier would be impressed.

The book concludes with an Appendix on additional measurement techniques for characterising the soil's bulk density, its water content and matric potential, the soil's water retention curve, plus the water stability of the soil's aggregates.

As we increasingly rely on the mechanistic modelling of soil and water processes for developing policies and practices to sustain our natural capital stocks of soil and water, we need to have confidence in the parameter values that drive these modelling schemes. This book outlines the suite of infiltration measurement techniques that can be used to characterise the soil's hydraulic properties. This book is an essential companion for all soil scientists and environmental engineers.

Plant and Food Research
Palmerston North, New Zealand

Brent Clothier

Contents

Chapter 1
An Introduction to Soil and Water Infiltration

1.1 General Characteristics of Soil

Soil is a three-phase system that includes a solid, a liquid and a gaseous phase (Hillel 1998). The solid particles of soil originate from climatic and biological processes. Exposed rocks are physically and chemically disintegrated into small fragments. Some primary materials (e.g., quartz) resist chemical change and retain their original character even in a fragmented form (sand). Other more reactive minerals (feldspars or micas) are reconstituted after decomposition into different secondary minerals (clays). Organic matter provided by living organisms reacts with the mineral component. Soil is also a disperse, heterogeneous and porous system, susceptible to energy (heat) and matter (water) exchanges with the atmosphere.

A soil profile generally consists of a succession of horizons. The upper layer or O horizon mainly contains organic material either undecomposed or subject to decomposition processes. The eluvial A horizon has more organic matter and a darker color than the underlying soil and it is the zone of major biological activity. Clay and carbonates leached from the A horizon accumulate into the illuvial B horizon. Weathered and fragmented rock material constitutes the C horizon.

The soil matrix consists of particles varying in chemical composition, size and shape. Organic matter may bind the mineral grains in assemblages called aggregates. Water in the soil always contains dissolved substances. The gaseous phase is in contact with the atmosphere above the soil surface. Soil is disperse since the solid phase is subdivided into numerous particles. The consequent interfacial activity gives rise to adsorption of water and chemicals, capillary, ion exchange, swelling and shrinking, dispersion and flocculation. Soil is also heterogeneous because its properties differ not only between one phase and another, but also between the internal parts of each phase and the edges or interfaces along which the phase comes into contact with a neighboring phase or phases. The structure of the soil matrix determines the geometric characteristics of the pore spaces.

R. Angulo-Jaramillo et al., *Infiltration Measurements for Soil Hydraulic Characterization*,
DOI 10.1007/978-3-319-31788-5_1

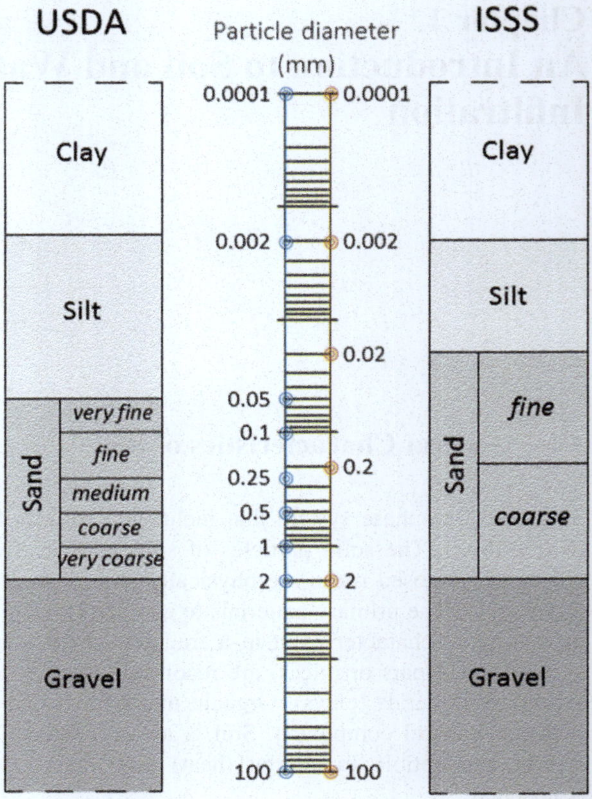

Fig. 1.1 Particle-size limits according to the USDA and ISSS soil classification schemes

The unsaturated zone of the soil, sometimes called the vadose zone, is the zone between the ground surface and the water table and represents the hydrological connection between the surface water component of the hydrologic cycle and the groundwater component (Haverkamp et al. 2006).

The term soil texture refers to the size range of the elementary soil particles and it denotes an intrinsic soil property that can be measured objectively. Conventionally, soil includes elementary particles of less than 2 mm in diameter whereas larger particles are generally referred to as gravel. The smallest soil particles are named clay, the largest particles constitute the sand and the particles of intermediate size are defined silt. In general, sand particles have quite jagged surfaces but can be represented as spherical. Silt particles are mineralogically and physically similar to sand particles but they are smaller, implying that silt has a greater surface area per unit mass. Clay particles, constituting the colloidal fraction, are plate-like or needle-like and they exhibit a very noticeable surface area per unit mass. The particle size limits of the three textural fractions vary with the adopted classification scheme (Fig. 1.1). The most common schemes are those introduced by the USDA (U.S. Department of Agriculture) and the ISSS (International Soil Science Society).

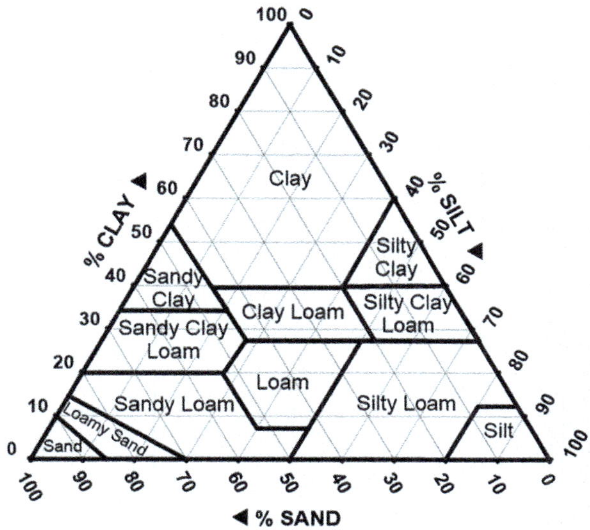

Fig. 1.2 USDA soil textural classification triangle

In both cases, the clay fraction includes particles not larger than 0.002 mm. Sand is defined as particles ranging in diameter from 2 mm to 0.05 mm according to the USDA classification and to 0.02 mm in the ISSS classification. Therefore, silt particles range in size between 0.05 and 0.002 mm and between 0.02 and 0.002 mm according to the USDA and ISSS schemes, respectively.

The clay, silt and sand percentages can be used to determine the soil textural class by the textural triangle that differs with the classification scheme of the three soil fractions (Fig. 1.2). Each side of the triangle corresponds to a particular textural fraction. For a given fraction, a straight line starting from the percentage of that fraction and parallel to the previous side, moving clockwise, is drawn. It is enough to carry out the procedure for two textural fractions. The intersection point of the two lines falls in a zone of the triangle identified by a particular textural class (e.g., silty-clay-loam), that is attributed to the soil sample.

More information on soil particle sizes is contained in the particle size distribution (PSD) curve (Fig. 1.3). The ordinate of the PSD graph indicates the percentage by weight of soil particles having diameters smaller than or equal to the diameter reported in the abscissa on a logarithmic scale. The PSD curve allows us to establish if the soil is composed of distinct groups of particles each of uniform size or of a more or less continuous array of particle sizes. In poorly graded soils, particles of one or several distinct sizes prevail. Well graded soils present a flattened and smooth distribution curve. The sand, silt and clay percentages are easily determined by reading on the PSD curve the percentages corresponding to the relevant diameters.

The measured PSD can be fitted on a PSD model, such as the one by Haverkamp and Parlange (1986) (Fig. 1.4):

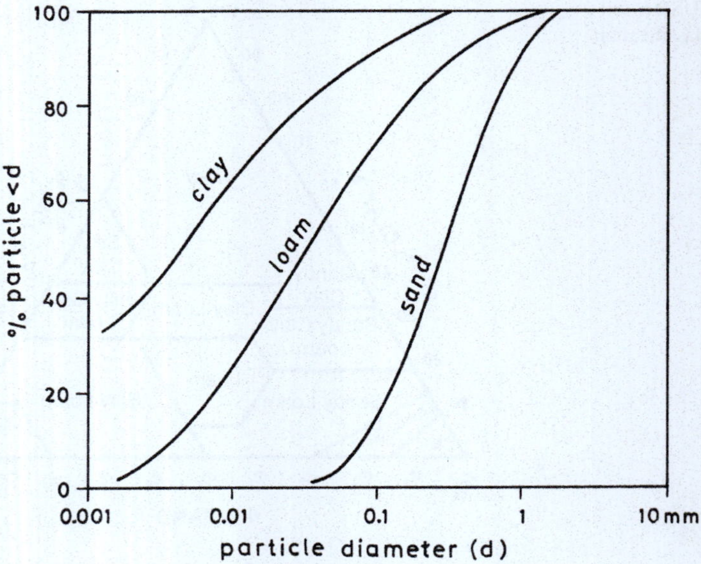

Fig. 1.3 Particle size distribution curves for various soil types

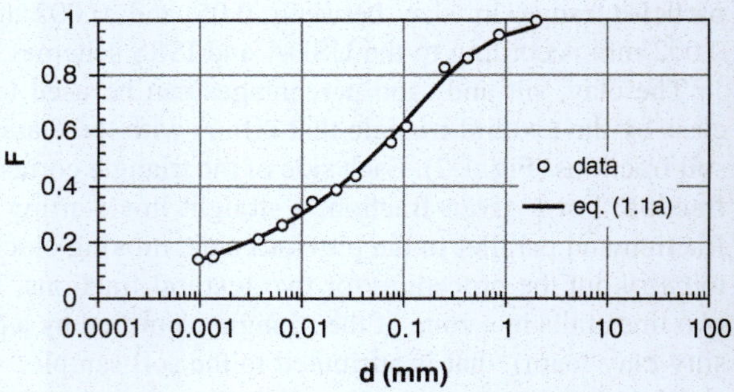

Fig. 1.4 Fitting of the measured particle size distribution on the theoretical model by Haverkamp and Parlange (1986) (From Bagarello and Iovino 2010)

$$F(d) = \left[1 + \left(\frac{d_g}{d}\right)^N\right]^{-M} \tag{1.1a}$$

$$M = 1 - \frac{1}{N} \tag{1.1b}$$

where $F(d)$, varying from 0 to 1 (or 0 to 100 % if expressed as a percentage) is the fraction by mass of particles passing a particular size, d (L), d_g (L) is a scale parameter and M and N are shape parameters. Parameters of Eq. (1.1a) can be determined by common least squares fitting techniques. Equation (1.1a) with $M = 1 - 2/N$ was used in the so-called BEST procedure of soil hydraulic characterization by Lassabatere et al. (2006), described in detail in Chap. 3. The PSD model by Fredlund et al. (2000) has been shown to perform well for most soil textural classes (Hwang et al. 2002; Hwang 2004; Bagarello et al. 2009):

$$F(d) = \frac{1}{\left[\ln\left(\exp(1) + \left(\frac{\alpha_f}{d}\right)^{n_f}\right)\right]^{m_f}} \left\{1 - \left[\frac{\ln\left(1 + \frac{d_f}{d}\right)}{\ln\left(1 + \frac{d_f}{d_m}\right)}\right]^7\right\} \quad (1.2)$$

in which d_m (L) is the diameter of the minimum allowable particle size, and d_f (L), α_f (L), n_f and m_f are fitting parameters. Fredlund et al. (2000) suggested to use a constant d_f value, equal to 0.001 mm. In other investigations, d_f was considered a fitting parameter but d_m was set at 0.001 mm (Hwang et al. 2002) or 0.0001 mm (Hwang 2004). Bagarello et al. (2009) set d_f at 0.001 mm and d_m at 0.0001 mm.

For a qualitative assessment of the textural characteristics of a soil, an attempt to prepare a donut is made using a stick of moistened soil of approximately 3 mm in diameter and 50–70 mm in length. The donut can easily be prepared if the soil contains at least 40 % of clay. The stick breaks easily if the soil is medium textured. A high sand content impedes preparing the stick.

Aluminosilicates are the most prevalent clay minerals in soils of temperate regions whereas hydrated oxides of iron and aluminum prevail in tropical regions. Therefore, clay particles differ mineralogically from silt and sand particles, mainly consisting of quartz. Clay is the fraction most influencing soil behavior since soil particles exhibit a great surface area per unit mass which implies a noticeable physico-chemical activity. Clay particles cause the soil to swell with wetting and to shrink with drying.

Two basic structural units compose the laminated microcrystals of the typical aluminosilicate clay minerals: a tetrahedron of four oxygen atoms surrounding a central cation, usually Si^{4+}, and an octahedron of six oxygen atoms or hydroxyls surrounding a somewhat larger cation of lesser valency, usually Al^{3+} or Mg^{2+}. In 1:1 layered aluminosilicate clay minerals, such as kaolinite, an octahedral sheet is attached by the sharing of oxygens to a tetrahedral sheet. In 2:1 minerals, such as montmorillonite, the octahedral sheet is attached in the same way to two tetrahedral sheets, one on each side. A clay particle is composed of lamellae, i.e. multiply stacked composite layers of this sort. The described structure is idealized. Typically, isomorphous replacements occur during crystallization, which means that some ions are substituted by other ions of approximately equal radii. In particular, Al^{3+} may take the place of Si^{4+} in the tetrahedral sheets and Mg^{2+} may take the place of Al^{3+} in the octahedral layer. Consequently, internally unbalanced negative charges occur at different sites in the lamellae and they are compensated by the adsorption of cations which concentrate near the external surfaces of the clay particle and occasionally penetrate into the interlamellar spaces. These adsorbed cations (Na^+, K^+, H^+, Mg^{2+}, Ca^{2+} and Al^{3+}) can be replaced by other cations in the soil solution. The cation exchange phenomenon affects retention and release of nutrients and other salts as well as flocculation-dispersion of soil colloids.

In a dry state, the neutralizing counter-ions are attached to the surface of the colloidal particle. In a hydrated colloidal particle, the adsorbed ions are spatially separated to a greater or lesser extent from the negatively charged particle. The term electrostatic double layer is used to denote the particle surface and the swarm of cations hovering about it. The double layer is thinner in dry than in wet conditions and cations are much more concentrated inside the double layer than in the external solution, i.e. outside the range of influence of the electrostatic force field of the particle. In addition, increasing valency of cations and concentration of the solution determine compression of the diffuse double layer. The greater the ion's hydration, the farther it is from the adsorbing surface and the weaker its adsorption. Replacement is easier for monovalent cations, that are attracted by a single charge, than divalent or trivalent cations. The cation exchange capacity depends on the clay type, and it is smaller for the kaolinite than the montmorillonite.

Prevalence of highly hydrated monovalent cations, such as Na^+, in the solution favors clay dispersion. Clay flocculation is favored by high solution concentrations and the prevalence of divalent and trivalent cations (i.e., Ca^{2+}, Al^{3+}), that reduce the thickness of the electrostatic double layer. In this case, two clay platelets are close together and their counterions form a unified layer of positive charges, attracting the negatively charged particles on both sides of it. When a dispersed clay dries, a dense and hard mass develops. Drying of a flocculated clay produces an ensemble of small aggregates.

Soil behavior depends on soil matrix or soil fabric, that denotes the manner in which the various particles are packed and held together to form a continuous spatial network. Soil structure denotes the arrangement of the particles in the soil, i.e. the internal configuration of the soil matrix. The forces acting within an aggregate are stronger than the forces acting between aggregates. Soil structure varies both in space and time. Soil structure depends on climate, biological activity and soil management practices, and it is susceptible to mechanical and physicochemical destructive forces. Soil structure affects total porosity as well as the shape and the size of the pores. Therefore, it influences the content and transmission of both air and water in the porous medium. Most coarse-textured soils have a single-grained structure because individual grains do not tend to form aggregates. If the clay content is appreciable, the primary particles can group themselves into aggregates. The visible aggregates have generally diameters of several millimeters to several centimeters and are often called macro-aggregates (>250 μm in diameter according to Tisdall and Oades 1982). Usually, a macro-aggregate is an assemblage of smaller groupings, or micro-aggregates (20–250 μm). In a micro-aggregate, the ultimate structural units, mainly composed by clay particles, are associated. Bundles of the latter units attach themselves to, and sometimes engulf, the larger primary particles of sand and silt.

Aggregate formation presupposes clay flocculation and the action of cementing agents. Organic matter must be supplied continually to steadily maintain aggregate stability because humic cements are susceptible to microbial decomposition. Soil micro-organisms bind aggregates by different mechanisms, such as adsorption, physical entanglement and envelopment, and cementation by excreted mucilaginous products. Some organic material is hydrophobic and this circumstance increases aggregate stability due to reduced wettability and swelling phenomena. Cementing agents can also have an inorganic nature. Cohesiveness between clay particles is the ultimate internal binding force within micro-aggregates. Calcium carbonate and iron and aluminum oxides increase aggregate stability.

Soil structure controls total soil porosity and pore size distribution. A distinction is frequently made between macropores and micropores. Macropores are mostly the inter-aggregate cavities, that control infiltration and drainage of water and aeration. The micropores are the intra-aggregate capillaries that control retention of water and solutes. However, the differentiation between macropores and micropores is often arbitrary. Cavazza (1981) suggested a limit of separation between macropores and micropores of 0.02 mm, that is the approximate diameter of the largest pores that remain full of water when the soil water content corresponds to the so-called field capacity condition. When empty of water, macropores act as a barrier to capillary flow, permitting only very slow film-creep along their walls. When filled with water, however, macropores permit very rapid, and often turbulent, flow.

Aggregate stability is a measure of the ability of soil aggregates to resist externally imposed destructive forces. This ability depends on the soil but also on the nature of the destructive force and the manner in which it is applied. Therefore, aggregate stability is not an intrinsic soil property. Aggregate stability to water has a noticeable practical interest. Wetting an aggregate may determine its collapse due to weakening of the bonding substances and clay swelling and possibly dispersion. Swelling is a slow process, that increases as the soil solution concentration decreases. Dispersion only occurs for concentrations lower than the flocculation value and it is a rapid process (Keren and Ben-Hur 2003). Non-uniform wetting may induce differential swelling of the aggregate. The resulting stress, compounded during subsequent shrinkage, may fracture the aggregate. A practically instantaneous wetting of an initially dry aggregate may determine slaking of the aggregate that can be particularly destructive. Water traps and compresses the originally present air. The cohesive strength of the outer part of the aggregate is reduced by swelling and the pressure of the entrapped air builds up in proportion to its compression. Eventually, the latter may exceed the former and the aggregate may actually explode. An increase in the exchangeable sodium percentage (ESP) promotes aggregate dispersion during wetting, particularly in presence of swelling clay and in the absence of other cementing agents. Abu-Sharar et al. (1987) suggested that soil hydraulic conductivity decreases if dispersion is associated with slaking since the dispersed clay particles alone cannot occlude the large conductive pores when swelling-shrinking phenomena play a minor role.

The volume of solids can generally be considered constant, whereas the volumes of the liquid and gaseous phases vary in time. Figure 1.5 shows a schematic representation of the relative volumes and masses of the three phases for a hypothetical soil. The mass of air, M_a, can be neglected in comparison to the masses of solids, M_s, and water, M_w. Total mass, M_t, is thus given by the sum of M_s and M_w. The sum of the air, V_a, and water, V_w, volumes gives the volume of pores,

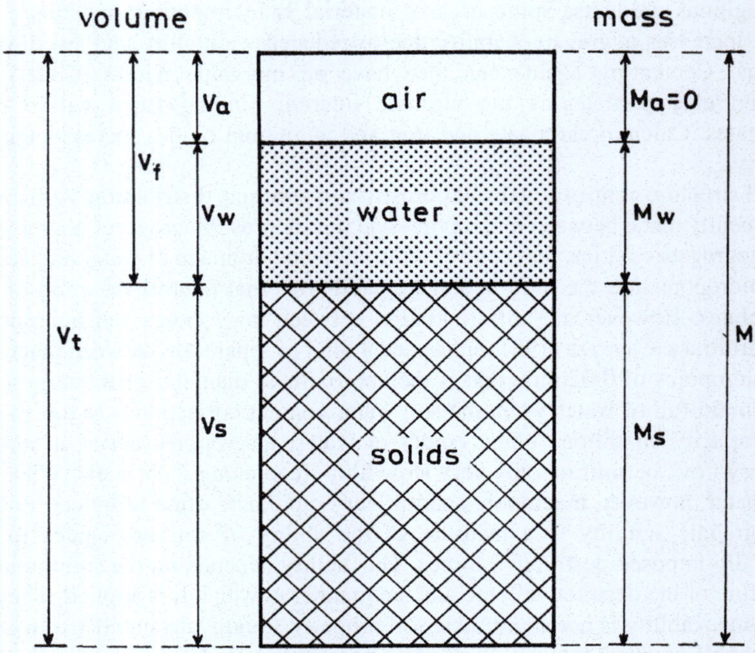

Fig. 1.5 Schematic diagram of the soil as a three-phase system

$V_f = V_a + V_w$. The volume of solids is V_s and the total volume of the sample is V_t. Different soil physical properties can be defined on the basis of this representation. In particular, density of solids, ρ_s (M L^{-3}), is given by:

$$\rho_s = \frac{M_s}{V_s} \tag{1.3}$$

For most mineral soils, ρ_s is typically constant and equal to 2600–2700 kg m^{-3}. Dry soil bulk density, ρ_b (M L^{-3}), is the ratio between the mass of solids and the total, or bulk, soil volume:

$$\rho_b = \frac{M_s}{V_t} = \frac{M_s}{V_s + V_a + V_w} \tag{1.4}$$

Obviously, ρ_b is always smaller than ρ_s. Total soil volume should be measured at a soil water pressure head of −1 m to account for possible soil shrink-swell behavior (Reynolds et al. 2009). The dry bulk density is affected by the soil structure and it may vary from 1600 kg m^{-3} for sandy soils to less than 1200 kg m^{-3} for aggregated loam and clay soils. Soil bulk density is used to calculate porosity if ρ_s is known, to convert weights to volumes and to estimate weights of soil volumes too large to weigh (Culley 1993). It is also an indicator of soil physical quality (Reynolds et al. 2007). Total porosity, f (L^3L^{-3}), is an index of the pore space in a soil:

$$f = \frac{V_f}{V_t} = \frac{V_a + V_w}{V_s + V_a + V_w} \tag{1.5}$$

Generally, f ranges from 0.3 to 0.6. Coarse-textured soils tend to be less porous than fine-textured soils, but individual pores are larger in the former soils than the latter ones. Size and shape of the soil pores have a noticeable impact on flow transport processes but f only gives an information on total pore space. Porosity and dry soil bulk density are related by the following relation:

$$f = 1 - \frac{\rho_b}{\rho_s} \tag{1.6}$$

When small stones and porous coarse fragments (diameter > 2 mm) occur in the sample, the dry soil bulk density measured over the whole soil sample (ρ_b) is different from the dry bulk density of the fine soil fraction (ρ_{bf}) (Haverkamp et al. 2006). The correction formula is:

$$\rho_{bf} = \frac{1 - a - b}{1/\rho_b - a/\rho_s - b/\rho_c} \tag{1.7}$$

where ρ_c is the density of the porous coarse fragments, and a and b are the stone and porous rock weight fractions, respectively, of the whole soil sample.

The specific surface is the total surface area of particles per unit mass or unit volume of particles or per unit bulk volume of soil. Specific surface depends on both size and shape of the soil particles. Flattened or elongated particles expose greater surface per unit mass or volume as compared with equidimensional (e.g., cubic or spherical) particles. The high specific surface of the clay particles depends on both their small size and their shape, since clay particles are generally platy.

1.2 Soil Water Content and Potential

The soil water content is commonly expressed relative to the mass of solids or to the total volume. The mass wetness, w ($M\ M^{-1}$), is the mass of water relative to the mass of dry particles:

$$w = \frac{M_w}{M_s} \tag{1.8}$$

The volume wetness, θ (L^3L^{-3}), or volumetric water content, is the volume of water relative to the total volume of the soil:

$$\theta = \frac{V_w}{V_t} \tag{1.9}$$

Calculations on a volumetric basis are directly applicable to the computation of fluxes and water volumes added to or extracted from the soil. The volumetric water content at saturation, θ_s (L^3L^{-3}), in theory equal to the porosity, varies approximately from 0.40 m^3m^{-3} in sandy soils to 0.60 m^3m^{-3} in clayey soils. Due to the presence of entrapped air, the field-saturated soil water content, θ_{fs} (L^3L^{-3}), is generally lower than porosity. For example, Rogowski (1971) reported $\theta_{fs} = 0.9 \times f$. The residual water content, θ_r (L^3L^{-3}), is the wetness of an air-dried soil. A drier condition, i.e. V_w approaching zero, implies placing the sample in an oven at a temperature higher than 100 °C. The physical meaning of θ_r is ambiguous (Haverkamp et al. 2006). Conceptually, the residual water content may be associated with the immobile water held by adsorptive forces within a dry soil profile in films on particle surfaces, in interstices between particles, and within soil pores. In practice, however, its value is generally estimated by fitting an equation to measured water retention data which reduces θ_r to an empirical fitting parameter restricted to the range of data points used. The relation between mass and volume wetness is:

$$\theta = w\frac{\rho_b}{\rho_w} \tag{1.10}$$

where ρ_w ($M\ L^{-3}$) is the density of water, equal to 1000 kg m^{-3} at standard temperature and pressure. In most hydrologic applications, volumetric soil water content is used in non-dimensional form (S_e):

$$S_e = \frac{\theta - \theta_r}{\theta_s - \theta_r} \tag{1.11}$$

When volumetric water content is expressed as the ratio to total pore space, it is referred to as the degree of saturation, S_d:

$$S_d = \frac{V_w}{V_w + V_a} = \frac{\theta}{f} \tag{1.12}$$

A few basic concepts, concerning water adsorption, surface tension, contact angle and capillary, need to be introduced before describing the soil-water relationships (Hillel 1998).

The adsorption of water on solid surfaces is an interfacial phenomenon, generally of electrostatic nature. The polar water molecules attach to the charged faces of the solids and the ions adsorbed on them. This mechanism causes the strong retention of water by clay soils.

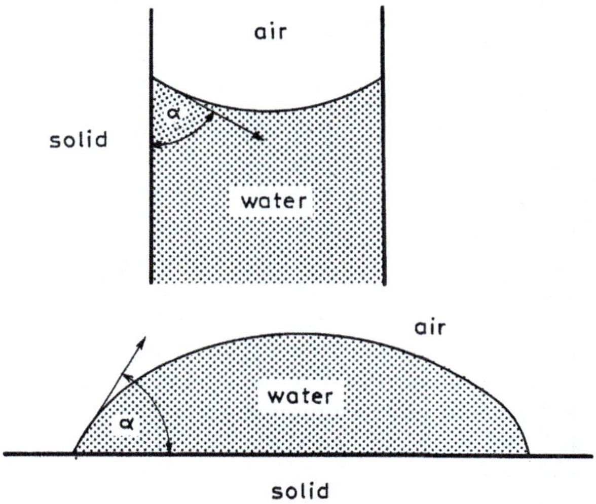

Fig. 1.6 The contact angle of a meniscus in a capillary tube and a drop on the surface of a solid

Surface tension occurs typically, but not exclusively, at the interface of a liquid and a gas. The liquid behaves as it was covered by an elastic membrane in a constant state of tension, which tend to cause the surface to contract. Increasing the surface area of a liquid requires work, which remains stored as potential energy in the enlarged surface. That potential energy can perform work if the enlarged surface is allowed to contract again. To explain the occurrence of surface tension, it should be considered that a molecule inside the liquid is equally attracted in all directions by the cohesive forces of neighboring molecules, while a molecule at the surface of the liquid is attracted into the relatively dense liquid phase by a net force greater than that attracting it toward the rarified gaseous phase. This unbalanced force draws the surface molecules inward into the liquid and results in the tendency for the surface to contract. Surface tension is associated with the phenomenon of capillary. A liquid with an interface that is convex toward the atmosphere is under a pressure greater than atmospheric. A liquid with an interface concave toward the gaseous phase is at a pressure smaller than atmospheric. A liquid with a flat interface is at the same pressure as the atmosphere.

If we place a drop of liquid on a dry soil surface, the liquid will usually spread over the surface to a certain extent and it will form a typical angle, named contact angle α, with the surface of the solid (Fig. 1.6). The angle is acute if the adhesive affinity between the solid and the liquid is strong relative to the cohesive forces inside the liquid itself and to the affinity between the gas and the solid. In this case, the liquid wets the solid. A contact angle of zero implies the complete flattening of the drop and the perfect wetting of the solid surface by the liquid. On the other hand, a contact angle of 180° implies a complete non-wetting or rejection of the liquid by the gas-covered solid. Therefore, neglecting gravity, the drop would retain its spherical shape. Surfaces on which water exhibit an obtuse contact angle are called water repellent or hydrophobic. The wetting angle of pure water on clean and smooth mineral surfaces is generally zero, but if the surface has a hydrophobic nature, the contact angle can be considerably greater than zero.

A capillary tube dipped in a body of free water will form a meniscus as the result of the contact angle of water with the walls of the tube. A greater curvature of this meniscus will occur (i.e., the radius of curvature will be smaller) in a narrower tube. Due to this curvature, a pressure difference will develop across the liquid-gas interface. A liquid with an acute contact angle, such as water on glass, will form a concave meniscus (Fig. 1.6). Therefore, the liquid pressure under the meniscus, P_1, will be smaller than the atmospheric pressure, P_0. For this reason, the water inside the tube and the meniscus will be driven up the tube from its initial position by the greater pressure of free water, that is water at atmospheric pressure under a horizontal air-water interface, outside the tube at the same level. This process will continue until the initial pressure difference between the water inside the tube and the water under the flat surface outside the tube is entirely countered by the hydrostatic pressure exerted by the water column in the capillary tube. The height of capillary rise, h_c (L), is:

$$h_c = \frac{2\gamma \cos \alpha}{\rho_w g r} \tag{1.13}$$

where γ is the surface tension between the liquid and the air, g is the acceleration of gravity and r is the radius of the capillary tube. At 20 °C and for $\alpha = 0$, Eq. (1.13) is written as:

$$h_c \approx \frac{0.149}{r} \tag{1.14}$$

where r and h_c are expressed in cm.

Kinetic energy of soil water is generally assumed negligible since water moves slowly in the porous medium. On the other hand, the potential energy of soil water, which is due to position or internal condition, is of primary importance in determining the state and movement of water in the soil. In fact, the spontaneous and universal tendency of all parcels of matter in nature is to equilibrate with their surroundings, moving from where the potential energy is higher to where it is lower. Therefore, soil water moves constantly in the direction of decreasing potential energy and the potential energy gradient with distance is the driving force determining flow.

For a rigid, unsaturated soil in which the gaseous phase is in equilibrium with the atmospheric pressure, the total soil water potential, ϕ, is given by:

$$\phi = \phi_c + \phi_g + \phi_o \tag{1.15}$$

where the matric or capillary potential, ϕ_c, expresses the energy status of water retained in a soil by capillary and surface adsorption, ϕ_g is the gravitational potential and ϕ_o is the osmotic potential. The total soil water potential measures the amount of work that must be done per unit quantity (volume, mass, weight) of pure water in order to transport reversibly and isothermically an infinitesimal quantity of water from a pool of pure water at a specified elevation at atmospheric pressure to the soil water at a specified point (Hillel 1998). The sum of the matric and the gravitational potentials is often referred to as the hydraulic potential. The osmotic potential has to be considered when there is a concentration gradient of dissolved substances. If the potential is expressed in terms of volume, the units are J m^{-3} or equivalently in units of pressure, i.e. Pa. If the potential is expressed per unit mass or per unit weight, the units are J kg^{-1} and meters (m), respectively. The latter option is conventionally used for most hydrological studies. When expressed per unit weight, the gravitational potential corresponds to the elevation head, z, and the capillary or matric potential is related to the soil water pressure head, h, by:

$$\phi_c = \rho_w g h \tag{1.16}$$

Therefore, the hydraulic head expressed per unit weight, H (L), is:

$$H = h + z \tag{1.17}$$

The pressure head, h, is positive below the water table and negative above it. In the unsaturated zone, the negative of the pressure is also called matric suction or tension which, therefore, is expressed as a positive number. In saturated conditions, the pressure head is positive (or null) and it is also called piezometric head. In dry soil conditions, the pressure head can reach extremely low values, i.e. very negative. For computational convenience, considering h in cm, its value is sometimes expressed in the logarithmic mode:

$$pF = \log|h| \tag{1.18}$$

The value of $pF = 4.2$, corresponding to $h = -150$ m of water, is referred to as the wilting point since plant transpiration does not occur for smaller, i.e. more negative, h values.

The relationship between the volumetric soil water content and soil water suction is referred to as the water retention characteristic curve. This characteristic, describing the soil's ability to store or release water, is a highly non-linear S-shaped curve (Fig. 1.7). The water retention curve changes strongly with soil texture and structure. In general, soils with more clay exhibit a greater water retention at any particular suction and a more gradual slope of the curve. In a sandy soil, most of the pores are relatively large, and once these large pores are emptied at a given suction, only a small amount of water is retained. The pore-size distribution is more uniform in a clayey soil. In this case, an increase of the matric suction determines a more gradual decrease in wetness.

Drying and wetting soils may have different characteristic curves. This phenomenon is referred to as hysteresis. For a given matric suction, more water is retained

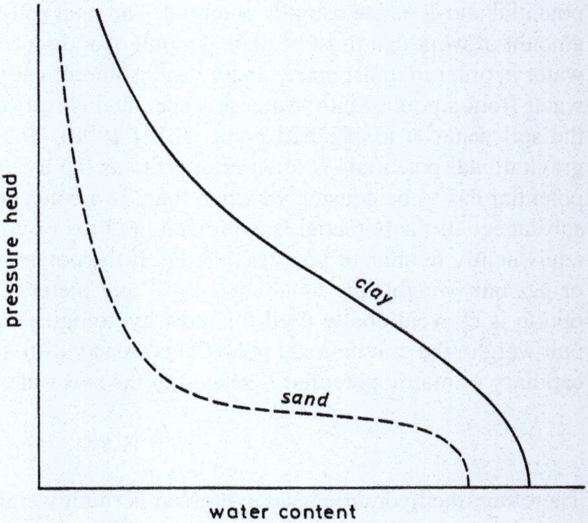

Fig. 1.7 The effect of texture on soil-water retention

by a drying soil than a wetting soil. Hysteresis is more pronounced for sands than for clay soils.

Several equations have been proposed to describe the soil water retention curve. Two well-known equations are that by Brooks and Corey (1964):

$$\frac{\theta - \theta_r}{\theta_s - \theta_r} = \left(\frac{h_{bc}}{h}\right)^{\lambda} \quad \text{for} \quad h \leq h_{bc} \tag{1.19a}$$

$$\theta = \theta_s \quad \text{for} \quad h_{bc} < h \leq 0 \tag{1.19b}$$

and van Genuchten (1980):

$$\frac{\theta - \theta_r}{\theta_s - \theta_r} = \left[1 + \left(\frac{h}{h_g}\right)^{n}\right]^{-m} \tag{1.20}$$

where the water pressure head, h, is usually taken to be negative and is expressed in cm of water, h_{bc} is the Brooks and Corey pressure scale parameter, h_g is the van Genuchten pressure scale parameter, and λ, m and n are water retention shape parameters. The water retention shape parameters m and n are frequently related according to:

$$m = 1 - \frac{k_m}{n} \quad \text{with} \quad n > k_m \tag{1.21}$$

where k_m is an integer value referred to as the user index (Haverkamp et al. 2005), taken equal to 1 or 2 according to the Mualem or the Burdine theory, respectively.

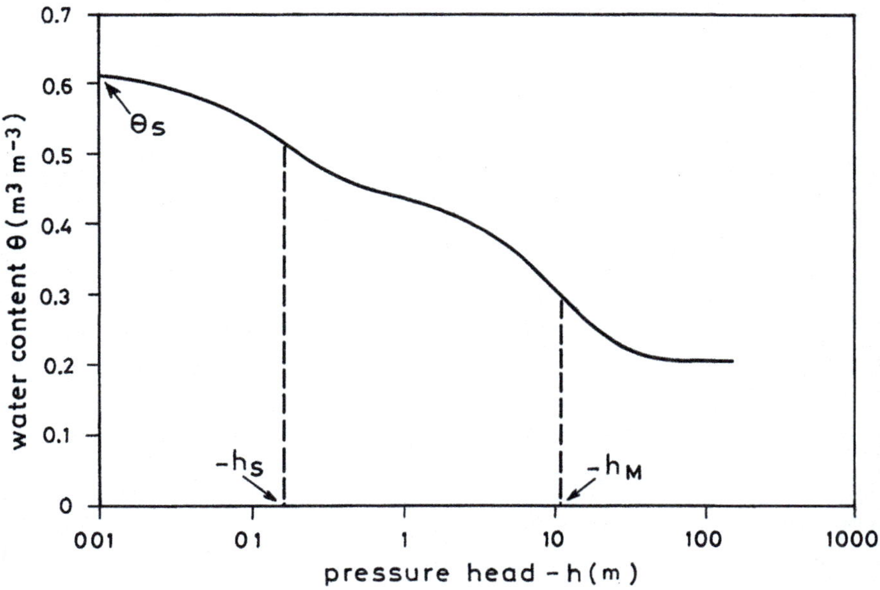

Fig. 1.8 Illustrative example of a bimodal soil water retention curve

A bimodal soil water retention curve may better describe retention for structured or two-domain soils with both matrix pores (small voids between individual soil mineral particles) and structure pores (relatively large inter-pedal cracks, biopores and inter-aggregate voids) (Reynolds et al. 2014). A convenient $\theta(h)$ function for soils composed of matrix pore and structure pore domains is that by Dexter et al. (2008):

$$\theta(h) = \theta_r + P_M e^{\left(-\frac{h}{h_M}\right)} + P_S e^{\left(-\frac{h}{h_S}\right)} \quad h \leq 0 \tag{1.22}$$

where θ_r (L^3L^{-3}) is the residual soil water content (as $h \rightarrow -\infty$), P_M (L^3L^{-3}) is the effective soil matrix porosity, P_S (L^3L^{-3}) is the effective soil structure porosity, h_M (L) is the pore water pressure head at the dry end or matrix inflection point on the $\theta(h)$ curve, and h_S (L) is the pressure head at the wet end or structure inflection point on the $\theta(h)$ curve (Fig. 1.8).

1.3 Flow of Water in Soils

Let us consider a steady flow of water from an upper reservoir to a lower one, in each of which the water level is maintained constant, through a horizontal column of a macroscopically uniform, saturated soil (Fig. 1.9). The discharge

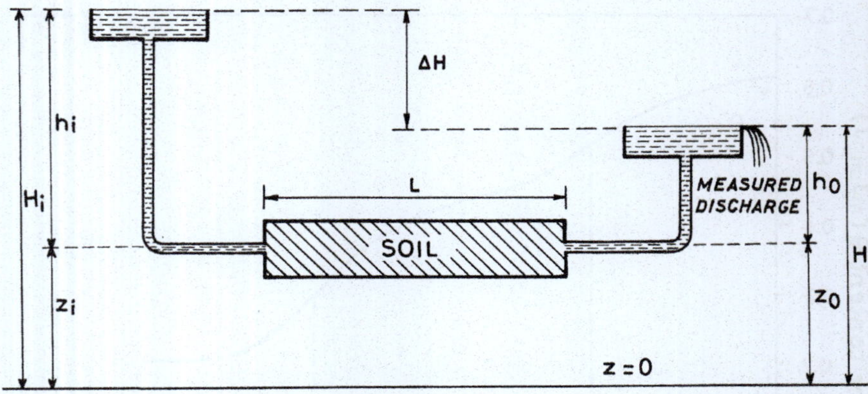

Fig. 1.9 Flow in a horizontal saturated column

rate, Q (L^3T^{-1}), i.e. the volume, V (L^3), flowing through the column per unit time, t (T), is directly proportional to the cross-sectional area of the column, A (L^2), and to the hydraulic head drop, ΔH (L), and inversely proportional to the length of the column, L (L). Measurement of the hydraulic head at the inflow, H_i (L), and the outflow, H_o (L), boundaries relative to some reference level yields ΔH. The gravitational head at a point, z (L), is given by the height of the point relative to the reference plane. For a horizontal soil column, z does not change between the two ends of the systems ($z_i = z_o$ in Fig. 1.9). Pressure of the water entering the column is the hydrostatic pressure plus the atmospheric pressure acting on the water surface in the reservoir. Taking into account that the atmospheric pressure does not change between the ends of the system, it can be disregarded and only the hydrostatic pressure needs to be considered. Accordingly, the water pressure at the inflow boundary is $P_i = \rho_w g h_i$. Since ρ_w and g are both nearly constant, this pressure can be expressed in terms of the pressure head, h_i. At the outflow boundary, the pressure head is h_o (Fig. 1.9). Therefore, $\Delta H = H_i - H_o = (z_i + h_i) - (z_o + h_o) = h_i - h_o$ for a horizontal soil column. The driving force is the head drop per unit distance in the direction of flow ($\Delta H/L$). Flux density, or simply flux, q ($L\ T^{-1}$), is the specific discharge rate, Q/A, i.e. the discharge rate flowing through a unit cross-sectional area. Thus, the flux is given by:

$$q = \frac{Q}{A} = K_s \frac{\Delta H}{L} \tag{1.23}$$

where the proportionality factor, K_s ($L\ T^{-1}$), is the hydraulic conductivity of the saturated porous medium. This equation, known as Darcy's law, establishes that the flow of a viscous liquid such as water through a porous medium is in the direction of, and at a rate proportional to, the driving force and also proportional to K_s, denoting the ability of the saturated medium to transmit the liquid.

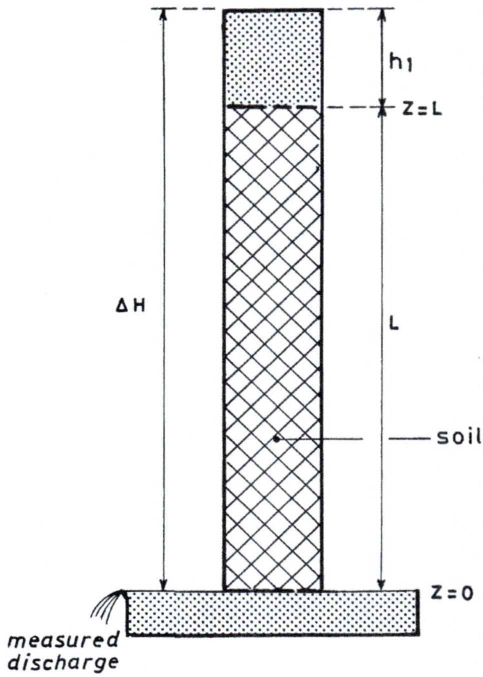

Fig. 1.10 Downward flow of water in a vertical column of saturated soil

For the one-dimensional flow in the vertical direction, z, the Darcy's law is written as:

$$q = -K_s \frac{dH}{dz} \tag{1.24}$$

in which dH/dz is the hydraulic gradient and the minus sign accounts for the fact that water moves from high to low H values. Equation (1.24) can be applied for both downward and upward steady flow in a vertical saturated soil column. Let us consider the case of an upper surface ponded under a constant head of water, h_1, and a bottom surface set in a lower, constant-level reservoir (Fig. 1.10). Flow takes place downward through a column of length L. The hydraulic head at inflow boundary, H_i, is h_1 (pressure head) $+L$ (gravity head). The hydraulic head at the outflow boundary is $H_o = 0$ because both the pressure head and the gravity head are zero. Therefore, the hydraulic head drop between the inflow and outflow boundaries of the column, ΔH, is equal to $h_1 + L$. The Darcy's equation for this case is:

$$q = K_s \frac{\Delta H}{L} = K_s \frac{h_1 + L}{L} = K_s \frac{h_1}{L} + K_s \tag{1.25}$$

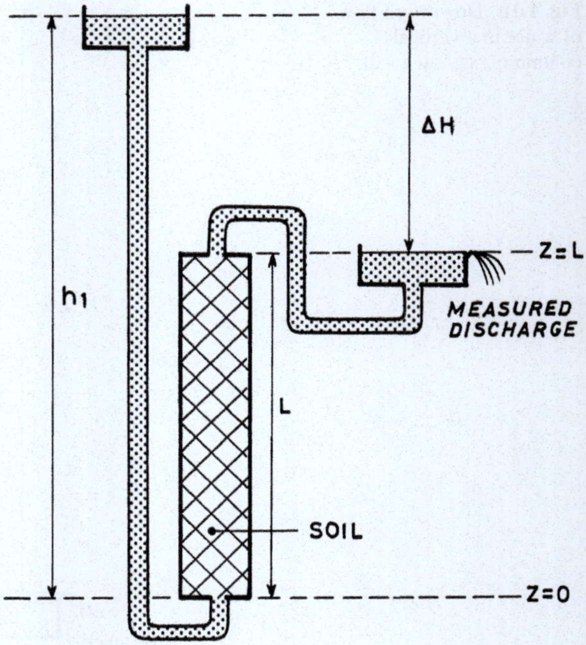

Fig. 1.11 Steady upward flow in a vertical column of saturated soil

If h_1 is small and hence negligible, $q \approx K_s$ since, in the absence of a pressure gradient, the only driving force is the gravitational head gradient which, in a vertical column, has the value of unity. In the case of steady upward flow in a vertical column (Fig. 1.11), the flow direction is opposite to the direction of the gravitational gradient. In this case, $H_i = h_1$ (pressure head) + 0 (gravity head) and $H_o = 0 + L$. Therefore, $\Delta H = h_1 - L$ and the Darcy's equation is:

$$q = K_s \frac{h_1 - L}{L} = K_s \frac{h_1}{L} - K_s \tag{1.26}$$

For a downward flow with the bottom end of the column exposed to the atmospheric pressure (Fig. 1.10), the hydraulic gradient is always ≥ 1. For the upward flow, it is also possible to establish hydraulic gradient values smaller than one, maintaining however $h_1 > L$. Let us now consider steady flow through a non-uniform soil column consisting of two distinct layers (Fig. 1.12), differing by both thickness (L_1 and L_2 for layer 1 and layer 2, respectively) and hydraulic conductivity (K_{s1} and K_{s2}). Layer 1 is at the inlet and layer 2 at the outlet side of the column. In this case, steady flux is given by (Hillel 1998):

$$q = \frac{\Delta H}{R_{h1} + R_{h2}} \tag{1.27}$$

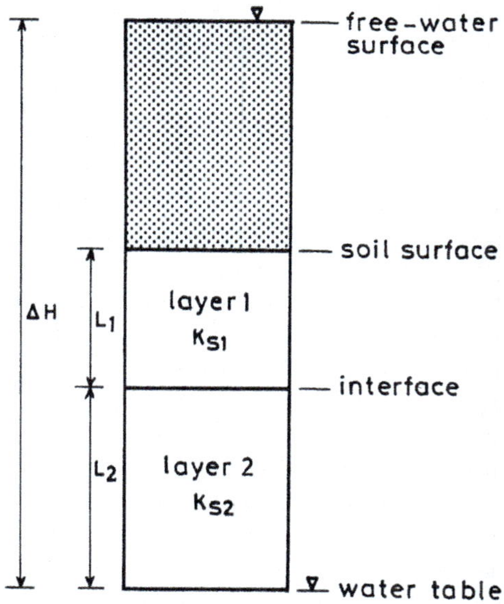

Fig. 1.12 Downward flow through a two-layer soil column

where ΔH is the total hydraulic head drop across the entire system and R_h (T) (R_{h1} for layer 1 and R_{h2} for layer 2), equal to L/K_s, is the hydraulic resistance per unit area. If the more conductive layer is placed on the top of the column ($K_{s1} > K_{s2}$), flow is impeded at the interface and there is a pressure buildup. If the upper layer is less conductive, pressure is dissipated through the top layer and a negative pressure may develop at the interface. If this negative pressure exceeds the sublayer's air entry value, the sublayer will become unsaturated.

Flux and velocity have the same dimensions ($L^3L^{-2}T^{-1} = L\ T^{-1}$) but the term flux is less ambiguous than flow velocity. Soil pores vary in shape, width and direction and wider pores conduct water more rapidly. Therefore, no more than an average velocity can be considered. Even this average velocity is not the flux because part of the entire cross-sectional area, A, is plugged by particles. Therefore, the real area through which flow takes place is smaller than A, which implies that the average velocity is greater than q. Furthermore, the tortuous nature of the soil pores implies an actual length of the path traversed by a parcel of liquid greater than the soil column length, L.

The saturated soil hydraulic conductivity is affected by both texture and structure since it depends on total porosity but also, and primarily, on the sizes of the conducting pores, i.e. on the soil's pore geometry. A gravelly or sandy soil with large pores can have a conductivity much greater than a clay soil with narrow pores, even though the total porosity of a clay soil is generally greater than that of a sandy soil.

The hydraulic conductivity is not a property of the soil alone since it also depends on density and viscosity of the fluid. It is possible to separate K_s (L T^{-1})

into the intrinsic permeability of the soil, k_s (L^2) and the fluidity of the permeating liquid, f ($L^{-1}T^{-1}$), that is related directly to density and inversely to viscosity:

$$K_s = k_s f = k_s \frac{\rho g}{\eta} \tag{1.28}$$

where ρ ($M L^{-3}$) is the fluid density, g ($L T^{-2}$) is the acceleration due to gravity and η ($M L^{-1}T^{-1}$) is the dynamic viscosity. Therefore, fluidity varies with composition of the fluid and temperature. On the other hand, the permeability is ideally an exclusive property of the porous medium. Equation (1.28) can be used to predict the expected K_s value at a different water temperature than the one used for a particular measurement. For example, if $K_s = 1.39 \times 10^{-6}$ m s^{-1} is obtained with water at 10 °C ($\rho = 999.73$ kg m^{-3}, $\eta = 1.307 \times 10^{-3}$ kg m^{-1}s^{-1}), a value of K_s of 2.04×10^{-6} m s^{-1} should be expected using water at 25 °C ($\rho = 997.08$ kg m^{-3}, $\eta = 8.90 \times 10^{-4}$ kg m^{-1}s^{-1}) (Bagarello and Iovino 2010).

Darcy's law applies only to laminar flow regime. As flow velocity increases, the occurrence of turbulent eddies or nonlinear laminar flow result in waste of effective energy so that more energy is dissipated. An increase of the hydraulic potential gradient determines a smaller increase of flux than predicted by the Darcy's law. Non-laminar flow conditions, impeding use of the Darcy's law, can occur in coarse sands and gravels and also in macropores such as wide cracks and worm holes.

The hydraulic conductivity may be uniform throughout the soil or the soil can be hydraulically inhomogeneous because K_s varies from point to point (Hillel 1998). In an isotropic soil, K_s does not change with the direction. Anisotropic conditions occur when K_s at a point varies with the direction (e.g., the horizontal conductivity is greater, or smaller, than the vertical conductivity). Anisotropy is generally due to the structure of the soil, determining development of micropores or macropores with a distinct directional bias.

The principles of water flow in saturated soil also apply in unsaturated soils. In this case, however, water is subject to a negative pressure potential, and the nature of the driving force and the effective geometry of the conducting pores may be very different (Hillel 1998).

Saturated and unsaturated soils can greatly differ by their hydraulic conductivity. In saturated conditions, all pores are water-filled and conducting, which implies continuity of the water phase and the highest possible conductivity. In unsaturated conditions, some soil pores are air-filled which implies a smaller conductive portion of the soil's cross-sectional area. The first pores that empty when the pressure head starts to decrease from $h = 0$ are the largest and the most conductive ones. Moreover, the large empty pores must be circumvented so that tortuosity increases with progressive desaturation. In soils with large pores (coarse-textured soils, aggregated soils), a high conductivity can be expected at saturation but, when emptied, these large pores act as barriers to liquid flow. Soil hydraulic conductivity may decrease by several orders of magnitude (sometimes down to 10^{-6} of its value at saturation) as suction increases from 0 to 1 MPa.

In saturated conditions, a sandy soil conducts water more rapidly than a clayey soil, and a well-aggregated soil conducts more than a poorly aggregated or dispersed soil. However, the opposite can occur in unsaturated conditions. In a soil with large pores, development of unsaturated conditions determines a quick emptying of these pores that become non-conductive, steeply reducing the initially high conductivity. In a soil with small pores, on the other hand, many pores remain saturated and conductive even at appreciable suction. Therefore, the hydraulic conductivity does not decrease as steeply and may actually exceed that of a soil with large pores subjected to the same suction.

For one-dimensional unsaturated flow in the vertical direction, the Darcy's law can be written in its generalized form (Haverkamp et al. 2006):

$$q = -K(h)\frac{dH}{dz} = -K(h)\left(\frac{dh}{dz} - 1\right) \quad (1.29)$$

where z (L) is depth taken positive downward, K (L T^{-1}) is the soil hydraulic conductivity and h (L) is the pressure head. Note that, when the z-axis is oriented downward, as commonly assumed in simulation of the water infiltration process, the hydraulic head [Eq. (1.17)] is defined as: $H = h - z$. For a given hydraulic gradient between two points of the porous medium, the flux is smaller in drier soil conditions since the soil hydraulic conductivity decreases with a decrease in h.

Various empirical equations have been proposed for expressing the relation between the hydraulic conductivity and the pressure head, $K(h)$, or the volumetric soil water content, $K(\theta)$. In particular, in the analysis of the infiltrometer data, the $K(h)$ function of Gardner (1958) is frequently applied:

$$K(h) = K_s \exp[\alpha(h - h_e)] \qquad 0 < \alpha < +\infty;\ h < h_e \quad (1.30a)$$

$$K(h) = K_s \qquad h_e \leq h \leq 0 \quad (1.30b)$$

where α (L^{-1}) is a slope parameter that depends primarily on soil texture and structure and h_e (L) is an entry pressure head which represents the air-entry value (h_a) for drainage from field saturation and the water-entry value (h_w) for wetting up to field-saturation. Another frequently used model is the one by Mualem (1976) that, under the hypothesis to describe the water retention curve according to Brooks and Corey (1966), is written as:

$$K(S_e) = K_s S_e^{\eta} \quad (1.31)$$

where η is a conductivity shape parameter. Using the van Genuchten (1980) relationship for the water retention curve with $m = 1 - 1/n$, the model by Mualem (1976) becomes:

$$K(S_e) = K_s S_e^{0.5}\left[1 - \left(1 - S_e^{1/m}\right)^m\right]^2 \quad (1.32)$$

Fig. 1.13 Schematization of the one-dimensional vertical flow

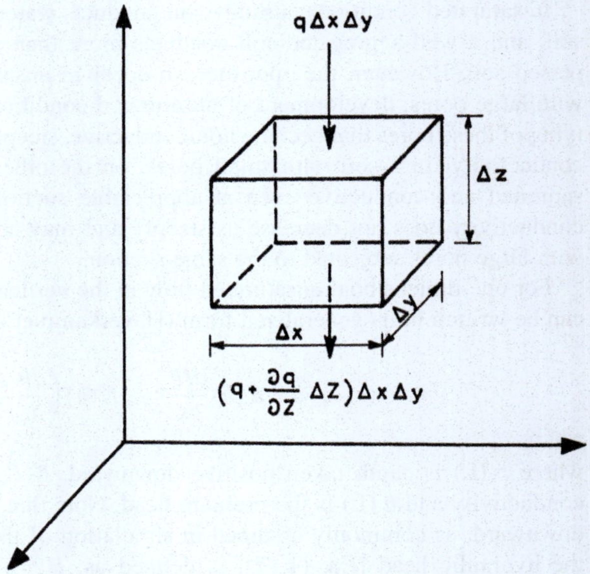

The relation of conductivity to suction depends on hysteresis, and is thus different in a wetting than in a drying soil. At a given suction, a drying soil is more conductive because it contains more water than a wetting one. The relation of conductivity to water content, however, appears to be affected by hysteresis to a much lesser degree.

Simulation of hydrological processes in unsaturated soil conditions requires a general flow equation that accounts for transient as well as steady flow. In the following analysis, the general flow equation is obtained for the case of isothermal water movement in a rigid porous material.

First, the continuity equation is derived. It is based on the law of mass conservation and simply states that, for a conserved substance such as water (i.e., a substance that is neither created nor depleted in the soil), the amount entering minus the amount exiting a given soil body must be equal to the change in water content of the same body. Considering a volume element of soil as a rectangular parallelepiped of sides Δx, Δy and Δz (Fig. 1.13), if the flux exiting from the bottom exceeds the flux entering from the top by the amount $\Delta q = (\partial q/\partial z)\Delta z$, then the net inflow discharge (volume per unit time) flowing through the two faces is given by $-(\partial q/\partial z)\Delta x\Delta y\Delta z$. This quantity expresses the rate of gain of water by the element volume of soil. On the other hand, the rate of gain of water can also be expressed in terms of the time rate of changes of the volumetric water content, θ, multiplied by the volume of the element: $(\partial\theta/\partial t)\ \Delta x\Delta y\Delta z$. Setting the two alternative expressions equal each other, the following expression for the equation of continuity in vertical direction can be obtained (Haverkamp et al. 2006):

$$\frac{\partial\theta}{\partial t} = -\frac{\partial q}{\partial z} \tag{1.33}$$

in which z (L) is the depth taken positive downward and the minus sign accounts for the fact that water content decreases when the flux density increases in the direction of flow. If the fluxes in the x and y directions are also considered, the following three-dimensional form of the continuity equation can be obtained:

$$\frac{\partial\theta}{\partial t} = -\left(\frac{\partial q_x}{\partial x} + \frac{\partial q_y}{\partial y} + \frac{\partial q_z}{\partial z}\right) \tag{1.34}$$

where q_x, q_y and q_z are the fluxes in the x, y and z directions, respectively. In shorthand mathematical notation, Eq. (1.34) can be written:

$$\frac{\partial\theta}{\partial t} = -\nabla \cdot q \tag{1.35}$$

where the three-dimensional gradient in space is indicated by the operator ∇ (del).

Combining Eq. (1.33) with Eq. (1.29), the following general equation for vertical water transfer into unsaturated soil can be derived, that is generally referred to as Richards equation:

$$\frac{\partial\theta}{\partial t} = \frac{\partial}{\partial z}\left[K(h)\left(\frac{\partial h}{\partial z} - 1\right)\right] \tag{1.36}$$

The corresponding three-dimensional (3D) expression is:

$$\frac{\partial\theta}{\partial t} = \nabla \cdot [K(h)\nabla H] \tag{1.37}$$

or, considering that ∇z is zero for horizontal flow and unity for vertical flow:

$$\frac{\partial\theta}{\partial t} = \frac{\partial}{\partial x}\left(K(h)\frac{\partial h}{\partial x}\right) + \frac{\partial}{\partial y}\left(K(h)\frac{\partial h}{\partial y}\right) + \frac{\partial}{\partial z}\left(K(h)\frac{\partial h}{\partial z}\right) - \frac{\partial K(h)}{\partial z} \tag{1.38}$$

When flow is axis-symmetric, flow equation can be written as follows in cylindrical coordinates:

$$\frac{\partial\theta}{\partial t} = \frac{1}{r}\frac{\partial}{\partial r}\left(rK(h)\frac{\partial h}{\partial r}\right) + \frac{\partial}{\partial z}\left(K(h)\frac{\partial h}{\partial z}\right) - \frac{\partial K(h)}{\partial z} \tag{1.39}$$

where r (L) is the radial coordinate which expresses the distance from the vertical symmetry axis.

Equation (1.36) can be expressed as a Fokker-Planck type equation, i.e. a θ-dependent equation, by introducing the concept of soil water diffusivity (Haverkamp et al. 2006):

$$\frac{\partial \theta}{\partial t} = \frac{\partial}{\partial z}\left[D(\theta)\frac{\partial \theta}{\partial z} - K(\theta)\right] \tag{1.40}$$

where $K(\theta)$ is the θ-dependent expression for the hydraulic conductivity function and $D(\theta)$ is diffusivity defined by:

$$D(\theta) = K(\theta)\frac{dh}{d\theta} \tag{1.41}$$

Similarly, Eq. (1.36) can be expressed in terms of soil water pressure head only by introducing the concept of specific capacity (Hillel 1998):

$$C(h)\frac{\partial h}{\partial t} = \frac{\partial}{\partial z}\left[K(h)\left(\frac{\partial h}{\partial z} - 1\right)\right] \tag{1.42}$$

where C is specific capacity given by $C(h) = d\theta/dh$.

Because of the highly non-linear $K(h)$ (or $K(\theta)$) and $\theta(h)$ relationships, analytical solutions of Eqs. (1.40) and (1.42) are not generally possible, except under very restricting assumptions regarding those relationships. Instead, numerical methods have been devised. However, the use of Eq. (1.40) with volumetric water content as independent variable causes computational difficulties when applied for numerical simulation of water movement in regions close to saturation. Therefore, the most commonly used numerical methods pertain to the h-based form of the Richards equation, Eq. (1.42), that can be applied to both saturated and unsaturated conditions, as well as to layered soils where h is generally continuous but θ may not be (Hillel 1998).

1.4 Infiltration Process and Equations

"Water infiltration is the term applied to the process of water entry into the soil, generally, by downward flow through all or part of the soil surface" (Hillel 2004). In hydrology and agricultural sciences, the study of infiltration is of great concern since water infiltration provides the stock of water available for plants and groundwater recharge and also defines water runoff at surface. Indeed, during rainfall events, an increase of water pressure head at surface induces the infiltration of water in soils, according to soil hydraulic conductivity. If rainfall intensity is relatively low (i.e., $\leq K_s$), all water will infiltrate with no runoff at surface. In opposite, when rainfall intensity exceeds the capability of soils to infiltrate water (infiltration capacity), a part of the water will runoff at surface. During the infiltration phase, wetting fronts will develop and move downwards by the combined effect of gravity and capillary, triggering the moistening of the soil profile. Once the rainfall stops, water will stop infiltrating at soil surface. Then, the stocks of water, previously added during the infiltration phase, will move downwards by gravity and capillary. Meanwhile, evapotranspiration may contribute to water depletion at surface and in the first horizons. In summer season, when evapotranspiration is high, most of water will evaporate before reaching the groundwater. In opposite, during the winter season, most of the water may reach the water table, thus contributing to groundwater recharge.

In this book, we focus on the methods used for the numerical and analytical resolution of flow equation and the quantification of water infiltration at soil surface. The presentation below, without being exhaustive in relation to the large existing literature, provides the necessary tools for understanding the infiltration-based methods illustrated in the following chapters. A detailed analysis of the infiltration theory can be found in Smith (2002). We mainly focus on water infiltration experiments that consist in infiltrating water under controlled conditions (i.e. water pressure heads, initial water contents, etc.) using specific devices (ponded or tension infiltrometers, infiltration pits, etc.). These experiments are designed to test and characterize the soil capability to infiltrate water as well as soil hydraulic properties. Water infiltration experiments are an interesting tool to get an insight on soil hydraulic properties since they are easy to conduct and non-invasive, allowing a proper characterization of soil hydraulic properties without any significant disturbance of the soil. Water infiltration experiments are designed to monitor infiltration rates, i.e. water flux over the infiltration surface, or cumulative infiltration which corresponds to the time integration of the infiltration rate, i.e. to the total volume of infiltrated water depth, I (L), at time t (T):

$$I(t) = \int_0^t i(\bar{t})\, d\bar{t} \tag{1.43}$$

where i stands for infiltration flux at soil surface ($z = 0$).

Several approaches have been proposed to model water infiltration as a function of initial and boundary conditions. These modeling approaches allow us to relate soil hydraulic properties to the response measured at surface, e.g. cumulative water infiltration. To build models, the first approach lies in the numerical resolution of Richards' equation [Eq. (1.38)], which can be adapted for the case of axis-symmetric flow (see [Eq. (1.39)]). An illustrative example of the use of numerical approaches to derive soil hydraulic properties will be provided in Sect. 3.1. Here, we present successively the development of analytical infiltration models for the cases of one-dimensional (1D) gravity-free water absorption, 1D gravity driven water infiltration, 1D gravity and capillary driven water infiltration and, finally, extensions to 3D for gravity and capillary driven water infiltration.

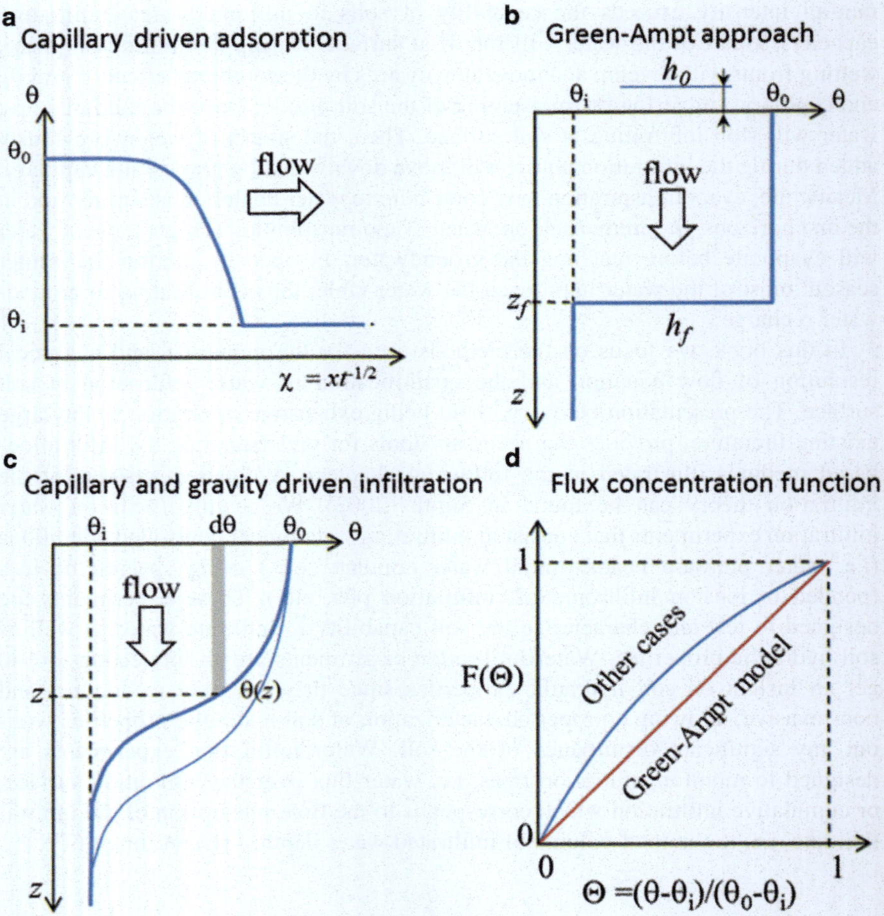

Fig. 1.14 Schematization of horizontal and vertical water infiltration, accounting for part or all mechanisms (capillary – gravity), and flux concentration function

1.4.1 Gravity Free Water Absorption

For one-dimensional approaches, we consider a semi-infinite column with only one boundary fixed in terms of water content θ_0 (L^3L^{-3}) or water pressure h_0 (L) (Fig. 1.14a). At any time, mass balance consideration allows the quantification of cumulative infiltration as the spatial integration of water content over the whole length column, leading to two expressions:

$$I(t) = \int_0^{+\infty} (\theta(x,t) - \theta_i)\, dx \tag{1.44a}$$

$$I(t) = \int_{\theta_i}^{\theta_0} x(\theta, t)\, d\theta \tag{1.44b}$$

where θ_i (L^3L^{-3}) is the initial volumetric soil water content. At time $t = 0$, we impose at the boundary, $x = 0$, a specific volumetric water content θ_0. Under these conditions, the flow equation [Eq. (1.40)] becomes:

$$\frac{\partial \theta}{\partial t} = \frac{\partial}{\partial x}\left[D(\theta)\frac{\partial \theta}{\partial x}\right] \tag{1.45}$$

Initial and boundary conditions are defined as follows:

$$t = 0 \qquad x \geq 0 \qquad \theta = \theta_i \tag{1.46}$$

$$t \geq 0 \qquad x = 0 \qquad \theta = \theta_0 \tag{1.47}$$

To solve the equation, the Boltzmann transformation is used as follows (Boltzmann 1894; Philip 1957a):

$$\chi(\theta) = x(\theta, t)\, t^{-1/2} \tag{1.48}$$

which allows us to define the position of the wetting front as a function of θ and t:

$$x(\theta, t) = \chi(\theta)\, t^{1/2} \tag{1.49}$$

The implementation of Eq. (1.48) into flow Eq.(1.45) leads to an ordinary differential equation:

$$\chi \frac{d\theta}{d\chi} = -2\frac{d}{d\chi}\left[D(\theta)\frac{d\theta}{d\chi}\right] \tag{1.50}$$

with the initial and boundary conditions:

$$\chi \rightarrow +\infty \qquad \theta = \theta_i \tag{1.51}$$

$$\chi = 0 \qquad \theta = \theta_0 \tag{1.52}$$

The integration of Eq. (1.50) with regards to variable χ leads to:

$$\int_{\theta_i}^{\theta} \chi(\theta)\, d\theta = -2D(\theta)\frac{d\theta}{d\chi} \tag{1.53}$$

It may be borne in mind that the Boltzmann transformation hypothesizes the similarity of profiles (Fig. 1.14a), meaning that different water profiles at several times have similar shapes and move forwards as time increases (Philip 1959). This hypothesis is rarely invalidated by experimental observations. The knowledge of χ fixes the position of the wetting front, $x(\theta,t)$ using Eq. (1.49), and allows us to calculate cumulative infiltration using Eq. (1.44b):

$$I(t) = \left[\int_{\theta_i}^{\theta_0} \chi(\theta)\, d\theta \right] t^{1/2} \tag{1.54}$$

The integral in Eq. (1.54) is the area under the advancing absorption front, and the integral coefficient is referred to as capillary sorptivity (Philip 1957b):

$$S(\theta_i, \theta_0) = \int_{\theta_i}^{\theta_0} \chi(\theta)\, d\theta \tag{1.55}$$

The sorptivity, which depends on the initial and boundary volumetric water contents, represents the soil capability to absorb water without gravity. Sorptivity has the dimension of L $T^{-1/2}$. Equation (1.55) leads to the following analytic model for the cumulative infiltration resulting from gravity-free capillary-driven absorption:

$$I(t) = S(\theta_i, \theta_0) t^{1/2} \tag{1.56}$$

This model requires no assumption on soil hydraulic properties.

1.4.2 Green and Ampt Model for Infiltration

For the case of vertical water infiltration, gravity must be accounted for. One of the models accounting for gravity solely was proposed by Green and Ampt (1911). The Green and Ampt model is widely used for hydrologic studies and pioneered several variables and concepts that are the basis of common more sophisticated models for hydrology. The Green and Ampt approach assumes a nominal wetting front in the form of a step function, with constant hydraulic conductivity $K(\theta) = K_0$ and water content $\theta = \theta_0$ above the wetting front, and a uniform water content $\theta = \theta_i$ and hydraulic conductivity, K_i, below (Fig. 1.14b). The shift in terms of water content and hydraulic conductivity at the wetting front corresponds to a specific water pressure head, h_f. In the original infiltration equation of Green and Ampt (1911), the initial value of hydraulic conductivity (K_i) was considered equal to zero. The equation for cumulative infiltration becomes:

$$I(t) = z_f(t)[\theta_0 - \theta_i] \tag{1.57a}$$

$$i(t) = [\theta_0 - \theta_i]\frac{dz_f}{dt} \tag{1.57b}$$

where z_f stands for the position of the wetting front. The infiltration flux can be written considering Darcy's law as follows:

$$i(t) = -K_0\frac{\left(h_f - z_f\right) - h_0}{z_f} \tag{1.58}$$

The integration of Eq. (1.57b) between times $t = 0$ and t gives:

$$I(t) = K_0 t + [\theta_0 - \theta_i]\left[h_0 - h_f\right]\ln\left\{1 + \frac{I(t)}{[\theta_0 - \theta_i]\left[h_0 - h_f\right]}\right\} \tag{1.59}$$

This formulation applies for very dry soils when initial water content and hydraulic conductivity can be neglected. Haverkamp et al. (2006) presented the generalized Green and Ampt infiltration equation with $K_i \neq 0$ (e.g., Swartzendruber 1987; Ross et al. 1996):

$$I(t) = K_0 t + \frac{[\theta_0 - \theta_i]\left[h_0 - h_f\right]K_0}{[K_0 - K_i]}\ln\left\{1 + \frac{[I(t) - K_i t][K_0 - K_i]}{[\theta_0 - \theta_i]\left[h_0 - h_f\right]K_0}\right\} \tag{1.60}$$

1.4.3 Philip Equation for Capillary and Gravity Driven Infiltration

Afterwards, the models that are presented account for both gravity and capillary. Philip (1957b) extended the Boltzmann transformation to the case of vertical infiltration with gravity and capillary (Fig. 1.14c). By analogy with Eq. (1.49), the author considered power series of $t^{1/2}$ to define the position of the water profile, $z(\theta,t)$ as follows:

$$z(\theta, t) = \chi_1(\theta)\ t^{1/2} + \chi_2(\theta)\ t + \chi_3(\theta)\ t^{3/2} + \ldots \tag{1.61}$$

In this equation, the first term represents the effect of capillary and the two next ones represent the effect of gravity. The following terms were neglected. The integration of this series considering the first two terms using the approach described above [Eqs. from 1.44b to 1.55] leads to the equation for cumulative infiltration. In this case, the quantification of the total cumulative infiltration by integrating soil water profile, by analogy with Eq. (1.44b), must be adapted to account for the quantity of water that infiltrates by gravity at initial water content θ_i, independently of any time evolution of water content. Total cumulative infiltration corresponds to water infiltration by gravity at water content θ_i plus the time variation of stock of water in the soil profile:

$$I(t) = K_i t + \int_{\theta_i}^{\theta_0} z(\theta, t)\, d\theta \tag{1.62}$$

The concatenation of Eqs. (1.61) and (1.62) leads to the following equations:

$$I(t) = S_1(\theta_i, \theta_0)t^{1/2} + [K_i + S_2(\theta_i, \theta_0)]t + S_3(\theta_i, \theta_0)t^{3/2} + \ldots \tag{1.63a}$$

$$S_i(\theta_i, \theta_0) = \int_{\theta_i}^{\theta_0} \chi_i(\theta)\, d\theta \tag{1.63b}$$

Considering small times, only the first two terms are accounted for. S_1 corresponds to the sorptivity as previously described and defined by Eq. (1.55). No simple approximation exists for S_2. Generally, the term S_2 is calculated from fitting onto numerically generated or experimental data with:

$$I(t) = A_1\sqrt{t} + [K_i + A_2]t \tag{1.64}$$

and for most soils, A_2 is confined into the following intervals:

$$\frac{1}{3}(K_0 - K_i) \leq A_2 \leq \frac{2}{3}(K_0 - K_i) \tag{1.65}$$

According to Talsma (1969), A_2 can be approached by 0.357 for most soils. The other terms of Eq. (1.63a) can be iteratively calculated, these calculations being quite complex and intensive (Haverkamp et al. 1990).

One of the main drawbacks of Eqs. (1.63a and 1.63b) is that its use must be limited in time, since convergence is not ensured for large times. Philip (1957c) proposed an asymptotic approximation based on the particular shape of the soil water profile. On this basis, Philip (1957d) proposed the following time limit for the validity of Eqs. (1.63a and 1.63b):

$$t_{grav} = \left[\frac{S(\theta_i, \theta_0)}{K_0 - K_i}\right]^2 \tag{1.66}$$

The so-called gravity time, t_{grav}, is the time limit between capillary driven and both gravity and capillary driven infiltration. Below this limit, the series with the two first terms give a good approximation of cumulative infiltration and Eqs. (1.63a and 1.63b) can be considered appropriate.

1.4.4 *Sorptivity and Flux Concentration Function*

The models presented above introduced the concept of sorptivity. The flux concentration function is another important concept often used for the analytical modelling of water infiltration. Philip (1973) introduced the concept of flux concentration function for the case of water absorption by capillary. He defined this function as the ratio of water flux at any location in the profile (i.e. at any water content θ), $i(\theta,t)$, to the infiltration rate at surface, $i_0(t)$. He also demonstrated that the flux concentration function mainly depends on θ and can be defined from the Boltzmann transformation as follows:

$$F(\theta) = \frac{i(\theta, t)}{i_0(t)} = \frac{\int_{\theta_i}^{\theta} \chi_1(\bar{\theta})\, d\bar{\theta}}{\int_{\theta_i}^{\theta_0} \chi_1(\theta)\, d\theta} \tag{1.67}$$

where $F(\theta)$ stands for the flux-concentration function. The flux concentration function can be used to estimate sorptivity, as follows (Philip and Knight 1974):

$$S^2(\theta_i, \theta_0) = 2\int_{\theta_i}^{\theta_0} \frac{[\theta - \theta_i]\, D(\theta)}{F(\theta)}\, d\theta \tag{1.68}$$

Such expression can also be written in terms of water pressure head, leading to:

$$S^2(h_i, h_0) = 2\int_{h_i}^{h_0} \frac{[\theta(h) - \theta(h_i)]\, K(h)}{F(\theta(h))}\, dh \tag{1.69}$$

In fact, Eq. (1.69) is more general and can be applied for the case of strictly positive values for water pressure head at surface ($h_0 > 0$). Its integration over the two distinct intervals $[h_i, 0]$ and $[0, h_0]$, considering that F equals unity over $[0, h_0]$, leads to the following expression for sorptivity:

$$S^2(h_0, h_i) = 2 \int_{\theta_i}^{\theta_s} \frac{[\theta - \theta_i]\, D(\theta)}{F(\theta)}\, d\theta + 2K_s h_0[\theta_s - \theta_i] \tag{1.70}$$

where θ_s is the saturated soil volumetric water content and K_s is the saturated soil hydraulic conductivity. The first term of the right side corresponds to the unsaturated component of sorptivity and the second term accounts for the saturated component of sorptivity and the influence of water pressure head at surface. Haverkamp et al. (1990) proposed an extension of Eq. (1.70) for the case of soils for which a certain air entry pressure is needed to start water desaturation.

Many estimates for sorptivity rely on the use of the flux concentration function, thus requiring its characterization. The flux concentration function is a priory unknown, and its estimation requires an iterative procedure (Philip and Knight 1974) or an accurate approximation for F function. It exhibits a specific concave shape (Fig. 1.14d) with concavity depending on the type of soil. Table 1.1 lists most of the approximations proposed in previous studies as a function of the type of soil. The Crank (1979) approximation concerns the case of linear soils for which diffusivity is constant (Table 1.1). The approximation of Philip and Knight (1974) addresses the case of soil with Dirac δ-function for diffusivity. In that case, infiltration model can be reduced to the Green and Ampt infiltration model. Other authors have proposed and compared several approximations using an exponential diffusivity relationship. Parlange (1975) proposed an equation that gives a good approximation for sorptivity for soils with strong non-linearity (Kutílek and Valentová 1986; Elrick and Robin 1981). For the soils with moderate non-linear behavior, the approximation of Brutsaert (1976) gives the best results. More recently, Parslow et al. (1988) proposed an asymptotic development of Eq. (1.55) using more orders than Parlange (1975) to define a more precise estimate of the flux concentration function. Finally, Parlange et al. (1994) proposed a simple method to estimate sorptivity on the basis of an arbitrary formulation of diffusivity.

1.4.5 General Infiltration Solution of Parlange

The flux concentration approach was developed to model wetting front and water infiltration at surface. As detailed above, the ratio between the water flux at any depth to the infiltration rate at surface is ruled by the flux concentration function through Eq. (1.67) for water infiltration driven by capillary only. When the flow is both capillary and gravity driven, flux concentration function and water flux in the profile are related through (Haverkamp et al. 2006):

Table 1.1 Main approximations of the flux-concentration function and sorptivity

Reference	$F(\theta)$	S^2 [L^2T^{-1}]	Comment
Crank (1979)	$\exp\left\{-\left[inverfc\left(\frac{\theta-\theta_i}{\theta_0-\theta_i}\right)\right]^2\right\}$	$2\,D\int_{\theta_i}^{\theta_0} \frac{\theta-\theta_i}{\exp\left\{-\left[inverfc\left(\frac{\theta-\theta_i}{\theta_0-\theta_i}\right)\right]^2\right\}}\,d\theta$	Linear soil (i.e. constant diffusivity)
Philip and Knight (1974)	$\frac{\theta-\theta_i}{\theta_0-\theta_i}$	$2(\theta_0-\theta_i)\int_{\theta_i}^{\theta_0} D(\theta)\,d\theta$	Soil with Dirac δ-function diffusivity (i.e. Green and Ampt model)
Parlange (1975)	$\frac{2(\theta-\theta_i)}{\theta_0+\theta-2\theta_i}$	$\int_{\theta_i}^{\theta_0} (\theta_0+\theta-2\theta_i)D(\theta)\,d\theta$	Soils with strong non-linear behavior
Brutsaert (1976)	$\left(\frac{\theta-\theta_i}{\theta_0-\theta_i}\right)^{1/2}$	$2(\theta_0-\theta_i)^{1/2}\int_{\theta_i}^{\theta_0} (\theta-\theta_i)^{1/2}D(\theta)\,d\theta$	Soils with moderate non-linear behavior

$\Theta = \frac{\theta-\theta_i}{\theta_0-\theta_i}$, non-dimensional form of volumetric soil water content

$$F(\theta) = \frac{i(\theta, t) - i_i}{i_0(t) - i_i} \tag{1.71}$$

where i_i corresponds to the infiltration rate at time zero (gravity flow at uniform initial water content).

Based on the flux concentration function, Parlange et al. (1982) developed a quasi-exact infiltration solution of the Fokker-Planck type equation [Eq. (1.40)] valid over the entire time range $t \in [0, +\infty]$. This work deals with zero and/or negative surface head condition ($h_0 \leq 0$). The main steps of the demonstration proposed by Parlange et al. (1982) are described below. Considering the case of infiltration into a soil with a uniform initial water pressure head profile, $h_i = h(\theta_i)$, and from a surface boundary source at either null or strictly negative water pressure head ($h_0 \leq 0$), double integration of Eq. (1.40) leads to the following rigorous expansions for the water content profiles, $z(\theta,t)$:

$$z(\theta, t) = \int_{\theta_i}^{\theta} \frac{D(\overline{\theta})}{F(\overline{\theta}, t)[i(t) - K_i] - [K(\overline{\theta}) - K_i]} d\overline{\theta} \tag{1.72}$$

where $\overline{\theta}$ is the integration variable, and $F(\theta,t)$ is the flux concentration function defined above. Integrating z by parts from θ_i to θ_0 and considering the relation between cumulative infiltration and water profile [Eq. (1.62)] yield:

$$I(t) - K_i t = \int_{\theta_i}^{\theta_0} \frac{[\overline{\theta} - \theta_i] D(\overline{\theta})}{F(\overline{\theta}, t)[i_0(t) - K_i] - [K(\overline{\theta}) - K_i]} d\overline{\theta} \tag{1.73}$$

For the integration of Eq. (1.73), Parlange et al. (1982) introduced an integral soil parameter β expressed as a function of conductivity and diffusivity. The relation was later slightly generalized by Haverkamp et al. (1990) to:

$$\frac{[K(\theta) - K_i]}{[K_0 - K_i]} = f(\theta) \left[1 - \frac{2\beta(\theta_i, \theta_0)}{S^2(\theta_i, \theta_0)} \int_{\theta}^{\theta_0} \frac{[\overline{\theta} - \theta_i]}{f(\overline{\theta})} D(\overline{\theta}) \, d\overline{\theta} \right] \tag{1.74}$$

where $f(\theta)$ is defined as the purely diffusivity driven flux concentration relation. Obviously, β is defined over the interval $0 \leq \beta \leq 1$, to ensure a positive value of the right member of Eq. (1.74), and depends on the soil type and initial and boundary conditions. The lower limit $\beta = 0$ corresponds to the Green and Ampt model for which $dK/d\theta$ increases much less rapidly with θ than the diffusivity, whereas the upper limit $\beta = 1$ corresponds to soils for which $dK/d\theta$ and the diffusivity behave similarly. Following Eq. (1.74), parameter β is defined as a function of volumetric water content ($\theta_i \leq \theta \leq \theta_0$); as a result, β is only influenced by the surface boundary condition when $\theta_0 \leq \theta_s$ or $h_0 \leq 0$. Ross et al. (1996) showed that β is slightly affected by changes in the surface boundary condition θ_0, especially when θ_0 stays close to θ_s ($\theta_0 \geq 0.75\ \theta_s$). Haverkamp et al. (1994) suggested the value of 0.6 for parameter β over the whole range for θ.

The integration of Eq. (1.73) was proposed using Eq. (1.74) and considering that the flux concentration function does not evolve with time and that $F(\theta,t) \cong f(\theta)$, leading to the following relationship:

$$I(t) = K_i t + \frac{S^2(\theta_i,\theta_0)}{2\beta[K_0 - K_i]} \ln\left(1 + \beta \frac{K_0 - K_i}{i_0(t) - K_i}\right) \tag{1.75}$$

The cumulative infiltration I is then defined as function of the infiltration flux at surface $i_0(t)$. The derivation in time of Eq. (1.75) leads to the following relation between time and the infiltration flux $i_0(t)$:

$$[K_0 - K_i]\, t = \frac{S^2(\theta_i,\theta_0)}{[1-\beta]\,[K_0 - K_i]} \left(\frac{1}{2\beta} \ln\left(1 + \beta \frac{K_0 - K_i}{i_0(t) - K_i}\right) - \ln\left(1 + \frac{K_0 - K_i}{i_0(t) - K_i}\right)\right) \tag{1.76}$$

The concatenation of the two equations [Eqs. (1.75) and (1.76)] leads to the following implicit relation between time and cumulative infiltration (Haverkamp et al. 1994):

$$\frac{2[K_0 - K_i]^2}{S^2(\theta_i,\theta_0)} t = \frac{1}{[1-\beta]} \frac{2[K_0 - K_i][I(t) - K_i t]}{S^2(\theta_i,\theta_0)} - \frac{1}{1-\beta} \ln\left(\frac{\exp\left(\frac{2\beta[K_0 - K_i][I(t) - K_i t]}{S^2(\theta_i,\theta_0)}\right) + \beta - 1}{\beta}\right) \tag{1.77}$$

Equation (1.77) may appear complex to apply. Some scaling procedures were proposed for Eq. (1.77) and studied by Varado et al. (2006), Ross et al. (1996) and Lassabatere et al. (2009a), leading to the following non-dimensional functions:

$$I^* = \frac{1}{\beta} \ln\left(1 + \frac{\beta}{i_0^* - 1}\right) \tag{1.78a}$$

$$t^* = \frac{1}{\beta(1-\beta)} \ln\left(1 + \frac{\beta}{i_0^* - 1}\right) - \frac{1}{1-\beta} \ln\left(1 + \frac{1}{1 - i_0^*}\right) \tag{1.78b}$$

$$t^* = \frac{1}{(1-\beta)}\left[I^* - \ln\left(\frac{\exp(\beta I^*) + \beta - 1}{\beta}\right)\right] \tag{1.78c}$$

with the following scaled variables:

$$I^* = \frac{2[K_0 - K_i]}{S^2(\theta_i, \theta_0)}[I(t) - K_i t] \tag{1.79a}$$

$$t^* = \frac{2[K_0 - K_i]^2}{S^2(\theta_i, \theta_0)}t \tag{1.79b}$$

$$i^* = \frac{i - K_i}{K_0 - K_i} \tag{1.79c}$$

Any infiltration data set may be calculated using non-dimensional functions (1.78) along with the scaling Eqs. (1.79). On the basis of this approach, Ross et al. (1996) addressed the case of water infiltration with ponding at soil surface (positive pressure head at surface). After mathematical derivation analogous to the development for the case of null or negative water pressure head, these authors obtained the following expansions:

$$I^* = \frac{\sigma}{i_0^* - 1} + \frac{1-\sigma}{\beta}\ln\left(1 + \frac{\beta}{i_0^* - 1}\right) \tag{1.80a}$$

$$t^* = \frac{1-\sigma}{\beta(1-\beta)}\ln\left(1 + \frac{\beta}{i_0^* - 1}\right) + \frac{\sigma}{i_0^* - 1} - \frac{1-\sigma\beta}{1-\beta}\ln\left(1 + \frac{1}{1 - i_0^*}\right) \tag{1.80b}$$

where the terms related to the parameter σ account for positive water pressure head at soil surface and corresponds to the relative contribution of the saturated part of sorptivity to the total sorptivity:

$$\sigma = \frac{2K_s h_0[\theta_s - \theta_i]}{S_{tot}^2} \tag{1.81a}$$

$$S_{tot}{}^2 = 2K_s h_0[\theta_s - \theta_i] + S^2(\theta_i, \theta_s) \tag{1.81b}$$

where h_0 stands for the value of water pressure head at surface.

1.4.6 Extension to 3D Water Infiltration Below a Disk Source

All the models presented above address the case of 1D water infiltration. 3D models may be needed, in particular for water infiltration experiments that makes use of disk or ring infiltrometers. Haverkamp et al. (1994) and Smettem et al. (1994)

proposed to relate 3D cumulative infiltration, $I_{3D}(t)$, from a disk source, to 1D cumulative infiltration, $I_{1D}(t)$ as follows:

$$I_{3D}(t) = I_{1D}(t) + \frac{\gamma S^2(\theta_i, \theta_0)}{r\Delta\theta} t \tag{1.82}$$

where $\Delta\theta$ (= $\theta_0 - \theta_i$) stands for the difference between the final and initial volumetric water contents θ_0 and θ_i, respectively, r is the radius of the disk source, and γ is a shape parameter. Despite Haverkamp et al. (1994) proposed an averaged value of 0.75 for γ, the proper value of this shape parameter is the subject of discussion because γ is known to be dependent on both the soil type and the initial water content (Lassabatere et al. 2009a). In line with the theoretical development by Haverkamp et al. (1994) and Smettem et al. (1994), Fuentes et al. (1992) proposed a direct formulation for the shape parameters γ and β as a function of the boundary and initial conditions and hydraulic functions:

$$\beta(\theta_i, \theta_0) = 2 - 2\frac{\int_{\theta_i}^{\theta_0} \frac{K(\theta) - K_i}{K_0 - K_i} \frac{\theta_0 - \theta_i}{\theta - \theta_i} D(\theta) d\theta}{\int_{\theta_i}^{\theta_0} D(\theta) d\theta} \tag{1.83a}$$

$$\gamma(\theta_i, \theta_0) = \sqrt{0.3}\frac{2(\theta_0 - \theta_i)\int_{\theta_i}^{\theta_0} D(\theta) d\theta}{\int_{\theta_i}^{\theta_0} (\theta_0 + \theta - 2\theta_i) D(\theta) d\theta} \tag{1.83b}$$

The analytical model described by the combination of Eqs. (1.77) and (1.82) is quite commonly used for water infiltration experiments using disk infiltrometers (Angulo-Jaramillo et al. 2000; Haverkamp et al. 2006; Lassabatere et al. 2009b). In fact, Eq. (1.77) may appear a little complex to compute since it is implicit and the determination of the function $I(t)$ requires to find the root of Eq. (1.77). This complex formulation is often replaced by its asymptotic behaviors close to zero and infinity. Haverkamp et al. (1994) derived the following direct time expansions leading to the following relationships for 3D cumulative infiltration for very short times, ${I_{3D}}^{O(1)}$, short times, ${I_{3D}}^{O(2)}$, and steady-state, ${I_{3D}}^{+\infty}$, respectively, (Lassabatere et al. 2009a):

$$I_{3D}^{O(1)}(t) = S\sqrt{t} \tag{1.84a}$$

$$I_{3D}^{O(2)}(t) = S\sqrt{t} + \left(\frac{(2-\beta)}{3}\Delta K + K_i + \frac{\gamma S^2}{r\Delta\theta}\right) t \tag{1.84b}$$

$$I_{3D}^{+\infty}(t) = \left(K_0 + \frac{\gamma S^2}{r\Delta\theta}\right)t + \frac{1}{2(1-\beta)}\ln\left(\frac{1}{\beta}\right)\frac{S^2}{\Delta K} \tag{1.84c}$$

where S stands for sorptivity ($S_0(\theta_i,\theta_0)$) and Δ refers to the operator difference between final and initial state. Time derivatives of Eqs. (1.84) provide corresponding models for infiltration rates for very short times, ${i_{3D}}^{O(1)}$, short times, ${i_{3D}}^{O(2)}$, and steady-state, ${i_{3D}}^{+\infty}$, respectively:

$$i_{3D}^{O(1)}(t) = \frac{S}{2\sqrt{t}} \tag{1.85a}$$

$$i_{3D}^{O(2)}(t) = \frac{S}{2\sqrt{t}} + \left(\frac{(2-\beta)}{3}\Delta K + K_i + \frac{\gamma S^2}{r\Delta\theta}\right) \tag{1.85b}$$

$$i_{3D}^{+\infty}(t) = K_0 + \frac{\gamma S^2}{r\Delta\theta} \tag{1.85c}$$

An illustrative example of the global analytical model [Eqs. (1.77) and (1.82)] along with its expansions Eqs. (1.84) and time derivatives is given in Fig. 1.15. The use of the approximate equations requires knowledge of their validity domains, but few studies have already focused on the definition of these time intervals and these are the subject of ongoing research (Lassabatere et al. 2009a).

Lassabatere et al. (2009a) studied the complete model [set of Eqs. (1.77) and (1.82)] against numerically generated data. These authors modelled 3D and 1D cumulative infiltrations for zero water pressure head at surface and several initial water contents over a disk with a radius of 10 cm and for different types of soil. The difference between 3D and 1D infiltration data was fitted to Eq. (1.82). The 1D infiltration data was fitted to Eq. (1.77). The results are presented in terms of scaled data (Fig. 1.16). Note that all cumulative infiltrations align on one sole curve after use of scaling equations. The fits of numerically generated data is particularly accurate and the values of parameters γ and β could be properly estimated. The

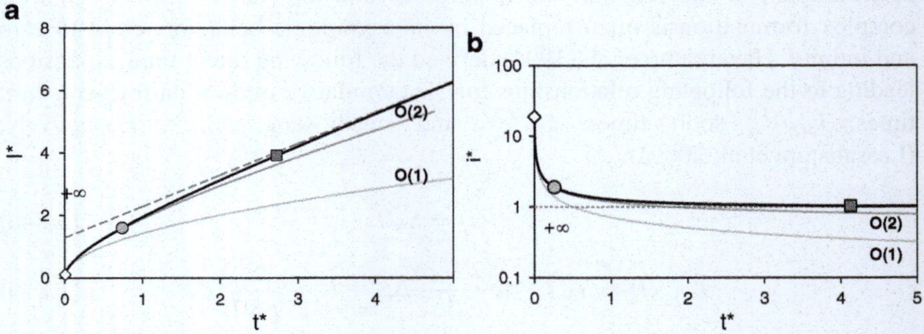

Fig. 1.15 Illustrative example of analytical models of water infiltration (Eqs. 1.77 and 1.84): (**a**) cumulative infiltration and (**b**) infiltration rate

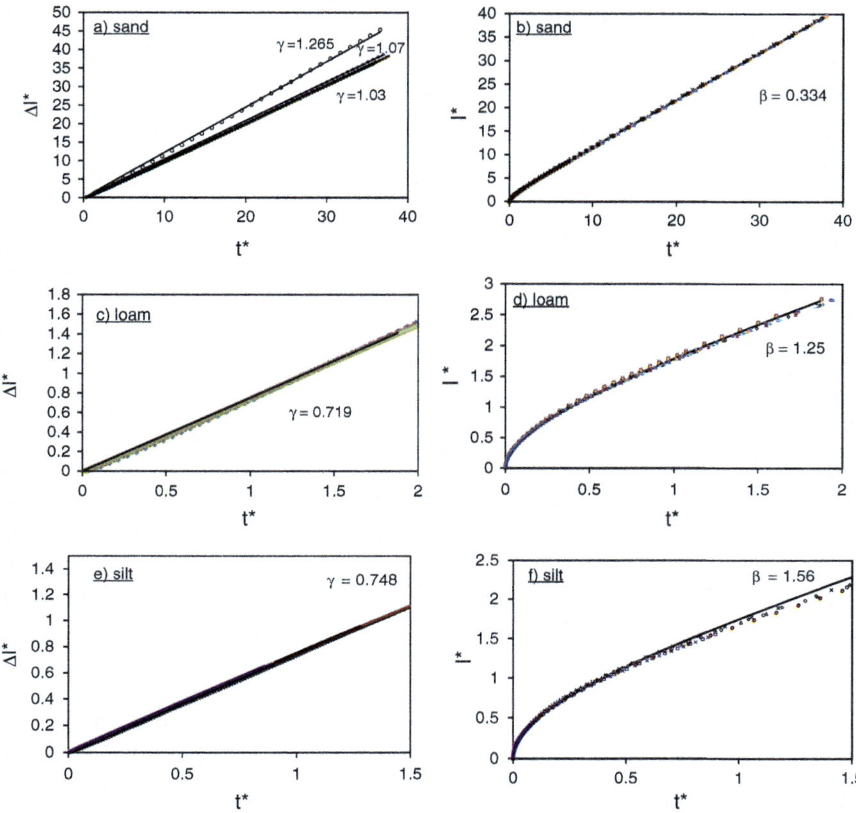

Fig. 1.16 Validations of Eqs. (1.77) and (1.82) with regards to numerical data for three types of soils. *Left*: relation between 3D and 1D cumulative infiltrations; *right*: 1D cumulative infiltration. The values of optimum parameters are indicated. Modelled and numerically generated data are presented as scaled data

values found for parameter γ remained in usual values and close to that derived from Eq. (1.83b) proposed by Fuentes et al. (1992). In opposite, the values of parameter β exceeds unity and differ from the values derived from Eq. (1.83a). Further research is still needed to understand and quantify these parameters with more precision.

References

Abu-Sharar, T. M., Bingham, F. T., & Rhoades, J. D. (1987). Stability of soil aggregates as affected by electrolyte concentration and composition. *Soil Science Society of America Journal, 51*, 309–314.

Angulo-Jaramillo, R., Vandervaere, J.-P., Roulier, S., Thony, J. L., Gaudet, J. P., & Vauclin, M. (2000). Field measurement of soil surface hydraulic properties by disc and ring infiltrometers. A review and recent developments. *Soil & Tillage Research, 55*, 1–29.

Bagarello, V., & Iovino, M. (2010). *Conducibilità idraulica del suolo – Metodi di misura nelle applicazioni idrologiche*. Milano: Ulrico Hoepli Editore S.p.A, 382 pp. ISBN 978-88-203-4411-5.

Bagarello, V., Provenzano, G., & Sgroi, A. (2009). Fitting particle size distribution models to data from Burundian soils for the BEST procedure and other purposes. *Biosystems Engineering, 104*, 435–441. doi:10.1016/j.biosystemseng.2009.07.008.

Boltzmann, L. (1894). Zur Integration Der Diffusionsgleichung Bei Variabeln Diffusionscoefficienten. *Annalen Der Physik, 289*(13), 959–964.

Brooks, R. H., & Corey, A. T. (1964). *Hydraulic properties of porous media* (Hydrology paper 3). Fort Collins: Colorado State University.

Brooks, R. H., & Corey, A. T. (1966). Properties of porous media affecting fluid flow. *Journal of Irrigation and Drainage Division, IR2*, 61–68.

Brutsaert, W. (1976). The concise formulation of diffusive sorption of water in a dry soil. *Water Resources Research, 12*(6), 1118–1124.

Cavazza, L. (1981). *Fisica del terreno agrario*. Torino: UTET, 589 pp.

Crank, J. (1979). *The mathematics of diffusion*. Oxford: Oxford University Press.

Culley, J. L. B. (1993). Chapter 50: Density and compressibility. In M. R. Carter (Ed.), *Soil sampling and methods of analysis* (pp. 529–539). Boca Raton: Canadian Society of Soil Science, Lewis Publishers.

Dexter, A. R., Czyż, E. A., Richard, G., & Reszkowska, A. (2008). A user-friendly water retention function that takes account of the textural and structural pore spaces in soil. *Geoderma, 143*, 243–253.

Elrick, D. E., & Robin, M. J. (1981). Estimating the sorptivity of soils. *Soil Science, 132*(2), 127–133.

Fredlund, M. D., Fredlund, D. G., & Wilson, G. W. (2000). An equation to represent grain-size distribution. *Canadian Geotechnical Journal, 37*, 817–827.

Fuentes, C., Haverkamp, R., & Parlange, J. Y. (1992). Parameter constraints on closed-form soil water relationships. *Journal of Hydrology, 134*(1), 117–142.

Gardner, W. R. (1958). Some steady state solutions of the unsaturated moisture flow equation with application to evaporation from a water table. *Soil Science, 85*(4), 228–232.

Green, W. H., & Ampt, G. A. (1911). Studies in soil physics. I. The flow of air and water through soils. *The Journal of Agricultural Science, 4*, 1–24.

Haverkamp, R., & Parlange, J. Y. (1986). Predicting the water-retention curve from particle-size distribution: 1. Sandy soils without organic matter. *Soil Science, 142*(6), 325–339.

Haverkamp, R., Parlange, J. Y., Starr, J. L., Schmitz, G., & Fuentes, C. (1990). Infiltration under ponded condition.3. A predictive equation based on physical parameters. *Soil Science, 149*(5), 292–300.

Haverkamp, R., Ross, P. J., Smettem, K. R. J., & Parlange, J. Y. (1994). 3-Dimensional analysis of infiltration from the disc infiltrometer. 2. Physically-based infiltration equation. *Water Resources Research, 30*(11), 2931–2935.

Haverkamp, R., Leij, F. J., Fuentes, C., Sciortino, A., & Ross, P. J. (2005). Soil water retention: I. Introduction of a shape index. *Soil Science Society of America Journal, 69*, 1881–1890.

Haverkamp, R., Debionne, S., Viallet, P., Angulo-Jaramillo, R., de Condappa, D. (2006). Chapter 6. Soil properties and moisture movement in the unsaturated zone. In J. W. Delleur (Ed.), *The handbook of groundwater engineering* (2nd ed., pp. 6-1–6.60). Boca Raton: CRC Press, Print ISBN: 978-0-8493-4316-2.

Hillel, D. (1998). *Environmental soil physics*. San Diego: Academic, 771 pp.

Hillel, D. (2004). *Introduction to environmental soil physics*. San Francisco: Elsevier Science.

Hwang, S. I. (2004). Effect of texture on the performance of soil particle-size distribution models. *Geoderma, 123*, 363–371.

Hwang, S. I., Lee, K. P., Lee, D. S., & Powers, S. E. (2002). Models for estimating soil particle-size distribution. *Soil Science Society of America Journal, 66*, 1143–1150.

Keren, R., & Ben-Hur, M. (2003). Interaction effects of clay swelling and dispersion and $CaCO_3$ content on saturated hydraulic conductivity. *Australian Journal of Soil Research, 41*, 979–989.

Kutílek, M., & Valentová, J. (1986). Sorptivity approximations. *Transport in Porous Media, 1*(1), 57–62.

Lassabatere, L., Angulo-Jaramillo, R., Soria Ugalde, J. M., Cuenca, R., Braud, I., & Haverkamp, R. (2006). Beerkan estimation of soil transfer parameters through infiltration experiments – BEST. *Soil Science Society of America Journal, 70*, 521–532.

Lassabatere, L., Angulo-Jaramillo, R., Soria-Ugalde, J. M., Simunek, J., & Haverkamp, R. (2009a). Numerical evaluation of a set of analytical infiltration equations. *Water Resources Research, 45*, 12415.

Lassabatere, L., Peyrard, X., Angulo-Jaramillo, R., & Simunek, J. (2009b). *Effects of the hydraulic conductivity of the matrix/macropore interface on cumulative infiltrations into dual-permeability media.* AGU Fall Meeting Abstracts, December, A764.

Mualem, Y. (1976). A new model for predicting the hydraulic conductivity of unsaturated porous media. *Water Resources Research, 12*, 513–522.

Parlange, J. Y. (1975). On solving the flow equation in unsaturated soils by optimization: Horizontal infiltration. *Soil Science Society of America Proceedings, 39*, 415–418.

Parlange, J. Y., Lisle, I., Braddock, R. D., & Smith, R. E. (1982). The three parameter infiltration equation. *Soil Science, 133*, 337–341.

Parlange, J. Y., Barry, D. A., Parlange, M. B., Lockington, D. A., & Haverkamp, R. (1994). Sorptivity calculation for arbitrary diffusivity. *Transport in Porous Media, 15*(3), 197–208.

Parslow, J., Lockington, D., & Parlange, J. Y. (1988). A new perturbation expansion for horizontal infiltration and sorptivity estimates. *Transport in Porous Media, 3*(2), 133–144.

Philip, J. R. (1957a). Numerical solution of equations of the diffusion type with diffusivity concentration dependent. II. *Australian Journal of Physics, 10*(1), 29–42.

Philip, J. R. (1957b). The theory of infiltration: 1. The infiltration equation and its solution. *Soil Science, 83*(5), 345–358.

Philip, J. R. (1957c). The theory of infiltration: 2. The profile at infinity. *Soil Science, 83*(6), 257–264.

Philip, J. R. (1957d). The theory of infiltration: 4. Sorptivity and algebraic infiltration equations. *Soil Science, 84*(3), 257–264.

Philip, J. R. (1959). The early stage of absorption and infiltration. *Soil Science, 88*(2), 91–97.

Philip, J. R. (1973). On solving the unsaturated flow equation: 1. The flux-concentration relation. *Soil Science, 116*, 1–7.

Philip, J. R., & Knight, J. H. (1974). On solving the unsaturated flow equation: 3. New quasi-analytical technique. *Soil Science, 117*(1), 1–13.

Reynolds, W. D., Drury, C. F., Yang, X. M., Fox, C. A., Tan, C. S., & Zhang, T. Q. (2007). Land management effects on the near-surface physical quality of a clay loam soil. *Soil & Tillage Research, 96*, 316–330.

Reynolds, W. D., Drury, C. F., Tan, C. S., Fox, C. A., & Yang, X. M. (2009). Use of indicators and pore volume-function characteristics to quantify soil physical quality. *Geoderma, 152*, 252–263.

Reynolds, W. D., Drury, C. F., Yang, X. M., Tan, C. S., & Yang, J. Y. (2014). Impacts of 48 years of consistent cropping, fertilization and land management on the physical quality of a clay loam soil. *Canadian Journal of Soil Science, 94*, 403–419.

Rogowski, A. S. (1971). Watershed physics: Model of soil moisture characteristics. *Water Resources Research, 7*, 1575–1582.

Ross, P. J., Haverkamp, R., & Parlange, J.-Y. (1996). Calculating parameters of infiltration equations from soil hydraulic functions. *Transport in Porous Media, 24*, 315–339.

Smettem, K. R. J., Parlange, J. Y., Ross, P. J., & Haverkamp, R. (1994). 3-Dimensional analysis of infiltration from the disc infiltrometer 1. A capillary-based theory. *Water Resources Research, 30*(11), 2925–2929.

Smith, R. E. (2002). *Infiltration theory for hydrologic applications* (Water resources monograph 15). Washington, DC: American Geophysical Union, 212 p. ISBN 0-87590-319-3.

Swartzendruber, D. (1987). Rigorous derivation and interpretation of the Green and Ampt equation. In Y.-S. Fok (Ed.), *Infiltration development and application.* Honolulu: Water Resources Research Center, University of Hawaii.

Talsma, T. (1969). In situ measurement of sorptivity. *Soil Research, 7*(3), 269–276.

Tisdall, J. M., & Oades, J. M. (1982). Organic matter and water-stable aggregates in soils. *Journal of Soil Science, 62*, 141–163.

van Genuchten, M. T. (1980). A closed-form equation for predicting the hydraulic conductivity of unsaturated soil. *Soil Science Society of America Journal, 44*, 892–898.

Varado, N., Braud, I., Ross, P. J., & Haverkamp, R. (2006). Assessment of an efficient numerical solution of the 1D Richards' equation on bare soil. *Journal of Hydrology, 323*(1–4), 244–257.

Chapter 2
Saturated Soil Hydraulic Conductivity

2.1 Well or Borehole Permeameters

Well or borehole permeameter methods allow subsurface measurement of soil hydrodynamic properties in a vadose zone with a deep water table. Early analysis of flow out of a borehole was carried out by Glover (in Zangar 1953), but the success of these methods has largely to be attributed to D.E. Elrick and W.D. Reynolds. Starting in the 1980s, these Canadian scientists improved substantially the theoretical treatment of the flow process out of the well. They also developed analytical solutions of great utility for a reliable soil hydraulic characterization, explicitly considering pressure, capillarity and gravity effects on flow from the water source.

Permeameter experiments are commonly carried out by maintaining a constant depth of water in an uncased borehole until near steady flow conditions have been reached. The commercially available Guelph permeameter (GP) allows relatively simple borehole experiments to be performed in the field. Soil hydrodynamic characteristics are then obtained by analytical relationships requiring one or more measurements of steady-state flow. Recently, alternative experimental and analytical procedures, that consider transient flow conditions and/or cased boreholes, have also been developed. In particular, advances mainly concern steady flow under constant head in cased boreholes, transient flow under constant head in both cased and uncased boreholes, and falling head infiltration from cased boreholes. These developments have still received little field testing but they appear promising to extend measurement of soil hydrodynamic properties below the soil surface to conditions that differ from those originally considered for the GP method.

R. Angulo-Jaramillo et al., *Infiltration Measurements for Soil Hydraulic Characterization*,
DOI 10.1007/978-3-319-31788-5_2

2.1.1 *Steady Flow Under Constant Head in Uncased Boreholes*

2.1.1.1 Description of the Method

Measuring steady flow out of an uncased borehole where a constant depth of water is established allows us first of all to determine the field-saturated soil hydraulic conductivity, K_{fs} (L T^{-1}), but also the matric flux potential, ϕ_m (L^{2}T^{-1}), the soil sorptivity, S (L T$^{-1/2}$), the so-called α^* (L^{-1}) parameter or sorptive number or macroscopic capillary length parameter, and the effective Green and Ampt (1911) wetting front pressure head, h_f (L) (Reynolds and Elrick 2002a, c). The Guelph permeameter (GP) is the device and the method, commonly named well or borehole permeameter method, is also called the shallow well pump-in method in the engineering literature (Reynolds and Elrick 2005).

By the GP, a positive head of water, H (L), is established in a small well having a flat bottom excavated to the desired depth (Fig. 2.1a), and a three-dimensional infiltration process through the walls and the base of the well into the initially unsaturated soil is activated. This process determines the development of a field-saturated soil region or bulb that extends outwards from the infiltration surface (Elrick and Reynolds 1992b). Enveloping the field-saturated bulb is an unsaturated wetting zone that extends from the bulb surface to the wetting front (Fig. 2.2a). Constant head flow into unsaturated soil goes through an initial transient phase and then approaches steady-state. During the transient phase, both the field-saturated bulb and the wetting zone increase in size by migrating downward and outward from the infiltration surface. After steady flow is attained, the field-saturated bulb remains essentially constant in size and shape, while the wetting zone continues to increase in size by outward migration of the wetting front.

In terms of design, the GP comprises of a support tube, with the base placed in contact with the bottom of the well, a reservoir assembly, and a system to control the ponded depth of water into the well. In the device marketed by the Soilmoisture Equipment Corp. (Santa Barbara, CA, U.S.A.), all parts are in plastics (Fig. 2.1b). The support tube supports the reservoir assembly and conducts water in the well. Outflow from the device occurs by a water outlet tip. Ribbed vents at the bottom of the tip disperse the energy of outflowing water to minimize soil particle detachment from the walls and the bottom of the well during the run. The reservoir assembly is made up of two concentric cylindrical reservoirs, providing a means of storing water and measuring the outflow rate while the GP is in use. With the same objective, a scale stamped on the inner reservoir is used to read water levels at given times and hence to determine the rate of fall in water level in the device during the run. A three-way valve at the bottom of the concentric reservoir allows selection of the reservoir assembly configuration to be used for a run (inner reservoir alone, inner and outer reservoirs) on the basis of the sampled soil characteristics, to provide in any case adequate resolution of outflow rate when making a reading. The reservoirs are filled through a fill port located on the

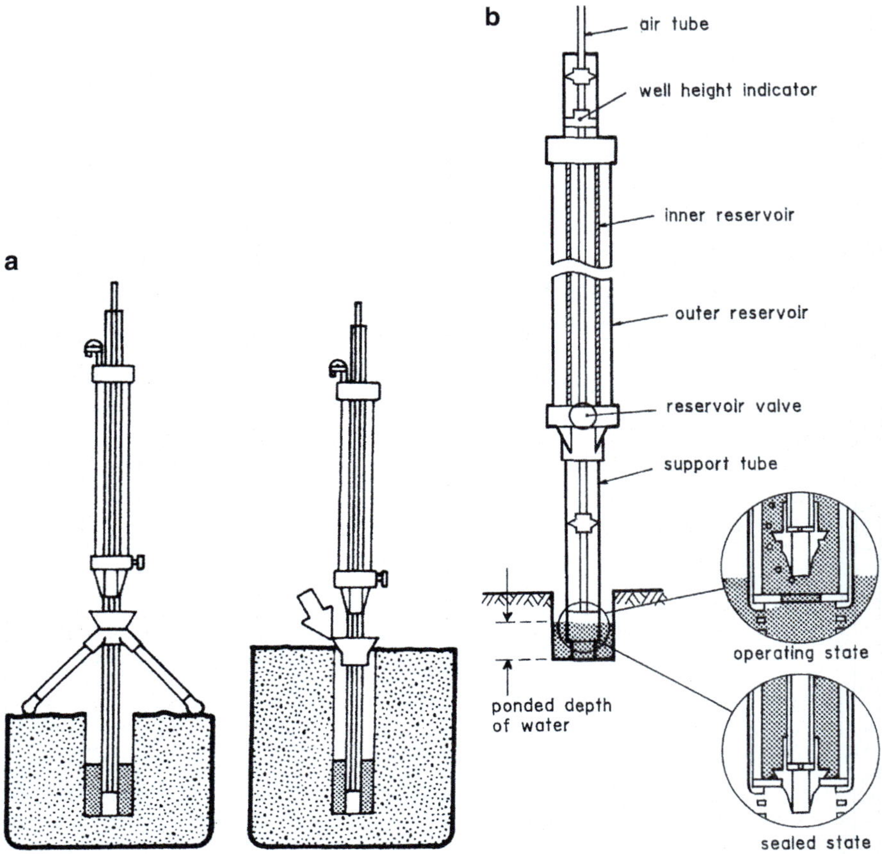

Fig. 2.1 Layout (**a**) and parts (**b**) of the commercial Guelph permeameter (cross-sectional area = 35.39 cm^2 for the combined reservoirs and 2.14 cm^2 for the inner reservoir; maximum permeameter capacity = 3.18 L; well height range = 2.5–25 cm) (From Soilmoisture Equipment Corp. 1987, reprinted with permission)

reservoir cap that remains sealed with a plug when the device is used to make a measurement. The ponded depth of water in the well is regulated by an air tube, open at both ends, sliding along the axis of the device within the inner reservoir and the support tube. On the reservoir cap, a scale allows easy regulation of the head established in the well, H, that can be changed by raising or lowering the air tube, grasped at the upper end. A tripod is used to support the permeameter in wells down to approximately 38 cm in depth (Fig. 2.1a). For use in wells deeper than 38 cm, the operating instructions of the device suggest that the tripod bushing alone, i.e. a truncated cone-shaped fitting sliding along the support tube, provides the functions of centering and stabilizing the permeameter. In reality, this fitting alone does not assure firmness of the device during the run. The GP comes as a kit and it can be

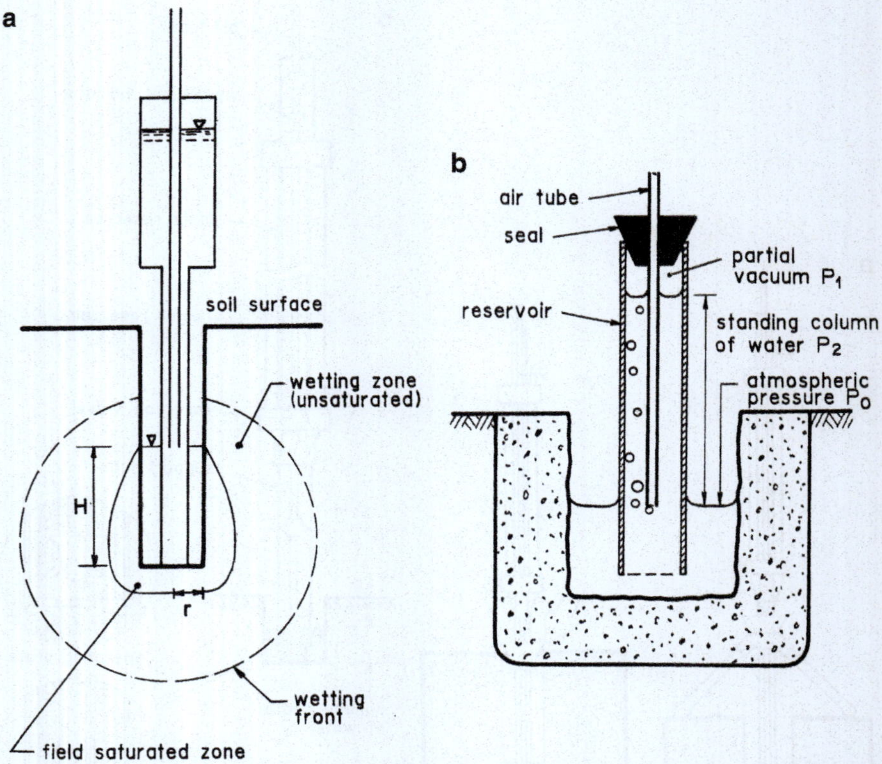

Fig. 2.2 Schematic of the (**a**) flow field outwards from the infiltration surface and (**b**) operating principle of the Guelph permeameter according to Soilmoisture Equipment Corp. (1987, reprinted with permission)

broken down into several segments for convenient storage in an easily transportable carrying case. Auxiliary tools included in the carrying case are the devices to prepare 6-cm-diam. wells, although the device is usable in 2- to 10-cm-diam. wells (Reynolds and Elrick 2002c). The established ponded depths of water in the well may vary from 2.5 to 25 cm. With the commercially available device, the range of measurable K_{fs} values is $10^{-8}-10^{-4}$ m s^{-1}. Obviously, the GP can only be used for determining subsurface soil properties. Sampling depths can also be noticeable considering, for example, that Gwenzi et al. (2011) applied the method for sampling soil of an artificial ecosystem at 15 cm intervals up to a depth of 3 m. According to Reynolds and Elrick (2005), however, the most common form of the well permeameter method uses wells that are 0.04–0.1 m in diameter by 0.1–1.0 m deep.

Figure 2.2b shows a flow chart of the operating principle of the device. The ponded water depth in the well depends on the level of the bottom of the air tube. As the water level in the reservoir falls, a vacuum is created in the air space above the water. On the basis of the Mariotte principle, the sum of this partial vacuum, P_1, and the pressure, P_2, exerted by the water column from the water surface in the reservoir to the water surface in the well has to be equal to the atmospheric pressure, P_0. Whenever the water level in the well begins to drop below the level of the lower end of the air tube, due to the infiltration process into the soil, the system tends to return to normality. In particular, air bubbles entering the air tube rise into the reservoir air space. The vacuum is then partially relieved and water flows from the reservoir into the well, re-establishing the fixed depth of ponding. Since these processes occur almost simultaneously, the ponded depth of water in the well remains practically constant.

Excavating a cylindrical well having a reasonably flat bottom, as required, is in practical terms difficult to achieve but is a critical step in the experimental GP procedure (Reynolds 1993). The implements included in the commercial GP kit to prepare a 6-cm-diam. well are a soil auger, a sizing auger and a well prep brush (Fig. 2.3). The soil auger is used to remove bulk amounts of soil up to the surface of the soil layer that has to be sampled. The well hole is augered by rotating the handle, maintained in a vertical position to avoid undesired enlargement of the well, in a clockwise direction while applying a steady, somewhat firm, downward pressure on the handle. When the bucket is full, the auger is extracted from the hole and it is inverted so that the collected soil slips out of the open end of the bucket. The sizing auger is used as a finishing tool to produce a proper sized well hole of uniform geometry and to clean debris off the bottom of the well hole. Generally, the sizing

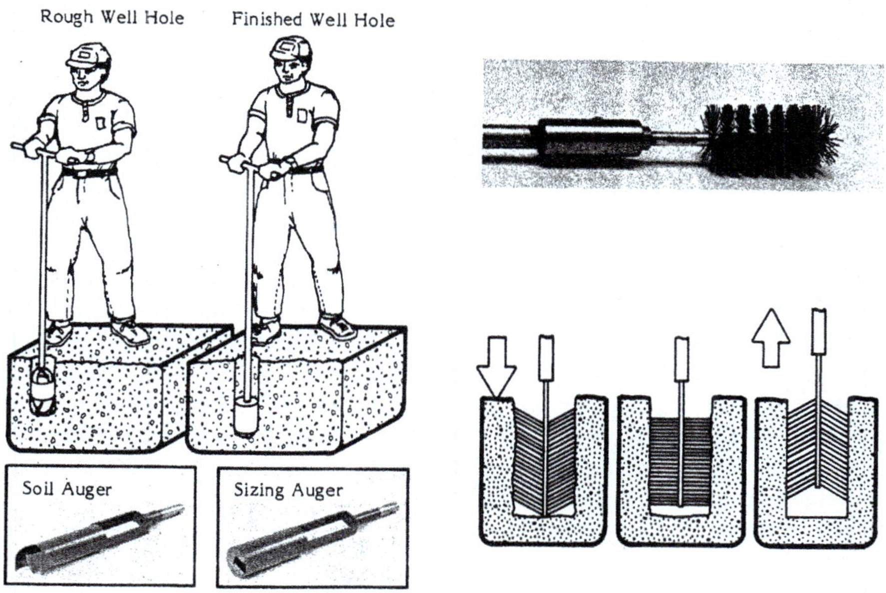

Fig. 2.3 Implements included in the Guelph permeameter kit to prepare the well (From Soilmoisture Equipment Corp. 1987, reprinted with permission)

auger is used for excavating the last 15 cm of the well, constituting the measurement zone. In gravelly soils, it may be necessary to use the soil auger to excavate all the way to the bottom of the well hole. The sizing auger is then used afterwards to clean loose debris off the bottom of the well hole. According to the manual of the device, in moist soils or in medium to fine-textured soils the procedure of augering the hole may create a smear layer, which can block the natural flow of water out of the well into the surrounding soil. In order to obtain reliable data, this smear layer has to be removed and the well prep brush, having an outside diameter slightly larger than the diameter of the well, can be used for this purpose. The brush is pushed into the well until reaching the bottom and then it is extracted. When the direction of the brush is reversed, the bristles dig into the sides of the well, roughening the surface and scouring the smear layer. This operation should not be done more than once or twice, to avoid enlarging the well diameter beyond the desirable limits needed to obtain accurate results. The difficulty of removing the smear layer increases with increasing wetness of the soil. Therefore, fine-textured soils should not be augered when their water content is very high. A flashlight can be used to check the well before applying the instrument.

When the well is ready, the procedure for the commercial GP manufactured by Soilmoisture Equipment Corp. (1987) prescribes connecting the inner and outer reservoirs to each other and filling the permeameter with water. It is important to verify that the air tube is completely lowered and that water does not drip from the base of the device. The well height indicator has to be correctly placed at $H=0$ before starting the run (sealed state preventing water infiltration). In addition, the fill plug must be fully seated against the reservoir cap. The permeameter is then placed into the well (Fig. 2.1a) and the air tube is slowly lifted out of the outlet tip to establish and maintain the desired depth of water in the well. Raising the air tube too quickly can cause turbulence and erosion in the well and a potential surging effect that could cause the well to temporarily overfill. The device is operating properly when air bubbles rise regularly up through the permeameter and into the reservoir. A drop of water in the reservoir without any bubble in the Mariotte-based permeameter may indicate air (vacuum) leak in the reservoir (Reynolds and Elrick 2002c). At the beginning of the run, the appropriate reservoir assembly (inner and outer reservoirs or inner reservoir only) and the time interval between two successive readings at the reservoir of the device, Δt (T), have to rapidly be chosen. The operating instructions suggest observing the initial fall in water level in the combined reservoir (inner and outer) having a total cross-sectional area of approximately 35 cm^2. Considering that the experimental information to be used in calculations is the steady-state flow rate, it is not strictly necessary to record the rate of fall in the reservoir of the device starting from the initial stages of the infiltration process. If this rate of fall is too slow to easily distinguish the drop in level between consecutive readings taken at a reasonable time interval, it is suggested that the outer reservoir be excluded by turning the valve at the base of the reservoir assembly. In this case, the difference between two successive readings will be larger for a given Δt because water is provided by the smaller reservoir, having a cross-sectional area of approximately 2 cm^2. Once the appropriate reservoir is selected, the reservoir valve must not be changed. The standard Δt value is 2 min but different values of Δt can be chosen, depending on the sampled soil characteristics. For example, a longer time interval (e.g., $\Delta t \geq 15$ min) can be more appropriate when investigating clay soils or situations where the rate of fall is so slow that a 2-min time interval does not allow us to detect a measurable change in the reservoir water level. When the rate of fall is very fast, such as in sandy soils, $\Delta t = 15$ s may be more appropriate. To improve flow measuring accuracy of Mariotte-based permeameters, some practitioners recommend timing intervals that produce a ≥ 1 cm drop in reservoir water level between readings (Reynolds and Elrick 2002c). A few preliminary applications of the device in the area to be sampled may provide data on the expected rates of fall, although these data are unavoidably approximate due to the high spatial variability of the infiltration process in the unsaturated soil. Water level readings collected with an automated pressure transducer-datalogger system are simpler and potentially less uncertain than the visual ones.

The difference of two successive readings of water level in the reservoir of the device divided by the time interval between the readings is the rate of fall of the water level in the reservoir, R (L T^{-1}). The cross-sectional area of the reservoir used for the run, A_s (L^2), multiplied by R gives the flow rate, Q (L^{3}T^{-1}), i.e. the volume of water entering the soil per unit time. The rate of fall is monitored until a practically constant value of R (R_s) is obtained in three to five consecutive time intervals. The product of R_s and A_s yields an estimate of the steady-state flow rate, Q_s (L^{3}T^{-1}), that is used to calculate K_{fs} and the other soil properties. According to Reynolds and Elrick (2002c), the late-time flow from constant head well permeameters has to be considered as quasi-steady, as true steady flow will occur only in completely homogeneous soil, which probably never exists in natural environments. Hence, a reasonable estimate of steady flow, i.e. the quasi-steady flow, is considered acceptable.

2.1.1.2 Calculation of Soil Parameters

The original well permeameter analysis was based on the Glover relationship (in Zangar 1953) that simply yields an estimate of K_{fs} using the steady-state flow rate measured, for a constant head of water, in a well of given radius. However, this relationship neglects the capillarity and gravity components of flow out of the well and it makes use of a shape factor that is estimated on the basis of the Laplace equation for saturated flow around the well (Reynolds and Elrick 2002c, 2005). As a result, the Glover equation can overestimate K_{fs} by an order of magnitude or more in dry, fine-textured soils, although the degree of overestimation can be reduced somewhat by establishing large depths of water in wells of small radius to reduce errors in K_{fs} resulting from lack of account for gravity and capillarity (Reynolds and Elrick 2002c).

More recent and realistic theoretical treatment of the flow process out of the well has allowed us to improve soil hydraulic characterization using one or more H-

levels in the well. Measurement of quasi steady-state flow rate for a single H value (One-Ponding-Depth, OPD, approach) allows us determination of K_{fs}, ϕ_m and S. Measuring quasi steady-state flow rates for two or more H-levels (Two-Ponding-Depth, TPD, and Multiple-Ponding-Depth, MPD, approaches, respectively) allows us to determine K_{fs}, ϕ_m, S, α^* and h_f. In particular, applying the TPD approach requires measuring quasi-steady flow rate for two H values (H_1 and H_2, $H_2 > H_1$), established in rapid succession to avoid drainage processes in the passage from H_1 to H_2. In practice, after having concluded the first measurement of Q_s (Q_{s1}, corresponding to H_1), the air tube is suddenly raised to establish a new ponded depth of water into the well (H_2) and the infiltration process is monitored until a new quasi steady-state condition is attained (Q_{s2}). The MPD approach involves measurement of quasi steady-state flow rate for several H values; H_1 is ponded first with $H_1 < H_2 < H_3 < \ldots$, and the water level in the well is not allowed to fall when switching from one head to the next higher head (Reynolds and Elrick 2002c).

Steady flow from a cylindrical, uncased well into homogeneous, isotropic, rigid and initially uniformly unsaturated soil can be represented by the following relationship (Reynolds and Elrick 1985b, 1987):

$$Q_s = \frac{2\pi H}{C}\left(K_{fs}H + \phi_m\right) + \pi r^2 K_{fs} \tag{2.1}$$

where Q_s (L^3T^{-1}) is the steady-state flow rate, H (L) is the constant height of ponded water in the well, C is a dimensionless shape factor, K_{fs} ($L\,T^{-1}$) is the field-saturated soil hydraulic conductivity, r (L) is the radius of the well, and ϕ_m (L^2T^{-1}) is the matric flux potential, defined by (Gardner 1958):

$$\phi_m = \int_{h_i}^{0} K(h)\ dh \qquad h_i \leq h \leq 0 \tag{2.2}$$

where h (L) is soil water pressure head, h_i (L) is the initial value of h and K ($L\,T^{-1}$) is the soil hydraulic conductivity. The first term on the right of Eq. (2.1) represents the hydraulic push of water into the soil due to the hydrostatic pressure of the water in the well, the second term represents the matric pull of water out of the well due to capillary forces in the soil, while the third term represents the gravitational pull of water through the bottom of the well. The first and the third terms of Eq. (2.1) may be thought of as the field-saturated component of flow out of the well, and the second term as the unsaturated or capillarity flow component. The C factor, depending primarily on the H/r ratio and, to a lesser degree, on soil type, can be estimated with the relationships obtained by Reynolds and Elrick (1987) using numerically simulated data and depicted in Fig. 2.4. These curves apply for $1 \leq r \leq 5$ cm, $0.5 \leq H \leq 20$ cm and $0.25 \leq H/r \leq 20$ (Reynolds and Elrick 2002c). As an alternative, relationships interpreting the curves of Fig. 2.4 can be used. In particular, Bosch and West (1998) suggested the following polynomial relationship:

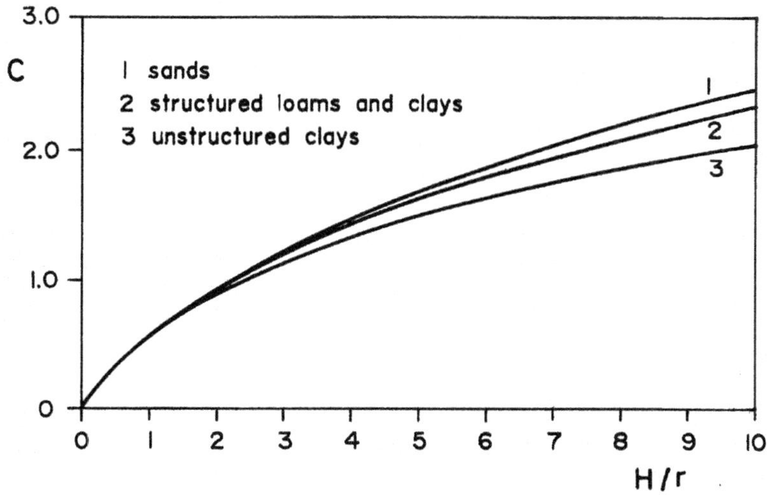

Fig. 2.4 Shape factor, C, versus the ratio between the ponded depth of water into the well, H, and the well radius, r, for different soil types (From Soilmoisture Equipment Corp. 1987, reprinted with permission)

$$C = A1 + A2\frac{H}{r} + A3\left(\frac{H}{r}\right)^2 + A4\left(\frac{H}{r}\right)^3 \tag{2.3}$$

where the $A1$, $A2$, $A3$ and $A4$ coefficients were obtained through fitting numerical results by Reynolds and Elrick (1987) for the same soil types. Zhang et al. (1998) proposed the following relationships for sands (Eq. 2.4a), structured loams and clays (Eq. 2.4b), and unstructured clays (Eq. 2.4c), respectively:

$$C = \left(\frac{H/r}{2.074 + 0.093 \times H/r}\right)^{0.754} \tag{2.4a}$$

$$C = \left(\frac{H/r}{1.992 + 0.091 \times H/r}\right)^{0.683} \tag{2.4b}$$

$$C = \left(\frac{H/r}{2.102 + 0.118 \times H/r}\right)^{0.655} \tag{2.4c}$$

It should be noted that Eq. (2.4c) is written in a slightly different form in Reynolds and Elrick (2005). In particular, 0.672, 2.081 and 0.121 replace 0.655, 2.102 and 0.118, respectively. According to Reynolds and Elrick (2002c), Eqs. (2.4a, 2.4b and 2.4c) are more accurate than Eq. (2.3). The relationships shown in Fig. 2.4 and described by Eqs. (2.3) and (2.4a, 2.4b or 2.4c) are usable under the hypotheses that the water table or the impermeable layer do not influence the infiltration process established with the GP. Using numerically simulated data, Hayashi and

Quinton (2004) suggested that the C values by Reynolds and Elrick (1987) can be used when the distance between the bottom of the well and the underlying impermeable layer is larger than $2H$. In practice, the bottom of the well should be at least 20–50 cm above the water table or the capillary fringe to avoid possible interference caused by mounding of the water table up into the well (Reynolds and Elrick 2002c). Moreover, the curves in Fig. 2.4 are expected to be sufficiently accurate for practical applications as long as the wetting front from the well does not appear on the soil surface before Q_s is attained (Reynolds and Elrick 1987). Therefore, methods for describing steady flow out of a well excavated in the unsaturated soil have progressed from the traditional Glover analysis based solely on saturated flow and hydrostatic pressure contribution to further analyses that allow a more realistic description of the physics of well permeameter flow since they include saturated-unsaturated flow and hydrostatic pressure, gravity and capillarity contributions to flow (Reynolds and Elrick 1987).

Equation (2.1), containing two unknowns (K_{fs} and ϕ_m), cannot be solved directly using a single measurement of Q_s. Therefore, appropriate experimental and data analysis procedures have to be applied to estimate the soil hydraulic parameters. In particular, these procedures differ by the number of the established H values with a run and, hence, the number of the collected Q_s data.

Measurement of quasi steady-state flow rate for a single H value allows determination of K_{fs} by the OPD approach, also named single-height approach or updated single-head approach (Reynolds and Elrick 1987, 2002c; Elrick and Reynolds 1992a; Reynolds 1994), making use of the soil hydraulic conductivity function by Gardner (1958):

$$K = K_{fs}\exp(\alpha h) \qquad h \leq 0 \tag{2.5}$$

where α (L^{-1}) is the slope of ln(K) versus h ($0 < \alpha < \infty$). Substituting Eq. (2.5) into Eq. (2.2) and integrating it yields:

$$\alpha = \frac{K_{fs} - K_i}{\phi_m} \tag{2.6}$$

where K_i ($L\ T^{-1}$) is the soil hydraulic conductivity corresponding to the initial pressure head, h_i. For many soils at field capacity and drier, K_i is small relative to K_{fs} and, consequently, Eq. (2.6) can be approximated by:

$$\alpha \approx \frac{K_{fs}}{\phi_m} = \alpha^* \tag{2.7}$$

Substituting Eq. (2.7) into Eq. (2.1) and finding K_{fs} yields:

$$K_{fs} = \frac{C\ Q_s}{2\pi H^2 + \pi r^2 C + \frac{2\pi H}{\alpha^*}} \tag{2.8}$$

Table 2.1 Texture-structure categories for selecting capillarity category and representative α* parameter when calculating field-saturated soil hydraulic conductivity, K_{fs}, using the single-head calculation procedure

Texture-structure category	Soil capillarity category	Representative α* (m^{-1})
Compacted, structureless, clayey or silty materials such as landfill caps and liners, lacustrine or marine sediments, etc.	Very strong	≤1
Porous materials that are both fine textured and massive; includes unstructured clayey and silty soils, as well as very fine to fine structureless sandy materials	Strong	4
Most structured and medium textured materials; includes structured clayey and loamy soils, as well as medium single-grain sands. This category is generally the most appropriate for agricultural soils	Moderate	12
Coarse and gravelly single-grain sands; may also include some highly structured soils with large and/or numerous cracks and biopores	Weak	36
Gravels, very coarse sands, etc. containing negligible amounts of coarse/medium/fine/very fine sand, silt and clay	Negligible	≥100

The soil capillary categories assume that antecedent pore water pressure head is sufficiently negative to produce near-maximum soil capillarity for that category
(From Reynolds and Lewis 2012, reprinted with permission; adapted from Elrick and Reynolds 1992a)

The α* (L^{-1}) parameter can be estimated on the basis of soil textural and structural characteristics, using for example Table 2.1 listing five values for different soil texture and structure categories. According to Reynolds and Elrick (2002c), Eqs. (2.4a), (2.4b) and (2.4c) should be used to calculate C for $\alpha^* \geq 9$ m^{-1}, $\alpha^* = 4$ m^{-1} and $\alpha^* = 1$ m^{-1}, respectively. A single C vs. H/r curve applies for all $\alpha^* \geq 9$ m^{-1}, including most natural soil textures and structures, because the impact of capillarity on flow out of the well decreases as α* increases. The OPD approach also allows estimation of ϕ_m and S by the following relationships (Philip 1957b; Reynolds and Elrick 2002a, c):

$$\phi_m = \frac{C\ Q_s}{\left(2\pi H^2 + C\pi r^2\right)\ \alpha^* + 2\pi H} \tag{2.9}$$

$$S = \left[\gamma_w\left(\theta_{fs} - \theta_i\right)\ \phi_m\right]^{1/2} \tag{2.10}$$

where γ_w is a dimensionless constant (White and Sully 1987) related to the shape of the wetting (or drainage) front ($\gamma_w \approx 1.818$ for a wetting front, but may be smaller for a drainage front), θ_{fs} (L^3L^{-3}) is the field-saturated volumetric soil water content and θ_i (L^3L^{-3}) is the initial or background volumetric soil water content.

The two unknowns of Eq. (2.1) can be estimated simultaneously by the TPD approach, also called simultaneous equations approach or two-head approach (Reynolds and Elrick 1985a, b, 1986, 1987, 2002c). In this case, the K_{fs} and ϕ_m values are obtained on the basis of the following two simultaneous equations:

$$\left(2\pi H_1^2 + C_1 \pi r^2\right) K_{fs} + 2\pi H_1 \phi_m = C_1 Q_{s1} \tag{2.11a}$$

$$\left(2\pi H_2^2 + C_2 \pi r^2\right) K_{fs} + 2\pi H_2 \phi_m = C_2 Q_{s2} \tag{2.11b}$$

where C_1 and C_2 are the values of C corresponding to H_1/r and H_2/r, respectively. In particular, K_{fs} and ϕ_m are given by:

$$K_{fs} = G_2 Q_{s2} - G_1 Q_{s1} \tag{2.12a}$$

$$\phi_m = J_1 Q_{s1} - J_2 Q_{s2} \tag{2.12b}$$

where G_2 (L^{-2}), G_1 (L^{-2}), J_1 (L^{-1}) and J_2 (L^{-1}) are equal to:

$$G_2 = \frac{H_1 C_2}{\pi\left[2H_1 H_2 (H_2 - H_1) + r^2 (H_1 C_2 - H_2 C_1)\right]} \tag{2.12c}$$

$$G_1 = G_2 \frac{H_2 C_1}{H_1 C_2} \tag{2.12d}$$

$$J_1 = \frac{\left(2H_2^2 + r^2 C_2\right)\ C_1}{2\pi\left[2H_1 H_2 (H_2 - H_1) + r^2 (H_1 C_2 - H_2 C_1)\right]} \tag{2.12e}$$

$$J_2 = J_1 \frac{\left(2H_1^2 + r^2 C_1\right)\ C_2}{\left(2H_2^2 + r^2 C_2\right)\ C_1} \tag{2.12f}$$

The K_{fs} and ϕ_m values obtained by Eqs. (2.12a, 2.12b, 2.12c, 2.12d, 2.12e and 2.12f) can be used to estimate α^* by Eq. (2.7) and S by Eq. (2.10). The h_f parameter can be estimated by the following relationship (Reynolds and Elrick 2002a, c):

$$h_f = -\frac{1}{\alpha^*} \tag{2.13}$$

Reynolds and Elrick (1986) suggested applying the following relationships to determine K_{fs} and ϕ_m by the MPD approach, also called multiple-head analysis (Reynolds and Elrick 2002c):

$$K_{fs} = \frac{\sum_{i=1}^{N} H_i^2 \sum_{i=1}^{N} C_i Q_{si}\left(\frac{C_i r^2}{2} + H_i^2\right) - \sum_{i=1}^{N} H_i C_i Q_{si} \sum_{i=1}^{N} H_i \left(\frac{C_i r^2}{2} + H_i^2\right)}{2\pi\left\{\sum_{i=1}^{N} H_i^2 \sum_{i=1}^{N}\left(\frac{C_i r^2}{2} + H_i^2\right)^2 - \left[\sum_{i=1}^{N} H_i\left(\frac{C_i r^2}{2} + H_i^2\right)\right]^2\right\}} \tag{2.14a}$$

$$\phi_m = \frac{\sum_{i=1}^{N} C_i Q_{si}\left(\frac{C_i r^2}{2} + H_i^2\right) \sum_{i=1}^{N} H_i \left(\frac{C_i r^2}{2} + H_i^2\right) - \sum_{i=1}^{N} H_i C_i Q_{si} \sum_{i=1}^{N}\left(\frac{C_i r^2}{2} + H_i^2\right)^2}{2\pi\left\{\left[\sum_{i=1}^{N} H_i\left(\frac{C_i r^2}{2} + H_i^2\right)\right]^2 - \sum_{i=1}^{N} H_i^2 \sum_{i=1}^{N}\left(\frac{C_i r^2}{2} + H_i^2\right)^2\right\}} \tag{2.14b}$$

where N is the number of the applied ponded depths of water for a run, Q_{si} (L^3T^{-1}) is the quasi steady-state flow rate corresponding to a ponded depth of H_i (L), and C_i is the shape factor corresponding to H_i/r. An alternative multiple-head analysis involves least squares regression fitting (Reynolds and Elrick 2002c, 2005):

$$C_iQ_{si} = P_1H_i^2 + P_2H_i + P_3 \qquad i = 1, 2, 3, \ldots, n \qquad n \geq 2 \tag{2.15a}$$

to the CQ_s vs. H data. Equation (2.15a) was derived from Eq. (2.1) and therefore the coefficients P_1, P_2 and P_3 have the following expressions:

$$P_1 = 2\pi K_{fs} \tag{2.15b}$$

$$P_2 = 2\pi\phi_m \tag{2.15c}$$

$$P_3 = C_i\pi r^2 K_{fs} \tag{2.15d}$$

Equations (2.15b) and (2.15c) can be solved for K_{fs} and ϕ_m, respectively, while the P_3 coefficient is not used (W.D. Reynolds, personal communication, 12 June 2013). The α^* parameter, S and h_f can then be calculated by Eqs. (2.7), (2.10) and (2.13), respectively. The two-head and multiple-head analyses give the same results when two ponded heads are used (Reynolds and Elrick 2002c).

2.1.1.3 Issues of Practical Interest

In the last 30 years, several investigations have increased our knowledge of the GP method showing its potential and also its limits. Well preparation may affect appreciably the representativeness of the K_{fs} value determined with the GP method (Campbell and Fritton 1994). The standard procedure of preparing a well, using the implements included in the commercial GP kit, can determine smearing and compaction of the well surfaces resulting in artificially low K_{fs} and ϕ_m values even if the well prep brush is applied before the run (Reynolds 1993). This phenomenon is particularly noticeable in fine-textured, initially wet soils, but it can also occur in soils with an appreciable percentage of coarse-textured particles (Bagarello and Provenzano 1996; Bagarello 1997). Auger induced smearing and compaction of the well surfaces can be minimized by not augering when the material is very wet (above field capacity), using a very sharp auger, applying very little downward pressure on the auger, and taking only small scoops with the auger before empting it out (Reynolds 1993). In particular, the two-finger/two-turn rule for augering within the measurement zone presupposes that, once the top of the measurement zone is reached, only two fingers on each hand are used to apply downward pressure on the auger (i.e., the weight of the auger provides most of the downward pressure), and only two complete turns of the auger are made before emptying it out. If inspection of the well reveals smearing within the measurement zone, i.e. a smooth, polished surface under the light of a flashlight, steps should be taken to remove it before performing the GP measurement. Bagarello et al. (1999)

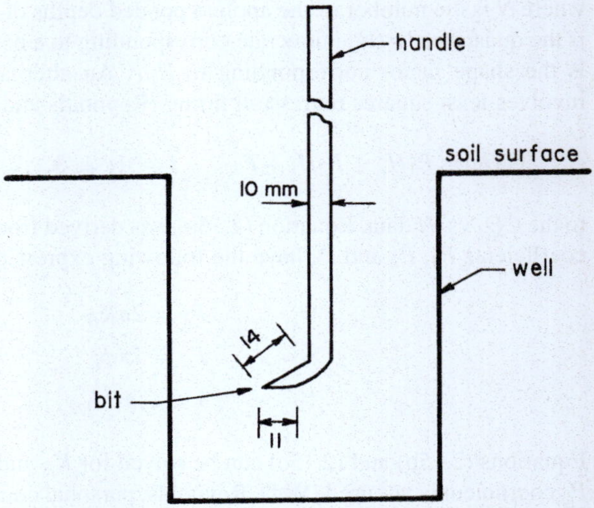

Fig. 2.5 Desmearing apparatus for the constant-head well permeameter method proposed by Bagarello et al. (1999, reprinted with permission)

removed smeared and compacted surfaces from GP wells by the plucking instrument shown in Fig. 2.5 with satisfactory results. The pointed bit is pushed into the well wall and base, and then it is lifted via the handle to pull off smeared and compacted soil. Alternatively, a desmearing apparatus comprising a spiked roller has been suggested by Reynolds (1993). Bird et al. (1996), working in soils with high clay content, applied the GP several days after excavating the well, to facilitate drying and fissuring of the soil at the infiltration surface. A method to prepare unsmeared soil surfaces using a quick setting epoxy resin was described by Koppi and Geering (1986). None of these desmearing procedures are completely effective under all conditions (Reynolds 1993). Clearly, as removing a smeared layer determines enlargement of the well, its radius has to be measured again before applying the device, since the well radius influences calculations. Boreholes can also be difficult to prepare in stony soils, where less invasive measurement methods should be used (Verbist et al. 2013).

The influence of soil water content at the time of augering the well on smearing can help explain an inverse relationship between K_{fs} and the corresponding initial soil water content values, such as the one shown in the example of Fig. 2.6. However, this type of relationship can also be expressive of the soil structural characteristics at the sampled field site. For example, in the investigation by Reynolds and Zebchuk (1996), the antecedent soil water content was low in sampling wells intersecting extensive soil structure where the individual pores and spaces comprising the structure (especially the macrostructure) were largely air filled, i.e. empty of water. When water was applied in these wells by the GP, the high flow rates through those same soil pores and spaces yielded high K_{fs} values. The reverse occurred when the wells intersected less structured soil conditions, i.e. higher soil water contents at the beginning of the run and lower measurements of K_{fs}. Jabro (1996) also determined the effect of the initial soil water

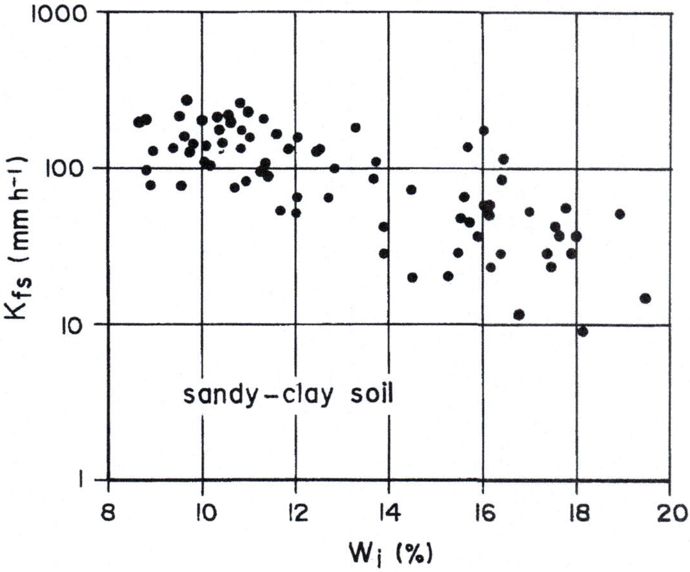

Fig. 2.6 Relationship between the field-saturated soil hydraulic conductivity, K_{fs}, measured with the Guelph permeameter and the initial gravimetric soil water content, w_i (From Bagarello and Provenzano 1996, reprinted with permission)

content, θ_i (L^3L^{-3}), on the K_{fs} data obtained with the GP in a structured clayey soil. Changes in K_{fs} were noticeable since swelling and shrinking of the clay associated with variations in the soil water content during the sampling period determined modifications in soil structure and macroporosity. In particular, data were described by the following relationship:

$$K_{fs} = \frac{0.0023W}{1 + W} \tag{2.16}$$

where K_{fs} is expressed in mm s^{-1} and $W = \exp(18.539 - 67.647 \times \theta_i)$. According to Eq. (2.16), in the range of the experimental θ_i values (0.16–0.35 m^3m^{-3}), an increase of the initial soil water content determines a decrease of K_{fs} by more than two orders of magnitude.

The well diameter and the established ponded depths of water can influence the measured variables. For example, several authors noted that the measured conductivity increased with the well diameter or the established ponded depth of water (Reynolds and Elrick 1985a; Bagarello 1993). This result was attributed to a greater difficulty of preparing small wells and to the circumstance that a larger soil volume sampled with an individual run implies a higher probability to intercept preferential flow paths. In an investigation by Wang et al. (2008), the water depth in the well was fixed at 15 cm since this depth was considered to be deep enough to provide a meaningful data interpretation and shallow enough to maintain the integrity of the borehole walls against slumping.

At the beginning of the run, turbulence of water filling the well to the desired depth can determine detachment of soil particles that are transported towards the infiltration surface, possibly occluding exposed pores. This phenomenon, determining an underestimation of K_{fs}, can be particularly noticeable if dispersion of clay and silt particles also occurs. To reduce the probability of an erroneous measurement of K_{fs}, the air tube should be raised slowly. In addition, native or tap water should be used, since they have chemical characteristics appropriate to prevent appreciable dispersion phenomena. In soils with high silt content, addition of flocculant to the water may be advisable (Reynolds 1993).

Solar heating of the device during the run can result in an anomalous infiltration process (Reynolds 1993). Thermal expansion of the air in the head space above the water surface can prevent bubbling and extreme heating of the water in the permeameter can cause a significant reduction in water viscosity. Especially in low-permeability soils, where the permeameter bubbles slowly, it is advisable to shade the permeameter reservoir from direct, hot sunlight. The saturated conductivity at the reference temperature (e.g. 20 °C), $K_{fs,ref}$, can be obtained by the following relationship (Bagarello and Iovino 2010; Rienzner and Gandolfi 2014):

$$K_{fs,ref} = K_{fs,T} \frac{\mu_T}{\mu_{20}} \tag{2.17a}$$

where $K_{fs,T}$ is the measured K_{fs} value, μ_{20} is the water viscosity at $T = 20$ °C (=0.001 Pa s^{-1}), and μ_T is the water viscosity corresponding to the water temperature in the well, T (°C), that can be calculated by (Likhachev 2003):

$$\mu_T = \left(2.42 \times 10^{-5}\right) \times 10^{[247.7/(T+273.15-139.86)]} \tag{2.17b}$$

The water outlet tip of the device should not sink into the base of the well during measurement. The occurrence of this phenomenon triggers an underestimation of K_{fs} because the actual value of H is smaller than the assumed one for the calculations and water flowing out of the device may be obstructed. Subsidence of the permeameter can be prevented by clamping the GP reservoir to a rigid tripod so that the weight of the instrument is carried by the legs of the tripod rather than by the outflow tip (Reynolds 1993). The GP subsidence can also be prevented by placing a layer of highly permeable pea gravel of known height on the bottom of the well, i.e. under the tip, and recalibrating the head scale of the device. Alternatively, a wire screen insert connected to a rigid surface support can be used (Fig. 2.7).

The equilibration time, t_s (T), i.e. the time required for a well permeameter run to reach quasi-steady flow, is determined primarily by the permeability of the material being tested, but also by the antecedent soil water content, the radius of the well, and the depth of water ponding in the well. Generally speaking, t_s increases with decreasing soil permeability, decreasing antecedent soil water content, increasing well radius, and increasing depth of water ponding (Reynolds and Elrick 1986,

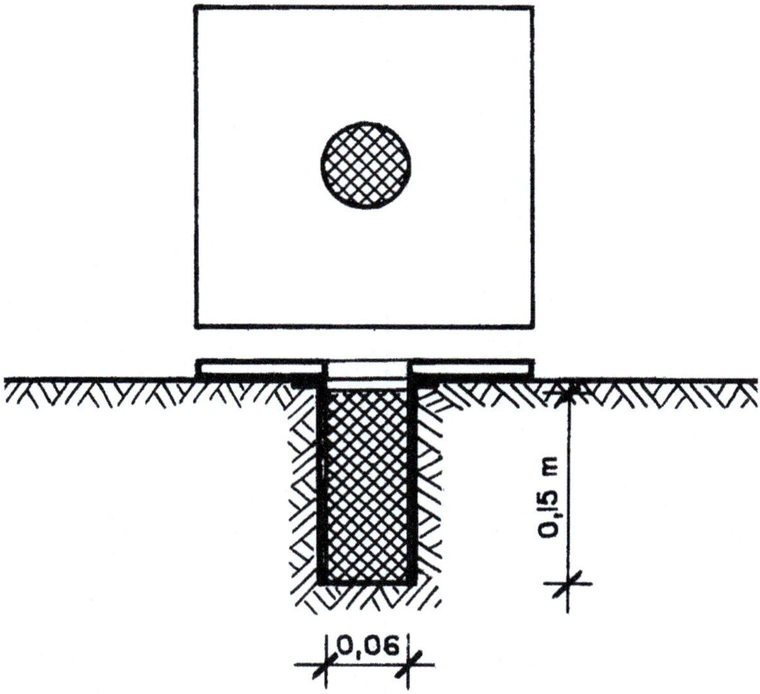

Fig. 2.7 Wire screen insert connected to a wooden surface support used by Bagarello (1997) to prevent subsidence of the permeameter at the base of the well (Reprinted with permission)

2002c). Equilibration times are expected to range from about 5–60 min in moderate to highly permeable soils ($K_{fs} \geq 3.6$ mm h^{-1}) to as much as two or more hours in low permeability materials ($K_{fs} \leq 0.036$ mm h^{-1}). Values of t_s of this order of magnitude presuppose that the well diameter and the ponding head are kept small (e.g. 0.04–0.1 m well diameter; 0.05–0.5 m head; Reynolds and Elrick 2005), which also implies that water consumption per measurement is generally limited to a few litres. Rapid detection of quasi steady-state conditions is common in practice (Bagarello et al. 1997, 2006a). The fact that steady-state is not reached may lead to an overestimation of flow rate and thus of saturated hydraulic conductivity. In any case, the applied procedure to detect quasi steady-state conditions influences the estimate of Q_s and hence it has consequences on the calculation of K_{fs} and the other variables of interest. Bagarello and Giordano (1999) suggested applying a Cumulative Drop (CD) procedure, consisting of plotting the cumulative drop in water level in the GP reservoir, L (L), against time, t (T), and checking on this plot when the slope of L versus t becomes constant (i.e. linear L vs. t relationship), which indicates steady flow (Fig. 2.8). The steady rate of the water level fall in the GP reservoir, R_s (L T^{-1}), is then determined as the least-squares regression slope of the

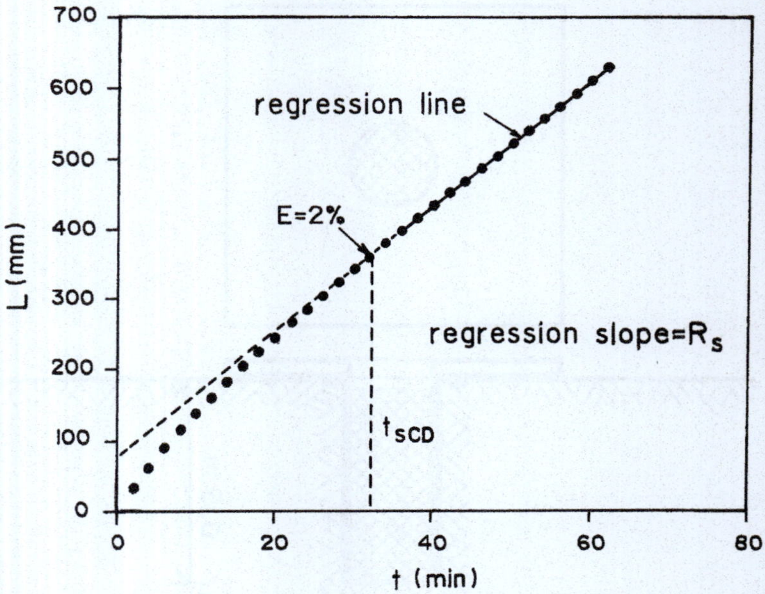

Fig. 2.8 Cumulative Drop (CD) procedure for estimating equilibration time and steady flow rate from Guelph permeameter data. *L* is the cumulative drop in water level in the GP reservoir (From Bagarello et al. 1999, reprinted with permission)

linear portion of the *L* vs. *t* curve (Ankeny et al. 1990; Bagarello et al. 1999). The t_s value is determined as the first *t* value for which:

$$\left|\frac{L - L_{reg}}{L}\right| \leq E \tag{2.18}$$

where *L* is the water level reading in the GP reservoir at time *t*, L_{reg} is the regression-based estimate of *L* at time *t*, and *E* is a criterion for establishing the onset of linearity in *L* versus *t*. It should be noted that the numbering increases downward in the scale stamped on the inner reservoir and therefore *L* increases with the infiltrated water volume. Equation (2.18) is applied sequentially until the criterion is met, starting from $t = 0$ and excluding progressively the first data points. A value of 2 % was selected for *E* (Bagarello et al. 1999). Bagarello and Giordano (1999) also showed that, for a given time interval between two successive readings at the instrument, Δt, procedures using three or four consecutive values of the rate of fall of the water level in the GP reservoir, *R* (L T^{-1}), equal or differing by not more than a fixed, small percentage (3 % or 5 %) gave rise to an overestimation of K_{fs} as compared to the CD procedure. The overestimation was higher in low permeability soil conditions and it decreased with an increase in Δt. However, detecting linearity of the cumulative infiltration curve, or the corresponding $L(t)$ curve, in a later stage of the run does not assure real flow steadiness conditions and the errors in the predictions of K_{fs} obtained in this case can still be appreciable (Logsdon and Jaynes 1993; Smettem et al. 1995).

The OPD, TPD and MPD approaches differ by both the run duration and the water volume necessary for the test, that are obviously minimized with the OPD approach and increase with the number of water heads successively established in the well. On the other hand, only the approaches using two or more H levels allow a direct determination of α^* and h_f. With the OPD approach, it seems reasonable to presume that the choice of α^* among the suggested values (Table 2.1) could be erroneous by not more than one category. For example, a non-correct value of $\alpha^* = 36\ \text{m}^{-1}$ is used instead of the correct $\alpha^* = 12\ \text{m}^{-1}$. This in turn introduces an error into the K_{fs}, ϕ_m and S calculations which is generally less than a factor of two to three and often less than $\pm 25\ \%$ (Elrick and Reynolds 1992a; Reynolds et al. 1992). This is a sufficient level of accuracy for many practical applications, given the inherent variability of these parameters (Reynolds and Elrick 2002c). Bosch (1997) suggested that the risk of erroneous determinations of K_{fs} is higher in low permeability soils. In addition, the sensitivity of K_{fs} to the choice of α^* decreases as H increases, while the sensitivity of ϕ_m and S to α^* decreases as H decreases (Reynolds et al. 1992; Reynolds and Elrick 2002c). Consequently, if one is interested primarily in K_{fs}, then the H level used for the single-head approach should be as large as possible. If interest is primarily in ϕ_m and/or S, on the other hand, then the H level should be as small as possible. If the technician is uncertain in the choice of α^*, the α^* value of first approximation, valid for a wide range of soils, should be $12\ \text{m}^{-1}$. The TPD and MPD approaches can yield negative, and hence meaningless, estimates of K_{fs} and ϕ_m due to soil heterogeneity in the form of, for example, layering, horizonation, cracks, worm holes or root channels (Reynolds and Elrick 1985a, 2002c). This occurs because both the infiltration surface and the wetted bulb around the well increase with increasing H, which increases the likelihood of encountering soil heterogeneities. In addition, the coefficient matrices in the two-head and multiple-head analyses are ill conditioned, which further increases sensitivity to soil heterogeneity (Philip 1985). Failure rates are of the order of 10 % in the structureless sandy soils, 30 % in the structured loam ones and 70–80 % in the highly structured clay soils (Elrick et al. 1990). Salverda and Dane (1993) obtained approximately 40 % of negative results when they applied the GP method in a sandy-loam, a loam and a loamy-sand soil, considering three sets of 16 wells with different diameter at each site and establishing four H-levels in each well. According to Reynolds and Elrick (1986), the risk of failure of the GP run is reduced if relatively similar H levels are established in sequence so that both H levels sample as nearly as possible the same volume of porous medium. However, in the performed checks, this suggestion was not found to be really advantageous (Bagarello 1993). For example, Wilson et al. (1989) used $H_1 = 3$ cm and $H_2 = 5$ cm, instead of the 5- and 10-cm heads typically recommended (Soilmoisture Equipment Corp., 1987), in two forested subsoils, but more than 50 % of the TPD calculations failed. Numerical simulation of flow out of a borehole appears suitable to better establish the dependence of the K_{fs} and ϕ_m calculations made with the multi-level experiment on various forms of heterogeneity. Using this approach, Wu et al. (1993) showed that unrealistic TPD results can be obtained due to macropores intersecting the base or walls of the well and also to soil layering. The single-height analysis was considered the most suitable analysis procedure in stony soils whereas the TPD approach was not appropriate due to the high failure rates in an investigation by Verbist et al. (2013). When the TPD or MPD approaches yield a negative result for either K_{fs} or ϕ_m, both values have to be discarded. For example, calculation of a negative ϕ_m value by Eq. (2.12b) makes the estimate of K_{fs} obtained by Eq. (2.12a) unusable even if positive. In this case, the OPD approach, that does not yield negative results, should be applied to each H value and the resulting K_{fs}, ϕ_m and S values averaged (Reynolds and Elrick 2002c). The same procedure should be applied when a multi-level run gives an estimate of α^* falling outside a realistic range of values, i.e. $1 \leq \alpha^* \leq 100\ \text{m}^{-1}$ (Reynolds and Elrick 2002c).

The so-called regression-based Richards analysis has been suggested as an alternative means to avoid negative K_{fs} and ϕ_m values (Vieira et al. 1988; Reynolds et al. 1992; Reynolds and Zebchuk 1996). This analysis procedure is based on the successful multiple-head well permeameter calculations and on the estimate of K_{fs} (K_L) obtained with the Laplace analysis by Reynolds and Elrick (1985a, 1987):

$$K_L = \frac{CQ_s}{2\pi H_m^2 + C\pi r^2} \tag{2.19}$$

where H_m (L) is the largest of the heads ponded in the well and C corresponds to H_m. The choice of the largest head of ponding is related to the fact that the Laplace analysis is expected to provide a good estimate of K_{fs} when the field-saturated component of flow dominates flow out of the well (Reynolds and Elrick 1987). Combining Eq. (2.19) with Eq. (2.8) yields:

$$K_{fs} = \left[\frac{2H_m^2 + Cr^2}{2H_m^2 + Cr^2 + \frac{2H_m}{\alpha^*}}\right] K_L \tag{2.20}$$

with $K_L \geq K_{fs}$ for $\phi_m \geq 0$ in Eq. (2.1) and for $0 < \alpha^* \leq \infty$ in Eq. (2.8), given that K_L is obtained by neglecting capillary effects. Equation (2.20) shows the functional relationship between K_{fs}, soil capillarity and K_L. The K_{fs} vs. K_L relationship is direct, but the relationship between K_{fs} and capillarity is inverse, i.e. K_{fs} decreases as ϕ_m increases. Equation (2.20) was generalized using the empirical form:

$$K_{fs} = \beta {K_L}^{\gamma'} \tag{2.21}$$

where β and γ' are dimensionless empirical parameters. Also, Eq. (2.20) was rearranged to give:

$$\frac{1}{\alpha^*} = \frac{2H_m^2 + Cr^2}{2H_m}\left(\frac{K_L}{K_{fs}} - 1\right) \tag{2.22}$$

The values of β and γ' are determined by least squares fitting Eq. (2.21) in its logarithmic form to plots of ln K_{fs} vs. ln K_L data. The (K_L, K_{fs}) data pairs are obtained from the wells where multiple-head well permeameter calculations were successful. Equations (2.19), (2.21) and (2.22) are then applied to all wells, regardless of whether or not the multiple-head calculations were successful, to produce estimates of K_{fs} and α^*, and ϕ_m via Eq. (2.7). The use of Eq. (2.21) in the regression provides a means of determining for any particular data set if α^* in Eqs. (2.20) and (2.22) is constant or variable. If the least squares fitting of Eq. (2.21) produces $\gamma' = 1$, then α^* is a constant in Eqs. (2.20) and (2.22). If the fitting produces $\gamma' \neq 1$, then α^* is not constant. It is also possible to conduct the above analysis in terms of ϕ_m rather than K_{fs} (Reynolds et al. 1992). The analysis in terms of K_{fs} produces the best results when H_m is large ($H_m \geq 0.09$ m).

Calculations of K_{fs} avoiding the risk of obtaining negative values could also be made by simply neglecting the effect of soil capillarity on steady flow out of the well. The impact of this assumption on the reliability of the K_{fs} data depends on the soil characteristics and the initial soil water content. According to Jabro and Evans (2006), for example, use of solutions that neglect capillarity should be considered possible in coarse-textured soils where the capillarity effects are minimal and the initial soil water content is near or at field capacity. For a silt-loam soil, the authors obtained, with Eq. (2.19), a mean K_{fs} value that was 1.4 times higher than that calculated by Eq. (2.8).

For completeness, it should also be said that not all soil scientists agree with the importance of a data analysis procedure based on Eq. (2.1). For example, Amoozegar (1989) stated that using the Glover solution with the constant-head well permeameter technique is superior to determine the conductivity in the vadose zone when compared to the simultaneous equations approach for the GP method. Moreover, Amoozegar (1993) concluded that declaring the Glover solution an inadequate analytical tool is questionable because models and approaches used for comparative purposes are also based on approximations and assumptions.

The single-height calculation of K_{fs} and ϕ_m is directly proportional to Q_s for a fixed α^*. Thus, standard-deviation calculations of K_{fs} and ϕ_m based on multiple measurements at a field site using the single-height approach give results that are proportional to the steady-state Q_s values and are, therefore, representative of the field variability. Standard deviations based on the simultaneous equations calculations generally give larger numbers that are artificially increased because of solving simultaneous equations where K_{fs} and ϕ_m are not precisely constant within the measurement volumes (Elrick and Reynolds 1992a). In other terms, the coefficient of variation of the K_{fs} values obtained with the TPD approach expresses the circumstance that the conductivity varies from a sampling point to another but also in correspondence of a particular sampling point. Probably, additional investigations on the estimated variability of K_{fs} are advisable, considering that, as shown by Eq. (2.1), K_{fs} is not the only variable controlling Q_s for given H and r.

According to Reynolds and Elrick (2002c), the main advantage of the MPD approach over the TPD one is that the deleterious effects on parameter accuracy of small scale soil heterogeneity and measurement error can be reduced by ponding more than two heads. The main disadvantages of multiple heads, on the other hand, are greater time requirements, greater water consumption, and greater chance of invalid results due to the crossing of major heterogeneities (e.g., soil horizon boundaries) as H is increased.

Independently of the data analysis procedure, the hypotheses of homogeneity, isotropy, rigidity and uniformity of the initial soil water content are not generally respected. In some cases, this violation can be detected since the decrease of the water level in the reservoir of the instrument does not evolve as expected, showing for example increasing R values for a part or the totality of the run (Bell and Schofield 1990), or the calculated variables are negative and hence meaningless. In other cases, measuring the infiltration process is not enough to signal a departure from the ideal conditions because the process evolves as expected notwithstanding that the soil is layered or it contains preferential flow paths (Jabro and Fritton 1990).

Using the GP in an anisotropic soil is expected to yield an estimate of K_{fs} that is intermediate between the vertical and the horizontal conductivity (Reynolds and Elrick 1985a), probably because the soil process established with the GP occurs through the well walls (horizontal direction) and the well bottom (vertical direction). The greater the ponding depth, the more the transmission parameters are weighted towards the horizontal direction (Reynolds and Elrick 2005).

The GP technique is mainly used to determine K_{fs} but it has also been applied to obtain an estimate of the unsaturated soil hydraulic conductivity function (Ragab and Cooper 1993) since this relationship can easily be determined with a multi-level experiment, setting $\alpha = \alpha^*$ in Eq. (2.5). However, using the GP to predict unsaturated soil hydraulic properties is expected to yield unreliable results (Paige and Hillel 1993; Reynolds 1994). In particular, ϕ_m, S, α^* and h_f can be of low accuracy when obtained using ponded infiltration techniques, regardless of the analysis procedure, because of the usual dominance of the pressure and gravity components of flow over the capillary component (Reynolds and Elrick 2002c).

According to Hayashi and Quinton (2004), Eq. (2.8) is also usable when the bottom of the well coincides with an impermeable substrate. However, in this case the C shape factor should be estimated by the following relationship:

$$C = u - s \exp(-p\ \alpha^*) \tag{2.23a}$$

where α^* is expressed in m^{-1} and p (m), s and u are coefficients depending on the H/r ratio:

$$p = p_1 \left(\frac{H}{r}\right)^{p_2} \tag{2.23b}$$

$$s = s_1 + s_2 \exp\left(-s_3 \frac{H}{r}\right) \tag{2.23c}$$

$$u = u_1 + u_2 \frac{H}{r} \tag{2.23d}$$

Table 2.2 Parameter values for Eqs. (2.23a, 2.23b, 2.23c and 2.23d). From Hayashi and Quinton, 2004, reprinted with permission

Initial soil water content	p_1	p_2	s_1	s_2	s_3	u_1	u_2
Low	0.00862	0.532	0.607	1.40	0.380	2.30	0.0813
High	0.00999	0.398	0.530	3.61	0.357	2.71	0.0848

where p_1 (m), p_2, s_1, s_2, s_3, u_1 and u_2 are coefficients varying with the initial soil water content (Table 2.2). Equations (2.23a, 2.23b, 2.23c and 2.23d) were developed because, when soil is underlain by an impermeable boundary, the C values obtained according to Reynolds and Elrick (1987) determine an underestimation of K_{fs}, although the difference with the true value may be relatively small compared to other uncertainties including measurement errors, the estimate of α^*, and soil heterogeneity (Hayashi and Quinton 2004). The inability or impossibility to detect the impermeable layer should in general only have small implications in terms of calculated K_{fs} values. The investigation by Hayashi and Quinton (2004) also showed that using the GP in a well with an impermeable layer at the bottom yields a measurement of the horizontal field-saturated soil hydraulic conductivity independently of the anisotropic characteristics of the sampled soil. In comparison with unrestricted, three-dimensional flow, two-dimensional flow occurring in presence of the impermeable layer takes more time to approach steady-state, which implies that infiltration runs have to be longer. Perhaps, the analysis by Hayashi and Quinton (2004) requires additional developments given that in Eq. (2.8), used to calculate K_{fs}, steady-state flow rate also depends on the gravitational pull of water through the bottom of the well but this effect should not be considered when the base of the well coincides with an impermeable substrate.

Many investigations have established comparisons between the K_{fs} values obtained with the GP and the corresponding values (K_{fs} or K_s, i.e. completely saturated soil hydraulic conductivity) measured with other field and laboratory methods. Table 2.3 gives a non-exhaustive overview of these comparisons, with the objective to frame the GP method into the ensemble of the K_{fs} or K_s measurement methods. This information may help a technician or a professional to establish if the GP is the most appropriate method to be applied in a particular circumstance. Other investigations yielding additional information on the reliability of the K_{fs} predictions obtained with the GP can also be found in literature. For example, Jabro (1992) showed that the K_{fs} values obtained by the GP in a silt-loam soil were predictable by using a relationship, developed on an independent database, making use of silt and clay percentages and soil bulk density. In other terms, the GP results were in line with the data collected by other methods in terms of influence of soil physical properties. A very good reproducibility of the GP test was recognized with reference to a wide range of K_{fs} values (from less than 0.01 m day^{-1} to more than 100 m day^{-1}) in a recent investigation by MacDonald et al. (2012).

The GP has been used to obtain the necessary data for interpreting and simulating hydrological processes, also explicitly considering temporal and spatial variability of K_{fs}. The few examples of these investigations summarized below attest to

Table 2.3 Synthesis of some comparisons between the saturated (K_s) or field-saturated (K_{fs}) soil hydraulic conductivities determined with the Guelph permeameter (GP) and other field and laboratory methods

References	Soils	Other methods	Results	Notes
Lee et al. (1985)	Loamy-sand, Fine sandy-loam, Silt-loam, Clay	Air-entry permeameter (AEP), Falling-head permeameter on small (47-mm-diam. × 50-mm-high) undisturbed soil cores (SC)	Soils differed according to a common sequence with all methods. A site-method interaction was detected	Site-method interaction interpreted in terms of the influence of macropores and air entrapment on each measurement technique. AEP the best choice to estimate the macropore dominated K_s of a structured soil. SC method suggested to obtain a soil matrix dominated K_s of a structured soil. GP the best for obtaining an average of the macropore- and matrix-dominated K_s value
Kanwar et al. (1989)	Silt-loam	Velocity permeameter (VP), constant-head permeameter (CHP) on undisturbed soil cores (75-mm-diam. × 75-mm-high)	GP and VP yielded similar values that were 10–800 times lower than the results of the CHP	Differences attributed to continuous macropores connecting the exposed surfaces and to complete saturation of the cores for the laboratory method
Dorsey et al. (1990)	Silt-loam, Silty-clay-loam	Pumping test, Auger hole, Velocity permeameter	Hydraulic conductivity based on GP runs equal to 15–20 % of the values found by the other methods	Differences between the GP and the other methods interpreted in terms of temperature and pressure fluctuations during long GP runs, smearing of the well walls, especially in the silty-clay-loam soil, and differences among the sampled soil volumes with the various methods
Gallichand et al. (1990)	Textures ranging from clay to loamy-sand	Falling-head permeameter (FHP), Auger hole (AH)	Lower values with the FHP (0.033 cm h^{-1}) than the GP (1.36 cm h^{-1}). Lower values	FHP < GP attributed to the smaller soil sample volume of the FHP method, soil structural changes within the core, and

(continued)

Table 2.3 (continued)

References	Soils	Other methods	Results	Notes
			with the GP ($0.13\ cm\ h^{-1}$) than the AH ($0.77\ cm\ h^{-1}$)	soil anisotropy. GP < AH attributed to the larger volume of soil sampled by the AH method, layering of the soil profile, direction of water flow
Gupta et al. (1993)	Sand = 44–66 % and clay = 5–19 % in the upper 50 cm of soil	Double ring infiltrometer (DRI), Rainfall simulator (RS), Guelph infiltrometer (GI)	RS and GI yielded 2.5–3.0 times higher results than DRI and GP. GP and GI yielded two times more variable results than RS and DRI	RS, GI > DRI, GP: attributed to differences in the depths of the experimental installations and to the relatively large surface area sampled by the RS. Variability differences attributed to the differences in the cross-sectional area of soil sampled with different techniques
Paige and Hillel (1993)	Fine sandy-loam, Silt-loam	Instantaneous profile (IP), Constant-head permeameter (CHP) on 3-cm-high soil cores	K_{fs} values obtained with the GP one to three orders of magnitude lower than K_s results obtained with the IP and CHP	Differences attributed to effects of entrapped air, anisotropy, smearing of the well walls, antecedent wetness conditions, discontinuous macropores in the soil
Mohanty et al. (1994a)	Loam	Velocity permeameter, Disk permeameter, Double-tube method, Constant-head permeameter on 7.6-cm-diam. × 7.6-cm-long soil cores	The GP gave the lowest average K_{fs} values (2.55×10^{-4} mm s^{-1} for the GP, $\geq 1.61 \times 10^{-3}$ mm s^{-1} with the other methods)	Low values with the GP attributed to small sample size, wall smearing, and air entrapment
Bagarello and Provenzano (1996)	Sandy-clay	Constant-head permeameter	Statistical similarity between the two methods (means = 3.74 - $\times 10^{-5}$–$3.90 \times 10^{-5}\ m\ s^{-1}$)	The similarity was detected when the field experiment was carried out in relatively dry soil (initial gravimetric soil water content ≤ 0.12) and relatively large soil cores (0.085-mm-diam. × 0.11-m-high) were used in the laboratory

(continued)

Table 2.3 (continued)

References	Soils	Other methods	Results	Notes
Reynolds and Zebchuk (1996)	Silty-clay	Auger-hole (AH)	Not significant differences between the GP and AH results (2.15×10^{-6}–3.30×10^{-6} m s^{-1}). Similar semivariograms	The GP method was capable of providing accurate estimates of K_{fs} in fine-textured soils having a stable and spatially variable structure
Noshadi et al. (2012)	Silt-loam	Auger hole, Original Porchet, Saturated Porchet, Drainage system (reference) method	GP results in the absence of a water table differing by 9.3 % from the reference method	The GP was the recommended method in regions with a low water table

possible applications of, and problems with, the method for practical and research purposes. Asare et al. (1993) determined seasonal variability of K_{fs} in a silt-loam soil. The K_{fs} value obtained in the fall season (1.19×10^{-8} m s^{-1}) was 29 % lower than the spring value (1.67×10^{-8} m s^{-1}), due to the relatively loose structure in spring resulting from the freeze-thaw cycles. These authors also noted that, in wetter soil conditions (spring), smearing and compaction in the borehole determined a homogenizing effect that reduced the range of variation of the K_{fs} data. In another silt-loam soil, the summer K_{fs} values differed from the field-saturated conductivity of the other seasons by up to a factor of 15 (Jabro 1996). These variations were attributed primarily to changes in soil structure and macroporosity caused by swelling and shrinking of clay associated with variation in water content throughout the year. Bosch and West (1998) measured K_{fs} to a depth of 2 m along a transect established in a 1-ha field including a loamy-sand and a sandy soil. Statistical separation of horizons and soil types according to soil hydraulic characteristics was plausible on the basis of the detected variations in soil textural characteristics. Therefore, the GP data were considered reliable, and hence usable for hydrological modeling purposes. Mohanty et al. (1998) implemented the estimates of K_{fs} obtained with different techniques, including the GP, in a two-dimensional model to predict water flow into a subsurface tile drain. The GP performance was poorer compared to the disk permeameter that was a more suitable method for reproducing measured data. This discrepancy was attributed to the limitations of the GP method including smearing, clogging of the soil pore structure by sedimentation, and artifacts involved in the original two-head analysis. On the other hand, GP data collected at different stages of maize growth in an investigation by Xu and Mermoud (2003) allowed them to define a time-dependent saturated soil hydraulic conductivity that was usable for modeling the soil water balance. The GP was also found to be usable in the description of dual permeability soils and to account for preferential water flow in structured soils (Kodesŏvá et al. 2010). In particular this device, combined with the tension disk infiltrometer, micromorphological images and dye tracer experiments, was used in a Haplic Luvisol to determine the soil hydraulic properties of matrix and macropore domains and mass exchange between these domains for the HYDRUS 2D/3D dual-permeability model. The GP was applied to measure the cumulative water flux under surface ponding conditions ($H = 5$ cm, $r = 3$ cm). Preferential flow occurred at the sampled site due to the presence of gravitational pores and the GP tests were considered appropriate to estimate the saturated conductivity of the macropore domain using the dual-permeability model. In a recent field-scale investigation, Rienzner and Gandolfi (2014) used the relation between the mean and the standard deviation of the log-transformed GP data to distinguish the matrix from the macropores (fracture system in dual porosity media). In more detail, the mean and the standard deviation of the log-transformed GP data can be linearly related according to the following relationship (Rienzner and Gandolfi 2014):

$$s_{\log} = a \times m_{\log} + b = a\left(m_{\log} + \frac{b}{a}\right) \tag{2.24}$$

where $s_{\log}$ is the standard deviation of the log-transformed K_{fs} values, $m_{\log}$ is the logarithm of the geometric mean of K_{fs}, and a and b are parameters that were estimated by the maximum likelihood approach. As the mean increases, the spread of the corresponding distribution is amplified to such an extent that not only very high but also very low values of log-conductivity have an increasingly high probability to occur. According to Rienzner and Gandolfi (2014), Eq. (2.24) allows us to distinguish between soil matrix and macropore effects on K_{fs}. When $m_{\log} = -b/a$, $s_{\log}$ becomes zero, indicating that $-b/a$ is a lower limit of the log-transformed conductivity. This value can be interpreted as the conductivity of the undisturbed soil matrix, when other physical, chemical or biological factors do not affect the pore structure. Defining the matrix conductivity, $m_m = -b/a$ and the macropore factor $m_p = m_{\log} - m_m$, where m_p incorporates all the additional factors that influence the actual value of K_{fs}, allows us to reformulate Eq. (2.24) as:

$$s_{\log} = a \times \left(m_{\log} - m_m\right) = a \times m_p \tag{2.25}$$

which shows that the spatial variability at the field scale can be explained by m_p only since the matrix conductivity, m_m, is constant. The analysis by Rienzner and Gandolfi (2014) could be interesting for improving field-scale characterization procedures for hydrological modeling purposes. Points that could be developed include, among others, the spatial extent to which a practically constant soil matrix saturated conductivity can be defined and the dependence of the empirically derived a and b parameters of Eq. (2.24) on both the considered sample sizes and the applied measurement method.

Table 2.4 Data (H = established ponded depth of water, t = time, L = water level in the reservoir of the Guelph permeameter, R = rate of fall of the water level in the GP reservoir) for the Example 2.1

$H_1 = 3$ cm			$H_2 = 5$ cm		
t (min)	L (cm)	R (cm min^{-1})	t (min)	L (cm)	R (cm min^{-1})
0	0.1		25	7.2	
2	0.9	0.4	27	8.3	0.55
4	1.4	0.25	29	9.1	0.4
6	1.9	0.25	31	9.9	0.4
8	2.3	0.2	33	10.6	0.35
10	2.8	0.25	35	11.3	0.35
12	3.2	0.2	37	12	0.35
14	3.7	0.25	39	12.6	0.3
16	4.1	0.2	41	13.2	0.3
18	4.5	0.2	43	13.8	0.3
20	4.9	0.2	45	14.4	0.3
22	5.3	0.2			

In summary, the well permeameter method allows measurement of three-dimensional K_{fs} and α^* values employing simple and robust equipment and procedures (Reynolds and Elrick 2005). All things considered, it is expected that K_{fs} measured by the GP would be accurate to within at least a factor of 2 in sandy soils with little structure, 2–3 in loams and 3–5 in structured clays (Elrick et al. 1990). Measurements are relatively rapid, with the exception of those in low permeability porous media, and generally require small water volumes. The method allows easy and rapid depth profiling and spatial-temporal replication of measurements. Smearing and compaction of the soil along the well walls during the augering process and progressive siltation of the soil along the infiltration surface by suspended silt and clay during the run are two important weaknesses of the method, but procedures to reduce their impact on the collected GP data can be applied. Use of small diameter boreholes may enhance horizontal flow thus adversely affecting the reliability of the data to describe vertical flow processes, even if small H-levels are applied for a run. Soil capillarity parameters estimated with the GP are expected to have a low accuracy but an extensive field testing of the predicted unsaturated hydraulic conductivity function by the device is lacking.

Example 2.1

A commercial GP manufactured by Soilmoisture Equipment Corp. (Santa Barbara, CA, U.S.A.) (cross-sectional area of the reservoir, $\sigma_R = 0.003579$ m^2) was applied in a soil visually belonging to the structured loams and clays category. A well of radius, r, of 0.03 m was excavated to a depth of 0.15 m below the soil surface. Two ponded depths of water, $H_1 = 0.03$ m and $H_2 = 0.05$ m, were successively established in the well and the water level, L (L), in the reservoir was monitored at $\Delta t = 2$ min time intervals until the rate of fall in water level in the reservoir, $R = \Delta L/\Delta t$ (L T^{-1}), did not vary during four successive intervals (Table 2.4). The

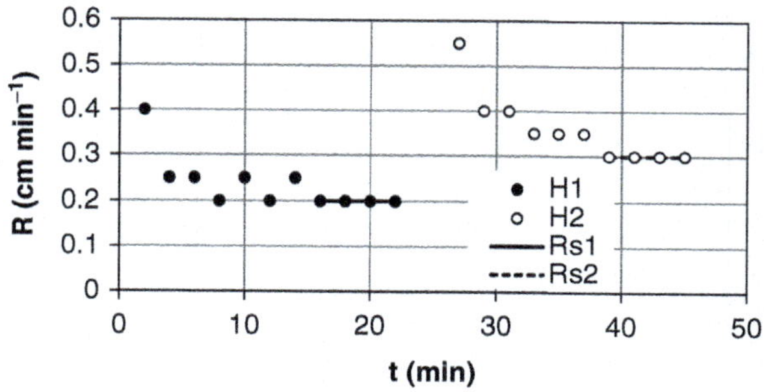

Fig. 2.9 Rate of fall of the water level in the reservoir of the Guelph permeameter, R, vs. time, t, for the Example 2.1

quasi-steady rate of fall in water level in the GP reservoir, R_s (L T^{-1}), was equal to 0.002 m min^{-1} for H_1 (R_{s1}) and to 0.003 m min^{-1} for H_2 (R_{s2}, Fig. 2.9). The corresponding quasi-steady flow rates (Q_{s1} and Q_{s2}, respectively) were equal to $Q_{s1} = 0.003579 \times 0.002/60 = 1.19 \times 10^{-7}$ m^{3}s^{-1} and $Q_{s2} = 1.79 \times 10^{-7}$ m^{3}s^{-1}. Equation (2.4b), applied to calculate the shape factors for both H_1/r (C_1) and H_2/r (C_2), yielded:

$$C_1 = \left(\frac{0.03/0.03}{1.992 + 0.091 \times 0.03/0.03}\right)^{0.683} = 0.606$$

and $C_2 = 0.842$. The TPD approach, i.e. Eqs. (2.12a, 2.12b, 2.12c, 2.12d, 2.12e and 2.12f), was then used to calculate K_{fs} and ϕ_m:

$$G_2 = \frac{0.03 \times 0.842}{\pi\left[2 \times 0.03 \times 0.05(0.05 - 0.03) + 0.03^2(0.03 \times 0.842 - 0.05 \times 0.606)\right]} = 144.95 \ \text{m}^{-2}$$

$$G_1 = 144.95\frac{0.05 \times 0.606}{0.03 \times 0.842} = 173.90 \ \text{m}^{-2}$$

$$J_1 = \frac{\left(2 \times 0.05^2 + 0.03^2 \times 0.842\right)\ 0.606}{2\pi\left[2 \times 0.03 \times 0.05(0.05 - 0.03) + 0.03^2(0.03 \times 0.842 - 0.05 \times 0.606)\right]} = 10.01 \ \text{m}^{-1}$$

$$J_2 = 10.01\frac{\left(2 \times 0.03^2 + 0.03^2 \times 0.606\right)\ 0.842}{\left(2 \times 0.05^2 + 0.03^2 \times 0.842\right)\ 0.606} = 5.67 \ \text{m}^{-1}$$

$$K_{fs} = 144.95 \times 1.79 \times 10^{-7} - 173.90 \times 1.19 \times 10^{-7} = 5.26 \times 10^{-6}\text{m s}^{-1}$$

$$\phi_m = 10.01 \times 1.19 \times 10^{-7} - 5.67 \times 1.79 \times 10^{-7} = 1.76 \times 10^{-7}\text{m}^2\text{s}^{-1}$$

Equation (2.7) yielded the following estimate of the α^* parameter:

Table 2.5 Water level readings (L, cm) at the Guelph permeameter reservoir recorded every minute for the Example 2.2 (t = time)

1.0 (t = 0), 2.2, 3.2, 4.2, 5.0, 5.8, 6.6, 7.3, 8.0, 8.7, 9.4, 10.1, 10.7, 11.4, 12.0, 12.6, 13.3, 14.0, 14.4, 15.0, 15.7, 16.3, 17.0, 17.5, 18.1, 18.7, 19.2, 19.8, 20.4, 21.0, 21.6, 22.1, 22.8, 23.4, 24.0, 24.5, 25.1, 25.6, 26.2, 26.7, 27.3, 27.9, 28.5, 29.0, 29.6, 30.1, 30.6, 31.2, 31.9, 32.4, 32.8, 33.4, 34.0, 34.6, 35.2, 35.7, 36.1, 36.7, 37.2, 37.9, 38.4, 39.0, 39.6, 40.1, 40.6, 41.2, 41.6, 42.3, 42.9, 43.3, 43.9, 44.5, 45.1, 45.5, 46.1, 46.7, 47.2, 47.8, 48.3, 48.9, 49.4, 49.9, 50.5, 51.0, 51.6, 52.1, 52.6, 53.2, 53.7, 54.2, 54.8, 55.3, 55.9, 56.5, 57.1, 57.7, 58.2, 58.8, 59.4, 59.9, 60.5, 61.0, 61.6, 62.1, 62.6, 63.2, 63.7, 64.3, 64.9, 65.3, 65.9, 66.5, 67.0, 67.6, 68.1, 68.7, 69.3, 69.8, 70.4, 71.0, 71.5, 72.0, 72.6, 73.2, 73.8, 74.3, 74.9, 75.5 (t = 127 min)

$$\alpha^* = \frac{5.26 \times 10^{-6}}{1.76 \times 10^{-7}} = 29.9 \ \text{m}^{-1}$$

Both K_{fs} and ϕ_m were positive and α^* fell within the $1 \le \alpha^* \le 100$ m^{-1} range. Therefore, the TPD calculations were considered successful although the estimated α^* value suggested that Eq. (2.4a) should have been used instead of Eq. (2.4b) to obtain C_1 and C_2. The OPD approach was also applied to the first ponded depth of water to obtain another estimate of K_{fs} with an assumed α^* parameter equal to 12 m^{-1}, i.e. the value of first approximation:

$$K_{fs} = \frac{0.606 \times 1.19 \times 10^{-7}}{2\pi \times 0.03^2 + \pi \times 0.03^2 \times 0.606 + \frac{2\pi \times 0.03}{12}} = 3.13 \times 10^{-6} \text{m s}^{-1}$$

This last estimate of K_{fs} was 1.7 times lower than the one obtained with the TPD approach. This difference is not substantial (Elrick and Reynolds 1992a) and it could be further reduced by a more appropriate choice of α^* for the sampled location ($K_{fs} = 5.73 \times 10^{-6}$ m s^{-1} with $\alpha^* = 36$ m^{-1}, a difference by a factor of 1.1). Another point to be considered is that the TPD and OPD approaches applied in this example might also differ by intrinsic soil permeability characteristics since a larger soil volume was sampled with the TPD approach (two ponded depths of water) than the OPD one (a single ponded depth of water).

Example 2.2
A commercial GP ($\sigma_R = 0.003579$ m^2) was used to establish a constant ponded depth of water, $H = 0.07$ m, in a well of radius, $r = 0.03$ m. Water level in the reservoir, L (L), was monitored at $\Delta t = 1$ min time intervals for a relatively long time, as shown in Table 2.5. Different estimates of the steady-state rate of fall in water level in the GP reservoir, R_s (L T^{-1}), were considered to obtain an estimate of K_{fs} by the OPD approach. Figure 2.10a shows the rate of fall in water level in the reservoir, $R = \Delta L/\Delta t$ (L T^{-1}), during the run. This figure shows that the precision of the measurement on L makes a piecewise function for R. The cumulative drop of water level in the reservoir against time is represented in Fig. 2.10b. Assuming that

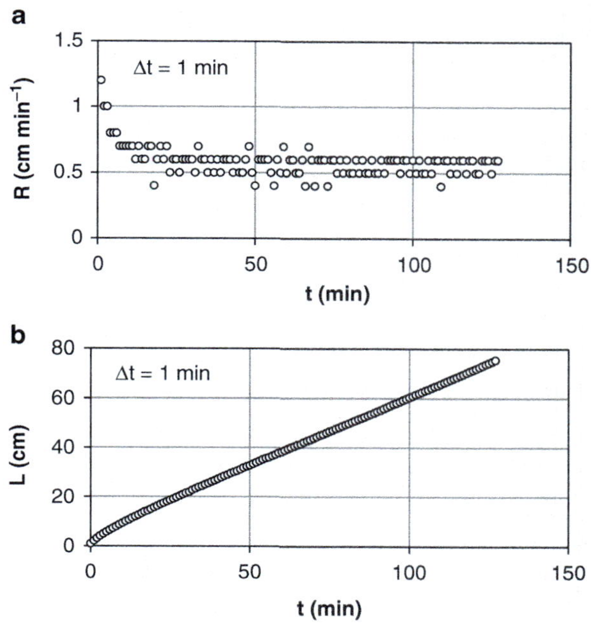

Fig. 2.10 Data for the Example 2.2 collected with a time interval between readings, Δt, of 1 min: (**a**) rate of fall of the water level in the reservoir of the Guelph permeameter, R, vs. time, t, and (**b**) water level readings in the reservoir of the device, L, plotted against time (From Bagarello and Iovino 2010)

flow steadiness occurs when three (3U criterion), four (4U criterion) or five (5U criterion) consecutive R values are equal, the estimated R_s values are of 0.8, 0.7 and 0.7 cm min^{-1}, respectively. Figure 2.10b suggests a linear L reading vs. t relationship at least for $t \geq 60$ min. Linear interpolation of the (L, t) data pairs for $t \geq 60$ min (CD criterion) yields a slope, i.e. an estimate of R_s, of 0.551 cm min^{-1} (coefficient of determination, $R^2 = 0.99996$). Assuming $\alpha^* = 12$ m^{-1} and using Eq. (2.4a) to estimate C (=1.014), Eq. (2.8) yields an estimate of $K_{fs} = 24.8$ mm h^{-1} with the 3U criterion, 21.7 mm h^{-1} with the 4U and 5U criteria, and 17.1 mm h^{-1} with the CD criterion, with a difference between the maximum and the minimum estimate of K_{fs} by a factor of approximately 1.5.

Excluding a water level reading every two measurements, i.e. assuming that data were recorded at $\Delta t = 2$ min time intervals, the R vs. t relationship of Fig. 2.11 is obtained. In this case, the R_s and K_{fs} values are equal to 0.65 cm min^{-1} and 20.1 mm h^{-1}, respectively (3U criterion), 0.60 cm min^{-1} and 18.6 mm h^{-1} (4U criterion), 0.55 cm min^{-1} and 17.0 mm h^{-1} (5U criterion), 0.552 cm min^{-1} ($R^2 = 0.99996$) and 17.1 mm h^{-1} (CD criterion), with a difference between the maximum and the minimum estimate of K_{fs} by a factor of approximately 1.2. Therefore, the increase in the time interval between successive readings at the reservoir did not influence the R_s estimate obtained with the CD criterion, as expected, and it determined smaller differences between the tested alternative approaches to establish steady flow conditions. However, this last result is not general because, as Δt increases, the prescribed condition (e.g., four consecutive, equal values of R) could not be detected although that the run appears clearly steady. For example, with $\Delta t = 5$ min, no more than three consecutive R values are equal ($R_s = 0.54$ cm min^{-1}, $K_{fs} = 16.7$ mm h^{-1}), even if the

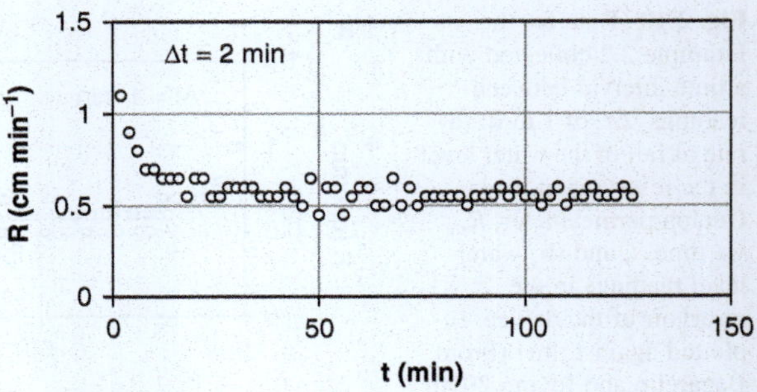

Fig. 2.11 Rate of fall of the water level in the reservoir of the Guelph permeameter, R, vs. time, t, corresponding to a time interval between readings of $\Delta t = 2$ min for the Example 2.2 (From Bagarello and Iovino 2010)

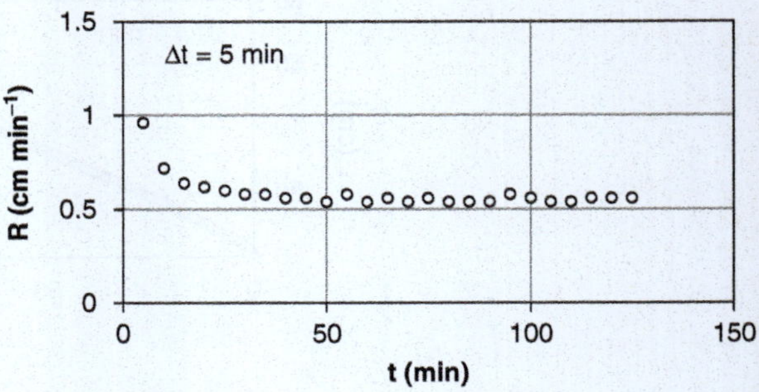

Fig. 2.12 Rate of fall of the water level in the reservoir of the Guelph permeameter, R, vs. time, t, corresponding to a time interval between readings of $\Delta t = 5$ min for the Example 2.2 (From Bagarello and Iovino 2010)

quasi steadiness character of the process is clear (Fig. 2.12). In this case, therefore, a rigid criterion, such as five consecutive equal values of R, cannot be applied to obtain a reasonable estimate of R_s. Alternatively, the CD criterion can be used since it does not reveal the above mentioned problem.

2.1.2 *Steady Flow Under Constant Head in Cased Boreholes*

2.1.2.1 Configurations and Principles

Constant head permeameter analyses for steady flow in cased boreholes were developed by Reynolds (2010) for in situ measurement of field-saturated soil hydraulic conductivity, K_{fs} (L T^{-1}) and matric flux potential, ϕ_m (L^{2}T^{-1}), in the vadose zone. According to this author, a cased borehole refers to any augered, drilled or driven hole in which a packer assembly, piezometer or well tube, well point, or tight-fitting liner has been installed (Fig. 2.13). Water exits the borehole into the surrounding unsaturated soil only through the permeable or open section (i.e., the "screen"), which may consist of the borehole segment below an inflatable packer or between two packers, the gravel-sand pack of a piezometer or cased well, a well point (Fig. 2.13a), or an unlined portion of the borehole (Fig. 2.13b). The water table is sufficiently deep so as not to affect steady discharge through the borehole screen. The analysis takes into account the radial pressure-capillary flux

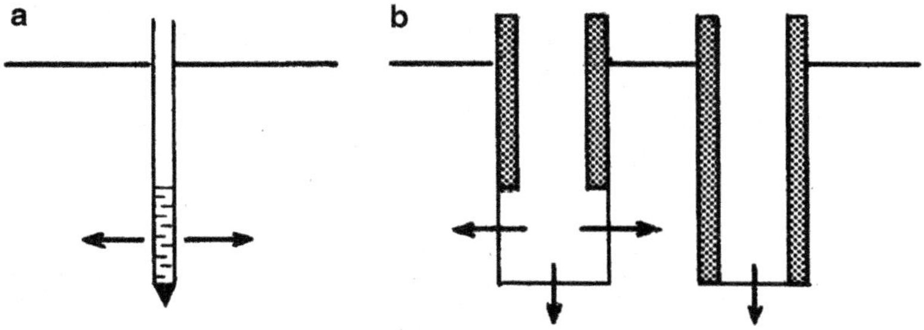

Fig. 2.13 Schematic view of a (**a**) driven pipe casing with well point attached to base, and (**b**) tightly lined borehole with an open section or base. The *arrows* on the infiltration surface show the flow direction

through the wall of the screen, the vertical pressure-capillarity flux trough the base of the screen and the vertical gravitational flux through the base of the screen. The pressure-capillarity fluxes represent the combined "hydrostatic push" of the ponded water in the borehole and the "capillarity pull" of the unsaturated soil. Representation of the pressure-capillarity and gravity components of flow as independent and additive terms is an approximation because such approach neglects gravity-capillarity interactions.

2.1.2.2 Modelling Flow Out of the Borehole

For a combined vertical and radial flow out of the borehole screen (Fig. 2.13b on the left) into rigid, homogeneous, isotropic, uniformly unsaturated soil, total steady flow out of the borehole, Q_s (L^3T^{-1}), is given by (Reynolds 2010):

$$Q_s = \frac{2\pi L K_{fs}}{C_{A-VR}} H + \frac{2\pi L \phi_m}{C_{A-VR}} + \pi r^2 K_{fs} \tag{2.26}$$

where C_{A-VR} is a dimensionless analytical shape function given by:

$$C_{A-VR} = \sinh^{-1}\left(\frac{L}{r}\right) - \sqrt{\left(\frac{r}{L}\right)^2 + 1} + \frac{r}{L} \tag{2.27}$$

L (L) is the length of the screen, H (L) is the steady water head in the casing, and r (L) is the radius of the borehole screen. Equation (2.26) is valid for $H \geq L$. The first term on the right of Eq. (2.26) describes discharge due to the hydrostatic pressure of the ponded water in the borehole, the second term describes discharge due to capillarity of the unsaturated soil surrounding the borehole, and the third term describes gravitational discharge out of the base of the borehole. Replacing L with H in Eqs. (2.26) and (2.27) produces the original Guelph permeameter analysis for steady, constant-head infiltration in an uncased borehole (Reynolds et al. 1985).

For radial discharge only (Fig. 2.13a), gravitational discharge is null in Eq. (2.26) and the formulation for Q_s is given by:

$$Q_s = \frac{2\pi L K_{fs}}{C_{A-R}} H + \frac{2\pi L \phi_m}{C_{A-R}} \tag{2.28}$$

where the dimensionless analytical shape function for C_{A-R} coincides with Eq. (2.27), i.e. $C_{A-R} = C_{A-VR}$.

For vertical discharge only (Fig. 2.13b on the right), the formulation for steady flow can be derived from the general Eqs. (2.26) and (2.27) considering that L tends to zero. This leads to:

$$Q_s = \frac{r K_{fs}}{C_{A-V}} H + \frac{r \phi_m}{C_{A-V}} + \pi r^2 K_{fs} \tag{2.29}$$

where the dimensionless shape factor constant, C_{A-V}, is equal to $1/(2\pi) = 0.159$. Equation (2.29) is equivalent to the steady-state form of the transient infiltration solution presented by Hinnell et al. (2009) for a fully lined borehole (Reynolds 2010).

2.1.2.3 Estimation of K_{fs} and ϕ_m from Experimental Flow Rate

Calculation of K_{fs} and ϕ_m can be made using a single head (one H level), dual heads (two H levels) or multiple ponded heads (two or more H levels). The dual- and multiple-head calculations are based on the fact that Eqs. (2.26), (2.28) and (2.29) are linear in H, which allows K_{fs} to be determined from the Q_s vs. H slope and ϕ_m from the Q_s axis intercept.

The single-head calculations have the form:

$$K_{fs} = \frac{Q_s}{(1/E_1)[H + (1/\alpha^*)] + \pi r^2 F} \tag{2.30a}$$

$$\phi_m = \frac{Q_s}{(1/E_1)(H\alpha^* + 1) + \pi r^2 \alpha^* F} \tag{2.30b}$$

where, for combined vertical and radial discharge:

$$E_1 = \frac{C_{A-VR}}{2\pi L}, \quad F = 1 \tag{2.31a}$$

for radial discharge only:

$$E_1 = \frac{C_{A-R}}{2\pi L}, \quad F = 0 \tag{2.31b}$$

and for vertical discharge only:

$$E_1 = \frac{C_{A-V}}{r}, \quad F = 1 \tag{2.31c}$$

and Q_s is the steady flow rate corresponding to the single ponded head, H. In this case, the sorptive number, α^* (L^{-1}), is selected a priori from the soil texture-structure categories (Table 2.1).

The dual-head calculations are given by:

$$K_{fs} = E_1 \left(\frac{Q_{s2} - Q_{s1}}{H_2 - H_1} \right) \tag{2.32a}$$

$$\phi_m = E_1 \left[\frac{Q_{s1}H_2 - Q_{s2}H_1}{H_2 - H_1} - E_2 \left(\frac{Q_{s2} - Q_{s1}}{H_2 - H_1} \right) \right] \tag{2.32b}$$

where E_1 is defined by Eqs. (2.31a, 2.31b or 2.31c), Q_{s1} and Q_{s2} are the steady-state flow rates (Q_s) corresponding to H_1 and H_2, respectively, that are ponded in ascending order ($H_2 > H_1$) and without intervening drainage and, for combined vertical and radial discharge:

$$E_2 = \frac{r^2 C_{A-VR}}{2L} \tag{2.33a}$$

for radial discharge only:

$$E_2 = 0 \tag{2.33b}$$

and for vertical discharge only:

$$E_2 = \pi r C_{A-V} \tag{2.33c}$$

The multiple-head calculations take the form:

$$K_{fs} = E_1 \frac{dQ_s}{dH} \tag{2.34a}$$

$$\phi_m = E_1 \left(I - E_2 \frac{dQ_s}{dH} \right) \tag{2.34b}$$

where E_1 and E_2 are defined by Eqs. (2.31a, 2.31b or 2.31c) and (2.33a, 2.33b or 2.33c), respectively. In Eqs. (2.34a and 2.34b), dQ_s/dH and I are the slope and the intercept, respectively, of the linear least-squares regression line through a plot of Q_s vs. H and the H levels are ponded continuously (no intervening drainage) and in ascending order (i.e., H_1 ponded first, ..., $H_3 > H_2 > H_1$) to obtain the corresponding Q_s values (i.e., ..., $Q_{s3} > Q_{s2} > Q_{s1}$).

2.1.2.4 Validation of Methods with Numerically Generated Data

Taking into account that the analytical solutions are approximate, the HYDRUS-2D numerical simulation model (Šimůnek et al. 1999) was used to assess and enhance these solutions (Reynolds 2010). The borehole parameters included screen radius, $r = 3.0$, 5.0 and 7.5 cm; screen length, $L = 0$–45 cm; L/r ratio $= 0$–6; ponding depth, $H = 3$–200 cm. The soil capillarities included moderate to negligible capillarity, i.e. $\alpha^* = 14.45\ \text{m}^{-1}$; high capillarity, $\alpha^* = 1\ \text{m}^{-1}$; and extreme capillarity, $\alpha^* = 0.1\ \text{m}^{-1}$.

According to the numerical study, the appropriate shape factor for vertical discharge, C_{N-V}, was found to be equal to 0.154, deviating by only 3.1 % from C_{A-V}. For combined vertical and radial discharge and for radial discharge alone, the following shape function, C_{N-VR}, was suggested as an alternative to Eq. (2.27):

$$C_{N-VR} = b_1\left(\sinh^{-1}\left(\frac{L}{r}\right) - \sqrt{\left(\frac{r}{L}\right)^2 + 1} + \frac{r}{L}\right) + b_0 \qquad (2.35)$$

where the b_0 and b_1 parameters are equal to 0.1479 and 0.8430, respectively.

The multiple-head and dual-head calculations were found to provide accurate K_{fs} determinations ($\leq$26 % error) for the full range of tested soil capillarities, screen or open section geometries, and ponded heads. The multiple-head and dual-head calculations of ϕ_m were similarly accurate ($\leq$25 % error) for high capillarity soils ($\alpha^* \leq 1\ \text{m}^{-1}$) and for $L/r \leq 1$ in soils with moderate to negligible capillarity. The ϕ_m calculations were not accurate, however, in moderate- to negligible-capillarity soils with $L/r > 1$, where ϕ_m was increasingly underestimated with increasing L/r ratio. The single-head calculation with estimated α^* was accurate only for K_{fs} determination ($\leq$23 % error) in soils with moderate to negligible capillarity. For ϕ_m determination and for K_{fs} determination in high-capillarity soils, the single-head calculation requires α^* to be known accurately. On average, the numerical shape factors for vertical discharge, C_{N-V}, and combined vertical and radial discharge, C_{N-VR}, yielded slightly more accurate K_{fs} and ϕ_m calculations than their analytical counterparts, i.e. C_{A-V} and C_{A-VR}, respectively. For most practical applications, however, the K_{fs} and ϕ_m calculations are expected to be effectively equivalent regardless of whether the numerically derived shape relationships or the analytical shape relationships are used. This occurs because the numerical and analytical shape relationships are similar in form and magnitude, and because the K_{fs} and ϕ_m calculations are not highly sensitive to the borehole shape relationships.

2.1.2.5 Drive Point Guelph Permeameter

Recently, constant-head, radial steady flow was further considered by Reynolds and Lewis (2012) on the premise that, with the classical Guelph permeameter (GP) technique, practical difficulties can arise in gravelly materials, where augering

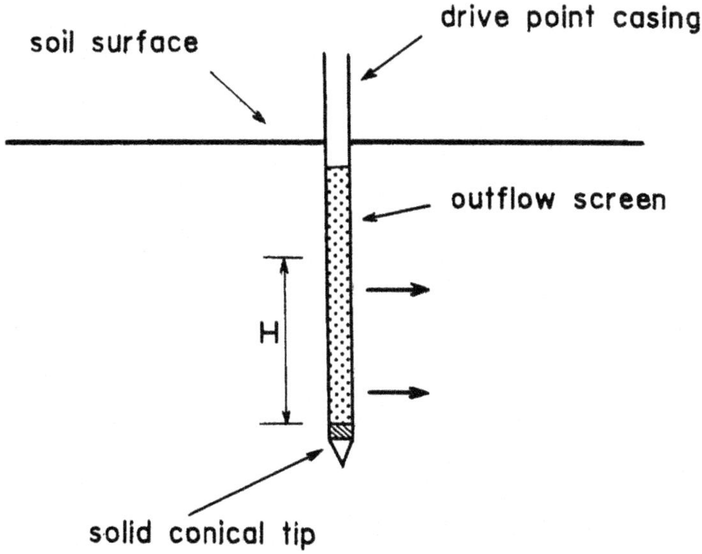

Fig. 2.14 Schematic of a drive point casing inserted into soil. The *block arrows* indicate radial direction of water discharge through outflow screen (From Reynolds and Lewis 2012, reprinted with permission)

can be problematic, and in uncohesive materials, where the well may collapse during the augering process or upon wetting. One convenient and rapid way designed to get around these problems was to form the well using a "drive point" casing rather than an auger. A stout pipe with a screened section and a solid basal "drive point" is pushed, driven, rotated or vibrated into the gravelly and/or uncohesive material to the desired depth (Fig. 2.14). The GP outflow tube is then simply inserted into the drive point casing. Alternatively, with some equipment modifications, the drive point casing itself can serve as the outflow tube to produce a drive point GP. This device is expected to be most applicable to porous materials where hydraulic parameters are not seriously altered by the disturbance of a drive point installation (e.g., smearing, compaction, loosening).

In the drive point application of the GP, using the same approach as for the original GP and neglecting vertical discharge, that is using Eq. (2.1), leads to:

$$Q_s = \frac{2\pi K_{fs}}{C_{DP}} H^2 + \frac{2\pi K_{fs}}{\alpha^* C_{DP}} H \tag{2.36}$$

where C_{DP} is the dimensionless drive point GP shape factor, H (L) is the ponding depth in the drive point screen ($H \leq$ length of drive point screen), and the gravitational discharge is not considered because the cone-shaped base of the drive point is impermeable. An estimate of C_{DP} can be obtained by the following relationship:

Table 2.6 Fitting constants (P_1, P_2, P_3 and P_4) of Eq. (2.37) for different soil capillarity conditions. From Reynolds and Lewis 2012, reprinted with permission

Soil capillarity conditions	α* parameter (m^{-1})	P_1	P_2	P_3	P_4
Negligible-Moderate	≥9	1.7905	0.0589	0.6671	0.0648
Strong	≈4	1.7491	0.0394	0.5817	0.0869
Very strong	≈1	1.4593	0.0282	0.4890	0.1872

$$C_{DP} = \left[\frac{(H/r)}{P_1 + P_2(H/r)}\right]^{P_3} - P_4 \tag{2.37}$$

where r (L) is the outside radius of the drive point screen, and P_1, P_2, P_3 and P_4 are dimensionless fitting constants varying with the soil capillarity conditions (Table 2.6). Equation (2.37) was specifically developed for the drive point GP, with procedures similar to Reynolds (2010, 2011), making use of numerically simulated data. As with the original GP analysis, K_{fs} and/or α* can be determined for the drive point application using single-head, dual-head and multiple-head calculations. In particular, the single-head calculation applies a single Q_s vs. H measurement to Eq. (2.36) re-arranged as:

$$K_{fs} = \frac{C_{DP} Q_s}{2\pi H(H + 1/\alpha^*)} \tag{2.38}$$

where α* is chosen on the basis of soil texture-structure characteristics (Table 2.1). The dual-head calculation is made by recasting Eq. (2.36) into its simultaneous equations form to produce:

$$K_{fs} = \frac{C_1 Q_{s1}(C_1 Q_{s1} H_2 - C_2 Q_{s2} H_1)}{2\pi H_1 (C_1 Q_{s1} H_1 H_2 - C_1 Q_{s1} H_2^2)} \tag{2.39a}$$

$$\alpha^* = \frac{C_1 Q_{s1} H_2 - C_2 Q_{s2} H_1}{C_2 Q_{s2} H_1^2 - C_1 Q_{s1} H_2^2} \tag{2.39b}$$

where Q_{s1} and Q_{s2} are the measured Q_s values for H_1 and H_2, respectively, and C_1 and C_2 are the C_{DP} values for H_1/r and H_2/r, respectively. The multiple-head calculation involves re-writing Eq. (2.36) as a quadratic equation, i.e.:

$$C_i Q_{si} = AH_i^2 + BH_i \qquad i = 1, 2, 3, \ldots m \tag{2.40a}$$

and then least squares regression fitting to two or more successive Q_s vs. H measurements, where $A = 2\pi K_{fs}$ and $B = 2\pi K_{fs}/\alpha^*$ are treated as regression constants; C_i, Q_{si} and H_i are the corresponding C_{DP}, Q_s and H values, respectively; m (≥2) is the number of ponded heads; and $H_i < H_{i+1}$. The K_{fs} and α* values are then determined from the fitted A and B coefficients using:

$$K_{fs} = \frac{A}{2\pi} \tag{2.40b}$$

$$\alpha^* = \frac{A}{B} \tag{2.40c}$$

Equation (2.40a) must be forced through the origin when fitted to the data. For $m = 2$, the regression approach is equivalent to the simultaneous equations approach. An alternative multiple-head calculation involves an iterative curve-fitting procedure (Reynolds and Lewis 2012).

In a method comparison carried out on a medium-coarse glacial outwash sand, the drive point GP method yielded two-three times higher α* and K_{fs} values as compared to the original GP method (Reynolds and Lewis 2012). The detected differences were thought to be a consequence of flow impedance in the original GP wells, due to progressive collapse of the unlined well wall during the flow measurements, and/or gradual sinking of the GP outflow tip into the unprotected base of the well due to the weight of the GP apparatus. An implication of this interpretation was that the drive point GP results were more reliable than the ones obtained by the original GP.

In principle, the drive point GP appears usable, in combination with other field techniques such as the Simplified Falling Head (SFH, Sect. 2.2.4) one (Bagarello et al. 2004), to determine in the field the anisotropy of field-saturated soil hydraulic conductivity. The drive point GP establishes a radial flow process, and therefore it could be expected to yield a determination of K_{fs} in the horizontal direction. The SFH technique establishes a vertical infiltration process and therefore yields an estimate of the vertical conductivity. The screened section for the drive point GP and the ring for the SFH technique can be chosen in such a way that the same soil layer is sampled and also a similar soil surface is explored by the two techniques. Both the SFH technique and the drive point GP with the single-head approach need an independent estimate of α*. Therefore, there is a methodological consistency between the two techniques. On the other hand, applying a transient (SFH) and a steady-state (drive point GP) technique could be a reason for concern since there is evidence to suggest that the duration of the field run may influence the measured conductivity (e.g., Bagarello et al. 2012). This factor should have a reduced impact in rigid and stable soils that are not altered appreciably by the wetting process.

2.1.3 Transient Flow Under Constant Head in Cased and Uncased Boreholes

Borehole or well permeameter methods are particularly suitable for subsurface measurement of soil hydraulic properties in the vadose zone with a deep water table. Establishing and maintaining a constant depth of water in the well is relatively simple and this is one of the reasons for the success of the original Guelph permeameter (GP) method, together with the availability of a theoretically robust analysis procedure of the data (Reynolds et al. 1985; Reynolds and Elrick 1986, 1987). In recent years, constant-head flow from boreholes has been reconsidered by several authors. In particular, methods to analyze transient infiltration data in both unlined, or uncased, and lined, or cased, boreholes were developed by Hinnell et al. (2009).

The starting point of the analysis by these authors was the model by Haverkamp et al. (1994), originally developed with reference to infiltration from a disk infiltrometer. This model describes three-dimensional cumulative infiltration, I_{3D} (L) (flow volume per unit area), from the surface disk source as the sum of a term accounting for one-dimensional infiltration, I_{1D} (L), and a second term accounting for multidirectional flow from the disk edge:

$$I_{3D} = I_{1D} + \frac{\gamma S_0^2 t}{r(\theta_0 - \theta_i)} \tag{2.41}$$

where t (T) is time, r (L) is the radius of the disk, γ is a dimensionless constant, S_0 (L $T^{-0.5}$) is the soil sorptivity, θ_0 (L^3L^{-3}) is the volumetric soil water content at the disk source and θ_i (L^3L^{-3}) is the initial volumetric soil water content. The last term in Eq. (2.41) was described by Hinnell et al. (2009) as the cumulative inflow per unit length of the edge of the wetted soil, Q_{edge} (L^3L^{-1}), divided by a characteristic length, L (L), equal to the ratio between the area across which flow occurs and the corresponding perimeter of the infiltration area:

$$I_{3D} = I_{1D} + \frac{1}{L} Q_{edge} = I_{1D} + \frac{\gamma S_0^2 t}{2L(\theta_0 - \theta_i)} \tag{2.42}$$

For the disk source, flow occurs across a surface of πr^2 and the appropriate edge length is $2\pi r$, leading to a ratio $L = r/2$ for which Eq. (2.42) corresponds to Eq. (2.41). Equation (2.42) was also found to be usable for both lined and unlined boreholes, with $L = r/2$ in the former case, r being the borehole radius in this case, and $L = r/2 + H$ in the latter case, H (L) being the constant ponded depth of water in the borehole. As noted by Hinnell et al. (2009), the lined borehole is analogous to the surface disk but the flow at the edge of the infiltration surface is more complex for the lined borehole due to the upward capillary uptake into the surrounding soil. For soil sorptivity, S_0, Hinnell et al. (2009) considered the Green and Ampt (1911) approximation for ponded infiltration [e.g., Haverkamp et al. (1988), their equation (14)]:

$$S_0 \approx \sqrt{2K_{fs}(\theta_{fs} - \theta_i)(H - h_f)} \tag{2.43}$$

where K_{fs} (L T^{-1}) is the soil hydraulic conductivity at field saturation since infiltration occurs under ponded conditions, θ_{fs} (L^3L^{-3}) is the field-saturated soil water content and h_f (L) is the pressure head corresponding to the wetting front, that can be defined as follows (Warrick 2003):

$$h_f = -\frac{1}{K_{fs}} \int_{h_i}^{0} K(h)dh \tag{2.44}$$

where h (L) is the soil water pressure head and h_i (L) is the initial value of h.

The values of γ were determined by Hinnell et al. (2009) using numerically simulated infiltration data. Simulations were carried out for six soil classes, i.e. sand, loamy-sand, sandy-loam, loam, silt-loam and silty-clay-loam, characterized with the soil hydraulic parameters by Carsel and Parrish (1988), r values ranging from 1.25 to 15 cm, H values varying between 0 and 50 cm and initial saturation values varying between 0.05 and 0.4. The values of γ for lined boreholes were similar to those for a disk source and practically independent of soil texture, radius, ponded water depth and initial saturation (Fig. 2.15). Assuming a constant value of γ, equal to 1.06, was found to be appropriate for lined boreholes. On the other hand, the values of γ estimated for the unlined boreholes covered a large range, varying from 1.02 to 3.16 (Fig. 2.15), showing that using a single average value is inappropriate in this case. In other terms, a case specific γ value has to be selected for the unlined borehole. We can note that the values of γ for lined borehole are in the order of the value proposed by Haverkamp et al. (1994) for the case of a disk source, i.e. 0.75. In addition, Eq. (2.41) was developed for null or negative pressure heads at the surface. For positive pressure heads like in boreholes, the value of γ may be a way of adapting Eq. (2.41) proposed for cases of unponded conditions to cases of ponded depths of water in the borehole.

The use of Eq. (2.42) requires the knowledge of 1D cumulative infiltration, I_{1D}. Hinnell et al. (2009) investigated several options for the estimation of I_{1D}: numerical resolution of Richards equation, Green and Ampt model, Philip's model. For instance, linking Eq. (2.42) with the cumulative one-dimensional infiltration model by Philip (1957b), leads to:

$$I_{3D}(t) = S_0 t^{1/2} + \left(a K_{fs} + \frac{\gamma S_0^2}{2L(\theta_0 - \theta_i)} \right) t \tag{2.45}$$

where a is a coefficient, often taken to be 0.5 (Warrick 2003; Hinnell et al. 2009), and θ_0 (L^3L^{-3}) coincides with the field-saturated soil water content. If θ_0, θ_i and γ are known, Eq. (2.45) can be used with a time series of early-time data to estimate soil hydraulic properties by a curve fitting procedure.

The transient analysis by Hinnell et al. (2009) has practical interest since it allows soil hydraulic characterization without waiting for steady-state flow conditions. The procedure appears more easily usable for lined boreholes, given that an approximate value of γ is available in this case. For unlined boreholes, γ should be predictable on the basis of the borehole radius, the height of water in the borehole

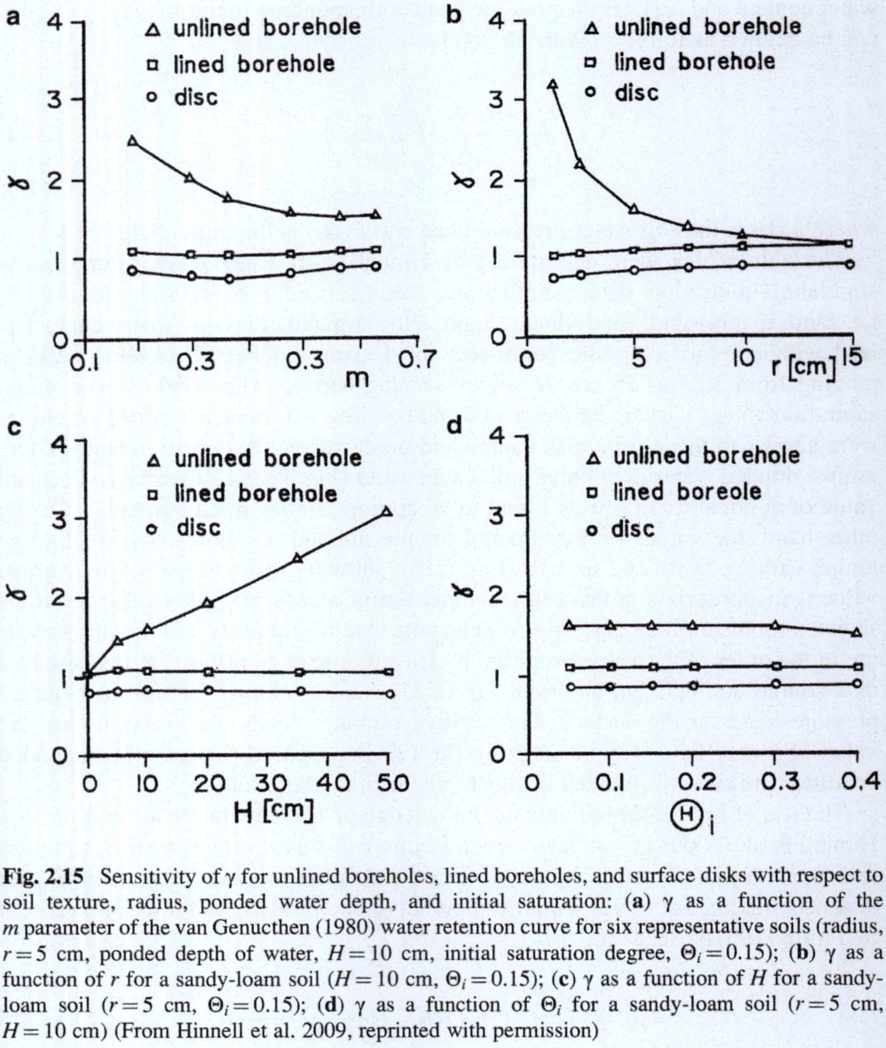

Fig. 2.15 Sensitivity of γ for unlined boreholes, lined boreholes, and surface disks with respect to soil texture, radius, ponded water depth, and initial saturation: (**a**) γ as a function of the *m* parameter of the van Genucthen (1980) water retention curve for six representative soils (radius, $r = 5$ cm, ponded depth of water, $H = 10$ cm, initial saturation degree, $\Theta_i = 0.15$); (**b**) γ as a function of *r* for a sandy-loam soil ($H = 10$ cm, $\Theta_i = 0.15$); (**c**) γ as a function of *H* for a sandy-loam soil ($r = 5$ cm, $\Theta_i = 0.15$); (**d**) γ as a function of Θ_i for a sandy-loam soil ($r = 5$ cm, $H = 10$ cm) (From Hinnell et al. 2009, reprinted with permission)

and an estimate of soil texture (Hinnell et al. 2009), but γ estimating procedures for this specific case have still to be developed. In any case, there is the need to check the applicability of the suggested analytical tool in real field situations. Taking into account that steady flow analysis procedures have been developed for both lined and unlined boreholes (e.g., Reynolds and Elrick 1986, 1987; Reynolds 2010), measuring both transient and steady flow in a single borehole might allow us to compare alternative procedures to analyze experimental data. These comparisons should preferably be carried out in soils where particle arrangement does not change appreciably upon wetting to be reasonably sure that different procedures are applied to practically the same porous medium. A point that should also be taken into account is the size of the sampled soil volume with the alternative procedures.

2.1.4 *Falling-Head Flow in Cased Boreholes*

2.1.4.1 Principles

The falling-head lined, or cased, borehole permeameter, also known as the Philip-Dunne permeameter, consists of a circular tube with an internal radius equal to r (L), that extends to the base of a vertical borehole realized to a certain depth in an unsaturated soil, and ideally fitting the borehole sides with zero gap (Fig. 2.16, Philip 1993). The device can be used for below surface measurements. Performing a test with this permeameter implies determining a falling-head infiltration process through the bottom of the borehole, representing the infiltration surface. To this end, water is suddenly introduced into the tube to a depth of D_0 (L) at time $t = 0$. The process continues up to $t = T$ (T), that is the time when all applied water disappears, i.e. the borehole empties. During the infiltration process, the water depth on the infiltration surface, D (L), is measured repeatedly to obtain the experimental $D(t)$ drawdown curve. A sensor at the bottom of the borehole can be used to register the depth of water over the base. According to Muñoz-Carpena et al. (2002), a 2-mm-mesh plastic screen can be glued to the lower edge of the tube to contrast soil erosion when filling the permeameter quickly at $t = 0$ with the aid of

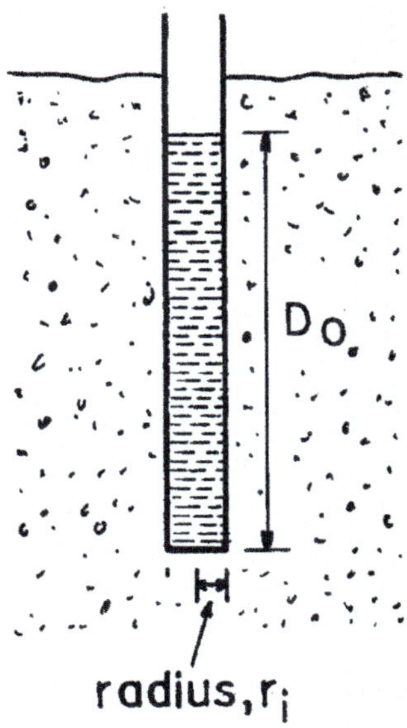

Fig. 2.16 Schematic view of the falling-head lined borehole permeameter with a water depth, D_0, at time $t = 0$ (From Philip 1993, reprinted with permission)

a funnel. The device was initially used by T.Dunne and E. Safran, at the University of Washington, in extensive pioneering hydrologic investigations of soils in the Amazon basin, where portability of equipment and water supply is an important constraint in the field when roads are scarce (Philip 1993).

2.1.4.2 Analytical Modeling of Water Drop

Philip (1993) developed an approximate analysis procedure of the data, based on the Green and Ampt (1911) (GA) model, with the flow treated as a gravity-perturbed spherically symmetrical pressure-capillarity-driven flow. This analysis allows us to obtain an estimate of both the field-saturated soil hydraulic conductivity, K_{fs} (L T^{-1}), and the GA wetting front potential, denoted by the symbol $-C$ (L) in the paper by Philip (1993) and indicated as h_f elsewhere in this book. The soil is assumed to be homogeneous and isotropic, with an initially uniform volumetric water content, θ_i (L^3L^{-3}). The disk-shaped water supply surface of radius r is replaced by a spherical supply surface of equal area, with radius, r_0 (L), equal to $r/2$ to keep the same surface for water infiltration. A factor of $8/\pi^2$ enters the analysis to take into account that flow through the actual system does not coincide with the spherically symmetrical flow (Fig. 2.17). More precisely, this factor is an *ad hoc* "flow efficiency correction" accounting for the fact that actual flow out of the borehole base may be less efficient than flow out of the assumed equivalent sphere

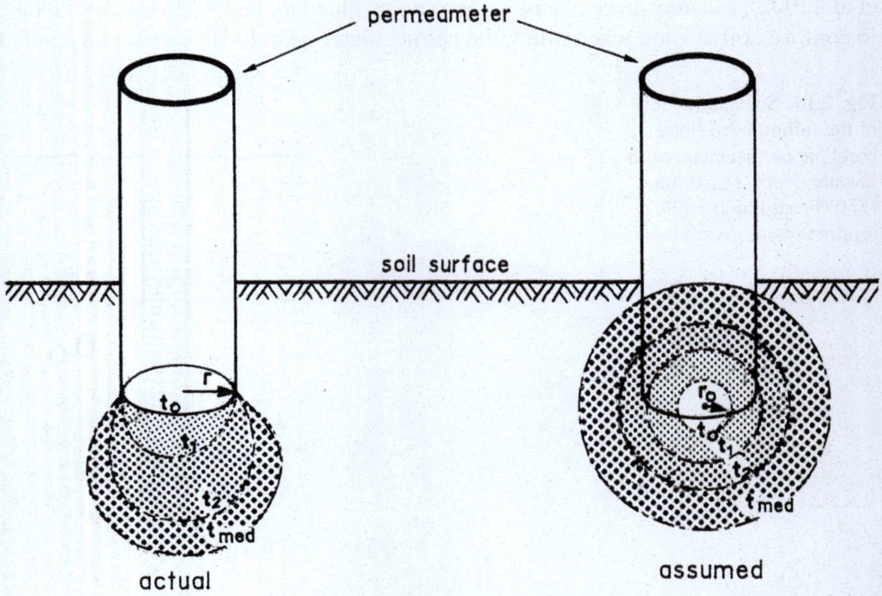

Fig. 2.17 Wetted bulb at different times, t, for the falling-head permeameter (r = radius of the disk-shaped water supply surface; r_0 = radius of the spherical supply source)

surface (Reynolds 2011). The wetted bulb is field-saturated and therefore it has a volumetric soil water content equal to θ_{fs} (L^3L^{-3}) and a hydraulic conductivity of K_{fs}. The bounding surface of the wetted bulb sharply separates wet soil ($\theta = \theta_{fs}$) from that yet to be wetted ($\theta = \theta_i$).

For his analysis, Philip (1993) defined the scaled time, τ, and the scaled radius of the wetted bulb, ρ, as follows:

$$\tau = \frac{8K_{fs}t}{\pi^2 r_0} \tag{2.46a}$$

$$\rho = \frac{R}{r_0} \tag{2.46b}$$

where R (L) is the wetted bulb radius. On the basis of the GA approach, Philip (1993) demonstrated that the scaled time and bulb radius were related through the following equation:

$$\begin{aligned} \tau = & \left(1 + \frac{1}{2A}\right) \ln\left(\frac{A^3 - 1}{A^3 - \rho^3}\right) \\ & - \frac{3}{2A} \ln\left(\frac{A-1}{A-\rho}\right) + \frac{\sqrt{3}}{A}\left[\tan^{-1}\left(\frac{A+2\rho}{\sqrt{3}A}\right) - \tan^{-1}\left(\frac{A+2}{\sqrt{3}A}\right)\right] \end{aligned} \tag{2.47}$$

where the A term is given by:

$$A = \left[\frac{3\left(C + D_0 + \frac{\pi^2 r_0}{8}\right)}{r_0 \Delta\theta} + 1\right]^{1/3} \tag{2.48}$$

with $\Delta\theta = \theta_{fs} - \theta_i$. This equation is valid for any time between the beginning ($t = 0$) and the end ($t = T$) of the experiment. The r_0 term in the numerator of Eq. (2.48) accounts for the effect of gravity that was assumed to be maximum when the radius of the wetted zone around the discharge surface is much larger than r_0 (Reynolds 2011). These equations may be adapted to the case of zero gravity flow assuming that the term $\pi^2 r_0/8$ is zero in Eq. (2.48).

Meanwhile, the fall of water in the tube, $D_0 - D$, and the scaled radius of the wetted bulb, ρ, are related through the following equation, derived from mass balance consideration:

$$D_0 - D = \frac{\Delta\theta \ r_0}{3}\left(\rho^3 - 1\right) \tag{2.49}$$

Equation (2.49) allows us to derive the maximum scaled radius of the wetted bulb at the end of the experiment, ρ_{max}, i.e. the ρ value corresponding to $D = 0$:

$$\rho_{max} = \left(\frac{3D_0}{\Delta\theta\ r_0} + 1\right)^{1/3} \tag{2.50}$$

The wetted bulb radius when $D = 0$ and $t = T$, i.e. the experiment ceases, R_{max} (L), is given by:

$$R_{max} = \rho_{max}\ r_0 \tag{2.51}$$

2.1.4.3 Procedure to Derive K_{fs} and C

For known values of D_0 and r_0, Philip (1993) suggested measuring infiltration time at two pre-established water levels to obtain K_{fs} and C. In particular, the proposed algorithm uses the t values corresponding to $D = D_0/2$ ($t = t_{05}$) and $D = 0$ (end of the run, $t = T$) and related dimensionless times. These times correspond to the scaled radius defined by Eq. (2.50) for $t = T$ and the following value of ρ corresponding to $D = D_0/2$, ρ_{05}:

$$\rho_{05} = \left(\frac{\rho_{max}^3 + 1}{2}\right)^{1/3} \tag{2.52}$$

The algorithm by Philip (1993) implies determining at first the relationship between C and the τ_{max}/τ_{05} ratio, where τ_{max} is the τ value calculated by Eqs. (2.47) and (2.50) ($\rho = \rho_{max}$) and τ_{05} is the τ value obtained with Eqs. (2.47) and (2.52) ($\rho = \rho_{05}$). With this aim, a sequence of C values with a given step is established. For a value of C, A is calculated by Eq. (2.48) and then Eq. (2.47) with the appropriate values of ρ (ρ_{max}, ρ_{05}) is applied to obtain τ_{max} and τ_{05} and hence τ_{max}/τ_{05}. This procedure is repeated for each C value. Taking into account that τ_{max}/τ_{05} must be equal to T/t_{05}, the developed $C(\tau_{max}/\tau_{05})$ relationship allows us to determine the value of C corresponding to the experimental T/t_{05} ratio. To determine K_{fs}, the τ_{max} vs. τ_{max}/τ_{05} relationship is plotted to individuate the value of τ_{max} corresponding to $\tau_{max}/\tau_{05} = T/t_{05}$. This τ_{max} value is then used in the following relationship, derived from Eq. (2.46a):

$$K_{fs} = \frac{\pi^2\ r_0\ \tau_{max}}{8T} \tag{2.53}$$

We should note that this method requires the knowledge of Δθ, which can be deduced from the measurement of water content before the experiment and soil porosity or measurement of water content in the wetted bulb at the end of the experiment.

As $t \rightarrow 0$, the flow from the permeameter base is one-dimensional and the developed analysis, based on a three-dimensional (3D) flow field, cannot be applied. In practice, the flow may be considered 3D when $D_0 - D \geq 4\Delta\theta r_0$ (Philip 1993). This circumstance reinforces the suggestion of making use of the experimental information collected in a temporally advanced stage of the experiment ($t = t_{05}$ and $t = T$) to obtain C and K_{fs}. According to Reynolds (2008), soil water transmission parameters obtained with the falling-head lined borehole permeameter are mainly expressive of water transport in the vertical direction given that flow occurs exclusively through the bottom of the borehole.

Example 2.3

This example, taken from Philip (1993), describes in detail the analysis of an experiment carried out with a tube having an internal radius $r = 0.03$ m ($r_0 = 0.015$ m) in a soil characterized by a $\Delta\theta = \theta_{fs} - \theta_i$ value of 0.2 m^3m^{-3}. The established water depth at time $t = 0$ was $D_0 = 0.3$ m. The water depth over the base of the borehole was equal to $D_0/2 = 0.15$ m at $t = t_{05} = 3600$ s and the borehole emptied ($D = 0$) at $t = T = 10{,}000$ s. These data have to be used for determining C and K_{fs}.

The term $4\Delta\theta r_0$ is equal to 0.012 m. Therefore, the 3D analysis can be applied starting from the instant in which the water depth over the infiltration surface, D, is equal to $D_0 - (4\Delta\theta r_0) = 0.288$ m. This last value is greater than $D_0/2$ and therefore the analysis by Philip (1993) can be applied. At first, Eqs. (2.50) and (2.52) are used to calculate ρ_{max} and ρ_{05}, respectively:

$$\rho_{max} = \left(\frac{3 \times 0.3}{0.2 \times 0.015} + 1\right)^{1/3} = 6.702$$

$$\rho_{05} = \left(\frac{301 + 1}{2}\right)^{1/3} = 5.325$$

Then, a sequence of possible C values is established ($0.037 \leq C \leq 0.3$ m, step $= 0.001$ m in this example) and A, τ_{max} and τ_{05} are calculated for each C. As an example, the following calculations are made for $C = 0.25$ m:

$$A = \left[\frac{3\left(0.25 + 0.3 + \frac{\pi^2 0.015}{8}\right)}{0.015 \times 0.2} + 1\right]^{1/3} = 8.289$$

$$\tau_{max} = \left(1 + \frac{1}{2 \times 8.289}\right)\ln\left(\frac{569.51 - 1}{569.51 - 301}\right) - \frac{3}{2 \times 8.289}\ln\left(\frac{8.289 - 1}{8.289 - 6.702}\right) +$$

$$+ \frac{\sqrt{3}}{8.289}\left[\tan^{-1}\left(\frac{8.289 + 2 \times 6.702}{\sqrt{3} \times 8.289}\right) - \tan^{-1}\left(\frac{8.289 + 2}{\sqrt{3} \times 8.289}\right)\right] = 0.5957$$

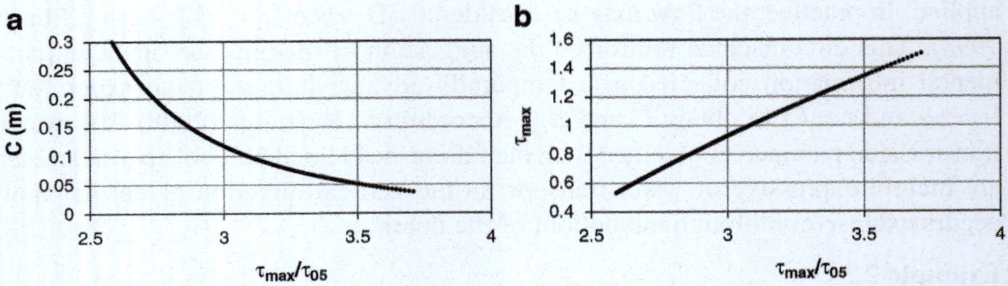

Fig. 2.18 Relationships between wetting front suction, C, dimensionless time at borehole empting, τ_{max}, and dimensionless time as the ponding depth of water in the borehole is 0.5 times the initial water depth, τ_{05}, for the data of the Example 2.3: (**a**) C vs. τ_{max}/τ_{05} relationship, and (**b**) τ_{max} vs. τ_{max}/τ_{05} relationship

$$\tau_{05} = \left(1 + \frac{1}{2 \times 8.289}\right)\ln\left(\frac{569.51 - 1}{569.51 - 151}\right) - \frac{3}{2 \times 8.289}\ln\left(\frac{8.289 - 1}{8.289 - 5.325}\right) +$$

$$+\frac{\sqrt{3}}{8.289}\left[\tan^{-1}\left(\frac{8.289 + 2 \times 5.325}{\sqrt{3} \times 8.289}\right) - \tan^{-1}\left(\frac{8.289 + 2}{\sqrt{3} \times 8.289}\right)\right] = 0.2247$$

Therefore, the τ_{max}/τ_{05} ratio is equal to 2.651. Figure 2.18a shows the relationship between C and the calculated τ_{max}/τ_{05} ratios within the range of selected C values. The corresponding relationship between τ_{max} and τ_{max}/τ_{05} is plotted in Fig. 2.18b. The T/t_{05} ratio is equal to 2.78. For $\tau_{max}/\tau_{05} = 2.78$, Fig. 2.18a yields $C = 0.185$ m and Fig. 2.18b yields $\tau_{max} = 0.722$. Using Eq. (2.53), the following estimate of K_{fs} is finally obtained:

$$K_{fs} = \frac{\pi^2 \times 0.015 \times 0.722}{8 \times 10000} = 1.34 \times 10^{-6} \text{ m s}^{-1}$$

According to Eq. (2.51), the radius of the wetted bulb at the end of the experiment is equal to:

$$R_{max} = 6.702 \times 0.015 = 0.1005 \text{ m}$$

Let us now suppose that there were errors of 5 % in the opposite sense of each time reading; in particular, t_{05} was underestimated (the reading was $t_{05} = 3420$ s) and T was overestimated (reading = 10,500 s). In this case, $C = 0.104$ m is obtained, with a percentage difference, Δ, between this estimate and the true value (i.e., no reading errors) equal to −43.8 %, and K_{fs} is 1.84×10^{-6} m s^{-1} ($\Delta = 37.5$ %). Therefore, a relatively small reading error had an appreciable impact on the calculations. If the errors were in the same direction (e.g., overestimation of both t_{05} and T), they did not have any impact on the calculated C and K_{fs} values because T/t_{05} did not vary.

2.1.4.4 Alternative Mathematical Methods

As also shown in the last part of the Example 2.3, small errors in the data can give large errors in the estimates of C and K_{fs} since making use of two (D, t) data pairs alone implies sensitivity to random measurement errors and lack of data-model fit (Philip 1993; Reynolds 2011). This limit has also been suggested with reference to the data analysis procedure by De Haro et al. (1998) (Muñoz-Carpena et al. 2002), avoiding graphical methods but still using two data pairs (Reynolds 2011). Moreover, Gómez et al. (2001) suggested that the application of the method to some soils of medium to fine texture generally requires a long measurement time and often shows non-convergence with the numerical scheme proposed by Philip (1993). To avoid this problem, these authors preferred to measure the whole $D(t)$ curve and they calculated C and K_{fs} by numerical fitting of the data on the theoretical $D(t)$ curve, obtained from Eqs. (2.46a and 2.46b) and (2.47). A similar suggestion was given by Reynolds (2008) and Philip (1993) also stated that a partial safeguard against large errors in the estimates of C and K_{fs} by his procedure is to measure the whole $D(t)$ curve and to compare it with the curve of Eq. (2.47), using the estimates of C and K_{fs}. If the two curves differ grossly in shape, skepticism is recommended.

According to Reynolds (2008), both the experimental and the analytical procedures can be simplified if the permeameter is applied to only determine K_{fs}. In this case, r_0 and D_0 are known and a measurement of $\Delta\theta$ and T is enough. In particular, the following relationship can be applied to estimate C using an estimate of the so-called α^* (L^{-1}) parameter (e.g., Table 2.1):

$$C = \frac{1}{\alpha^*} \tag{2.54}$$

The ρ_{max} term is calculated by Eq. (2.50), and Eq. (2.48) is applied to calculate A using the estimated C value. Then, Eq. (2.47) is used to calculate τ_{max} and, finally, Eq. (2.53) yields K_{fs}.

Regalado et al. (2005) used the Philip's (1993) method to analyze approximately 300 runs carried out in eight texturally different soils (from clay to sand) with a tube having an internal radius, $r = 0.018$ m, and an initial depth of water over the bottom of the borehole, $D_0 = 0.3$ m. These authors recognized a small sensitivity of the calculations to $\Delta\theta$, in agreement with the findings by other authors (Muñoz-Carpena et al. 2002; Reynolds 2008). Moreover, the suction at the wetting front, C, violated the positivity condition for $T/t_{05} > 5.4$. Finally, for T/t_{05} close to 5.4, the estimate of C was very sensitive to small variations (<2 %) in T, t_{05} and D_0. On the basis of their findings, Regalado et al. (2005) suggested that calculation of T/t_{05} quickly allows us to validate the run for calculating K_{fs} and C ($T/t_{05} < 5$). On the basis of their study, these authors proposed the following statistical relationships for the estimation of τ_{max} and C (within the range from 0.01 to 1 m):

$$\tau_{max} = 0.731\frac{T}{t_{05}} - 1.112 \tag{2.55}$$

$$\ln(C) = -13.503 + 19.678\left(\frac{T}{t_{05}}\right)^{-1/2} \tag{2.56}$$

Then, K_{fs} can be obtained with Eq. (2.53) and the value of τ_{max}. The practical interest for the empirically derived Eqs. (2.55) and (2.56) is related to the fact that they allow us to obtain an estimate of the two unknowns without the need to measure $\Delta\theta$.

2.1.4.5 Practical Issues and Comparison with Other Experimental Methods

An incomplete contact between the external surface of the tube and the soil may determine upflow of water in the hollow space during the run. Therefore, the tube has to be inserted snugly in order to prevent this risk. With this aim, Muñoz-Carpena et al. (2002) suggested boring a flat bottom hole with a slightly smaller diameter than the one of the tube (3.8 and 4.0 cm, respectively, in the investigation by these authors). Soil has then to be pressed around the tube at the surface to ensure a close fit. Reynolds (2008) suggested spreading some grease on the outer side of the tube before inserting it into the hole. In any case, the tube has to be long enough to line the hole up to the bottom. Depending on the objective of the experiment (for example, near-surface measurements with a relatively large value of D_0), a tube extending above the soil surface can be used. Stoniness or other forms of soil heterogeneity can influence appreciably the reliability of the collected data due to the generally small, and difficult to inspect, infiltration surface used by this technique. For example, Vanderlinden et al. (1998) suggested that even a small stone located right below the bottom of the borehole can cause an anomalous behavior in the measured times.

Gómez et al. (2001) compared the falling-head lined borehole permeameter with ring, tension and rainfall infiltrometers in a clay-loam soil, finding similar results for K_{fs} but not for C. These Authors could not give a definitive explanation to justify the high C values obtained with the borehole permeameter. In a loamy-sand soil, Muñoz-Carpena et al. (2002) also obtained large C and K_{fs} values as compared with the ones determined with the Guelph permeameter. The detected differences were interpreted in terms of representativeness and anisotropy of the sampled soil volume and flow field geometry established by the two techniques. In an investigation on an artificial ecosystem, the mean K_{fs} value obtained with the Philip-Dunne permeameter was 1.4 times higher than the saturated hydraulic conductivity measured in the laboratory with the constant-head permeameter (Gwenzi et al. 2011).

Therefore, the method appears simple and useful, especially as it allows determination of K_{fs} in low permeability soils with a run of reasonable duration, shorter than that of a steady-state, constant-head infiltration experiment. The lining of the exposed soil at the borehole walls appears to be a critical factor influencing the applicability of the method since upflow of water in the space between the soil and the outer side of the tube during the run determines overestimation of K_{fs} (Reynolds 2008). In particular, lining can be difficult or even impossible in gravelly soils, due to a possibly unavoidably incomplete contact between the tube and the walls of the hole. In addition, only a few comparisons with other more established methodologies have been carried out, suggesting that the reliability of the collected data has still to be considered more or less uncertain. According to Reynolds (2011), the adoption of the falling-head lined borehole permeameter has been limited because (i) the method applies only for vertical discharge of water through the base of the borehole, (ii) it uses a two-head graphical calculation procedure that requires nomographs and specific head vs. time measurements, (iii) its accuracy has not been rigorously assessed, and (iv) there is some evidence that the method may systematically overestimate K_{fs} and provide low-accuracy or unstable estimates of C. This Author used theoretical analysis of the flow process and data for strong- (sorptive number, $\alpha^* = 1\ \text{m}^{-1}$), moderate- ($\alpha^* = 11.7\ \text{m}^{-1}$) and weak- ($\alpha^* = 26.3\ \text{m}^{-1}$) capillarity soils, simulated by an *ad hoc* "well version" of HYDRUS-2D, to demonstrate that inaccuracies in the K_{fs} and α^* values, i.e. overestimation by amounts ranging from a few percentage points in strong-capillarity soils to several orders of magnitude in weak-capillarity soils, were due to the choice of the flow efficiency correction and the treatment of the gravity component of borehole discharge, that was overestimated in the original analysis by Philip (1993).

2.1.4.6 Alternative Setups and Mathematical Methods for Falling-Head Flow in Cased Boreholes

Reynolds (2011) developed an extended analysis for determining K_{fs} and α^* in a homogeneous, isotropic, uniformly unsaturated soil, with the aim to improve the accuracy and utility of the falling-head lined borehole permeameter. The extended analysis allows a range of discharge geometries and it introduces different values of the parameters accounting for flow efficiency and gravity effects. In particular, the considered discharge geometries were vertical discharge only, as in Philip (1993) (Fig. 2.19a), radial discharge only (permeable wall section of length L (L) but impermeable base) (Fig. 2.19b), and combined vertical and radial discharge where the permeable section or screen has length L and radius, a (L), that can be different from the borehole radius, r_1 (L) (Fig. 2.19c, d). A flow efficiency correction equal to 1.0 is used in the extended solution instead of $\pi^2/8$ (=1.2337) used by Philip (1993) because radial and combined radial-vertical discharge through the borehole screen should be about as efficient as flow through the equivalent sphere surface. Finally, a zero gravity flow is considered, which implies that the gravity component of discharge through the borehole screen is small

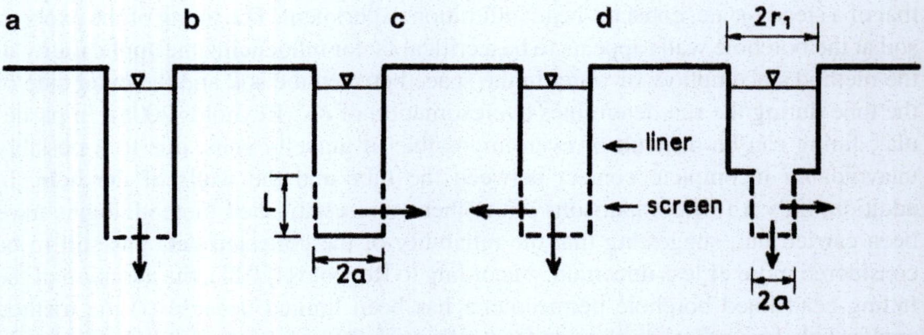

Fig. 2.19 Examples of discharge geometries for falling-head lined borehole permeameters: (**a**) vertical discharge only, (**b**) radial discharge only, (**c**) combined vertical–radial discharge with screen radius a equal to borehole radius r_1, and (**d**) combined vertical–radial discharge with $a < r_1$. The *arrows* on the infiltration surface show the flow direction

relative to hydrostatic pressure and capillarity. Reynolds (2011) proposed the following equations for falling-head flow in cased boreholes:

$$t = \frac{r_1^2}{4r_0 K_{fs}} \tau_E \tag{2.57a}$$

where:

$$\tau_E = \left(1 + \frac{1}{2A_E}\right) \ln\left(\frac{A_E^3 - 1}{A_E^3 - \rho_E^3}\right) - \frac{3}{2A_E} \ln\left(\frac{A_E - 1}{A_E - \rho_E}\right) + \frac{\sqrt{3}}{A_E}\left[\tan^{-1}\left(\frac{A_E + 2\rho_E}{\sqrt{3}A_E}\right) - \tan^{-1}\left(\frac{A_E + 2}{\sqrt{3}A_E}\right)\right] \tag{2.57b}$$

$$A_E^3 = \frac{3r_1^2(H_0 + \alpha^{*-1})}{4r_0^3 \Delta\theta} + 1 \tag{2.57c}$$

$$\rho_E^3 = \frac{3r_1^2(H_0 - H_t)}{4r_0^3 \Delta\theta} + 1 \tag{2.57d}$$

where H_t (L) is the effective pressure head in the borehole screen at time t ($H_t = H_0$ at $t = 0$) and the equivalent sphere radius, r_0 (L), for the extended solution is equal to:

$$r_0 = \frac{a}{2} \tag{2.58a}$$

for vertical discharge only (Fig. 2.19a),

$$r_0 = \left(\frac{aL}{2}\right)^{1/2} \tag{2.58b}$$

for radial discharge only (Fig. 2.19b), and

$$r_0 = \left(\frac{a^2}{4} + \frac{aL}{2}\right)^{1/2} \tag{2.58c}$$

for combined vertical and radial discharge (Fig. 2.19c, d). The H_t term is related to the water depth in the borehole at time t, D_t, with D_t measured from the base of the screen, by the following relationship:

$$H_t = D_t - E \tag{2.59a}$$

where

$$E = 0 \tag{2.59b}$$

for vertical discharge only,

$$E = \frac{L}{2} \tag{2.59c}$$

for radial discharge only, and

$$E = \frac{L^2}{a + 2L} \tag{2.59d}$$

for combined vertical and radial discharge. Equations (2.59a, 2.59b, 2.59c and 2.59d) account for the fact that pressure head is independent of elevation along the screen base but changes linearly with elevation along the screen wall. Note that $H_t = D_t$ for vertical discharge through a basal circular screen but $H_t < D_t$ for both radial discharge and combined vertical-radial discharge through a cylindrical screen. The minimum allowable D_t and H_t values, D_{min} (L) and H_{min} (L), respectively, are given by:

$$D_{\min} = L \tag{2.60a}$$

and

$$H_{\min} = L - E \tag{2.60b}$$

because it is implicitly assumed in Eqs. (2.57a, 2.57b, 2.57c and 2.57d) that the borehole screen is always submerged.

To reduce sensitivity of the K_{fs} and α^* determinations to random measurement error and lack of data-model fit, an iterative, nonlinear, curve-fitting approach was implemented via Excel solver (Frontline Systems, Incline Village, NV), which is independent of borehole, reservoir and screen dimensions, and can be used for two or more (H_t, t) data pairs falling anywhere between zero drawdown and maximum drawdown. The solution is achieved, through iterative adjustment of K_{fs} and α^*, by:

$$\min\left[\sum_{i=1}^{n}\left(t_i^{Data} - t_i^{Analy}\right)^2\right] \tag{2.61}$$

where n is the number of (H_t, t) values, t_i^{Data} (T) are the measured times when convenient H_t values are reached and t_i^{Analy} (T) are the corresponding times predicted by the analytical solution. The initial guess and constraint for α^* was set to:

$$\alpha^* \geq \alpha^*_{\min} = 0.001\ \text{cm}^{-1} \tag{2.62}$$

where α^*_{min} represents an extreme capillarity situation. The initial guess and constraint for K_{fs} was:

$$K_{fs} \leq \frac{1}{n}\sum_{i=1}^{n} K_i \tag{2.63a}$$

$$K_i = \frac{r_1^2}{4r_0 t_i}\ln\left(\frac{H_0}{H_i}\right) \tag{2.63b}$$

and (H_i, t_i) are the corresponding (H_t, t) values. Equation (2.63b) applies for a saturated soil with zero capillarity and thereby yields the average maximum K_{fs} if all other parameters remain unchanged. The precision and convergence criteria for the iterative solution were arbitrarily set at 10^{-9}. Solver options selected to improve solution uniqueness, stability, and accuracy included automatic scaling of inputs and outputs, quadratic extrapolation of parameter estimates, central difference estimation of partial derivatives, and use of the quasi-Newton method to determine solution search direction. To avoid possible fitting bias, it was suggested to base model-data fit on 11 (H_t, t) data pairs, equally spaced between H_0 and H_{min}.

The assessment of the extended solution carried out by Reynolds (2011) showed that Eqs. (2.57a, 2.57b, 2.57c and 2.57d) remained valid and accurate ($\leq$20 % error in the K_{fs} and α^* calculations) for almost the entire $\Delta\theta$ range, including near-saturated conditions (i.e., θ_i close to θ_{fs}). Moreover, this equation was found to meet the adopted accuracy criterion (i.e., error <20 %) for all considered borehole discharge geometries except with reference to α^* under the discharge geometry of Fig. 2.19d in a weak-capillarity environment, where the error (underestimate) was about 25 %. Accurate determinations of K_{fs} were obtained for all tested capillarities and for all practical ranges and combinations of a, L and L/a ($a = 2$, 3 and 5 cm; $0 \leq L/a \leq 15$). The accuracy of the α^* determination, on the other hand, was variable and depended on both screen dimensions and soil capillarity although the 20 % accuracy criterion was met for all tested capillarities and $1 \leq L/a \leq 4$.

In conclusion, the falling-head lined borehole permeameter appears a versatile tool for determining K_{fs} and α^* below soil surface and the extended analysis by Reynolds (2011) shows considerable promise for improving the accuracy and utility of this device, taking into account the theoretical improvements and the details in the assessment of the new analytical procedure. A more definitive assessment of the method requires checking K_{fs} and α^* determinations in field conditions, which generally differ from idealized porous medium.

2.2 Ring Infiltrometers

2.2.1 *Single-Ring Pressure Infiltrometer*

The single-ring pressure infiltrometer (PI) technique allows measurement of the soil hydrodynamic properties with particular, but not exclusive, reference to the field-saturated soil hydraulic conductivity, K_{fs} (L T^{-1}). Typically, a ring (or a cylinder) is inserted into the initially unsaturated porous medium to a short depth and one or more constant depths of water are established on the confined soil surface. For a given ponded depth of water, H (L), the rate at which water infiltrates the soil is monitored until the infiltration rate becomes quasi steady because the most common methods to analyze the PI data are based on the assumption of steady flow (Reynolds and Elrick 1990; Reynolds 1993). However, alternative data analysis procedures, considering the transient phase of the constant-head infiltration process (Wu et al. 1999) were also developed. Ring infiltrometers essentially measure vertical soil water transmission parameters since rings establish downward flow (Reynolds and Elrick 2005), but flow divergence has to be considered in the analysis of field data to avoid overestimation of K_{fs}. Use of the PI technique has been greatly favored by the theoretical, three-dimensional (3D) analysis of Reynolds and Elrick (1990), that takes into account the three main components of steady flow from the ring infiltrometer, i.e. hydrostatic pressure of the ponded water in the cylinder, capillarity of the unsaturated soil under and adjacent to the cylinder, and gravity.

2.2.1.1 A Review of Devices and Apparatus

Various terms have been used in the literature to denote the single-ring pressure infiltrometer technique. Reynolds and Elrick (1990) used the phrase *single-ring approach* to describe steady infiltration from within a single ring in which a constant head of water is established. Accordingly, the term *single-ring infiltrometer* was used by some authors to denote the technique (Ciollaro and

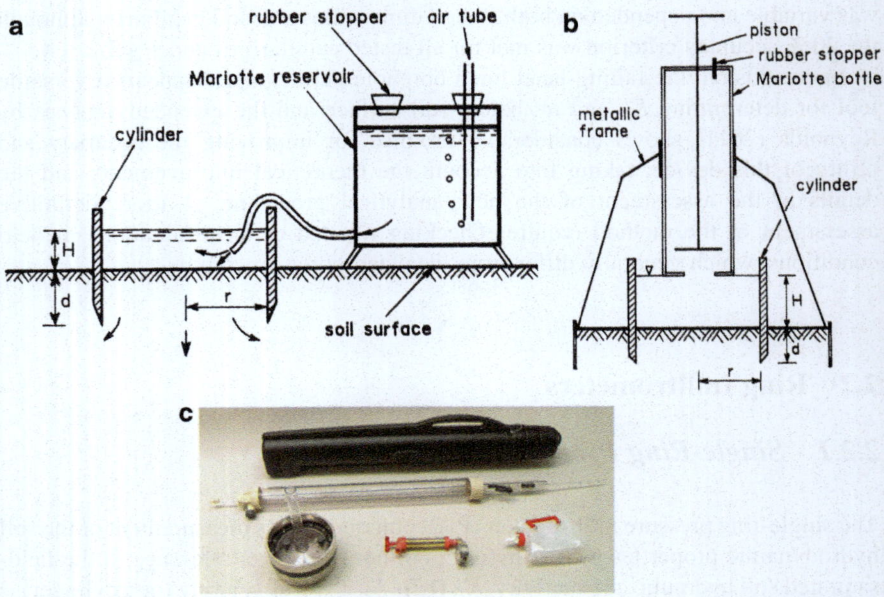

Fig. 2.20 Schematic of the (**a**) single-ring infiltrometer; (**b**) device by Ciollaro and Lamaddalena (1998, reprinted with permission); and (**c**) commercially available Guelph pressure infiltrometer (From Soilmoisture Equipment Corp. 2009, reprinted with permission)

Lamaddalena 1998; Bagarello and Sgroi 2004). However, Reynolds also used a different terminology such as *constant head pressure (single-ring) infiltrometer* (Reynolds 1993) and *single-ring pressure infiltrometer* (Reynolds et al. 2000). Other expressions are *surface* or *pressure permeameter* (Elrick and Reynolds 1992b), *Guelph pressure infiltrometer* (Fallow and Elrick 1996), *pressure single-ring infiltrometer* (Angulo-Jaramillo et al. 2000) and *pressure infiltrometer* (Bagarello et al. 2006a). Single-ring infiltrometer and pressure infiltrometer were explicitly considered synonymous by Wu et al. (1999). In the book *Methods of Soil Analysis, Part 4, Physical Methods*, Reynolds et al. (2002a, b) and Reynolds and Elrick (2002b) established a distinction between the *single-ring infiltrometer* and the *pressure infiltrometer*. The single-ring infiltrometer (Fig. 2.20a) is used to measure the cumulative infiltration, I (L), and the infiltration rate, $i = dI/dt$ (L T^{-1}), t (T) being the time, and to determine the field-saturated soil hydraulic conductivity, K_{fs}. A constant, small depth of water, H ($5 \leq H \leq 20$ cm), is established on the infiltration surface by using a Mariotte bottle, a float valve arrangement or simply by manually adding water when infiltration rates are low. A single H value is established for the run. The pressure infiltrometer allows us to determine at the same time the following parameters: the field saturated soil hydraulic conductivity, K_{fs}, the matric flux potential, ϕ_m (L^{2}T^{-1}), the soil sorptivity, S (L T$^{-1/2}$), the so-called α^* (L^{-1}) parameter, the effective wetting front pressure head of the Green and Ampt (1911) infiltration model, h_f (L), the air-entry pressure head, h_a

(L), and the water-entry pressure head, h_w (L) (Fallow and Elrick 1996). A constant depth of ponding is established on the infiltration surface by using a Mariotte bottle, such as the one provided with the Guelph permeameter. One, two or more H values are established on the infiltration surface, mainly depending on the variables that have to be determined. Therefore, the distinction between the single-ring and pressure infiltrometers essentially depends on how the constant depth of water is maintained on the infiltration surface and the number of H values that can be established. In other words, in the description by Reynolds et al. (2002a, b) and Reynolds and Elrick (2002b), the single-ring infiltrometer can be viewed as a simplified version of the pressure infiltrometer. However, even a handcrafted device, such as the one realized by Ciollaro and Lamaddalena (1998) (Fig. 2.20b), can be used to measure quasi-steady flow for different depths of water established in succession. This circumstance lessens the need to distinguish between the single-ring infiltrometer and the pressure infiltrometer. In this chapter, the denomination "single-ring pressure infiltrometer" and the abbreviation PI (pressure infiltrometer) are used to generically denote the constant-head technique involving measurement of a 3D infiltration process through a soil surface delimited by a ring or a cylinder inserted to a short depth into the initially unsaturated soil.

The device makes use of a thin-walled (thickness < 5 mm), metallic or plastic open-ended cylinder with a sharp outside-beveled cutting edge at the base to facilitate its insertion into the soil. Typically, the inner diameter of the ring varies between 0.10 and 0.50 m, the height is of 0.10–0.20 m, and the depth of ring insertion into the soil is of 0.03–0.10 m. Occasionally, larger rings are used (diameter also equal to 1 m) but a ring diameter of 0.10 m is considered appropriate for most field applications (Reynolds and Elrick 2002b). A mallet and a wood tablet to be placed on the top of the ring can be enough to insert the ring into the soil. Alternative systems, more suited to minimize soil disturbance during ring insertion, can be used (Reynolds and Elrick 2002b).

In the commercially available *Guelph pressure infiltrometer* (Fig. 2.20c, Soilmoisture Equipment Corp. 1992, 2005, 2009), the reservoir of the Guelph permeameter is connected to a 0.10 or 0.20 m diameter ring covered by a cap allowing a watertight seal. Therefore, the device is constituted by a Mariotte bottle directly attached to a ring of radius, r (L), inserted to a depth, d (L), into the soil. The air tube, sliding in the vertical direction, allows us to establish a wide range of positive pressure heads, H, on the infiltration surface. The Mariotte bottle provides the necessary water to maintain the constant-head infiltration process, and the rate of fall in the water level in this bottle can be used to calculate infiltration rates. In particular, for a given time interval, the water level drop (L) times the cross-sectional area of the bottle (L^2) is the infiltrated water volume (L^3) in that time interval. This volume is divided by the area of the infiltrating surface (L^2) and duration of the time interval (T) to obtain the infiltration rate ($L\ T^{-1}$). A standpipe attached to the cap of the ring can be used for measuring established pressure head on soil surface. According to Reynolds and Elrick (2002b), small diameter Mariotte reservoirs (0.02–0.03 m) should be used in low permeability soils for an increased measurement accuracy of slow flow rates; large diameter reservoirs (0.05–0.10 m or more) allow an increased capacity for use of the technique in high permeability soils where flow rates are high.

The handcrafted device by Ciollaro and Lamaddalena (1998) (Fig. 2.20b) uses a stainless steel cylinder having an inner diameter of 0.15 m and a height of 0.20 m. A graduated Mariotte reservoir 1.20 m high with an inner diameter of 0.11 m is used to establish the water ponding depth on the infiltration surface and to monitor infiltration rates. The reservoir is filled through a fill port located on the reservoir cap that is hermetically sealed by a rubber stopper. The water outlet hole, realized at the bottom of the reservoir, is relatively large (diameter = 0.042 m) to rapidly establish the ponding depth of water at the beginning of the run. Such a relatively large size may be responsible for some turbulence at the sampled soil's surface, especially during the initial stages of the run (Bagarello et al. 2014c). Soil disturbance may however be reduced by placing a wire net on the soil surface or gently pouring water on the infiltration surface to establish ponding conditions before starting the run. The water outlet hole can be hermetically closed by a piston provided with an o-ring. The reservoir is supported by a metallic frame that can easily be taken to pieces for transport in the field. The run presupposes placing the filled reservoir, with the piston completely lowered (no flowing water), above the infiltration surface confined by the ring in such a way as to maintain a distance between the soil and the bottom of the reservoir equal to first head of water, which is to be established. The infiltration run begins when the piston is raised. Water flows out of the Mariotte bottle until the prescribed H value is established, i.e. the space between the soil surface and the reservoir is filled with water. Soil water infiltration determines a raising of air bubbles inside the reservoir that empties progressively. Reading the water level at the graduated scale of the reservoir at given time intervals allows calculation of the infiltration rates. After quasi steady-state conditions have been reached for the established ponded head, the reservoir is easily raised, so that a new H value, higher than the previous one, can be established, and another infiltration process can be monitored.

The single-ring pressure infiltrometer can be equipped with pressure transducers connected to a data-logger. Infiltrated water volumes are calculated from the pressure transducer readings at pre-established time intervals and the transducer calibration curve (Wu et al. 1999). This methodology determines an increase in the cost of the device but also an easier experiment in the field and, potentially, a better monitoring of the infiltration process. The infiltrometer data collection system is based on the concept that tension (negative pressure) in the air pocket at the top of the water reservoir is linearly related to the height of water in the column. A millimeter decrease in water column height means a millimeter decrease in tension in the air pocket (Ankeny 1992). Thus, cumulative infiltration can be monitored by recording tension changes measured by a transducer situated at the top of the air pocket. In the automated single-ring pressure infiltrometer by Prieksat et al. (1992), a four-wire full bridge pressure transducer is located at the top of the water reservoir and another transducer is located at the base. Using two transducers nullifies the bubbling induced pressure variations within the Mariotte reservoir and therefore improves precision of water flow measurement (Prieksat et al. 1992). The data-

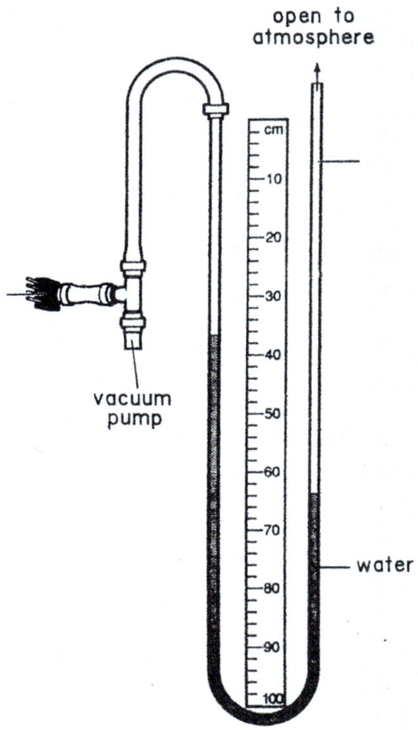

Fig. 2.21 Schematic of the manometer/manifold for pressure transducer calibration

logger is programmed to record paired readings of top and bottom transducers at regular time intervals. Water flow from the infiltrometer is calculated by monitoring change over time of the tension difference between the top and bottom pressure transducers since the change in this tension difference is directly related (1:1) to the change in the water reservoir height (Ankeny 1992). This height change is multiplied by the cross-sectional area of the reservoir and divided by the infiltration surface area to obtain the infiltration rate.

Calibration of pressure transducers before use is carried out taking into account that their voltage output is linearly proportional to tension (Ankeny 1992). In the setup for calibrating pressure transducers (Fig. 2.21), the right end of the manometer tubing is open to the atmosphere and the closed left end is connected to a manifold holding the pressure transducer and having a septum for changing tensions using a vacuum gun or syringe. Initially, the difference between the left and right heights of water column in the manometer is zero. Then vacuum is applied by pulling air out of the closed end of the water manometer and the tension, i.e., the difference between the two water levels, and the corresponding voltage output for the transducer are recorded. Ten points with approximately 100 mm increments are adequate for a good calibration since pressure transducers are generally calibrated for a tension range of 0–1 m water tension. The slope and the intercept of the voltage – tension least squares regression line are the parameters needed for transducer calibration. However, the intercept is not important for determining infiltration rates because the rate calculation is a difference method.

A practically constant-head infiltration run can even be carried out by manually pouring water into the ring (Reynolds et al. 2002b; Braud et al. 2005). The simplest means is to mark a given level on the cylinder wall and to add a pre-established water volume whenever the water level in the ring reaches the mark. The average infiltration rate is calculated using the volume of water added and the time interval between additions. If a constant water volume is poured each time, steady-state flow rate is practically signaled by an approximately constant time interval between additions. By this approach, used when transporting a Mariotte bottle is complicated and soils have a low permeability, the established head is not perfectly constant and an approximately constant head is estimated by averaging the water levels corresponding to the mark and the mark plus the added water volume with an individual pouring. Obviously, the approximation on the *H* value used in the calculations is reduced if the lower and the upper water levels are close to each other.

2.2.1.2 Physics of Flow Under the Single-Ring Pressure Infiltrometer

As schematically shown in Fig. 2.22, flow into unsaturated soil from within a ring containing a constant depth of ponded water determines development of a saturated

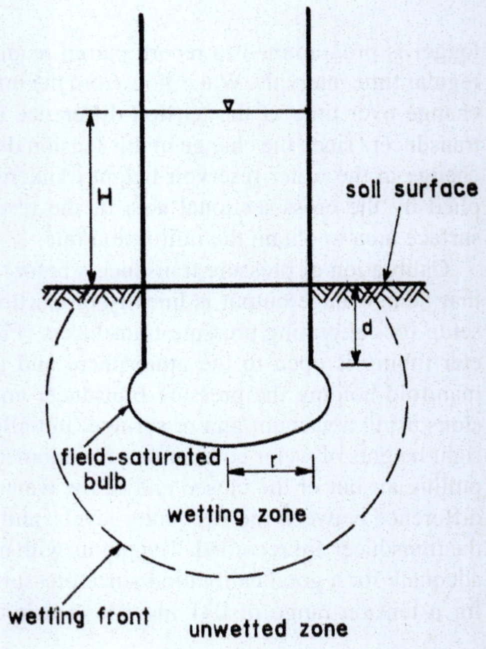

Fig. 2.22 Schematic of the flow field established with a single-ring pressure infiltrometer run

soil bulb which extends outwards from the infiltration surface (Elrick and Reynolds 1992b). The soil water pressure head varies from H along the infiltration surface to 0 at the surface of the saturated bulb, assuming a water entry pressure head equal to zero. The saturated bulb is surrounded by an unsaturated wetting zone extending to the wetting front. In this zone, the soil water pressure head varies between 0 and h_i at the wetting front, h_i (L) being the initial or antecedent soil water pressure head. The infiltration rate is initially large and decreases with time to approach a quasi steady-state value. During the transient phase, the saturated bulb and the unsaturated zone increase in size in both the vertical and the horizontal directions. Under steady-state conditions, both the shape and the size of the saturated bulb remain essentially constant. The wetting zone continues to increase in size due to the expansion of the wetting front (Elrick and Reynolds 1992b).

The equilibration time, i.e. the duration of the transient phase, generally increases with finer soil texture, drier soil conditions, decreasing soil structure, and increasing depth of water ponding, depth of ring insertion, and ring radius (Reynolds 1993; Reynolds and Elrick 2002b; Reynolds et al. 2002b). Steady flow is generally attained within 5–30 min using small rings ($\leq$0.1 m in diameter) and small ring insertion depths ($\leq$0.1 m) in soils of moderate to high permeability ($K_{fs} = 0.36$–360 mm h^{-1}). Equilibration times of 60–120 min or more are expected for larger rings, larger ring insertion depths, and low permeability soils. With large cylinders ($\geq$0.30 m in diam.) and in very low permeability soils, equilibration times as long as several hours or days may be required. Reynolds et al. (2000) assumed that flow rates were quasi-steady when the rate of fall of water level in the reservoir was constant for at least 10 min. These authors also suggested using relatively small ponding depths for faster flow rates to reduce water consumption. According to Reynolds and Elrick (2002b), quasi steady-state flow conditions are detectable with the infiltration run because true steady flow will only occur in completely homogeneous soil, which probably does not exist in natural environments. Hence, a reasonable estimate of steady flow, or quasi steady-state flow, is considered acceptable. In practice, with this technique, a rapid attainment of near steady-state conditions appears common (e.g., Bagarello and Iovino 1999; Reynolds et al. 2000).

Taking into account that the soil is wetted by downward infiltrating water under ponding conditions, the saturated bulb generally contains entrapped air bubbles and, for this reason, it is more properly denoted as a field-saturated bulb. Air entrapment implies that the water content of a field-saturated soil is lower than at complete saturation. Depending on the amount of entrapped air, K_{fs} can be a factor of two or more below the hydraulic conductivity of fully saturated soil, K_s (Bouwer 1978; Reynolds and Elrick 1987). For many unsaturated zone applications, K_{fs} is considered more appropriate than K_s because most natural and man-made infiltration processes result in significant air entrapment within the porous medium (Reynolds 1993).

The range of K_{fs} values that can be measured using relatively small heads ($H \leq 0.50$ m) and the quasi steady flow measurements is on the order of 0.036–1800 mm h^{-1} (Reynolds and Elrick 2002b). Measuring lower K_{fs} values implies higher ponded heads (>1.0 m) and alternative analyses, using early-time transient flow data (e.g., Odell et al. 1998, Sect. 4.3). The PI can be applied at both the soil surface and some depth below the surface, in expressly excavated pits.

2.2.1.3 Analysis of Steady-State Flow

The analytical expression for steady, ponded flow out of the ring into rigid, homogeneous, isotropic, uniformly unsaturated soil is (Reynolds and Elrick 1990):

$$Q_s = \frac{r}{G}(K_{fs}H + \phi_m) + \pi r^2 K_{fs} \tag{2.64}$$

where Q_s (L^3T^{-1}) is the steady-state flow rate, r (L) is the ring radius, K_{fs} ($L\,T^{-1}$) is the field-saturated soil hydraulic conductivity, H (L) is the ponded head of water on the infiltration surface, ϕ_m (L^2T^{-1}) is the matric flux potential and G is a dimensionless shape factor expressing the complex interactions between ring radius, depth of ring insertion, d (L), depth of ponding in the ring, soil capillarity and gravity. According to Eq. (2.64), steady-state flow rate out of the ring is the sum of three components, i.e. flow due to the hydrostatic pressure of the established depth of water on the infiltration surface (first term on the right of the equation), flow due to the capillarity of the unsaturated soil under and adjacent to the ring (second term), and flow due to gravity (third term). Equation (2.64) was developed under the assumption that ponding does not occur around the outside of the ring during a measurement. The appearance of a wetting front at the soil surface, however, is both admissible and even expected when the depth of ring insertion is small and/or soil permeability is low (Reynolds 1993; Reynolds and Elrick 2002b). Reynolds and Elrick (1990) used numerical simulations of the Richards equation for three-dimensional, saturated-unsaturated flow to determine G for pre-established values of K_{fs}, α^* (L^{-1}) ($\alpha^* = K_{fs}/\phi_m$), d, r and H. The G values depend significantly on both d and r but are nearly independent of soil hydraulic properties (i.e., K_{fs}–α^* combination). For this reason, Reynolds and Elrick (1990) also determined the values, $\overline{G}$, of the shape factor by averaging, for a given H-r-d combination, the G data corresponding to the four considered porous media. In addition, the shape factor is nearly independent of H for $H \geq 0.05$ m. Therefore, for practical application of the PI technique, Reynolds and Elrick (1990) suggested using the following relationship, valid for $0.03 \leq d \leq 0.05$ m, $0.05 \leq r \leq 0.10$ m and $0.05 \leq H \leq 0.25$ m, to obtain an estimate of G (G_e):

$$G_e = 0.316\frac{d}{r} + 0.184 \tag{2.65}$$

According to Youngs et al. (1995), Eq. (2.65) is usable for practical purposes within wider ranges of both d ($0 < d \leq 0.10$ m) and H ($0.05 \leq H \leq 1$ m) without substantially compromising the reliability of the estimates (Reynolds and Elrick 2002b).

Table 2.7 Values of the i_s/K_{fs} ratio and percentages of total steady flow due to hydrostatic pressure of the ponded water, capillarity of the unsaturated soil and gravity for different combinations of H, d, r and α^*

					Percentage (%)		
H (m)	d (m)	r (m)	α^* (m^{-1})	i_s/K_{fs}	Hydrostatic pressure	Capillarity	Gravity
0.05	0.05	0.05	12	2.698	24	39	37
0.05	0.05	0.10	12	2.241	21	35	45
0.05	0.05	0.15	12	1.978	18	31	51
0.05	0.05	0.20	12	1.807	17	28	55
0.05	0.05	0.40	12	1.475	12	20	68
0.05	0.05	0.60	12	1.336	9	16	75
0.05	0.03	0.30	12	1.656	15	25	60
0.05	0.05	0.30	12	1.598	14	23	63
0.05	0.10	0.30	12	1.489	12	21	67
0.05	0.15	0.30	12	1.414	11	18	71
0.05	0.20	0.30	12	1.358	10	16	74
0.05	0.05	0.30	12	1.598	14	23	63
0.10	0.05	0.30	12	1.822	25	20	55
0.15	0.05	0.30	12	2.046	33	18	49
0.20	0.05	0.30	12	2.270	40	16	44
0.40	0.05	0.30	12	3.167	57	12	31
0.05	0.05	0.30	36	1.349	17	9	74
0.05	0.05	0.30	12	1.598	14	23	63
0.05	0.05	0.30	4	2.345	10	48	42
0.05	0.05	0.30	1	5.707	4	79	17

i_s (L T^{-1}) steady-state infiltration rate, K_{fs} (L T^{-1}) field-saturated soil hydraulic conductivity, H ponding depth of water, d depth of ring insertion, r ring radius, α^* parameter varying with the soil textural and structural characteristics

The following relationship is obtained by combining Eqs. (2.64) and (2.65):

$$\frac{i_s}{K_{fs}} = \frac{H}{G_e \pi r} + \frac{1}{\alpha^* G_e \pi r} + 1 \tag{2.66}$$

where $i_s = Q_s/(\pi r^2)$ (L T^{-1}) is the steady-state infiltration rate. Table 2.7 shows the relative importance of the three flow components at steady-state for common H, d, r and α^* values. In all cases, i_s/K_{fs} is greater than one, showing that the measured steady-state infiltration rate cannot be set equal to the field-saturated soil hydraulic conductivity. The i_s/K_{fs} ratio decreases as the ring radius and the ring insertion depth increase. Moreover, gravity flow contribution to total flow increases with r and d. An increase in H determines higher i_s/K_{fs} values, a more noticeable contribution of pressure flow and a reduced contribution of both capillarity and gravity flow. The ratio between i_s and K_{fs} increases as the α^* parameter decreases. Moreover, a decrease in α^* determines a larger contribution of capillarity flow and a decreasing contribution of both pressure and gravity flow. In a field investigation

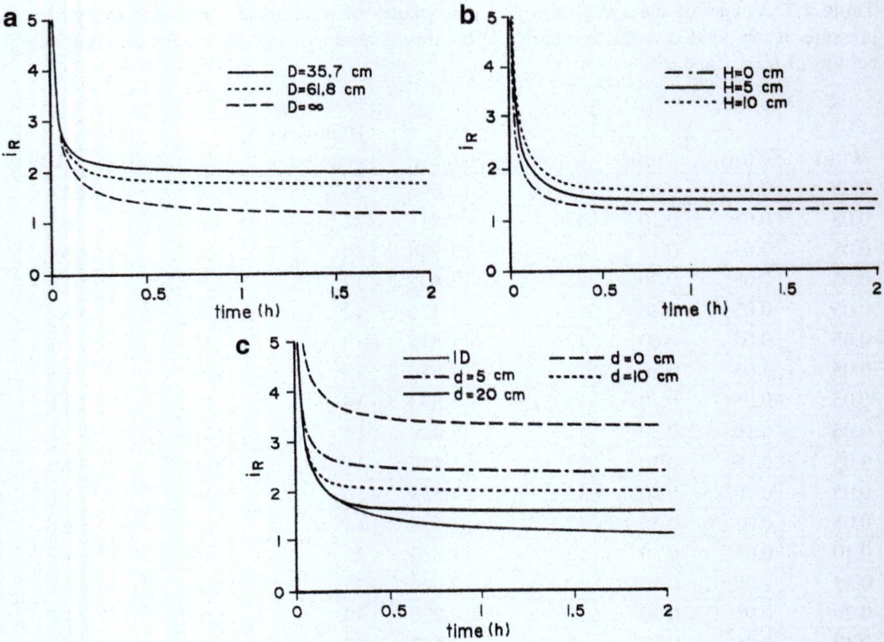

Fig. 2.23 Ratio, i_R, between the infiltration rate from a single-ring infiltrometer and the saturated soil hydraulic conductivity plotted against time for different (**a**) ring diameters, D; (**b**) values of the ponded head of water on the infiltration surface, H; and (**c**) depths of ring insertion, d (From Dušek et al. 2009, reprinted with permission)

carried out by Verbist et al. (2010) in stony soils, neglecting capillarity and pressure flow (i.e., simply assuming $i_s = K_{fs}$), overestimated K_{fs} by a factor of 2.0–2.4. The dependence of the infiltration rates on ring diameter, ponded depth of water and depth of ring insertion was also assessed by Dušek et al. (2009), using numerically simulated data for a homogeneous, coarse sandy-loam soil. One-dimensional infiltration rates approached asymptotically the saturated hydraulic conductivity, K_s, but, for the 3D process, the infiltration rates were higher than K_s and steady-state infiltration rates increased as the ring diameter decreased (Fig. 2.23a). A higher water level maintained in the ring produced higher infiltration rates due to a greater pressure gradient (Fig. 2.23b). Appreciable deviations among the simulated infiltration rates with different depths of ring insertion were predicted given that, for the considered scenario, the simulations with zero depth of insertion yielded a steady-state infiltration rate higher than three times K_s (Fig. 2.23c).

The two unknowns, K_{fs} and ϕ_m, contained in Eq. (2.64) can be differently determined, depending on the number of ponding depths of water established during the experiment. In particular, the approach named *One Ponding Depth* (OPD) by Reynolds and Elrick (1990) and *single-head analysis* by Reynolds and Elrick (2002b) implies measurement of a single Q_s value corresponding to an established ponding depth of water, H, and the estimation of the α^* parameter on the basis of the visually determined soil textural/structural characteristics (Table 2.1) (Elrick and Reynolds 1992a; Reynolds et al. 2002a). The following relationships are then used to obtain the two unknowns:

$$K_{fs} = \frac{\alpha^* \ \gamma_G \ Q_s}{r(\alpha^* H + 1) + \gamma_G \alpha^* \pi r^2} \tag{2.67}$$

$$\phi_m = \frac{\gamma_G \ Q_s}{r(\alpha^* H + 1) + \gamma_G \alpha^* \pi r^2} \tag{2.68}$$

where γ_G is the shape factor, determined by using the values given by Reynolds and Elrick (1990) (G or $\overline{G}$) or Eq. (2.65) (G_e). The ϕ_m calculation allows us to obtain an estimate of the soil sorptivity, S, by Eq. (2.10). The OPD approach was developed because the range of α^* is usually about 1–50 m^{-1} although K_{fs} and ϕ_m can range over many orders of magnitude. The reduced variability of α^* and its connection with soil texture and structure was considered to make it a very useful parameter in simplified analyses for determination of K_{fs} and ϕ_m.

The *Two Ponding Depth* (TPD, Reynolds and Elrick 1990) approach or *two-head analysis* (Reynolds and Elrick 2002b) requires the steady-state flow rates, Q_{s1} and Q_{s2}, corresponding to two ponding depths of water, H_1 and H_2 ($H_2 > H_1$) respectively, consecutively established on the infiltrating surface, i.e. without occurrence of a drainage phase in the passage from H_1 to H_2. If the dependence of the shape factor on the ponding depth is taken into account, K_{fs} and ϕ_m are calculated by the following relationships:

$$K_{fs} = \frac{F_1 \ Q_{s2} - F_2 \ Q_{s1}}{E_2 \ F_1 - E_1 \ F_2} \tag{2.69}$$

$$\phi_m = \frac{E_2 \ Q_{s1} - E_1 \ Q_{s2}}{E_2 \ F_1 - E_1 \ F_2} \tag{2.70}$$

$$E_1 = \frac{r \ H_1}{\overline{G_1}} + \pi r^2 \tag{2.71a}$$

$$E_2 = \frac{r \ H_2}{\overline{G_2}} + \pi r^2 \tag{2.71b}$$

$$F_1 = \frac{r}{\overline{G_1}} \tag{2.71c}$$

$$F_2 = \frac{r}{\overline{G_2}} \tag{2.71d}$$

where $\overline{G_1}$ and $\overline{G_2}$ are the $\overline{G}$ values corresponding to H_1 and H_2, respectively (Reynolds and Elrick 1990). If only the effect of the ring radius and the insertion depth on the shape factor is accounted for by Eq. (2.65), the following relationships can be applied to calculate K_{fs} and ϕ_m:

$$K_{fs} = \frac{G_e}{r}\left(\frac{Q_{s2} - Q_{s1}}{H_2 - H_1}\right) \tag{2.72}$$

$$\phi_m = \frac{G_e}{r}\left(\frac{H_2\ Q_{s1} - H_1\ Q_{s2}}{H_2 - H_1} - \pi r G_e \frac{Q_{s2} - Q_{s1}}{H_2 - H_1}\right) \tag{2.73}$$

The simultaneous calculation of K_{fs} and ϕ_m allows us to determine the α^* (L^{-1}) parameter by Eq. (2.7), the soil sorptivity by Eq. (2.10), and the effective Green and Ampt (1911) wetting front pressure head, h_f, by Eq. (2.13).

The approach named *Multiple Ponding Depth* (MPD, Reynolds and Elrick 1990) or *multiple-head analysis* (Reynolds and Elrick 2002b) determines K_{fs} and ϕ_m using two or more consecutively established ponded heads, starting from the smallest water level. With the shape factor estimated by Eq. (2.65), that is, assuming that it is unaffected by either the value of H or the kind of soil, the following relationships are applied to determine the two unknowns:

$$K_{fs} = \frac{G_e}{r} \times b_1 \tag{2.74}$$

$$\phi_m = \frac{G_e}{r}\left(b_0 - \pi r^2 K_{fs}\right) \tag{2.75}$$

where b_1 and b_0 are the slope and the intercept, respectively, of the linear least squares regression line through a plot of Q_s vs. H data. Obviously, Eqs. (2.7), (2.10) and (2.13) can be used to determine α^*, S and h_f, respectively. If two ponding depths are used, Eqs. (2.74) and (2.75) give the same results to Eqs. (2.72) and (2.73), respectively.

2.2.1.4 Analysis of Transient Flow

Especially in soils with a low or relatively low water permeability, where equilibration times are long, using methodologies analyzing the transient phase of the infiltration process may be advisable to maintain the duration of the field experiment within sustainable limits from a practical point of view. By applying scaling theory, Wu and Pan (1997) developed a generalized solution to infiltration from single-ring pressure infiltrometers and Wu et al. (1999) used this solution to determine K_{fs} from measurements of cumulative infiltration. In particular, considering an infiltration surface of radius, r (L), delimited by a ring inserted into the soil to a short depth, d (L), the cumulative infiltration, I (L), can be described by a relationship formally identical to the two-term infiltration model by Philip (1957a):

$$I = A_w t + B_w \sqrt{t} \tag{2.76}$$

where A_w (L T^{-1}) and B_w (L $T^{-0.5}$) are coefficients and t (T) is the time. The A_w coefficient has the following expression:

$$A_w = a'K_{fs}\ f \tag{2.77}$$

where a' is a dimensionless constant equal to 0.9084 and f is a correction factor that depends on soil initial and boundary conditions and ring geometry:

$$f = \frac{H + \phi_m^{'}/K_{fs}}{G^*} + 1 \tag{2.78}$$

where $\phi_m^{'}$ (L^2T^{-1}) is the matric flux potential calculated by a modified Mualem-van Genuchten hydraulic conductivity – pressure head function and G^* (L) is equal to:

$$G^* = d + \frac{r}{2} \tag{2.79}$$

According to Wu et al. (1997), f represents the ratio between the final infiltration rate of a single-ring pressure infiltrometer and the corresponding one-dimensional infiltration for the same soil. The expression of B_w is:

$$B_w = 2\ b'f\ K_{fs}\ \sqrt{T_c} \tag{2.80}$$

where b' is a dimensionless constant equal to 0.1682 and T_c (T) is approximately equal to:

$$T_c = \frac{\Delta\theta}{K_{fs}^2}\phi_m^{'} \tag{2.81}$$

where $\Delta\theta$ (L^3L^{-3}) is the difference between the field-saturated volumetric soil water content, θ_{fs} (L^3L^{-3}), and the initial one, θ_i (L^3L^{-3}). There are two ways to calculate K_{fs} by applying the generalized infiltration equation to the measured infiltration curves from a single-ring pressure infiltrometer (Wu et al. 1999). With method 1, using the entire cumulative infiltration curve, the A_w and B_w coefficients are estimated by fitting Eq. (2.76) to the data collected from the beginning of the run. These coefficients, together with the measurement of $\Delta\theta$, can be used to calculate K_{fs}:

$$K_{fs} = \frac{\lambda_c \Delta\theta}{T_c} \tag{2.82}$$

where:

$$\lambda_c = \frac{1}{2}\left[\sqrt{(H + G^*)^2 + 4G^*C} - (H + G^*)\right] \tag{2.83}$$

$$C = \frac{1}{4\ \Delta\theta}\left(\frac{B_w}{b'}\right)^2 \frac{a'}{A_w} \tag{2.84}$$

$$T_c = \frac{1}{4}\left(\frac{B_w a'}{b' A_w}\right)^2 \tag{2.85}$$

in which λ_c (L) is the macroscopic capillary length. Method 1 seems appropriate to also estimate α^* since Wu et al. (1999) showed that:

$$\alpha^* = \frac{K_{fs}}{\phi_m} \cong \frac{K_{fs}}{\phi'_m} \tag{2.86}$$

The ϕ'_m term of Eq. (2.86) is obtained by solving Eq. (2.81) for ϕ'_m, using the K_{fs} and T_c values obtained with Eqs. (2.82) and (2.85), respectively:

$$\phi'_m = \frac{K_{fs}^2}{\Delta\theta}\ T_c \tag{2.87}$$

Method 2 is based on the assumption that the last part of the infiltration run has reached steady-state. In this case, the following linear relationship:

$$I = A_w\ t + c = a'\ f\ K_{fs}\ t + c \tag{2.88}$$

is fitted to the data corresponding to the steady-state phase of the infiltration process, where A_w is the slope and c is the intercept. Therefore, K_{fs} is given by:

$$K_{fs} = \frac{A_w}{a'\ f} \tag{2.89}$$

where f, taking into account Eqs. (2.78) and (2.86), can be estimated by the following relationship:

$$f \cong \frac{H + 1/\alpha^*}{G^*} + 1 \tag{2.90}$$

The methods by Wu et al. (1999) should have a wide applicability taking into account that the generalized infiltration equation was successfully tested for soils ranging in texture from sand to clay and with reference to various ring radii ($0.06 \le r \le 0.2$ m), insertion depths ($0.02 \le d \le 0.1$ m), and ponding depths of water ($0 \le H \le 0.2$ m) (Wu and Pan 1997). Method 2 by Wu et al. (1999), which is conceptually similar to the OPD approach by Reynolds and Elrick (1990), is expected to give less variable predictions of K_{fs} when compared to method 1 since it uses a fixed α^* value for the whole field whereas variations of this parameter exist in the field. With method 2, the erroneous inclusion of the last transient infiltration data into the assumed steady-state phase of the process determines an overestimation of K_{fs} that, however, should not be greater than a few percentage units (Wu et al. 1999).

2.2.1.5 Validation and Use of the Single-Ring Pressure Infiltrometer

Using numerically simulated infiltration data for a sand, a sandy-clay-loam and a clay soil, Wu et al. (1999) concluded that the differences between the real K_{fs} values and the estimated conductivity by their method 1 were practically negligible since the estimates were 87–130 % of the real K_{fs}, depending on the soil. The authors also established a comparison, for a limited number of sampling points ($N = 8$) on a sandy-loam soil, between the field measured conductivity (0.20 m diameter ring) and the K_s data obtained by using the laboratory falling-head method with small undisturbed soil cores (0.05 m × 0.05 m). The two sets of conductivity data differed by a factor of approximately 2.2. However, this comparison did not provide strong support for the use of method 1, due to large differences in the two experimental methods. More interesting is the fact that methods 1 and 2 yielded K_{fs} values that were close to those obtained by the OPD approach (differences by 7.8 % for method 1 and 4.6 % for method 2), supporting the methodological soundness of the proposed methods. Gómez et al. (2005) applied method 2 to sample twice 113 locations of a silty-clay-loam soil, each having a square shape, a surface of 1×1 m^2 and a depth of insertion of the septum of 0.1 m. The duration of each run with a constant ponded depth of water of 0.08 m was of three hours and the last 2 hours of data were used to calculate K_{fs} because quasi steady-state conditions were reached within an hour from the beginning of the infiltration process. To calculate G^* with Eq. (2.79), Gómez et al. (2005) set $r = 1$ m. Method 1 by Wu et al. (1999) has advantages despite the fact that soil water content has to be measured, because it does not imply a separation between the transient and the steady-state phase of the infiltration process. According to Vandervaere et al. (2000), non-linear fitting of Eq. (2.76) to the (I, t) dataset by least squares optimization technique offers no check to determine if the model fits the data adequately. To compensate for this, Smiles and Knight (1976) proposed linearizing Eq. (2.76) by dividing both sides by $\sqrt{t}$, giving:

$$\frac{I}{\sqrt{t}} = A_w \sqrt{t} + B_w \tag{2.91}$$

and then plotting $I/\sqrt{t}$ as a function of $\sqrt{t}$. With Eq. (2.91), the applicability of the infiltration model is easily established by evaluating the linearity of the data (Vandervaere et al. 2000). The A_w and B_w coefficients are equal to the slope and the intercept, respectively, of the linear regression line interpolating the ($I/\sqrt{t}$, $\sqrt{t}$) data pairs. The transient method by Wu et al. (1999) and the TPD approach were compared by Bagarello et al. (2009a) in a sandy-loam soil. The choice of the initial values of the A_w and B_w parameters of Eq. (2.76) did not affect the optimized values by non-linear fitting of the model to the data. However, the linearization of the infiltration curve yielded more reliable results than the non-linear optimization technique, essentially because the α* estimates were closer to the expected values with the former approach than the latter one. Method 1 gave both positive and reasonable K_{fs} and α* values for all infiltration tests, while the TPD approach yielded negative results or excessively low α* values in some cases. Using the mean soil water content for all tests instead of the locally measured Δθ values did not substantially affect calculations because the highest difference between two estimates of a variable (K_{fs} or α*) at a location was equal to 22 %. Finally, the mean results of the TPD approach and method 1 with the linearization technique were similar for both K_{fs} and α* (K_{fs} = 175–217 mm h^{-1}; α* = 3.3–3.9 m^{-1}). In an investigation carried out by Verbist et al. (2009) in degraded drylands of Chile, the mean K_{fs} values determined with the OPD approach differed by not more than a non-statistically significant 36 % from the values obtained with method 1 by Wu et al. (1999). The experiment was carried out both in the field and on the < 2 mm granulometric fraction packed in a container at the same bulk density as in the field. Differences between the two calculation methods were smaller in the latter case, suggesting that a more homogeneous soil structure and the absence of rock fragments improved similarity of the two tested methods. Another investigation was carried out at three field sites in Chile on loam and sandy-loam soils with approximately 10–60 % of stone fragment content, depending on the site (Verbist et al. 2010). A relatively large ring (inner diameter = 0.28 m) was used to perform ten randomly replicated runs at each field site with an established ponded depth of water of H = 0.05 m. The run duration was 60 min or until the infiltration rates of three successive 1 min time intervals remained constant. In this case, a comparison among the OPD approach and methods 1 and 2 by Wu et al. (1999) was established. The average K_{fs} values for each of the three sites were within the same order of magnitude for the three tested calculation methods (i.e., maximum difference among methods at a site by a factor of 1.6) and the associated coefficients of variation were similar at two of the three sites. The conclusion was that the steady-state and transient methods were equally usable with all datasets. Therefore, the available checks are encouraging since they suggest that the transient method by Wu et al. (1999) is a valid alternative to the steady-state approaches, allowing us to obtain positive results at each sampling point, which is particularly important when soil has to be sampled precisely at pre-established locations. The transient method should be preferred especially when time available for the test is limited and flow rates can easily be monitored at short time intervals from the beginning of the test. On the other hand, the TPD approach should be preferred when estimating soil water content is uncertain or impossible, and only steady flow rates can be detected. For example, a detailed monitoring of transient flow rates may be complicated when several instruments are used simultaneously by a single operator or when the transient phase of infiltration is impractically short due to high soil permeability, relatively small Δθ, or unfavorable ring geometry (e.g., small insertion depth and/or ring radius).

The PI method with a ring of 0.10 m in diameter and the TPD approach, the tension infiltrometer method and the classical undisturbed soil core (SC) method for measuring K_{fs} or the fully saturated soil hydraulic conductivity (K_s, SC method) were compared on structureless single-grain sand, structured loam and cracking clay-loam soils under different land management treatments by Reynolds et al. (2000). Of the 27 between-method correlations, only four were significant. However, the PI yielded conductivity values that were statistically equivalent to one or both of the other methods in all soil type-land management combinations. A possible limitation to the use of the PI method was only suggested for two treatments (conventional tillage, no-tillage) of the cracking clay-loam soil, probably because the sampled soil volume was too small.

In an investigation carried out at 10 Sicilian sites, the following trends appeared. The K_{fs} values obtained with the PI by Ciollaro and Lamaddalena (1998) and the OPD approach increased with both the soil organic matter content and the soil structural stability index by Pieri (1992), K_{fs} values decreased with an increase in soil bulk density and they were not significantly correlated with the clay, silt and sand content (Bagarello et al. 2014c). Therefore, the OPD approach was found to yield plausible results, since soil particle arrangement has a large influence on saturated hydraulic conductivity (e.g., Bouma and Dekker 1981) and the organic matter content should be positively correlated with K_{fs} because it stimulates soil aggregation, which lowers bulk density, increases porosity and hence elevates K_{fs} (Rawls et al. 2005; Agnese et al. 2011). The fact that K_{fs} was higher in more stable soils was viewed as a sign of the occurrence of some soil disturbance when the ponding depth of water was established on the infiltration surface by simply raising the piston of the device, but this disturbance was not enough to impede detection of plausible relationships with other soil properties.

In any case, it should be taken into account that the PI technique is primarily suited for determining K_{fs}, notwithstanding that it was also applied to estimate the unsaturated soil hydraulic conductivity function with plausible results (Vauclin et al. 1994; Angulo-Jaramillo et al. 2000). In a sandy-loam soil, the PI method with the TPD approach produced individual soil hydraulic conductivity, K, vs. pressure head, h, relationships that were substantially different from the ones determined by the combined use of the PI for determining K_{fs} and the tension infiltrometer method for measurement of unsaturated soil hydraulic conductivity (Fig. 2.24) (Bagarello et al. 2000). Therefore, this investigation supported the conclusion that the use of the PI to predict the unsaturated soil hydraulic conductivity should not generally be recommended because ponded flow tends to maximize the hydrostatic pressure and gravity components of flow while reducing the importance of the capillarity component (Elrick et al. 1995; Reynolds and Elrick 1990).

Yet, the pressure and tension infiltrometers can be combined. A single-ring pressure infiltrometer can first be used for saturated infiltration measurements. Then, the ring is filled with the contact material and the tension infiltrometer (TI) is applied for unsaturated infiltration measurements from low to high suction (Ankeny et al. 1991; Logsdon and Jaynes 1993) because the wet-to dry sequence reduces the antecedent negative head effects at low infiltration rates (Mohanty et al. 1994b). However, an ascending pressure head sequence with the TI is often

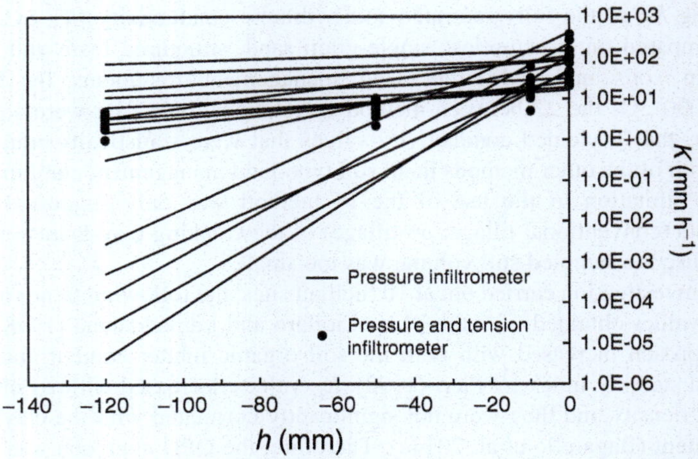

Fig. 2.24 Comparison between the hydraulic conductivity, K, vs. soil water pressure head, h, relationships obtained with the pressure and tension infiltrometer techniques and the pressure infiltrometer alone (From Bagarello et al. 2000, reprinted with permission)

recommended because a decreasing sequence may cause hysteresis, with progressive drainage occurring close to the disk while wetting continues at the infiltration front (Reynolds and Elrick 1991; Jarvis and Messing 1995), and it may enhance air entrapment and favor the establishment of upward gradients for initial pressure heads. The presence of a thin contact layer on the soil surface can introduce flow impedance effects at the high infiltration rates associated with ponded conditions (Reynolds et al. 2000) but it should not influence steady-state infiltration rates if the water permeability of the contact material is greater than that of the soil. Only in this last case, a positive head established either in presence or not of a layer of contact material should yield similar K_{fs} results and the thickness of the contact layer should not be included in the computation of the positive pressure head value at the soil surface. Bagarello et al. (2000) in a sandy-loam soil showed that the presence on the soil surface of a layer of different types of contact material having a saturated hydraulic conductivity greater than the field-saturated soil hydraulic conductivity did not alter appreciably ponded infiltration measurements. A consequence of this last result is that the PI can be used to establish a positive head at the end of an ascending pressure head sequence established with the TI without the need to interrupt the infiltration process for removing the contact material.

The PI is usable to determine sodicity effects on K_{fs}. For example, Bagarello et al. (2006a) tested the potential for reduction in hydraulic conductivity by applying a two-stage sodic water-deionized water infiltration procedure. Stage 1 infiltration consisted of using the PI to infiltrate a given volume of sodic water, having pre-established sodium adsorption ratio and cationic strength. Stage 2 infiltration, which followed immediately after stage 1 by maintaining ponding in the PI ring during the switch-over between the two infiltration stages, consisted of using

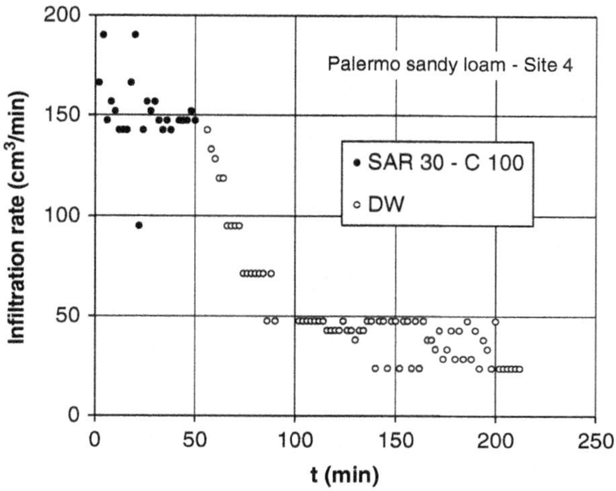

Fig. 2.25 Example of single-ring pressure infiltrometer measurements of infiltration (flow) rate vs. time, t, for stage 1 sodic water infiltration (*solid circles*) and stage 2 deionised water infiltration (*open circles*) (From Bagarello et al. 2006a, reprinted with permission)

the PI to displace the previously infiltrated sodic water by infiltrating an additional volume of deionized water. Plots of infiltration rate vs. time usually showed a decline in infiltration rate to a steady value for stage 1 infiltration, and then a second decline in infiltration rate to a second steady value for stage 2 infiltration (Fig. 2.25). The piecewise behavior of infiltration rates in this figure depends on the selected time interval between two readings from the device's reservoir (see also the Example 2.2) and the resolution of the visual reading with the used PI. The initial decline in infiltration rate during sodic water infiltration was an expected consequence of wetting of the initially unsaturated soil. The stage 2 early-time transient indicated that changes occurred in the soil during deionized water infiltration, such as aggregate slaking and clay dispersion, causing K_{fs} to decline. A conclusion by Bagarello et al. (2006a) was that the PI run may be useful for characterizing in the field the time scale of sodicity effects on saturated soil hydraulic conductivity.

2.2.1.6 Issues of Practical Interest

Preparing the site for a PI run is important to obtain good quality data, since it is necessary to minimize the risk of occurrence of smearing and siltation at the infiltration surface, compaction or shattering of the soil during ring insertion, and short-circuit flow at the contact zone between the soil and the inside surface of the ring wall (Reynolds 1993). The most effective way to prevent smearing phenomena is not to alter the infiltration surface when the site is prepared, especially in fine-textured and wet soil conditions, taking into account that, in any case, the technique cannot be applied in very wet conditions, i.e. close to saturation. In vegetated soils, the surface vegetation can be removed by shears while the roots remain in situ. Compaction and shattering, which are more probable in clay rich soils, can be reduced by avoiding the soil when it is excessively wet (compaction prone) or dry (shattering prone). Prewetting an initially dry and/or hard soil may allow easier ring insertion but it can also determine a change in the soil hydraulic properties to be measured (Reynolds and Elrick 2002b). However, one or more quasi steady-state flow conditions have to be reached with the PI run, and the equilibration times are longer in drier and finer soils. In other terms, it is reasonable to suppose that a change of the antecedent K_{fs}, ϕ_m, S, α^*, and h_f values is a practically unavoidable consequence of the run, at least in soils exhibiting some shrink-swell behavior, independent of the fact that the soil is prewetted to facilitate ring insertion. Short circuit flow can be minimized by lightly tamping the contact between the soil and the ring wall and by careful insertion of the ring (Reynolds and Elrick 2002b). In any case, the risk to disturb the soil volume to be sampled is reduced when a metal, thin walled cylinder with a sharp outside beveled cutting edge is inserted to a short depth into the soil. According to Reynolds (1993), no attempt should be made to straighten a ring if it tilts during the insertion process. This is because straightening will tend to compact the soil inside the ring on one side and to open up a gap between the ring and the soil on the other side. The effect of soil disturbance specifically due to ring insertion on the determination of K_{fs} has not been studied in detail but some attempt to go into this topic was carried out by Bagarello and Sgroi (2004). These authors monitored temporal change of surface K_{fs} in a clayey soil during a 1.5-year period with two different application procedures of the PI, i.e. using rings inserted permanently at the soil surface at the beginning of the study period (permanent sites, PSs) and rings installed immediately before taking the infiltrometer measurement (non-permanent sites, NPSs). The K_{fs} values were 1.0–3.5 times higher at the PSs than at the NPSs, and higher differences between two datasets were detected in relatively wet soil conditions. These differences were attributed primarily to the macrostructure dynamics following ring insertion. In particular, it was suggested that soil disturbance phenomena due to ring insertion into relatively wet soil (compaction, macrostructure collapse) were responsible for low conductivity values, and that soil structure re-organization processes occurring after ring insertion reduced the tendency to produce low values of K_{fs}.

The single-ring pressure infiltrometer, and also the double-ring infiltrometer, yielded higher estimates of K_{fs} as compared with other methods (inverse auger hole method, rainfall simulator, constant-head well permeameter and tension infiltrometer) in stony soils (Verbist et al. 2013). According to these authors, ring insertion into the compacted, stony and sealed arid soils opened up preferential pathways and connected to larger pores that were not active when the other techniques were applied. A general conclusion by Verbist et al. (2013) was that an estimated value for K_{fs} will always be dependent on the measurement method and the calculation technique used, which invalidates the idea of a well defined and unique soil characteristic saturated hydraulic conductivity based on field measurements.

The turbulence of the applied water at the beginning of the PI run may determine siltation phenomena, i.e. plugging of the exposed soil pores by resettlement of the soil particles detached from the infiltration surface. This phenomenon, that determines measurement of a lower K_{fs} than the actual one, may be enhanced by the occurrence of clay particle dispersion phenomena. Siltation can be reduced by slowly establishing the water depth of ponding on the infiltration surface, using a diffuser, and adding, if necessary, a flocculent to the water used for the run (Reynolds et al. 2002b). Distilled water, favoring deflocculation, should not be used (Bagarello et al. 2006a).

The measured infiltration process can contain uncertain information if solar heating of the device occurs during the run (Reynolds 1993). Thermal expansion of the air in the head space of the reservoir above the water surface can prevent bubbling whereas heating of the water in the instrument may determine a reduction in water viscosity that should be taken into account in the calculation of K_{fs} and ϕ_m.

The commercial Guelph pressure infiltrometer may be prone to a lift of the cylinder out of low permeability, initially wet soil as a consequence of the hydrostatic pressure. This phenomenon can be prevented by placing a counter weight on top of the infiltrometer cap (Reynolds and Elrick 2002b).

The choice of the ring diameter must be optimized to ensure a reasonable representativeness of the sample and permit satisfactory conditions for the infiltration experiment. The ring radius influences the steady state infiltration rate, i_s (L T^{-1}). In particular, the effect of r on i_s can be predicted for a soil of given characteristics using Eqs. (2.64) and (2.65) written in the following form:

$$i_s = \frac{Q_s}{\pi r^2} = \frac{K_{fs}H + \phi_m}{\left(0.316\frac{d}{r} + 0.184\right)\pi r} + K_{fs} \tag{2.92}$$

Figure 2.26 shows, for a clay ($K_{fs} = 1 \times 10^{-8}$ m s^{-1}, $\phi_m = 2.5 \times 10^{-9}$ m^{2}s^{-1}, $\alpha^* = K_{fs}/\phi_m = 4$ m^{-1}) and a sandy ($K_{fs} = 1 \times 10^{-4}$ m s^{-1}, $\phi_m = 2.8 \times 10^{-6}$ m^{2}s^{-1}, $\alpha^* = 36$ m^{-1}) soil, the relationship between i_s and r ($0.05 \leq r \leq 0.10$ m) determined with Eq. (2.92) for two values of H (0.05 and 0.25 m) and d (0.03 and 0.05 m) (Bagarello and Iovino 2010). Halving the ring radius from 0.1 to 0.05 m determines an increase of i_s by a factor varying between 1.15 and 1.42. In particular, the increase of i_s is larger for the higher H value and the smaller d value. For given H and d values, i_s increases more in the clay soil than in the sandy one (higher capillary driven flow). However, due to soil heterogeneity, the hydraulic conductivity measured in the field can vary, also appreciably, with the individual sample size (Youngs 1987; Lauren et al. 1988). For example, using a small ring in a soil with macropores implies a relatively high probability to sample only the soil matrix, which yields a measurement of K_{fs} that is not representative of the sampled system. A non-representative result can also be obtained if the sampled soil volume contains one or more macropores. In this case, the collected infiltration rates are mainly controlled by the soil macroporosity because the small sampled volume implies a small volume of the soil matrix and, hence, an imperfect representation of the interaction between the soil matrix and the soil macroporosity occurring in the field. The representativeness of the measurement increases, and the variability of the data decreases, as the size of the sampled soil volume increases. When an infiltrometer is used in the field, ring insertion can locally compact the soil or it can determine the

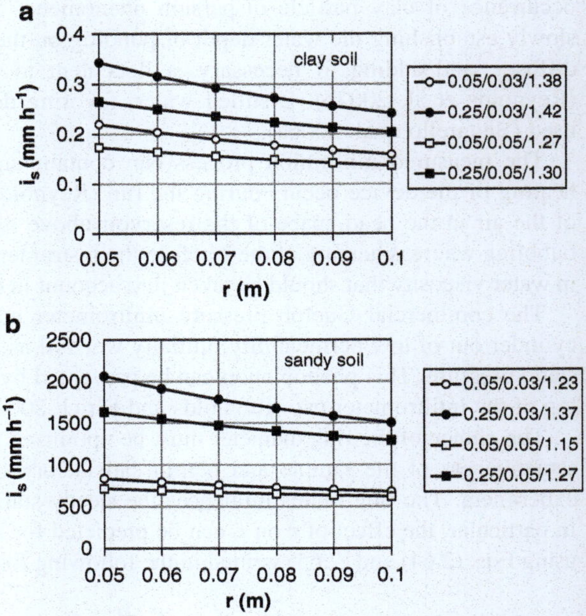

Fig. 2.26 Relationship between the steady-state infiltration rate, i_s, and the ring radius, r, for a (**a**) clay soil, and (**b**) sandy soil. Legend: ponded depth of water, H (m)/depth of ring insertion, d (m)/ratio between the highest and the lowest i_s values (Modified from Bagarello and Iovino 2010)

interruption of large, non-vertical pores (Shouse et al. 1994; Wuest 2005). Obviously, these phenomena have a larger impact on the measured infiltration rates, and hence on the calculated conductivities, when the sampled soil surface is small. On the other hand, a large soil volume sampled with an individual measurement implies a more complicated experiment. The relationship between the measurement of K_{fs} and the sampled soil volume can be expressed using the Representative Elementary Volume (*REV*) concept, which represents the smallest volume of soil usable for the measurement (Bouma 1983; Lauren et al. 1988). Using soil samples having a volume $V < REV$ may yield both higher and smaller conductivity values as compared with those obtained with $V = REV$. The mean conductivity obtained with $V > REV$ is expected to coincide with the one corresponding to $V = REV$ (Vepraskas and Williams 1995). The *REV* depends on the soil structural characteristics and it should contain approximately 20 elementary units of soil structure or a representative number of macropores (Bouma 1983, 1985; Lauren et al. 1988). For measurements carried out with infiltrometers, the representative elementary area, *REA*, or representative measurement area, *RMA*, is sometimes considered to define the smallest surface area to be sampled (Mallants et al. 1997; Haws et al. 2004). For example, if a vertical macropore is detected every 200 cm^2 of soil surface, the sampled area with an individual measurement should be a multiple of 200 cm^2 (Bouma 1983). The *REV* concept is considered rather vague by some authors (van Es 2002) and it is not easily usable in practice. Experimentally assessing the ring size effect on the measurement of K_{fs} represents a valuable source of information that can be used for improving experimental strategies. For example, Youngs

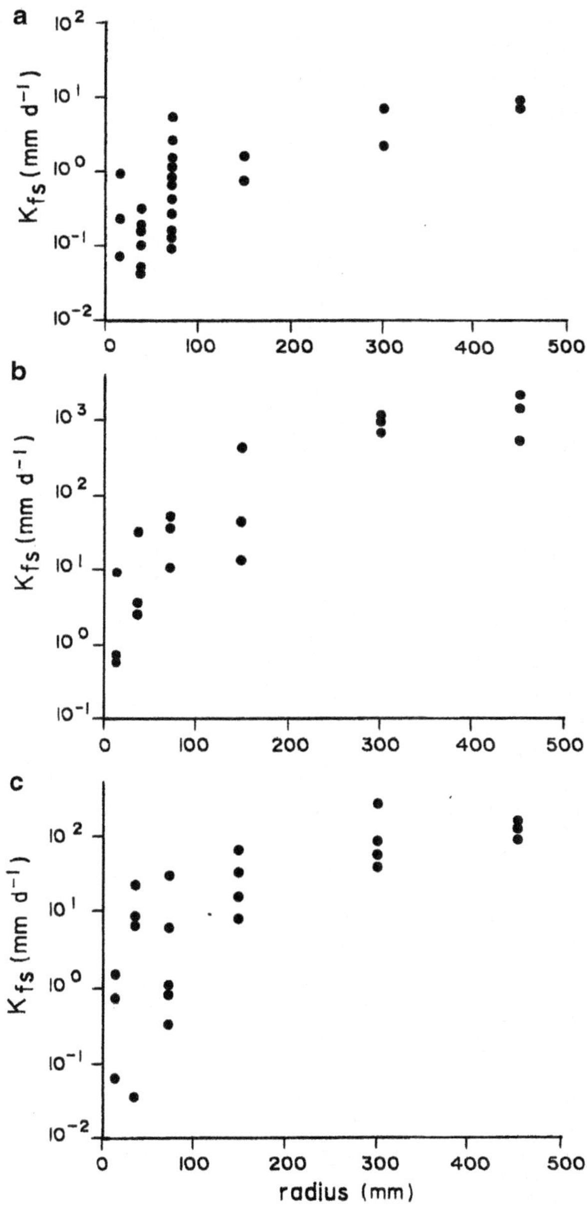

Fig. 2.27 Estimated field-saturated soil hydraulic conductivity, K_{fs}, from measurements with ring infiltrometers of different radii for a (**a**) stony sandy soil, (**b**) sandy-loam soil, and (**c**) silty-loam soil (Modified from Youngs 1987, reprinted with permission)

(1987) compared the K_{fs} values measured on a stony sandy soil, a sandy-loam soil and a silty-loam soil using rings with a radius varying from 17.5 to 456 mm (Fig. 2.27). The hydraulic conductivity values varied over several orders of magnitude with the smallest ring, but this variation was absent for the larger rings.

Generally, estimates of hydraulic conductivity tended to be smaller for the smaller ring sizes. The conclusion by the author was that ring radii above about 150 mm provide a representative sampling area. According to Reynolds and Elrick (2005), ring diameters for the PI method need to be ≥ 0.05–0.1 m for single-grain sands and uniform structureless materials, ≥ 0.3 m for stony/heterogeneous sands, structured sandy-loams and silt-loams, and ≥ 0.5 m for structured clays and clay-loams. Large rings appear more appropriate than small rings to give a signal of the occurrence of heterogeneous conditions in the field, such as invalid TPD calculations, maybe because smaller soil volumes are functionally more homogeneous than larger volumes (Bagarello et al. 2013a). However, it should be noted that this last conclusion was based on numerically simulated data and relatively small datasets and hence it should currently be considered with some caution.

According to Reynolds and Elrick (2002b), soil heterogeneity (e.g., extensive layering, cracks, worm holes, root channels) and extreme vertical antecedent water content gradients may explain the occurrence of invalid TPD and MPD results, i.e. negative K_{fs} or ϕ_m (or α^*) values. Other error sources that could cause negative values include non-attainment of steady-state conditions, errors in the experimental measurement of quasi-steady flow rates because of air bubble size and reading errors, and entrapment of air in the soil due to water redistribution during refilling of the infiltrometer and restarting of infiltration (Mertens et al. 2002). Reynolds and Elrick (1990) and Reynolds (1993) suggested that errors induced by local, small scale heterogeneity should be reduced with the PI method as compared with the Guelph permeameter method because the infiltration surface does not change with each ponded head of water in the former case. Reynolds and Elrick (2002b) also suggested that the PI method should not be highly susceptible to soil heterogeneity or antecedent water content gradients and hence that the expected success rates for the TPD and MPD approaches are generally acceptable. However, Mertens et al. (2002) showed that failure of the PI run can frequently occur (i.e., the 34 % of the 120 two-level runs carried out in a sandy-loam soil failed) even if measurement times are increased, small air inlet tubing is used and large reservoirs are used to make measurements at two heads without refilling. Air entrapment phenomena during refilling can also be avoided if a ponding condition is manually maintained on the infiltration surface when the reservoir of the instrument is refilled. In an investigation on different soils varying from clay to sandy-loam, the TPD calculation procedure yielded simultaneously positive K_{fs} and α^* values for 15 of the 40 runs with a failure rate, therefore, of 62.5 % (Bagarello et al. 2014c). In another investigation, higher failure rates (40 %) were obtained in two clay-loam soils than in a sandy-loam soil (25 %) and a similar result, i.e. fine-textured soils yielding higher failure rates than the coarse-textured one, was obtained by using numerically simulated infiltration data (Bagarello et al. 2013a). Soil heterogeneity and reading errors at the device were suggested to be factors influencing the field results. With the numerical data, allowing a less generic definition of soil heterogeneity, heterogeneous conditions did not impede to obtain success rates of 100 % with the TPD calculations but invalid results were occasionally obtained with the simultaneous occurrence of high values of the standard deviation factor, SD ($SD \geq 0.5$), expressive of the magnitude of variation of saturated conductivity, and the correlation length, L ($L \geq 20$ cm), indicating the structural correlation for this variable (i.e., the conductivity at one node is related with that at another node within a distance equal to L) (Lai and Ren 2007; Lai et al. 2010).

According to Reynolds and Elrick (1990), the TPD or MPD run has to be considered successful if K_{fs} and ϕ_m, and hence α^*, are positive. More recently, Reynolds and Elrick (2002b) suggested that the OPD approach should be applied to each H value, and the resulting K_{fs}, ϕ_m and S values averaged, when an α^* value that falls substantially outside the realistic range of $1 \leq \alpha^* \leq 100$ m^{-1} is obtained. However, extreme values of α^*, also falling outside a wider range ($0.1 \leq \alpha^* \leq 1000$ m^{-1}) than the one suggested by Reynolds and Elrick (2002b), can be found in literature (White and Sully 1992; Russo et al. 1997; Khaleel and Relyea 2001). In an investigation carried out by Bagarello et al. (2010b) on different soils, excluding from the analysis the infiltration runs yielding an α^* parameter lower than 1 m^{-1} or higher than 100 m^{-1} did not modify significantly the estimated means of K_{fs} and α^* but it determined an appreciable decrease of data variability. Therefore, if the objective of the field campaign is to characterize an area in terms of mean value of K_{fs} and α^*, excluding or not the infiltration runs yielding anomalous values of α^* could not influence the results. However, a risk of assuming that $\alpha^* < 1$ m^{-1} or $\alpha^* > 100$ m^{-1} denotes a failed run is that some information on the K_{fs} and α^* variability is lost when it should not be ignored. Calculation of K_{fs} and α^* can be made very quickly after concluding the field infiltration run. Maybe, an empirical and approximate criterion to assess the reliability of an "anomalous" α^* result at a given sampling point could be to rapidly repeat the PI run close to this point. The data point is excluded from the data set unless the repeated test gives support to the unexpected α^* value.

The TPD approach is sensitive to G (Reynolds and Elrick 1990). If α^* is known a priori, then the appropriate G value can be selected from the list of values given by Reynolds and Elrick (1990) and accurate results can be obtained. However, if α^* is already known, one would logically use the less labor-intensive OPD approach. The TPD approach based on G_e (Eq. (2.65)) determines K_{fs} more accurately than it determines ϕ_m. Simulations carried out with reference to a sandy soil and a clay cap/liner material showed that the K_{fs} value was determined within ± 17 % of the actual value whereas the error in ϕ_m reached a factor of more than three (Reynolds and Elrick 1990). The OPD approach prevents calculation of negative K_{fs} and ϕ_m values but it needs an independent estimate of α^*. Moreover, this approach has the advantage of speed and simplicity over the other procedures, because only one H level must be ponded. This requirement also results in minimum susceptibility to variations in K_{fs} and ϕ_m due to small scale soil heterogeneity. The accuracy of the OPD calculations depends primarily on the accuracy of the α^* value. In most instances, an improper site evaluation of α^* will be in error by at most only one category (Table 2.1) among the suggested values for α^* (i.e. $\alpha^* = 4$ m^{-1} instead of $\alpha^* = 12$ m^{-1}). The result is that K_{fs} and ϕ_m may be in error by a factor of nearly three (Reynolds and Elrick 1990). This error is considered acceptable for many practical applications, given that K_{fs} ranges from 10^{-9} m s^{-1} for "tight" clays to 10^{-4} m s^{-1} for coarse sands and given the extreme spatial variability of K_{fs} and ϕ_m found in the field (Elrick and Reynolds 1992a). The sensitivity of K_{fs} to the choice of α^* decreases as H increases, while the sensitivity of ϕ_m to α^* decreases as H decreases (Reynolds 1993). Consequently, if one is interested primarily in K_{fs}, H should be as large as possible; if the interest is primarily in ϕ_m, H should be as small as possible.

Mertens et al. (2002) analyzed the steady infiltration rates collected in a field by assuming that each sampling point may differ in terms of K_{fs} but not of α^*. In other words, the suggested procedure yields one overall "field" α^* and a different K_{fs} value for each measurement location. The assumption of a constant α^* for the sampled field introduces an approximation in the calculation of K_{fs} because, in reality, the α^* parameter varies in space. However, this approximation was considered practically negligible because the K_{fs} estimate is not strongly influenced by α^*. The procedure presupposes that N locations are sampled with the PI at the field site by establishing two ponding depths of water at each measurement point. The TPD approach is applied at each sampling point to obtain an estimate of K_{fs} and α^*. The runs yielding positive values of the two variables are then selected (sample size, $N' \leq N$). Using the data corresponding to these N' runs, an optimization technique is applied to calculate a K_{fs} value for each individual sampling point and a common α^* value for the N' sampling points (i.e., number of optimized parameters $= N' + 1$). This procedure minimizes the sum of the squared differences between the observed steady-state flow rates and the simulated values obtained by Eq. (2.64) (i.e., number of auxiliary variables $= 2N'$). The optimized value of α^* is then used in the N–N' sites excluded from the optimization procedure. For each of these sites, two estimates of K_{fs} are obtained, one for each applied pressure head, and the average of these two estimates is considered to be the in situ measured K_{fs} for that location. With an alternative procedure, the overall α^* parameter could be the average of the N' positive values of α^*. This parameter can then be used to obtain two K_{fs} values for each of the N sampling points and the average of the two K_{fs} values at a measurement location could be considered as the best estimate of K_{fs}. According to Mertens et al. (2002), however, the optimization procedure should be preferred to the alternative procedure because (i) considering the average of the positive α^* values as the overall field α^* is not very meaningful when considering the large uncertainty in α^*, and (ii) the optimization considers simultaneously all the usable experimental information.

In conclusion, the overall strengths of ring infiltrometer methods include accurate measures of vertical K_{fs}, simple and robust equipment and procedures, relatively easy and rapid spatial/temporal replication of measurements, ability to measure water transmission parameters at the soil surface, and widespread acceptance by the science and engineering community (Reynolds and Elrick 2005). The general weaknesses of ring methods include difficult use in stony soils (rings difficult to insert), potential disturbance/alteration of the measured soil volume during the ring insertion process, inconvenience for subsurface measurements given that large access pits have to be dug, measurement of only the vertical water transmission properties, and potentially reduced accuracy for determining the soil capillarity parameters. Strengths of the steady analyses include reasonably accurate and robust determination of vertical K_{fs}, extensive field testing, relatively simple measurements (i_s, H, d, r, α^*), and relatively large sample volumes. Weaknesses include potentially long equilibration times and/or extensive water consumption for large rings or highly structured soils. According to Angulo-Jaramillo et al. (2000), ring infiltrometers are very suitable for determining in situ soil hydraulic properties, with particular reference to the surface soil layer. However, additional efforts are necessary to establish the best ways of characterizing preferential flow or transport processes induced by biological activity, specific pedological conditions, and soil management practices.

Example 2.4
A relatively long, two-level infiltration run was carried out at the surface of a sandy-loam soil using the handcrafted device by Ciollaro and Lamaddalena (1998). The reservoir of the instrument, having an inner diameter of 11 cm (cross-sectional area, $A_{res} = 95.0$ cm^2), was refilled several times during the run. Air entrapment in the sampled soil volume during refilling was avoided by manually maintaining a ponding condition on the infiltration surface. A ring of radius, r, equal to 7.45 cm was inserted into the soil to a depth, d, of 12 cm. Therefore, the infiltration surface had an area, A_{inf}, of 174.4 cm^2 and the G_e shape factor, calculated by Eq. (2.65), was equal to 0.693. For the two ponded heads of water, i.e. $H_1 = 5.3$ cm and $H_2 = 11.0$ cm, Table 2.8 lists the water levels in the reservoir of the instrument, L (cm), at different times, t (min). The lack of data at a given time is indicative of refilling. The difference between the field-saturated and the initial volumetric soil water content, $\Delta\theta$, was equal to 0.417 m^{3}m^{-3}. The field-saturated soil hydraulic conductivity, K_{fs}, was calculated with the TPD and OPD approaches and also with methods 1 and 2 by Wu et al. (1999). The TPD approach and method 1 were also used to calculate the α^* parameter.

The rate of fall of the water level in the reservoir is calculated as $R = \Delta L/\Delta t$ (cm min^{-1}). The R vs. time data are plotted in Fig. 2.28. It has to be noted that, at $t = 2$ min, the calculated R value needs to be corrected to account for the water delivered by the device that accumulates on the infiltration surface to establish the pre-determined ponded depth of water. Given that the volume of water necessary for this is $A_{inf} \times H_1 = 174.4 \times 5.3 = 924.1$ cm^3, the water level in the reservoir decreased by $924.1/95.0 = 9.72$ cm to fill the space above the infiltration surface. Therefore, the rate of fall for the first time interval is given by (24-1-9.72)/(2–0) = 6.6 cm min^{-1}. Figure 2.28 suggests that steady-state conditions were reached for both ponded depths of water. For $H = H_1$, the water level fall rate can be considered practically constant (R_{s1}) for $t \geq 138$ min:

$$R_{s1} = \frac{91.4 - 4.4}{157 - 138} = 4.58 \text{ cm min}^{-1}$$

The corresponding steady-state flow rate, Q_{s1}, is:

Table 2.8 Readings at the reservoir of the PI device, L (cm), at different times, t (min), during the run with two ponded depths of water, H (cm), for the Example 2.4

$H_1 = 5.3$				$H_2 = 11.0$			
t	L	t	L	t	L	t	L
0	1	82	38.2	161	3	239	5.3
2	24	83	43	162	8.2	240	10.5
4	33.8	84	47.8	163	13.5	241	15.7
6	43.8	85	53	164	18.7	242	20.9
8	54.5	86	57.7	165	24	243	26
9	59.7	87	62.7	166	29.7	244	31.4
10	65	88	67.8	167	34.9	245	36.6
11	70.5	89	72.7	168	40	246	41.6
12	75.8	90	77.5	169	45.5	247	46.9
13	81	91	82.7	170	51	248	51.9
14	86.3	92	87.7	171	56.5	249	57
15	91.6	93	92.5	172	61.8	250	62.5
17		95		173	67.4	251	67.6
18	9	96	3	174	72.7	252	72.5
19	13.6	97	7.9	175	78	253	77.9
20	18.9	98	12.6	176	83.5	254	83
21	24	99	17.3	177	88.8	255	88.2
22	29.3	100	22	178	94.2	256	93.2
23	35	101	26.8	180		258	
24	40	102	31.7	181	6.5	259	4.8
25	45.4	103	36.5	182	12	260	10.1
26	50.7	104	41	183	17.4	261	15.2
27	56	105	46	184	22.8	262	20.4
28	61.4	106	50.5	185	28	263	25.3
29	66.5	107	55.4	186	33.6	264	30.5
30	72	108	60.1	187	39	265	35.5
31	77.4	109	65	188	44.4	266	40.5
32	82.8	110	70	189	49.6	267	46.1
33	88.2	111	74.5	190	54.8	268	51.1
34	93.6	112	79.5	191	60	269	56
36		113	84.5	192	65.4	270	61.4
37	6	114	89	193	70.6	271	66.5
38	11.1	116		194	76	272	
39	16.3	117	3.6	195	81.4	273	2
40	21.5	118	8.5	196	86.3	274	7
41	26.7	119	13.3	197	91.7	275	12.3
42	31.8	120	18.2	199		276	17
43	37	121	22.8	200	6.2	277	22.5
44	42.3	122	27.8	201	11.5	278	27.5
45	47.6	123	32.7	202	16.5	279	32.4
46	53	124	37.5	203	21.8	280	37.5

(continued)

Table 2.8 (continued)

$H_1 = 5.3$				$H_2 = 11.0$			
t	L	t	L	t	L	t	L
47	58	125	42	204	27.3	281	42.5
48	63.4	126	47	205	32.3	282	47.7
49	68.7	127	51.8	206	37.8	283	52.8
50	73.8	128	57	207	43	284	58
51	79	129	61.6	208	48	285	63.1
52	84.2	130	66.4	209	53.2	286	68.1
53	89.4	131	71.2	210	58.7	287	73.1
55		132	76	211	64	288	78.1
56	5.5	133	80.8	212	69.2	289	83.1
57	10.5	134	85.6	213	74.4	290	88.1
58	15.5	135	90.6	214	79.6	291	93.1
59	20.7	137		215	84.9		
60	25.7	138	4.4	216	90		
61	30.8	139	9	218			
62	35.9	140	13.7	219	5.7		
63	40.7	141	18.1	220	10.7		
64	45.6	142	22.6	221	16.2		
65	50.5	143	27.4	222	21.2		
66	55.8	144	32.2	223	26.5		
67	61	145	36.7	224	31.7		
68	65.7	146	41.2	225	36.7		
69	71	147	45.7	226	41.7		
70	76	148	50.4	227	47.1		
71	81	149	55	228	52.3		
72	85.7	150	59.7	229	57.5		
73	90.8	151	64	230	62.7		
75		152	68.8	231	67.7		
76	8.3	153	73.1	232	72.9		
77	13.2	154	77.8	233	78.4		
78	18	155	82.3	234	83.1		
79	23.1	156	86.7	235	88.6		
80	28.1	157	91.4	236	93.4		
81	33			238			

$$Q_{s1} = R_{s1} \times A_{res} \times 60 \times 1000 = 4.58 \times 95.0 \times 60 \times 1000$$
$$= 26109116 \left(2.61 \times 10^7\right) \text{ mm}^3 \text{ h}^{-1}$$

For $H = H_2$, flow equilibration is detected for $t \geq 273$ min, measured from the beginning of the run. The R_{s2} and Q_{s2} values are 5.06 cm min^{-1} and 28,858,409 (2.89×10^7) mm^{3}h^{-1}, respectively.

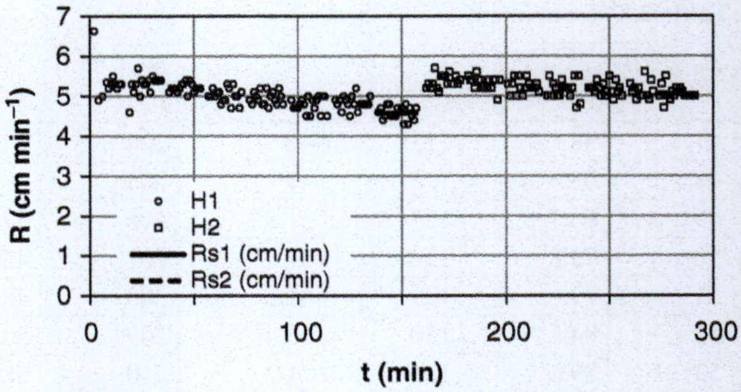

Fig. 2.28 Rate of fall of the water level in the reservoir of the instrument, *R*, vs. time, *t*, for the data of the Example 2.4

Using Eqs. (2.72), (2.73) and (2.7) (TPD approach), the following estimates of K_{fs}, ϕ_m and α^* are obtained:

$$K_{fs} = \frac{0.693}{74.5}\left(\frac{2.89 \times 10^7 - 2.61 \times 10^7}{110 - 53}\right) = 448.7 \ \text{mm h}^{-1}$$

$$\phi_m = \frac{0.693}{74.5}\left[\frac{110 \times \ 2.61 \times 10^7 - 53 \times 2.89 \times 10^7}{110 - 53} - \pi \times 74.5 \times 0.693 \frac{2.89 \times 10^7 - 2.61 \times 10^7}{110 - 53}\right] = 146315.6 \ \text{mm}^2 \ \text{h}^{-1}$$

$$\alpha^* = \frac{448.7}{146315.8} = 0.0031 \ \text{mm}^{-1}\left(\text{i.e., } 3.1 \ \text{m}^{-1}\right).$$

With the OPD approach applied to the first ponded head, Eq. (2.67) yields the following estimate of K_{fs} if the α^* value of first approximation ($\alpha^* = 0.012$ mm^{-1}, Table 2.1; Most structured and medium textured materials) is used in the calculations:

$$K_{fs} = \frac{0.012 \times 0.693 \times 2.61 \times 10^7}{74.5(0.012 \times 53 + 1) + 0.693 \times 0.012 \times \pi \times 74.5^2} = 813.5 \ \text{mm h}^{-1}.$$

This estimate of K_{fs} is 1.81 times higher than the K_{fs} value obtained with the TPD approach. For the sampled site, a more appropriate estimate of α^* would probably be $\alpha^* = 0.004$ mm^{-1}. In this case, Eq. (2.67) yields $K_{fs} = 522.1$ mm h^{-1}, that is appreciably closer (difference between the two K_{fs} estimates by a negligible 16 %), as expected, to the K_{fs} value obtained with the TPD approach.

An additional analysis of the data can be carried out by considering the first part of the run with $H = H_1$, i.e. until the reservoir emptied the first time (i.e., the first 15 min of the run). In this case, a cumulative infiltration curve can be obtained. At first, the readings at the reservoir of the instrument are corrected so that the reading at $t = 0$ is set to 0 and the volume of water delivered to fill the space above the infiltration surface is accounted for. The corrected reading at $t = 2$ min is 24-1-9.72 = 13.28 cm. The corrected readings at later times are obtained by

Table 2.9 Readings at the reservoir of the PI device, L, corrected readings, L', and cumulative infiltration, I, at different times, t, during the initial part of the infiltration run, i.e. before the first refilling of the reservoir, with a ponded depth of water of 53 mm for the Example 2.4

t (h)	L (cm)	L' (cm)	I (mm)
0	1.0	0	
0.033	24.0	13.3	72.4
0.067	33.8	23.1	125.8
0.1	43.8	33.1	180.3
0.133	54.5	43.8	238.6
0.15	59.7	49.0	266.9
0.167	65.0	54.3	295.8
0.183	70.5	59.8	325.8
0.2	75.8	65.1	354.7
0.217	81.0	70.3	383.0
0.233	86.3	75.6	411.9
0.25	91.6	80.9	440.8

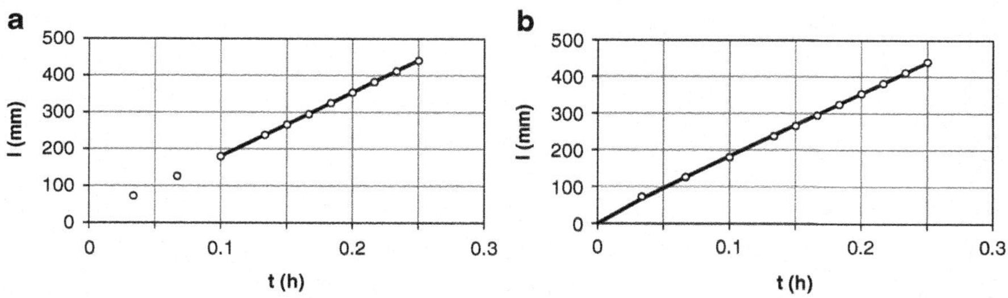

Fig. 2.29 Cumulative infiltration, I, vs. time, t, for the initial part of the infiltration run with a ponded depth of water of 53 mm for the data of the Example 2.4 and (**a**) linear regression line of I against t for the linear portion of the dataset, and (**b**) fit of Eq. (2.76) to the data

repeatedly adding to this corrected reading the difference between L at a given time and L at $t = 2$ min. The corrected readings are denoted by the symbol L' in Table 2.9. This table also lists the corresponding cumulative infiltration, I (mm), values, obtained as $I = L' \times 10 \times A_{res}/A_{inf}$, for different times expressed in hours. Plotting I vs. t (Fig. 2.29a) suggests that a steady-state condition, denoted by a linear relationship between the two variables, can be detected with reference to the last nine data points. The slope of this linear portion of the cumulative infiltration curve, calculated by a linear regression analysis of the last data points, is equal to 1737.4 mm h^{-1}. The corresponding estimate of the steady flow rate is $1737.4 \times 174.4 \times 100 = 30{,}294{,}125.1$ (3.03×10^7) mm^3h^{-1}. The OPD approach with this "early-time" estimate of steady-state flow rate and $\alpha^* = 0.004$ mm^{-1} yields a K_{fs} value of 605.7 mm h^{-1}. The percentage difference between 605.7 and 522.1 mm h^{-1}, equal to 16 %, quantifies the effect of assuming that steady-state conditions of the infiltration process occurred soon after starting the run.

Fitting Eq. (2.76) to the (I, t) data pairs listed in Table 2.9 by using the solver routine of Microsoft Excel yields $A_w = 1625.0$ mm h^{-1} and $B_w = 65.4$ mm $h^{-0.5}$

(Fig. 2.29b). Equations (2.79), (2.84), (2.83), (2.85), (2.82), (2.87) and (2.86) are then applied to determine K_{fs} and α* by method 1 of Wu et al. (1999):

$$G^* = 120 + \frac{74.5}{2} = 157.2 \text{ mm}$$

$$C = \frac{1}{4 \times 0.417} \left(\frac{65.4}{0.1682} \right)^2 \frac{0.9084}{1625.0} = 50.6 \text{ mm}$$

$$\lambda_c = \frac{1}{2} \left[\sqrt{(53 + 157.2)^2 + 4 \times 157.2 \times 50.6} - (53 + 157.2) \right] = 32.7 \text{ mm}$$

$$T_c = \frac{1}{4} \left(\frac{65.4 \times 0.9084}{0.1682 \times 1625.0} \right)^2 = 0.0118 \text{ h}$$

$$K_{fs} = \frac{32.7 \times 0.417}{0.0118} = 1157.7 \text{ mm h}^{-1}$$

$$\phi'_m = \frac{1157.7^2}{0.417} 0.0118 = 37894.6 \text{ mm}^2 \text{ h}^{-1}$$

$$\alpha^* \approx \frac{1157.7}{37894.6} = 0.031 \text{ mm}^{-1}$$

Therefore, method 1 yields an estimate of K_{fs} that is 2.6 times higher than the corresponding estimate obtained with the TPD approach. Much larger differences (i.e. by approximately an order of magnitude) are detected with reference to α*. A linear regression analysis of the ($I/t^{0.5}$, $t^{0.5}$) data pairs yields $A_w = 1565.8$ mm h^{-1} and $B_w = 90.0$ mm h$^{-0.5}$ (Fig. 2.30). In this case, $K_{fs} = 1009.2$ mm h^{-1} and α* = 0.017 mm^{-1} is obtained. These values are closer to, but still higher than, the ones obtained with the TPD approach (differences by a factor of 2.2 for K_{fs} and 5.6 for α*). A possible reason for the difference between the values of K_{fs} obtained with the TPD approach and method 1 is that a longer run (TPD approach) determined more noticeable pore clogging phenomena. In method 2 by Wu et al. (1999), A_w is the slope of the linear portion of the I vs. t curve. Using Eqs. (2.90) and (2.89) with α* = 0.004 mm^{-1} yields:

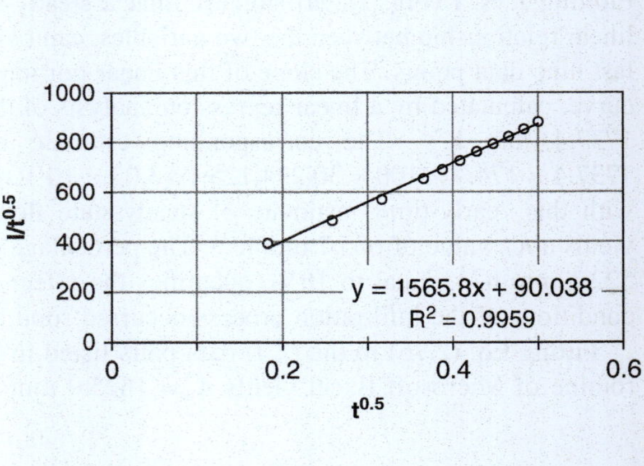

Fig. 2.30 Relationship between the cumulative infiltration, I (mm), and the time, t (h), on the $I/t^{0.5}$ vs. $t^{0.5}$ plot for the Example 2.4

$$f \approx \frac{53 + 1/0.004}{157.2} + 1 = 2.93$$

$$K_{fs} = \frac{1737.4}{0.9084 \times 2.93} = 653.5 \text{ mm h}^{-1}$$

This estimate of K_{fs} is very close to the one obtained by the OPD approach and an "early-time" estimate of steady-state flow rate (difference by 7.9 %), which is a plausible result given that the same part of the infiltration process was considered with these two approaches.

2.2.2 *Other Single-Ring Approaches*

Besides the classical single-ring pressure infiltrometer technique, other approaches use measurements of three-dimensional infiltration through a confined area by a ring to determine the hydrodynamic parameters of an initially unsaturated soil. In particular, the method by Touma et al. (2007) and the Simplified method based on the Beerkan Infiltration run (Bagarello et al. 2013b, 2014b) rely on the so-called Beerkan experimental protocol and they consider the transient phase of an infiltration process under a constant head of water on the soil surface, theoretically equal to zero. With the Steady-State Falling-Head method by Elrick et al. (1995) and the Bottomless Bucket method by Nimmo et al. (2009), a falling-head infiltration process has to be established. These methods are described in this section.

2.2.2.1 The Beerkan Experimental Protocol

The Beerkan experimental protocol was pioneered by Braud et al. (2005) and it was detailed more precisely in Lassabatere et al. (2006). The technique consists of measuring the infiltrated water with time after the supply of a constant volume of water of 120–150 mL in a ring of 120–150 mm in diameter driven 4–5 mm into the soil to prevent lateral loss of water. At time zero, the first volume is poured into the ring. When this volume has completely infiltrated, the time is measured and another equal volume of water is poured in the ring. The infiltration is terminated when the time elapsed between two supplies becomes nearly constant. The infiltration test is completed by the collection, at the end of infiltration, of a wet disturbed soil sample inside the ring and a dry disturbed sample in the area surrounding the ring. These samples allow us to determine the field-saturated, w_{fs} (M M^{-1}), and the initial, w_i (M M^{-1}), gravimetric water content. An undisturbed soil sample of known volume is collected to measure the soil's bulk density, ρ_b, and to convert the gravimetric to volumetric water contents. According to Lassabatère et al. (2006) and Touma et al. (2007), the advantage of this type of experiment is its simplicity and economy, also it can be carried out by a single operator. The time necessary for an infiltration test is of approximately 1–2 h and the total amount of water used is limited to a few liters for each test. The wetted depth of soil is small, and therefore it is realistic to consider the soil as homogeneous and isotropic.

2.2.2.2 The Method by Touma et al. (2007)

Touma et al. (2007) developed a method to analyze transient single-ring infiltrometer data for the special case of a negligible head of water on the infiltration surface, i.e. H (L) ≈ 0, and a very small insertion depth of the ring into the soil. The method allows the estimation of field-saturated soil hydraulic conductivity, K_{fs} (L T^{-1}), and soil sorptivity, S (L T$^{-1/2}$), on the basis of measured cumulative infiltration from the beginning of the process. It considers the following relationship between three-, I_{3D} (L), and one-dimensional, I_{1D} (L), cumulative infiltration (Smettem et al. 1994):

$$I_{3D} = I_{1D} + \frac{\gamma S^2}{r\Delta\theta}t \tag{2.93}$$

where γ is a constant equal to 0.75, t (T) is the time, r (L) is the radius of the source, and $\Delta\theta$ (L^{3}L^{-3}) is the difference between the final, θ_{fs}, and the initial, θ_i, volumetric soil water content. One-dimensional infiltration is modeled by the Brutsaert's (1977) relationship:

$$I_{1D} = K_{fs}t + \frac{S^2}{\beta_B K_{fs}}\left[1 - \left\{1 + \beta_B\left(\frac{K_{fs}\sqrt{t}}{S}\right)\right\}^{-1}\right] \tag{2.94}$$

where β_B is a constant that can be set equal to 2/3 for practical applications. This model was selected because it performed better than alternative models and also because it gives the cumulative infiltration explicitly as a function of time. Moreover, Eq. (2.94) applies for all infiltration times. Combining Eqs. (2.93) and (2.94) yields the following expression for I_{3D}:

$$I_{3D} = K_{fs}t + \frac{S^2}{\beta_B K_{fs}}\left[1 - \left\{1 + \beta_B\left(\frac{K_{fs}\sqrt{t}}{S}\right)\right\}^{-1}\right] + \frac{\gamma S^2}{r\Delta\theta}t \tag{2.95}$$

An estimate of S and K_{fs} is obtained by an optimization procedure using Eq. (2.95) and the measured (I_{3D}, t) data pairs. In particular, optimization is done by minimizing the following objective function, F:

$$F(S, K_{fs}) = \sum_{i=1}^{n_m}\left\{I_{3D,i}(S, K_{fs}) - I_{3D,i}\right\}^2 \tag{2.96}$$

where n_m is the number of measurements, $I_{3D,i}(S,K_{fs})$ is the calculated three-dimensional infiltration and $I_{3D,i}$ is the measured three-dimensional infiltration. Using Eq. (2.96) needs an initial estimate of S and K_{fs}. In a test of the method carried out with numerically simulated infiltration data for different soils, S was underestimated and K_{fs} was overestimated. However, vertical cumulative infiltration, considered as an appropriate indicator of the suitability of the method, was predicted satisfactorily, i.e. with errors ranging from 13.9 % for the sandy soil to 5.3 % for the clay soil. Testing the applicability of the method in real soils appears advisable to draw conclusions about its practical interest.

2.2.2.3 The Simplified Method Based on the Beerkan Infiltration Run

The Simplified method based on the Beerkan Infiltration run (SBI method; Bagarello et al. 2013b, 2014b) is another method yielding an estimate of K_{fs} by the same infiltration run performed according to the Beerkan experimental protocol (Lassabatere et al. 2006). Even with the SBI method, a cylinder of radius r (L) is inserted to a short depth into a soil with a uniform initial water content, θ_i (L^3L^{-3}), so to produce a minimal disturbance of the porous medium, and the infiltration time of a few small volumes of water repeatedly applied at the surface of the confined soil is measured. An experimental, three-dimensional (3D) cumulative infiltration, here denoted by the symbol I (L) for simplicity, vs. time, t (T), relationship including a given number of discrete points (8–15 according to Lassabatere et al. 2006) is then obtained and used to estimate K_{fs}. The infiltration data can be fitted to the following explicit transient two-term relationship (Haverkamp et al. 1994):

$$I(t) = S\sqrt{t} + \left(AS^2 + BK_{fs}\right)\, t \tag{2.97}$$

where S ($L\,T^{-1/2}$) is soil sorptivity, and A (L^{-1}) and B are constants defined as:

$$A = \frac{\gamma}{r\,\left(\theta_{fs} - \theta_i\right)} \tag{2.98a}$$

$$B = \frac{2-\beta}{3}\left[1 - \left(\frac{\theta_i}{\theta_{fs}}\right)^{\eta}\right] + \left(\frac{\theta_i}{\theta_{fs}}\right)^{\eta} \tag{2.98b}$$

where θ_{fs} (L^3L^{-3}) is the field-saturated volumetric soil water content, β and γ are coefficients equal to 0.6 and 0.75, respectively, for $\theta_i < 0.25\ \theta_{fs}$ (Smettem et al. 1994; Haverkamp et al. 1994) and η is a shape parameter. Equation (2.98b) assumes that the hydraulic conductivity function can be represented with the relationship proposed by Brooks and Corey (1964). If the soil is relatively dry at the beginning of the experiment, $K(\theta_i) \ll K_{fs}$ and B is equal to $(2-\beta)/3$. Dividing both sides of Eq. (2.97) by $\sqrt{t}$, as suggested by Vandervaere et al. (2000), and introducing Eqs. (2.98a and 2.98b) in the infiltration equation, the following linear relationship between $I/\sqrt{t}$ and $\sqrt{t}$ is obtained:

$$\frac{I(t)}{\sqrt{t}} = S + \left[\frac{\gamma\ S^2}{r(\theta_{fs} - \theta_i)} + \frac{2-\beta}{3}\ K_{fs}\right]\ \sqrt{t} = S + b_1\sqrt{t} \tag{2.99}$$

Therefore, the slope, b_1, of Eq. (2.99), equal to:

$$b_1 = \frac{\gamma\ S^2}{r(\theta_s - \theta_i)} + \frac{2-\beta}{3}\ K_{fs} \tag{2.100}$$

can be estimated by a linear regression analysis of the $(I/\sqrt{t},\ \sqrt{t})$ data.

Reynolds and Elrick (1990) and Elrick and Reynolds (1992a) expressed the relative importance of gravity and capillary forces during a ponding infiltration process by the so-called α^* (L^{-1}) parameter, given by Eq. (2.7), where ϕ_m is defined by Eq. (2.2). The relationship between S and ϕ_m can be written according to Eq. (2.10) (Philip 1957b; Reynolds and Elrick 2002a). Equation (2.10) with $\gamma_w = 1.818$ (i.e., the value for a wetting front) was considered by Reynolds and Elrick (2002b) to be suitable for estimating sorptivity with a ponded infiltration experiment from a single ring. Combining Eqs. (2.100), (2.7) and (2.10) and solving for K_{fs} gives:

$$K_{fs} = \frac{b_1}{\frac{\gamma\gamma_w}{r\alpha^*} + \frac{2-\beta}{3}} \tag{2.101}$$

Using Eq. (2.101) needs knowledge of the α^* parameter that, according to the literature, can be estimated on the basis of a general description of soil textural and structural characteristics (Elrick and Reynolds 1992a). In particular, four values of α^* (0.036, 0.012, 0.004 and 0.001 mm^{-1}) were suggested for practical use of permeameters and infiltrometers in soils varying from coarse sands to compacted clays (Table 2.1) and $\alpha^* = 0.012\ mm^{-1}$ was considered to be the value of first approximation for most field soils (Reynolds et al. 2002a). With the proposed Eq. (2.101), additional field and laboratory measurements, such as initial and final soil water content, particle size distribution, or bulk density, are not strictly necessary. A theoretical limit is that Eq. (2.97) is only valid for the transient phase of the infiltration process (Lassabatere et al. 2006). From a practical point of view, however, the duration of the infiltration run in the field does not seem to represent a crucial step of the data analysis procedure based on Eq. (2.97) (Bagarello et al. 2011a).

The effect of an erroneous choice of α^* on the predictions of K_{fs} with Eq. (2.101) was explored for each of the four K_{fs}–α^* combinations that define representative porous media according to Reynolds and Elrick (1990) (Bagarello et al. 2014b). In particular, Eq. (2.101) was used to calculate the b_1 value corresponding to a particular K_{fs}–α^* combination. Then, K_{fs} was re-calculated with the true b_1 value

Table 2.10 Factor of discrepancy (= maximum K_{fs} between the erroneous and the true value/minimum K_{fs} between the erroneous and the true value) obtained by calculating the field-saturated soil hydraulic conductivity, K_{fs}, with Eq. (2.101) and an erroneous choice of the α* parameter by a category, for two values of the ring radius, r. From Bagarello et al. 2014b, reprinted with permission

Porous medium		Erroneous α* (mm^{-1})	r= 75 mm	r= 150 mm
Sand soil	(K_{fs}=1.0×10^{-1} mm s^{-1}, α*=0.036 mm^{-1})	0.012	2.04	1.70
Loam soil	(K_{fs}=1.0×10^{-3} mm s^{-1}, α*=0.012 mm^{-1})	0.004–0.036	2.04–2.53	1.70–2.24
Clay soil	(K_{fs}=1.0×10^{-5} mm s^{-1}, α*=0.004 mm^{-1})	0.001–0.012	2.53–3.72	2.24–3.49
Clay cap/liner	(K_{fs}=1.0×10^{-6} mm s^{-1}, α*=0.001 mm^{-1})	0.004	3.72	3.49

and a value of α* differing by plus or minus one category from the correct one, according to the values of α* suggested by Elrick and Reynolds (1992a). For example, 0.012 mm^{-1} was replaced by either 0.036 or 0.004 mm^{-1}, whereas 0.036 mm^{-1} was only replaced by 0.012 mm^{-1}. The factor of discrepancy, f_D (maximum K_{fs} between the erroneous and the true value/minimum K_{fs} between the erroneous and the true value), associated with an erroneous choice of α* was determined. Calculations were carried out for $r = 75$ and 150 mm to assess the ring size effect on the results of this sensitivity analysis. The f_D values varied from 1.70 to 3.72 (Table 2.10), with higher values in low permeability porous media. For a given soil, larger errors were associated to the underestimation of α* as compared with those due to the overestimation of this parameter. The errors were slightly lower with the larger ring. An error in K_{fs} by a factor of two or three has been considered acceptable in many cases given that K_{fs} ranges from 10^{-9} m s^{-1} for tight clays to 10^{-4} m s^{-1} for coarse sands and given the extremely high spatial variability of K_{fs} commonly found in the field (Elrick and Reynolds 1992a). Assuming that a factor of more than three is indicative of an important error, this analysis suggested that an erroneous choice of α* by a category should not be expected to substantially compromise the reliability of the K_{fs} prediction for most field soils. In any case, the risk to be in error decreases as the soil permeability to water increases. If a technician has a doubt on which α* value has to be chosen, it is preferable to use a relatively high value because the expected error is lower. Finally, the use of a large ring for the experiments was recommended to reduce the K_{fs} estimation error due to an improper selection of α*.

Bagarello et al. (2013b, 2014b) tested the applicability of the SBI method on both Burundian ($N = 149$ infiltration runs) and Sicilian (south Italy, $N = 43$) soils. The field-saturated soil hydraulic conductivity determined with the original BEST procedure (Lassabatere et al. 2006), denoted by the symbol K_{fsB}, was considered to be the reference value in the comparison with the SBI method calculations (K_{fsS}). The comparison between K_{fsS} and K_{fsB} was carried out by simply setting

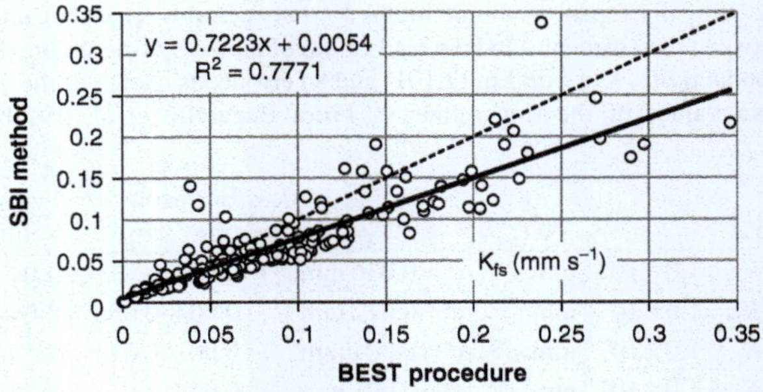

Fig. 2.31 Comparison between the field-saturated soil hydraulic conductivity, K_{fs}, values obtained by the BEST procedure of soil hydraulic characterization and the SBI method for the Sicilian and Burundian data. *Identity line* denoted by the *dashed line*; the *continuous line* is the linear regression line (sample size, $N = 192$; from Bagarello et al. 2013b, reprinted with permission)

$\alpha^* = 0.012$ mm^{-1} in Eq. (2.101) to test what happens if the value of first approximation is used to estimate K_{fs}. With reference to the complete dataset (Burundian and Sicilian soils), K_{fsB} (mean = 0.088 mm s^{-1}) was significantly greater than K_{fsS} (0.069 mm s^{-1}) (Bagarello et al. 2013b). A statistically significant correlation was detected between the two variables (coefficient of determination, $R^2 = 0.78$) but the regression line (Fig. 2.31) differed significantly from the identity line (95 % confidence intervals for the intercept and the slope equal to −0.0006 – 0.011 and 0.67 – 0.78, respectively). The difference between K_{fsS} and K_{fsB} did not exceed a factor of two and three in the 96.4 % and 98.4 % of the cases, respectively, and the maximum difference was by a factor of 6.4. This check suggested that the SBI method with a rough estimate of α^* can be used to obtain at least a first approximation value of K_{fs} since K_{fsB} and K_{fsS} differed by a factor of less than three for the large majority of the sampled points.

To improve the derivation of the α^* parameter for the SBI method and the subsequent estimation of K_{fs}, Bagarello et al. (2013b, 2014b) also tested the possibility to predict α^* (mm^{-1}) on the basis of the slope of the linearized cumulative infiltration equation, b_1 (mm s^{-1}). The following relationships were obtained for the Burundian and the Sicilian soils, respectively, by testing different functional relationships between α^* and b_1 and minimizing the differences between K_{fsS} and K_{fsB} for the considered datasets (Fig. 2.32):

$$\alpha^* = 0.0262 + 0.0035 \times \ln(b_1) \tag{2.102a}$$

$$\alpha^* = 0.0052 + 0.016 \times b_1 \tag{2.102b}$$

For both datasets, the K_{fsS} values calculated by Eqs. (2.101) and (2.102a) or (2.102b), depending on the dataset, were significantly correlated with the associated

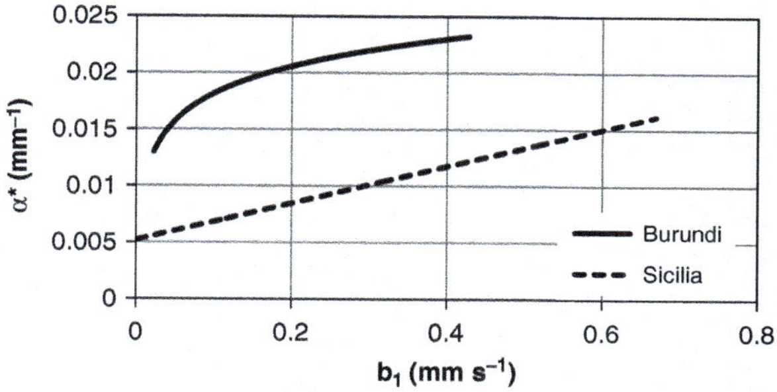

Fig. 2.32 Relationships between the α* parameter and the slope of the linearized cumulative infiltration curve, b_1, obtained for Burundian and Sicilian soils (From Bagarello et al. 2013b, reprinted with permission)

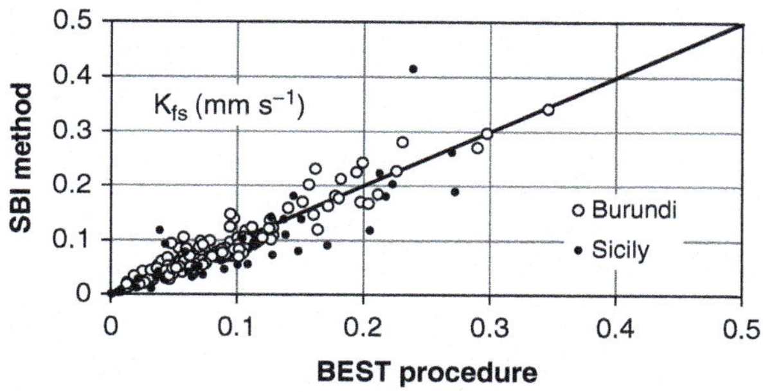

Fig. 2.33 Comparison between the field-saturated soil hydraulic conductivity, K_{fs} (mm s^{-1}), values obtained with the BEST procedure of soil hydraulic characterization and the SBI method using the slope of the linearized cumulative infiltration curve to estimate the α* parameter (sample size, $N = 149$ for Burundian soils and $N = 43$ for Sicilian soils; from Bagarello et al. 2013b, reprinted with permission)

K_{fsB} data, the two K_{fs} datasets did not show significant differences, and the linear regression line between K_{fsS} and K_{fsB} was not significantly different from the identity line (Fig. 2.33). The largest difference between K_{fsS} and K_{fsB} was by a factor of 3.2 that is very close to the highest factor of difference that can be considered negligible from a practical point of view in many cases (Elrick and Reynolds 1992a). Therefore, Eqs. (2.102a and 2.102b), suggesting that the measured infiltration curve contains the necessary information to estimate α*, allowed us to improve K_{fs} prediction as compared with the α* value of first approximation. The two Eqs. (2.102a and 2.102b) clearly differed, maybe because temperate and tropical soils show differences in their chemical and physical properties (Hodnett and Tomasella 2002; Tomasella and Hodnett 2004). One of the drawbacks of this method lies in the need to calibrate the α* vs. b_1 relationship for any soil database.

In the investigations by Bagarello et al. (2013b, 2014b), plotting $I/\sqrt{t}$ against $\sqrt{t}$ yielded in general the expected linear behavior for the entire infiltration run (Fig. 2.34a) or at least with the exclusion of the first few points (generally, one or two points, Fig. 2.34b). In some cases, however, a linear relationship between the two variables was undetectable (Fig. 2.34c). A perturbation of the run in the early

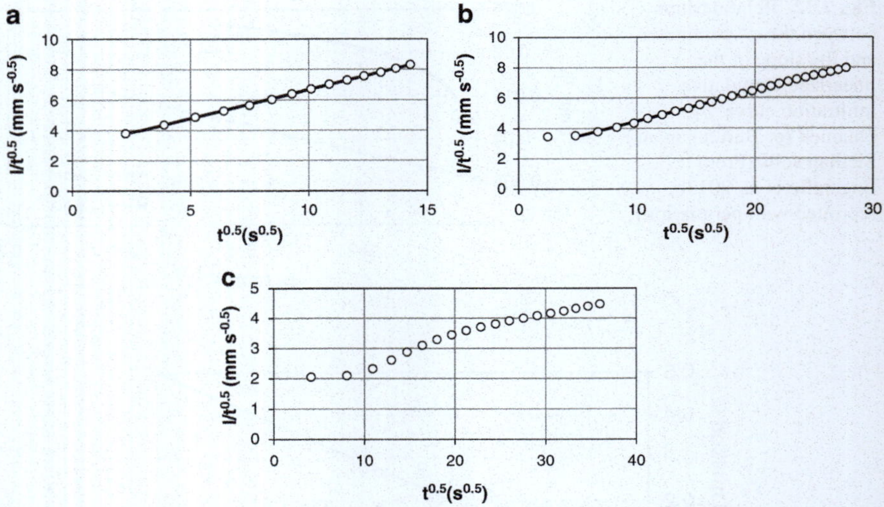

Fig. 2.34 Values of the ratio between the cumulative infiltration, I, and the square root of time, t, plotted against the square root of t: (**a**) run showing the expected linear relationship between the two variables for the entire infiltration process; (**b**) run showing the linear relationship with the exclusion of the first data point; and (**c**) run with an undetectable linear relationship between the two variables (From Bagarello et al. 2014b, reprinted with permission)

stage of the infiltration process has been reported in other investigations and it has been removed in the fitting of the selected model to the data (e.g., Wu et al. 1999). Therefore, the very early phase of the run denoting an anomalous process can be excluded from the fitting. However, when the data do not allow to detect a linear relationship between the two variables, the infiltration run has to be considered failed since the two-term infiltration model is not appropriate to describe the measured process. From a practical point of view, field data allow us to locally evaluate the applicability of the procedure on the basis of the detected linearity of the data points. However, it would be advisable to more clearly define factors determining the observed departures of the data from the expected linear behavior.

Equation (2.101) is theoretically sound because it combines a physically based infiltration model with basic relationships between soil variables. However, Eqs. (2.102a and 2.102b) were developed empirically, assuming that larger α^* values are expected in more permeable soils (Bagarello et al. 2014a). From a theoretical point of view, the α^* parameter, which is related primarily to the shape of the $K(h)$ relationship (Reynolds 2011) and it is indicative of the relative importance of the field-saturated and capillarity components of steady flow (Reynolds et al. 1992), is not functionally related to K_{fs}. In particular, α^* is independent of the K_{fs} value when the $K(h)$ function has K_{fs} as a multiplier, which is the case of the Gardner's (1958) relationship but also of the van Genuchten's (1980) model (Reynolds 2011). Experimental data indicating that α^* is independent of K_{fs} can be found in literature. For example, Fig. 2.35a shows the relationship between these

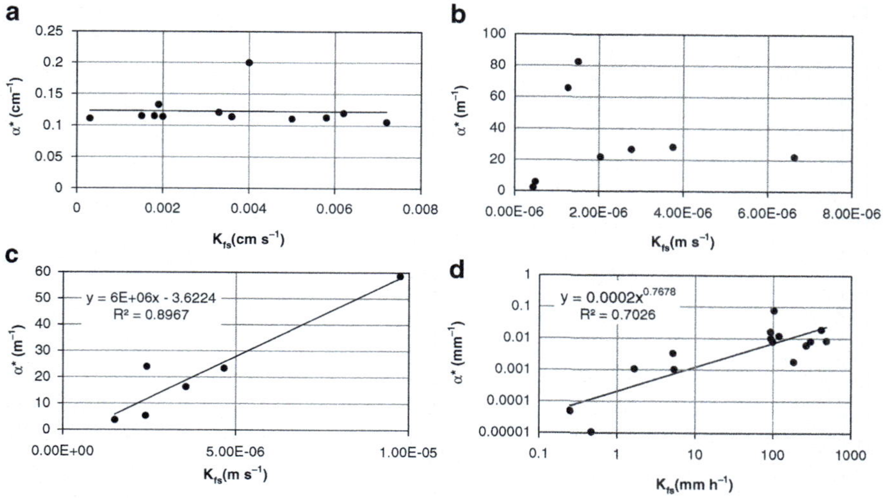

Fig. 2.35 Relationship between the α* parameter and the field-saturated soil hydraulic conductivity, K_{fs}, obtained by using (**a**) the Guelph permeameter (GP) data published by Paige and Hillel (1993) in their Tables 1 and 2; (**b**) the GP data listed by Reynolds and Elrick (1985b) in their Table 3 (well radius = 0.02 m); (**c**) the GP data listed by Reynolds and Elrick (1985b) in their Table 4 (well radius = 0.03 m); and (**d**) the single-ring pressure infiltrometer data obtained by Bagarello et al. (2014c) in several Sicilian soils (From Bagarello et al. 2014a, reprinted with permission)

two variables obtained with the Guelph permeameter data published by Paige and Hillel (1993, their Tables 1 and 2). However, low α* values are typical of soils where the proportion of steady flow due to capillarity is relatively high whereas high α* values occur in soils where this proportion is relatively small. The relative importance of the two flow components is expected to depend on soil texture and structure (Elrick and Reynolds 1992a). Therefore, a conceptual link between α* and K_{fs} can be expected because K_{fs} also depends on soil textural and structural characteristics. According to Reynolds and Elrick (1990), α* increases monotonically from 0.001 to 0.036 mm^{-1} as K_{fs} increases from 1×10^{-6} (clay cap/liner) to 1×10^{-1} mm s^{-1} (sand soil). This relationship appears logical from a physical point of view. Soils with a low capillarity (high α*) include "coarse and gravelly sands, may also include some highly structured soils with large and/or numerous cracks and biopores" (e.g. Table 1 by Reynolds 2010 and Table 2.1). In addition, a high α* value corresponds to an initially steep $K(h)$ relationship (Reynolds 1994), that is a signal of the fact that a small decrease in pressure head is enough to determine a noticeable pore emptying. This phenomenon can only occur if the pores are large or relatively large. Therefore, soils with high α* values have high K_{fs} values, as clearly stated by White and Sully (1992). On the other hand, soils with a high capillarity (low α*) include "porous materials that are both fine textured and massive; unstructured clayey and silty soils, as well as very fine to fine structureless sandy materials". In addition, a small α* value corresponds to an initially flat $K(h)$ relationship, that is a signal of a less appreciable pore emptying as the pressure head decreases. This can occur when pores are small. Therefore, small α^* values are associated with fine textured, fine structured or compacted soils (Reynolds 2011) that are expected to have small K_{fs} values. Another support to this reasoning can be found in Yitayew et al. (1998), who stated that higher λ_c (=1/α^*) values indicate heavier soils with lower hydraulic conductivity and vice versa. Data suggesting an increase of α^* with K_{fs} can also be found in the literature. For example, using the valid K_{fs} and ϕ_m calculations obtained by Reynolds and Elrick (1985b, Table 3, well radius = 0.02 m), a clear relationship between the two variables was not detected (Fig. 2.35b). However, $\alpha^* \leq 5.5$ m^{-1} was obtained for low K_{fs} values ($\leq 4.92 \times 10^{-7}$ m s^{-1}) whereas α^* values varying between 22 and 82 m^{-1} were obtained for high K_{fs} values ($\geq 1.26 \times 10^{-6}$ m s^{-1}). Using the data published in Table 4 of the above mentioned article (well radius = 0.03 m, Fig. 2.35c), an increasing relationship between the two variables was found. Moreover, figure 2 by Reynolds et al. (1992) shows a plot of α^* versus $\log_{10} K_{fs}$ for four different soils. According to these authors, α^* was essentially constant for a structureless loamy-sand soil, and it increased, mildly or substantially (in a single case), with K_{fs} for the other soils. Therefore, an increasing relationship was the most common result in that investigation. Finally, Fig. 2.35d shows the α^* vs. K_{fs} relationship obtained by considering single-ring pressure infiltrometer data collected in several Sicilian soils (Bagarello et al. 2014c). The fitted line suggests an increasing relationship between the two variables. Perhaps, a more scientifically exhaustive assessment of α^* should be carried in the near future, also considering that this parameter (i) seems to be directly comparable with parameters of the water retention curve (i.e., Mubarak et al. 2010), and (ii) has a noticeable practical interest since it was included in many other equations allowing rapid calculations of K_{fs} (Elrick and Reynolds 1992a; Reynolds and Elrick 1990; Bagarello et al. 2004; Nimmo et al. 2009; Wu et al. 1999).

The choice by Bagarello et al. (2013b, 2014b) to use K_{fsB} as a benchmark was due to the fact that comparing K_{fs} measurement methods is uncertain (Reynolds et al. 2000) and K_{fsB} was determined on the basis of the same experimental information used to derive K_{fsS}. Obviously, developing more confidence on the SBI method implies additional investigations that should be carried out with other datasets and also with independent measurements of both α^* and K_{fs} from different soils. A point that should be clarified is the reason why, for a given b_1 value, gravity was predicted to be relatively more important than capillarity (higher α^*) for the tropical soils. The experimental procedure used to collect a set of (t, I) data at a given sampling point also needs testing and maybe improvements. The reason is that the SBI method theoretically assumes that a null pressure head is steadily maintained on the infiltration surface of a rigid porous medium. Pouring water when the previously applied amount had completely infiltrated may promote air entrapment phenomena in the sampled soil volume and may also favor soil structure alteration phenomena at the infiltration surface. Therefore, the impact of the suggested procedure, that has the obvious advantage of being very simple, on the soil hydraulic characterization should specifically be taken into account. With this aim, comparisons of infiltration runs carried out by steadily maintaining a very small (i.e., close to zero) depth of water on the infiltration surface could be developed. Another factor of possible developments is the choice of the β and γ values. The reason is that $0 < \beta < 2$ and $0.6 < \gamma < 0.8$ are the feasible ranges of these two constants and their calibration as a function of the soil type has been suggested to potentially improve the estimates of the soil hydraulic properties with the BEST procedure (Nasta et al. 2012). The SBI method is cheap and rapid in terms of both the devices that have to be transported and the measurements that have to be carried out in the field. Therefore, it is a good candidate method for intensively sampling an area of interest with a practically sustainable experimental effort and, hence, it could practically simplify interpretation and simulation of soil hydrological processes, such as runoff generation.

Example 2.5

Infiltration data collected at a Burundian sampling point were analyzed according to the SBI method. A volume of water, V_w, of 150 mL was applied 18 times on an infiltration surface delimited by a ring having a radius $r = 75$ mm. The cumulative infiltration vs. time data are reported in Table 2.11. For this infiltration run, the original BEST procedure by Lassabatere et al. (2006) yielded an estimate of K_{fs} of 0.112 mm s^{-1}. Table 2.11 also lists the $\sqrt{t}$ and $I/\sqrt{t}$ values and Fig. 2.36 shows the linearized cumulative infiltration curve. The b_1 value, estimated by linear regression analysis of all $(I/\sqrt{t}, \sqrt{t})$ data pairs, was equal to 0.1706 mm s^{-1}. Equation (2.102a) was applied to obtain the following estimate of α^* (mm^{-1}):

Table 2.11 Cumulative infiltration, I, and associated time, t, defining the experimentally determined infiltration curve of the Example 2.5, and corresponding $\sqrt{t}$ and $I/\sqrt{t}$ values

No.	I (mm)	t (s)	$t^{0.5}$ ($s^{0.5}$)	$I/t^{0.5}$ (mm/s$^{0.5}$)
1	8.5	8	2.828	3.001
2	17.0	25	5.000	3.395
3	25.5	48	6.928	3.676
4	33.9	73	8.544	3.974
5	42.4	103	10.149	4.182
6	50.9	130	11.402	4.467
7	59.4	162	12.728	4.668
8	67.9	197	14.036	4.838
9	76.4	229	15.133	5.048
10	84.9	262	16.186	5.244
11	93.4	297	17.234	5.418
12	101.9	330	18.166	5.607
13	110.3	365	19.105	5.776
14	118.8	402	20.050	5.927
15	127.3	439	20.952	6.077
16	135.8	475	21.794	6.231
17	144.3	512	22.627	6.377
18	152.8	549	23.431	6.521

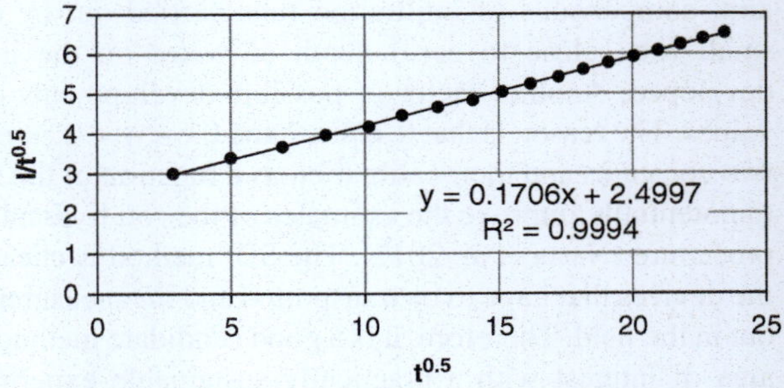

Fig. 2.36 Values of the ratio between the cumulative infiltration, I (mm), and the square root of time, t (s), plotted against the square root of t for the data of the Example 2.5

$$\alpha^* = 0.0262 + 0.0035 \times \ln(0.1706) = 0.020$$

Therefore, the following estimate of K_{fs} (mm s^{-1}) was obtained by Eq. (2.101):

$$K_{fs} = \frac{0.1706}{\frac{0.75 \times 1.818}{75 \times 0.020} + \frac{2 - 0.6}{3}} = 0.124$$

The percentage difference between the two estimates of K_{fs} (BEST, SBI method) was equal to 11.0 %.

2.2.2.4 Steady-State Falling-Head Method

Infiltration under steady-state falling-head conditions can provide information from which both K_{fs} and ϕ_m can be obtained (Elrick et al. 1995). Using this approach, particularly suited for permeable porous media, infiltration is initially allowed to come to steady-state under constant ponded head conditions and then the head, H, is allowed to fall. Values of K_{fs} and ϕ_m are calculated by fitting to the data the following relationship, that was developed by extending the steady flow equation by Reynolds and Elrick (1990) to transient falling-head conditions:

$$H(t) = \left(H_0 + \frac{\phi_m}{K_{fs}} + \pi r G\right)\exp\left(-\frac{rK_{fs}t}{XG}\right) - \frac{\phi_m}{K_{fs}} - \pi r G \tag{2.103}$$

where t (T) is the time, H_0 (L) is the ponded head at $t = 0$, G is the dimensionless shape factor by Reynolds and Elrick (1990), and X (L^2) is the cross-sectional area of the infiltrometer's cylindrical water reservoir. The iterative fitting procedure requires an initial guess value for each parameter, and measurement errors coupled with bad first guesses can give incorrect best fit values. Elrick et al. (1995) suggested that the slope, m, of the early data of ln[$H(t)$] versus t gives a guess value of K_{fs}:

$$K_{fs} = \frac{mXG}{r} \tag{2.104}$$

Equation (2.104) was derived from the pressure head component of falling-head infiltration and it is expected to provide a sufficiently accurate initial guess of K_{fs}, provided that m is obtained when the perturbing effects of soil capillarity and gravity, i.e. the second and third term, respectively, on the right hand side of Eq. (2.103), are minimal. Elrick et al. (1995) established a field comparison between the Two Ponding Depth (TPD) approach and the steady-state falling-head analysis by sampling a silty-clay soil at 12 locations. Initially, steady constant-head infiltration was obtained at two successively ponded heads (0.11 and 0.34 m). Then the permeameter reservoir was quickly refilled while maintaining ponded flow and the filler hole left open to the atmosphere so that the reservoir then produced falling head infiltration conditions. The choice of $t=0$ for the falling head event was arbitrary, and H_0 ranged from 1.30 to 1.61 m, depending on the measurement point. The maximum duration of the falling-head event was 11 min. The correlation between the two sets of K_{fs} data (TPD approach, steady-state falling-head analysis) was excellent (coefficient of correlation, $R=0.95$), producing coefficients of variation that differed by only 12 % and site averages that differed by only 1.5 %. The correlation between the ϕ_m values was lower ($R=0.65$), but the site averages still fell within 0.4 %. The α^* values did not correspond particularly well. Probably, other field tests of the steady-state falling-head procedure could yield more information on the practical usability of this method.

2.2.2.5 Bottomless Bucket Method

The Bottomless Bucket (BB) method by Nimmo et al. (2009) is another falling-head single-ring infiltration method, developed with the specific purpose of allowing characterization of large areas with limited resources in terms of time, equipment and personnel. Typical implementation is with a straight-walled or a bucket-sized infiltration ring, a small shovel, and a few litres of water. The ring is inserted into the soil surface to a short depth. Any remaining loose soil along the outside bucket edge is packed, if necessary, by adding small amounts of bentonite to prevent lateral leakage. A rubber mat is laid within the bucket to minimize surface disturbance during the initial stage of the infiltration run. A preselected amount of water is then poured within the ring to establish an initial ponding depth of 0.03–0.1 m and the mat is removed. The time from the start of pouring until the last patch of bulk liquid water within the bucket infiltrated the soil surface is measured. The following relationship is used to estimate K_{fs} on the basis of the measured ponded depths:

$$K_{fs} = \frac{L_G}{t} \ln\left(\frac{L_G + \lambda_c + H_0}{L_G + \lambda_c + H(t)}\right) \tag{2.105}$$

where L_G (L) is the so-called ring installation scaling length, t (T) is the time, λ_c (L) is the macroscopic capillary length of the soil (White and Sully 1987), H_0 (L) is the initially established ponded depth of water and $H(t)$ (L) is the ponded depth of water at time t. The L_G length, based on Eq. (2.65), is given by:

$$L_G = 0.316\pi d + 0.184\pi r \tag{2.106}$$

where d (L) is the ring insertion depth and r (L) is the ring radius. Equation (2.105) can be written:

$$L_G \ln\left(\frac{L_G + \lambda_c + H_0}{L_G + \lambda_c + H(t)}\right) = K_{fs} t \tag{2.107}$$

from which it follows that, if a set of $H(t)$ data has been collected, the slope of a plot of the so-called effective infiltration length, i.e. the left-hand side of Eq. (2.107), vs. t should equal K_{fs}. Equation (2.107) can be applied whether or not the test is continued until no water remains in the ring, as long as both H_0 and $H(t)$ have been measured. Nimmo et al. (2009) suggested that a moderate sensitivity of the conductivity calculations to λ_c has to be expected. Therefore, it is enough to choose a value from one of the four broad categories based on soil textural and structural characteristics (Elrick and Reynolds 1992a), taking into account that $\lambda_c = 1/\alpha^*$. In particular, for most soils with significant structural development, a λ_c value of about 0.08 m would be suitable. For extremely coarse and gravelly soils, a value of 0.03 m may be better and, for fine-textured soils without macropores, the suggested λ_c value is 0.25 m. If the head is allowed to fall to zero, and does so at time t_f (T), Eq. (2.105) simplifies to:

$$K_{fs} = \frac{L_G}{t_f} \ln\left(1 + \frac{H_0}{L_G + \lambda_c}\right) \tag{2.108}$$

A few tests of Eq. (2.107) yielded data reasonably close to a straight line through the origin, after deviations early in each test (Nimmo et al. 2009). These authors also applied Eq. (2.108) to obtain an estimate of K_{fs} of alluvial fan deposits using 1–4 L of water and $\lambda_c = 0.08$ m. Nimmo et al. (2009) used a steel bucket with its bottom removed, 0.22 m high, that tapered from 0.26 m diameter at the top to 0.21 m at the bottom, inserted to a depth of 0.05 m into the soil. Mirus and Perkins (2012) applied the BB method with both straight-walled (0.10 and 0.20 m diameter) and tapered (0.14 m diameter) buckets. According to Nimmo et al. (2009), the use of commonly available buckets has the advantages of cheap and easy acquisition and it allows us to obtain additional buckets even near remote field sites. A tapered bucket has a non-uniform diameter. Therefore, the average diameter of the initially filled portion of the bucket has to be considered for the K_{fs} calculations, but the authors suggested that the associated error should be small and practically negligible.

A new volume of water is poured after infiltration of the previously applied volume and the infiltration process is monitored again. These falling-head tests, each giving an estimate of K_{fs}, are repeated until the rate of head decline is of sufficient regularity to support the assumption that field-saturated conditions have been reached (Mirus and Perkins 2012). These last authors applied the new water volume into the ring before the water level reached the soil surface. In this case, Eq. (2.105) was used considering the time interval between H_0 and $H(t) > 0$. An expected advantage of this approach is that possible air entrapment phenomena into the soil are avoided.

Equation (2.105) is based on the analysis developed by Reynolds and Elrick (1990) with reference to 3D, steady, ponded infiltration from within a single ring. Moreover, it can be shown that Eq. (2.105) can be obtained from Eq. (2.103), developed by Elrick et al. (1995) for a steady-state falling-head process, under the assumptions that the cross-sectional area of the infiltrometer's cylindrical water reservoir coincides with the infiltration surface and the dimensionless shape factor can be estimated with the approximate relationship of Reynolds and Elrick (1990), i.e. $G \approx G_e = 0.316(d/r) + 0.184$. As noted above, with the steady-state falling-head analysis, infiltration is initially allowed to come to steady-state under constant ponded head conditions and then the head is allowed to fall. Therefore, the theoretical analysis by Nimmo et al. (2009) should be considered strictly usable for a falling-head infiltration run following attainment of steady-state flow conditions. In other words, using Eq. (2.108) to determine K_{fs} with a single water application cannot be suggested because a transient infiltration process is analyzed by a solution valid for steady flow conditions. According to Nimmo et al. (2009), however, measuring H at various times and plotting the effective infiltration length against t allows us to ascertain which early data should be excluded. Nimmo et al. (2009) repeated their measurements with the ring in the same place and they found fairly steady K_{fs} values after about 0.05–0.10 m of water had been applied. Therefore, these authors suggested discarding the earliest calculated K_{fs} values changing significantly from one measured interval to the next.

Mirus and Perkins (2012) modified the BB method to measure field-saturated hydraulic conductivity of bedrock outcrops. In this case, instead of inserting the lower BB rim into the subsurface, the BB is sealed to the surface using a nontoxic, quick drying silicone gel. Gaps between the bottom ring of the bucket and the bedrock greater than 5 mm should be avoided to prevent failure of the seal. If necessary, the bottom of the bucket can be cut with a saw or a knife to accommodate any irregularities in the bedrock microtopography. According to Mirus and Perkins (2012), the lack of ring insertion into the bedrock eliminates the impact of d on L_G. Therefore, Eqs. (2.105) and (2.106) are combined to yield:

$$K_{fs} = \frac{0.578r}{t} \ln\left(\frac{0.578r + \lambda_c + H_0}{0.578r + \lambda_c + H(t)}\right) \quad (2.109)$$

Equation (2.109) implicitly assumes that the $G_e = f(d/r)$ relationship can be used to estimate the dimensionless shape factor when d is equal to zero, although it was developed with specific reference to 0.03 m $\leq d \leq$ 0.05 m. However, some support to the usability of this relationship for d close to zero can be found in literature (Reynolds and Elrick 2002b).

Obviously, factors such as site preparation, water application, ring size or water quality, affecting a typical PI run, can also influence data collection with the alternative approaches described in this section.

Example 2.6

The Bottomless Bucket (BB) method was applied using a ring of radius $r = 0.075$ m inserted on the soil surface to a depth $d = 0.05$ m, giving a ring installation scaling depth, calculated with Eq. (2.106), equal to:

$$L_G = 0.316\pi \times 0.05 + 0.184\pi \times 0.075 = 0.093 \text{ m}$$

A volume of water of 1767 mL was poured on the infiltration surface at time $t = 0$ to establish an initial water level, $H_0 = 0.1$ m. When the ponding depth of water, H, was of 0.02 m, the time was recorded and another volume of water of 1414 mL was poured in the ring to raise the water level to 0.1 m. The time for H to reach 0.02 m was noted and another equal volume of water was applied. This procedure was repeated until the rate of the water level decline approached a practically constant value. Water was applied ten times in this example (Table 2.12). An estimate of K_{fs} was obtained for each step of the infiltration run using the measured time from $H = 0.1$ m to $H = 0.02$ m and setting $\lambda_c = 0.083$ m in Eq. (2.105). For example, the following K_{fs} value was obtained for the first volume of water:

Table 2.12 Data for the Example 2.6 and field-saturated soil hydraulic conductivity, K_{fs}, values obtained with the Bottomless Bucket method

Pouring no.	Volume (mL)	Progressive time (s)	Partial time (s)	K_{fs} (m s^{-1})	ΔK_{fs} (%)
1	1767.1	32	32	9.93×10^{-4}	
2	1413.7	97	65	4.89×10^{-4}	−50.8
3	1413.7	192	95	3.35×10^{-4}	−31.6
4	1413.7	318	126	2.52×10^{-4}	−24.6
5	1413.7	463	145	2.19×10^{-4}	−13.1
6	1413.7	616	153	2.08×10^{-4}	−5.2
7	1413.7	783	167	1.90×10^{-4}	−8.4
8	1413.7	961	178	1.79×10^{-4}	−6.2
9	1413.7	1143	182	1.75×10^{-4}	−2.2
10	1413.7	1326	183	1.74×10^{-4}	−0.5

$$K_{fs} = \frac{0.093}{32} \ln\left(\frac{0.093 + 0.083 + 0.1}{0.093 + 0.083 + 0.02}\right) = 9.93 \times 10^{-4} \text{m s}^{-1}$$

The last two K_{fs} values differed by approximately 0.5 % (Table 2.12), suggesting that the duration of the run was appropriate for obtaining a reliable estimate of K_{fs}, equal to 1.74×10^{-4} m s^{-1} in this particular case.

2.2.3 *Double-Ring Infiltrometer*

2.2.3.1 Description and Operation of the Double-Ring Infiltrometer

The double- or concentric-ring infiltrometer, DRI, consists of an open-ended measuring, or inner, cylinder or ring placed concentrically inside an open-ended buffering, or outer, cylinder or ring (Fig. 2.37). The device is used to measure cumulative infiltration, I (L), and infiltration rate, $i = dI/dt$ (L T^{-1}), t (T) being the time, and to determine field-saturated soil hydraulic conductivity, K_{fs} (L T^{-1}). The measuring cylinder is usually about 0.10–0.20 m in diameter by 0.10–0.20 m long, while the buffering cylinder is generally about 0.50 m in diameter and it has the same length with the measuring cylinder (Reynolds et al. 2002b). The cylinders are inserted into the soil to a depth of 0.03–0.10 m. The run is carried out by establishing a constant depth of water on the soil surface, usually on the order of 0.05–0.20 m (Reynolds et al. 2002b). The same water height should be ponded in the measuring and buffering cylinders, although it is not necessary to measure infiltration through this last cylinder to determine K_{fs}. Infiltration into the soil is determined by monitoring flow rates through the measuring cylinder. Quasi-steady flow in the near-surface soil under the measuring cylinder is assumed when the flow rate becomes practically constant. The time required to reach quasi-steady flow, i.e. the equilibration time, generally increases with finer soil texture, decreasing soil structure, and increasing depth of water ponding, H (L), depth of cylinder insertion, d (L), and cylinder radius, r (L).

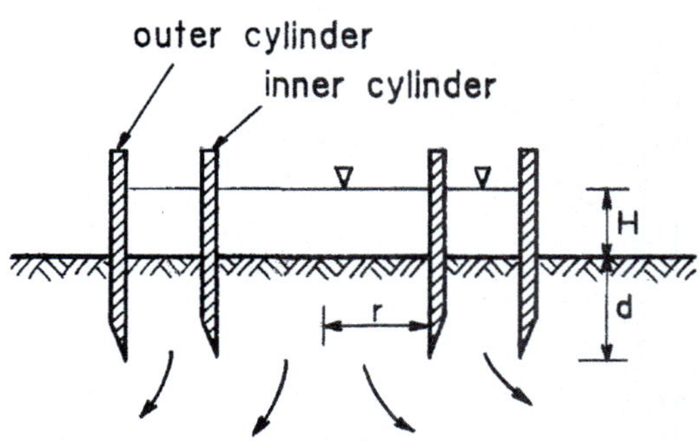

Fig. 2.37 Schematic view of the double- or concentric-ring infiltrometer

Lai and Ren (2007) used different DRIs, having inner cylinders of 0.20–1.20 m in diameter and outer cylinders of 0.70–1.40 m, depending on the device. The water level in the inner cylinder is maintained by a Mariotte tube made in polyvinyl chloride pipe, while the water level in the outer cylinder is adjusted manually to match that in the inner cylinder. Water is carefully added to the outer cylinder every few seconds, so that water level fluctuations are kept within 0.5 cm and have a negligible impact on the infiltration inside the inner cylinder. The flux in the inner cylinder is measured using a calibrated sight tube attached to the side of the Mariotte tube. One more cylinder with a base is nested and fixed inside the inner cylinder by four pins. This nested cylinder is about 0.20 m high, and its diameter is 0.03 m less than that of the inner cylinder. The nested cylinder is held several centimeters above the soil surface, so it does not affect the infiltration process. The operator fills in the inner cylinder while keeping water-free the nested cylinder. This last cylinder has two functions: (i) to reduce the upper free water area in the inner cylinder, improving measurement accuracy, and (ii) to minimize surface evaporative losses, especially for long-term infiltration. Lai and Ren (2007) suggested inserting the two concentric cylinders at least 0.05 m into the soil surface. To minimize the risk of altering the soil surface at the beginning of the infiltration process, water is carefully poured on the soil surface confined by the ring to a depth of approximately 0.05 m just before switching on the Mariotte tube to establish a constant 0.05 m head of water. Both the inner and outer cylinders are filled with water at the same time. The criterion used by Lai and Ren (2007) to establish attainment of steady-state flow conditions was that infiltration volumes measured at 5 min time intervals remained constant for a 30 min period.

In the DRI used by Frey et al. (2012) to assess the ability of the macropores to remain hydraulically active under saturated conditions, the inner and the outer cylinders, having a 0.30 and 0.55 m diameter, respectively, were hammered 0.05 m into the soil and a constant depth of water of 0.10 m was maintained in the inner and outer spaces of the infiltrometer by separate 40 L Mariotte bottles. Infiltration rates were measured by monitoring the water level in these bottles and quasi steady-state conditions were usually obtained in 2–3 h.

2.2.3.2 Calculation of Soil Parameters

The outer cylinder has the function of reducing the three-dimensional (3D) radial flow from the inner cylinder and hence establishing a practically one-dimensional (1D) infiltration process under the inner cylinder. Assuming that water infiltration below the inner cylinder is 1D and that steady flow only depends on gravity, the following assumption can be made (Reynolds et al. 2002b):

$$K_{fs} = i_s \tag{2.110}$$

where i_s (L T^{-1}) is the quasi-steady infiltration rate. Equation (2.110) is valid only if steady-state has been reached and the water pressure at the surface is close to zero.

However, neglecting the hydrostatic and capillarity components of total flow will overestimate K_{fs}. As pointed out by Reynolds et al. (2002b), the accuracy of Eq. (2.110) for determining K_{fs} increases with r, d and the α^* (L^{-1}) parameter (i.e., it is higher in coarse textured soils) and with decreasing H. Consequently, the cylinder diameter should be as large as possible and ponding depth as small as possible to use Eq. (2.110). However, Reynolds et al. (2002b) also showed that a radius of 2.6 m would be required to estimate K_{fs} within 5 % for $H = d = 0.05$ m and $\alpha^* = 12\ m^{-1}$. Large cylinders are not practical since they are difficult to install, highly consumptive of water, and may require an excessively long equilibration time. Large insertion depths also improve the accuracy of Eq. (2.110) but they are not suggested in practice since a large d value tends to increase the equilibration time and may cause excessive soil disturbance.

The analysis by Bouwer (1966, 1986) assumes one-dimensional vertical flow within and below the measuring cylinder and takes into account the hydrostatic pressure, capillarity and gravity components of infiltration:

$$\frac{i_s}{K_{fs}} = \frac{H}{L_f} + \frac{1}{\alpha^* L_f} + 1 \tag{2.111}$$

where L_f (L) is the distance from the infiltration surface to the wetting front, that has to be estimated or measured. The α^* parameter also must be estimated on the basis of the textural/structural soil characteristics, as suggested by Elrick and Reynolds (1992a) (e.g., Table 2.1), or measured independently. In any case, Eq. (2.111) does not account for the lateral flow divergence in the pressure and capillarity terms and, for this reason, it tends to overestimate K_{fs} under quasi-steady flow (Reynolds et al. 2002b).

As stated above, the DRI should physically prevent flow divergence under the measuring cylinder by the outer buffering cylinder. For this reason, the assumption of Eqs. (2.110) and (2.111) is that infiltration through the annular space between the buffering and measuring cylinders absorbs the flow divergence, leaving only vertical flow under the measuring cylinder. Wu et al. (1997), using an inner cylinder having a diameter of 0.20 m, showed that when the diameter of the outer cylinder was increased to 1.20 m, the measured infiltration rates were 20–33 % greater than the one-dimensional infiltration rates for their three test soils. As ring size increased, measurement error due to lateral flow decreased, and consequently the measured infiltration rates better approximated the one-dimensional vertical flow rates.

Reynolds et al. (2002b) concluded that the buffering cylinder is often not effective, with the quasi-steady infiltration rate from the measuring cylinder still being influenced by flow divergence. Consequently, these authors suggested estimating K_{fs} from DRI data by the following relationship, originally developed for the single-ring pressure infiltrometer technique (Reynolds and Elrick 1990):

$$K_{fs} = \frac{i_s}{\frac{H}{C_1 d + C_2 r} + \frac{1}{\alpha^*(C_1 d + C_2 r)} + 1} \tag{2.112}$$

where $C_1 = 0.316\pi$ and $C_2 = 0.184\pi$ are dimensionless constants that apply for $d \geq 0.03$ m and $H \geq 0.05$ m. The α^* parameter of Eq. (2.112) must be estimated or measured independently but lateral divergence of flow due to hydrostatic pressure and capillarity is accounted for implicitly in the $(C_1 d + C_2 r)$ term. An implication of the suggestion by Reynolds et al. (2002b) is that the single-ring infiltrometer technique should be used in practice instead of the more laborious DRI technique. In any case, a practical means to check flow divergence during a DRI run is to compare the infiltration rates measured in the inner and outer areas of the device. Theoretically, the two cumulative infiltration curves may differ appreciably with a 3D component for infiltration from the outer cylinder and 1D infiltration from the inner cylinder. Yet, in some circumstances, such differences may be insignificant. For example, when the soil has relatively high initial soil water content, field measurements should show small differences between buffered and unbuffered infiltration rates (Burgy and Luthin 1956).

Other DRI data analysis procedures can also be found in literature. For example, the capillarity component of total flow was neglected by Lai and Ren (2007) when a small depth of water was maintained on the surface of a soil having relatively high soil water content (>50 % of the saturated water content) at the time of the infiltration run. This last circumstance suggested a negligible lateral flow contribution to total infiltration. The Philip's (1957a) two term equation was fitted to the cumulative infiltration, I, vs. time, t, data collected by the DRI:

$$I = St^{0.5} + At \tag{2.113}$$

where S (L $T^{-1/2}$) is the soil sorptivity and A (L T^{-1}) is a constant. Taking into account that this constant represents the main part of the gravitational influence, $K_{fs} = A$ was assumed by Lai and Ren (2007). However, it should be noted that this assumption could only be made for very long times, i.e. when $t \rightarrow \infty$ (e.g., Reynolds 2010). For shorter times, A in Eq. (2.113) could have a lower value and theory suggests that $K_{fs}/3 < A < 2K_{fs}/3$ (Clothier and Scotter 2002; Verbist et al. 2010). Hinnell et al. (2009) set A at $0.5K_{fs}$ following Warrick (2003) and Gupta et al. (1993) suggested assuming $A = 2K_{fs}/3$ in the analysis of DRI data fitted to the Philip's (1957a) model.

Different methods of calculating K_{fs} from DRI data (Green and Ampt 1911; Horton 1939; Philip 1957a; Talsma and Parlange 1972; Brutsaert 1977; Swartzendruber 1987; Gupta et al. 1993; Wu et al. 1997; Reynolds et al. 2002b) were compared by Verbist et al. (2010) at three locations in Chile showing from moderate (15 %) to high (55 %) stoniness. The lowest and the highest means of K_{fs} for each calculation method, obtained on the basis of ten replicated runs, were separated by a factor of not more than four for each sampled plot. Therefore, the suggestion by the authors was that, if the order of magnitude of K_{fs} has to be determined for an area of interest, carrying out many infiltration runs is more important than attempting to find the most appropriate calculation method for the considered circumstance. This suggestion probably deserves additional consideration with testing on other soils, since an intensive soil sampling is a practical possibility whereas the choice of the calculation method is more or less subjective, due to the lack of a reference method to determine K_{fs} and the fact that any calculation method makes assumptions that are difficult or even impossible to check in the field.

2.2.3.3 Influence of Setup Design on DRI Results

Assessing the size dependency of measured hydraulic conductivity with the DRI is important for obtaining reliable data. With reference to seven sites, mostly established on a silt-loam soil, Lai and Ren (2007) measured K_{fs} with four devices having inner diameters varying from 0.20 m to 1.20 m. The mean hydraulic conductivity did not change significantly across the full range of inner cylinder diameters but the range and standard deviation of the measurements decreased appreciably with an increased size of the infiltrometer. According to Lai and Ren (2007), the stability of the mean was indicative of the fact that lateral flow did not cause significant differences between cylinder sizes. As this size increased, the representativeness of the area covered by the infiltrometer also increased, so that the measured conductivity became more representative and stable. In other words, the measurement scale effect was due to soil heterogeneity, confirming previous results (Rovey and Cherkauer 1995; Zhang 1997; Dirk et al. 1999).

The size dependency of the measured conductivity with the DRI due to soil heterogeneity was also investigated by Lai and Ren (2007) using numerically simulated data. In particular, inner cylinders having a diameter, d_i, varying from 0.10 to 2.0 m were considered, and K_{fs} was treated as a realization of a stationary, second-order, spatially random distributed field with correlation length, L ($0 \leq L \leq 2.0$ m), and standard deviation, SD ($0 \leq SD \leq 1.0$). The greatest scatter in the calculated K_{fs} values was found for small cylinders and large SD values. Large cylinders showed little scatter regardless of SD, and media with a small standard deviation showed little scatter regardless of cylinder size. For a homogeneous hydraulic conductivity field ($SD = 0$), K_{fs} was essentially constant across all cylinder diameters suggesting that, in almost homogeneous media, K_{fs} should not be expected to change appreciably with the measurement scale. The effect of L was similar to that of SD. Even a small ring allowed the authors to sample the full range of K_{fs} values when the correlation length was small, but a larger ring was necessary to integrate across greater correlation lengths. The practical conclusion by Lai and Ren (2007) was that a large ring, i.e. $d_i > 0.80$ m, is required to reliably quantify the soil hydraulic conductivity. Lai et al. (2010) investigated the effect of the outer cylinder diameter, d_o, and in particular the effect of the buffer index, $b = (d_o - d_i)/d_o$ ($0.2 \leq b \leq 0.71$) on the predictions of K_{fs}. When the soil heterogeneity was great (large L or SD), there was a limited effect on increasing the stability of the K_{fs}

measurements by only increasing the buffer index. A more stable and reliable measurement of K_{fs} was obtained by increasing the inner cylinder size than that of the outer cylinder. In practice, the inner cylinder size is more important than the buffer index to obtain a reasonably stable and representative measurement by DRI. An infiltrometer with $d_i \geq 0.80$ m and $b \geq 0.33$ (i.e. $d_o = 1.20$ m for $d_i = 0.80$ m) represents an efficient method for improving measurement accuracy and representativeness. More recently, Lai et al. (2012) suggested that the insertion depth of the inner cylinder plays a more important role than the insertion depth of the outer cylinder in K_{fs} measurement by DRI, and also that increasing the insertion depth of the inner cylinder could improve the measurement accuracy of K_{fs}. In practice, an insertion depth of 0.05–0.15 m, which has been adopted in most ring infiltrometer related field experiments, appears acceptable and it was recommended for K_{fs} measurement.

The majority of landscapes, natural or cultivated, are non-level (slopes > 1 %) but the DRI, such as most of the other infiltrometer techniques, is designed to apply on horizontal surfaces. Locally nearly flat areas can probably be found in the field when small cylinders are used for the run, but this last choice can have negative impacts on the representativeness of the individual measurement due to soil heterogeneity. The problem of non-horizontality of the sampled area is particularly important with the DRI since the total sampled area cannot be too small given that two concentric cylinders have to be installed at a sampling point. When the device is used on a sloping surface, the pressure head varies across the sampled area with the highest value at the downslope side and the lowest value at the upslope side. Therefore, the infiltration rate is lower at the upslope side than at the downslope side. Despite this, Bodhinayake et al. (2004) showed that, for a more or less homogeneous silt loam soil, the lower infiltration rate at the upslope side was offset by the higher infiltration rate at the downslope side, resulting in no substantial differences in K_{fs} among slopes with a 0–20 % gradient. In that investigation, K_{fs} was calculated by Eq. (2.112) and the α^* parameter was estimated by fitting the Wooding's (1968) equation to the steady-state infiltration rate vs. water pressure head, h ($-22 \leq h \leq -3$ cm), data collected by a tension infiltrometer. Bodhinayake et al. (2004) also reported that, according to the theoretical study by Philip (1991), downslope flow in sloping lands occurs as a result of the downslope component of gravity and, for homogenous and isotropic soils under constant flux boundary condition, the infiltration normal to the slope is expected to differ relatively little from infiltration on a horizontal surface for slope gradients not exceeding 58 %. Therefore, their conclusion was that the DRI is suitable for characterizing soil hydraulic properties in lands with slopes up to 20 %.

2.2.3.4 Sources of Errors in the DRI Results

Physical sources of error in the application of DRI are those generally expected when a ponded depth of water is established on an infiltration surface confined by a ring. In particular, soil compaction during ring insertion, short circuit flow along the cylinder walls, siltation of the infiltration surface, and gradual plugging of soil pores by deflocculated silt and clay particles can occur (Reynolds et al. 2002b). These errors can be prevented or reduced by (i) using small insertion depths and thin-walled cylinders with sharp cutting edges; (ii) tamping the soil adjacent to the walls and/or backfilling gaps between the soil and wall with powdered bentonite or fine clay; (iii) using diffuser devices to reduce surface soil disturbance during water application, and (iv) using water having major cation concentrations similar to those of the resident soil water or local tap water. Distilled or deionized water should not be used if deflocculation has to be avoided.

Another source of variability affecting steady-state infiltration rates measured with infiltrometers and permeameters is water viscosity, although this factor is seldom considered to interpret data. In a recent investigation carried out with specific reference to the DRI (Clancy and Alba 2011), i_s increased with initial infiltration water temperature according to a soil texture-dependent relationship. For a loamy-sand soil, changes in i_s with temperature were fully accounted for by viscosity changes. For a sandy soil, however, the effect of temperature on i_s was higher than predicted by viscosity. This phenomenon, called the *T effect* (Lin et al. 2003), appears to be linked to the fact that the pore spaces are not completely saturated because some air remains entrapped in the bulk soil under ponding conditions. Therefore, the viscosity of the total fluid system, i.e. air and water, within the soil matrix affects i_s. Clancy and Alba (2011) also suggested that interpreting the variability of the i_s data collected with the DRI could be improved by measuring air, soil and water temperatures.

2.2.4 Simplified Falling Head Technique

2.2.4.1 Principles and Application of the Technique

The Simplified Falling Head (SFH) technique by Bagarello et al. (2004) allows us to determine the field-saturated hydraulic conductivity, K_{fs} (L T^{-1}), of an initially unsaturated soil by a one-dimensional falling head infiltration process. The technique is based on the following relationship which describes one-dimensional cumulative infiltration, I (L), for falling head conditions using the Green and Ampt (1911) approach (Philip 1992):

$$t = \frac{\Delta\theta}{K_{fs}(1-\Delta\theta)}\left[\frac{I(t)}{\Delta\theta} - \frac{H_0 - h_f}{1-\Delta\theta}\ln\left(1 + \frac{(1-\Delta\theta)I(t)}{\Delta\theta(H_0 - h_f)}\right)\right] \quad (2.114)$$

where t (T) is the time, $\Delta\theta$ (L^{3}L^{-3}) is the difference between the field-saturated (θ_{fs}) and the initial (θ_i) volumetric soil water content, H_0 (L) is the height of the ponded head at $t = 0$, and h_f (L) is the soil water pressure head at the wetting front, with h_f negative. It may be born in mind that $I = H_0 - H$, where H (L) corresponds to the head of water into the ring. Equation (2.114) includes gravity and it is valid until the falling head drops to zero. The h_f term can be replaced by the so-called α* (L^{-1}) parameter by using Eq. (2.13) (Elrick et al. 2002). Equation (2.114) can then be written in the following form:

$$t = \frac{\Delta\theta}{K_{fs}(1-\Delta\theta)}\left[\frac{I(t)}{\Delta\theta} - \frac{H_0 + \frac{1}{\alpha^*}}{1-\Delta\theta}\ln\left(1 + \frac{(1-\Delta\theta)I(t)}{\Delta\theta\left(H_0 + \frac{1}{\alpha^*}\right)}\right)\right] \tag{2.115}$$

The SFH technique consists of applying quickly a small volume of water, V (L^3), on the soil surface confined by a ring inserted at a fixed distance, d (L), into the soil and in measuring the time, t_a (T), from the application of water to the instant at which the infiltration surface, having a cross-sectional area equal to A (L^2), is no longer covered by water. At $t = t_a$, $I(t_a) = H_0 = V/A$, i.e. the depth of water corresponding to V. Therefore, K_{fs} can be calculated from Eq. (2.115) as:

$$K_{fs} = \frac{\Delta\theta}{(1-\Delta\theta)t_a}\left[\frac{H_0}{\Delta\theta} - \frac{\left(H_0 + \frac{1}{\alpha^*}\right)}{1-\Delta\theta}\ln\left(1 + \frac{(1-\Delta\theta)H_0}{\Delta\theta\left(H_0 + \frac{1}{\alpha^*}\right)}\right)\right] \tag{2.116}$$

Estimation of K_{fs} by Eq. (2.116) also requires the measurement of Δθ and the estimation of α*, which can be carried out on the basis of the textural/structural soil characteristics (Table 2.1) (Elrick and Reynolds 1992a; Reynolds and Lewis 2012). Taking into account that Eq. (2.116) applies to one-dimensional flow, the wetting front should not emerge from the bottom of the ring, since three-dimensional flow commences in this case. In practice, a volume of water, V, less than or equal to the volume of voids, V_p (L^3), within the bulk soil volume confined by the ring, V_c (L^3), has to be used:

$$V \le V_p = V_c\ \Delta\theta = d\ A\ \Delta\theta \tag{2.117}$$

The SFH run begins with the exposition of the surface area to be sampled, that could only imply removal of the residuals or the surface vegetation while the roots remain in situ (Fig. 2.38) (Bagarello and Iovino 2010). A stainless steel cylinder with a thin wall and a sharp cutting edge is then inserted into the soil. The cylinder is at least 0.20 m high and it has an appropriate diameter in order to sample a representative surface area with an individual measurement (e.g., 0.15–0.30 m). The cylinder's insertion can be carried out by using a mallet and a wood tablet placed on the top of the cylinder, controlling repeatedly its verticality by a level. If the contact between the inner wall of the cylinder and the soil appears locally poor, it is advisable to at least gently press the soil along the edge of the sampled surface with the non-sharp end of a pencil. In some cases, such as investigations on the temporal variability of K_{fs} in natural and non-tilled soils, the cylinder remains installed into the soil until all sets of measurements have been collected (Bagarello and Sgroi 2004). The contact between the sampled soil volume and the cylinder has obviously to be checked before each run. Water should be poured very quickly on the infiltration surface

Fig. 2.38 Steps of the SFH run: (**a**) exposition of the infiltration surface; (**b**) insertion of the cylinder into the soil; (**c**) check of the contact between the sampled soil volume and the inner walls of the cylinder; (**d**) pouring of the prescribed water volume; (**e**) intermediate stage of the falling head infiltration process; (**f**) infiltration surface at the end of the run (From Bagarello and Iovino 2010)

because the theoretical analysis assumes instantaneous establishment of the corresponding depth of ponding. However, water also needs to be applied with caution to prevent alteration of the soil structure close to the infiltration surface. In practice, the water volume can be poured in approximately 5 s, possibly breaking the applied volume on a plastic or metal net that is temporarily held close to the soil surface. The time, t_a, from the beginning of the water application to the instant at which the surface area is no longer covered by water, which is used to calculate K_{fs}

by Eq. (2.116), is expected to vary from <5 min in highly permeable soils ($K_{fs} > 360$ mm h^{-1}) to more than 3.5 h in low permeable soils ($K_{fs} < 0.04$ mm h^{-1}). Small uncertainties in the measurement of t_a can occur since water application cannot be really instantaneous and disappearance of free water from the soil surface, denoting the end of the run, may not occur uniformly due to the presence of small irregularities on the sampled area. However, Bagarello et al. (2004) suggested that these uncertainties do not influence appreciably the calculation of K_{fs}. Moreover, small errors (i.e., $\pm 10\,\%$ of the true value) in the determination of H_0 and the soil water content have a practically negligible effect on the calculated conductivity.

The SFH technique is based on a simple and generally rapid experiment in the field, and it requires standard and easily usable laboratory equipment, such as an oven and a balance. A large ring (e.g. 0.30 m diam.) and a small volume of water (<1 L) can be used so that a representative elementary volume can be sampled. Vertical variations of K_{fs} can also be assessed with a high resolution, by carrying out infiltration tests at closely spaced depths, given that a small depth of soil is sampled. Especially in moderately permeable soils, many experiments can be conducted simultaneously by a single operator. Therefore, the SFH technique appears suitable for intensively sampling large areas in relatively short periods of time. As an example, Bagarello et al. (2010a) concluded the field work necessary to measure K_{fs} at 350 sampling points established on an approximately 100 m^2 area in 1 week. This investigation, carried out on a clay soil, yielded K_{fs} values ranging from slightly less than 1 mm h^{-1} to more than 5000 mm h^{-1}. Therefore, the SFH technique seems suitable for measuring in the field the saturated conductivity of both the soil matrix only and the soil with macropores and cracks.

The SFH technique has been applied to monitor temporal changes in K_{fs} at the surface of a sandy-loam soil (Bagarello and Sgroi 2007), to characterize a clay soil at the plot scale (Bagarello et al. 2010a, 2013c), and to establish a comparison between pasture and forest soils (Agnese et al. 2011). As an example, the possibility offered by the SFH technique to make an intensive sampling of K_{fs} with a relative simplicity allowed Bagarello et al. (2013c) to experimentally test the sample size influence on the relationship between K_{fs} and the initial water content of a clay soil. With reference to the complete dataset, including $N = 528$ values for K_{fs}, the range of K_{fs} did not vary appreciably with θ_i (Fig. 2.39a). However, a low θ_i value suggested a high probability to obtain high K_{fs} results, whereas a high θ_i value suggested an increased probability to obtain low K_{fs} data. Extracting a sub-sample ($\ll 528$) of K_{fs} values from the complete dataset modified the information provided by the data. Figures 2.39b, c show two possible K_{fs} vs. θ_i relationships identifiable with a reduced, but not small ($N = 90$), sample size for a sampling covering in practice the entire range of θ_i values. Both relationships were defensible on the basis of the existing literature (e.g., Reynolds and Zebchuk 1996; Das Gupta et al. 2006), but neither was appropriate to capture the effect of θ_i on K_{fs} detected at the field site with a more intensive sampling. Experimental investigations of this

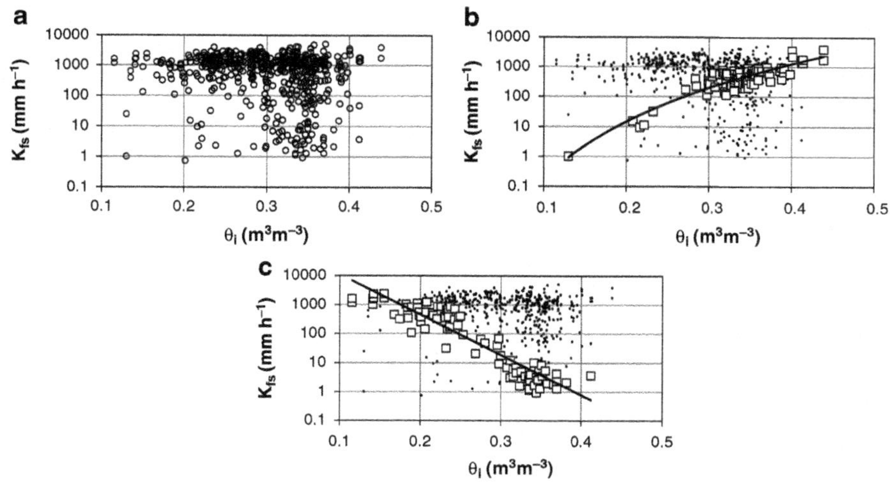

Fig. 2.39 Field-saturated soil hydraulic conductivity, K_{fs}, vs. initial volumetric soil water content, θ_i, measured at the clay soil of the Sparacia site, in Sicily: (**a**) complete dataset (sample size, $N = 528$); (**b**) example of a K_{fs} vs. θ_i relationship detectable with a smaller sample size ($N = 90$); (**c**) example of another K_{fs} vs. θ_i relationship detectable with $N = 90$ (From Bagarello et al. 2013c, reprinted with permission)

type, or even more intensive, could help to better establish how to sample soil for interpreting and simulating hydrological processes. The availability of simple methods to determine K_{fs} makes these investigations feasible in practical terms even with limited resources. Other investigations making use of the SFH technique have been carried out by Azam (2005), Shivakoti (2005), Rex and Dubé (2006), Waugh et al. (2006), Azam et al. (2008), Rojas et al. (2008), and Keller et al. (2012). The SFH technique has also been applied to determine the field-saturated hydraulic conductivity of an initially unsaturated soil column in the laboratory (Bagarello and Sgroi 2008).

Some of the above mentioned investigations also validated the applicability of the SFH technique. Indeed, they proved that the estimations obtained with this technique lead to physically plausible relationships between K_{fs} and other soil properties. For example, Agnese et al. (2011) showed that the mean K_{fs} at their sampled sites increased with the mean effective porosity, defined as the porosity minus $\theta_{-0.05m}$, i.e. the soil water content at a pressure head of -0.05 m (Fig. 2.40a). These authors also found that K_{fs} increased with the soil organic matter content (Fig. 2.40b) and decreased with the dry soil bulk density, ρ_b (Fig. 2.40c). A similar relationship between K_{fs} and ρ_b was obtained in an investigation carried out by Keller et al. (2012) in three Swedish fields. These last authors also concluded that the K_{fs} data obtained with the SFH technique can be used to explain crop yields.

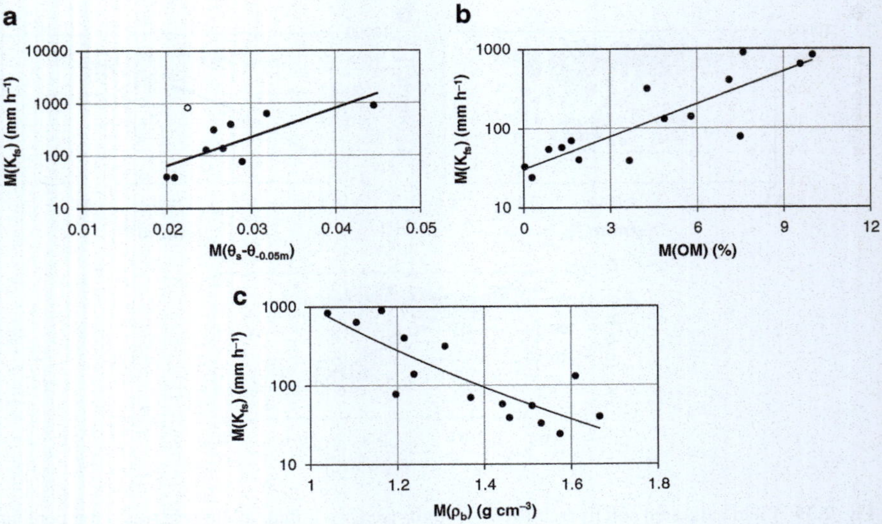

Fig. 2.40 Mean values of the field-saturated soil hydraulic conductivity, $M(K_{fs})$, obtained at different sites plotted against (**a**) mean effective porosity, $M(\theta_s - \theta_{-0.05\text{m}})$; (**b**) mean organic matter content, $M(OM)$; and (**c**) mean dry soil bulk density, $M(\rho_b)$. The *regression lines* are shown for each established relationship. The data point represented by an *open circle* in (**a**) was not included in the regression (Modified from Agnese et al. 2011, reprinted with permission)

2.2.4.2 Issues of Practical Interest

The choice of the insertion depth of the cylinder depends on both the thickness of the soil layer to be sampled and the fact that, according to Eq. (2.117), a short insertion depth implies the use of a small water volume for the run. Therefore, the order of magnitude of the H_0 term in Eq. (2.116) can be similar to the surface roughness, which makes the measurement of t_a uncertain. On the other hand, a deep insertion increases the risk of compacting or shattering the sampled soil volume. As a general suggestion, Bagarello et al. (2004) and Bagarello and Sgroi (2007) used $d = 0.12$ m without visually detecting the above mentioned problems. Possible alteration of a sandy-loam soil due to a relatively deep insertion of the cylinder was specifically tested by Bagarello et al. (2009b) by applying the SFH technique on differently isolated soil volumes. At 12 measurement points, a 0.15 m diam. cylinder was inserted to a depth of 0.15 m to enhance the risk of altering soil (Wuest 2005). Another 12 measurements were carried out on soil columns of the same size that were manually exposed and covered along their walls by a casing in polyurethane foam (Bagarello and Sgroi 2008). In particular, a soil column was exposed by digging a small trench (Fig. 2.41). A PVC cylinder (height = 0.25 m; diameter = 0.20 m) with opened ends was placed around the soil column and a stopper in polyurethane foam previously prepared in the laboratory was put on the surface of the column to prevent direct contact between the expanding foam and the

Fig. 2.41 Steps of the realization of a casing in polyurethane foam for applying the SFH technique: (**a**) exposition of the soil column; (**b**) placement of the PVC cylinder with opened extremities around the soil column; (**c**) placement of a stopper in polyurethane foam on the surface of the column; (**d**) filling of the space between the PVC cylinder and the soil column with polyurethane foam; (**e**) placement of a tablet and a small weight on the upper end of the PVC cylinder; and (**f**) detachment of the PVC cylinder from the casing (From Bagarello et al. 2009b)

upper end of the sampled soil volume. The 60–70 % of the space between the PVC cylinder and the soil column was filled with polyurethane foam and a tablet and a small weight of 1–2 kg were placed on the upper end of the PVC cylinder to confine foam expansion only partially. After the foam hardened, the PVC cylinder was detached along two previously realized cutting lines, and the stopper was removed to expose the soil surface for the SFH run. The mean K_{fs} values obtained on the two types of soil columns differed by a negligible and non-significant factor of 1.2. Taking into account that using polyurethane foam was thought to be appropriate to really minimize soil disturbance, Bagarello et al. (2009b) concluded that ring insertion did not affect the reliability of the measured conductivities. Despite this encouraging result, compaction or shattering phenomena during cylinder insertion (Reynolds 1993), preferential flow at the interface between the inner cylinder wall and the soil (Chappell and Ternan 1997), and blocking of non-vertical macropores near the wall edges (Shouse et al. 1994; Haws et al. 2004; Wuest 2005) cannot be excluded in general, due to the relatively deep insertion of the cylinder into the soil. The possible occurrence of these phenomena suggests that a large cylinder should yield more reliable K_{fs} data than a small cylinder. However, soil can also be expected to be more disturbed in the former case because more energy is required to insert a large cylinder than a small cylinder.

Attempts to better understand the performances of the SFH technique were carried out with reference to the size of the sampled soil volume, the usability of a literature estimate of α^*, and the information contained in a measurement of K_{fs} carried out with this transient technique. The effect of the soil volume sampled with an individual run on the K_{fs} determinations was tested in several investigations. For example, Sgroi (2005) compared the K_{fs} values obtained by sampling a sandy-loam soil with 0.15 and 0.30 m diam. cylinders randomly inserted to a depth of 0.12 m. Ten and 40 runs were carried out with the large and the small cylinders, respectively, to sample the same total area, equal to 0.71 m^2. The small cylinders yielded significantly higher and more variable results (mean of K_{fs}, $M = 1076$ mm h^{-1}; coefficient of variation, $CV = 43$ %) than the large cylinders ($M = 615$ mm h^{-1}; $CV = 25$ %), but the differences between the means were not substantial being equal to a factor of 1.7. Figure 2.42 shows that the lowest K_{fs} values did not differ appreciably between the two cylinder sizes, and an appreciable percentage of K_{fs} data obtained with the small cylinder was higher than the highest K_{fs} value measured with the large cylinder. A similar investigation was carried out by Bagarello et al. (2009b) in a silt-loam soil, where K_{fs} was determined by randomly sampling a total area of 0.71 m^2 with 40 cylinders of 0.15 m diam., 23 cylinders of 0.20 m diam., and 10 cylinders with a diameter of 0.30 m. The three cylinders yielded non significantly different mean values of K_{fs} (Table 2.13) and the

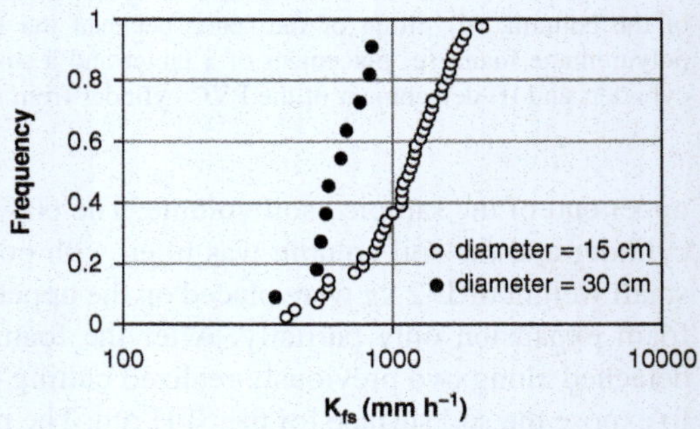

Fig. 2.42 Empirical cumulative frequency distribution of the field-saturated soil hydraulic conductivity values, K_{fs}, obtained by Sgroi (2005) with cylinders of two different diameters in a sandy-loam soil (Modified from Bagarello and Iovino 2010)

Table 2.13 Field-saturated soil hydraulic conductivity, K_{fs} (mm h^{-1}), obtained by Bagarello et al. (2009b) in a silt-loam soil with cylinders of different diameter

	Diameter of the cylinder (cm)		
Statistic	15	20	30
Minimum	1.7	2.2	16.6
Maximum	1648.7	1017.6	509.6
Mean	88.9	61.8	81.2
Coefficient of variation (%)	481.2	480.7	128.8

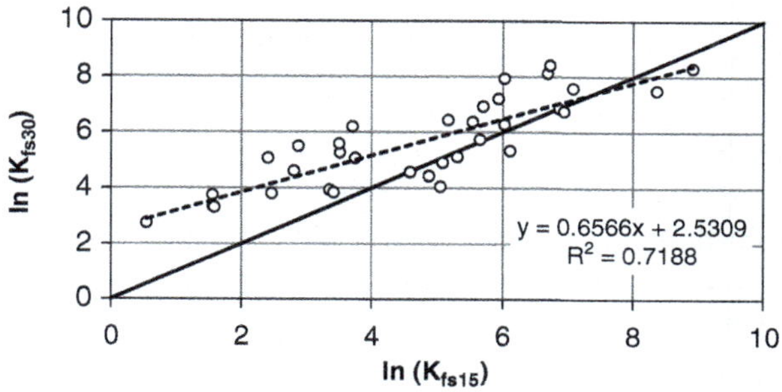

Fig. 2.43 Relationship between the field-saturated soil hydraulic conductivity measured with large (K_{fs30} in mm h^{-1}) and small (K_{fs15} in mm h^{-1}) diameter cylinders (0.30 and 0.15 m, respectively) at 34 Sicilian sites (From Bagarello et al. 2012, reprinted with permission)

difference between two cylinder sizes did not exceed a practically negligible factor of 1.4. However, the small cylinders (0.15 and 0.20 m diam.) yielded minimum and maximum values of K_{fs} lower and higher, respectively, than the large cylinders, which predicted the lowest variability of the individual K_{fs} data. Using relatively small cylinders probably implied a more appreciable impact of preferential flow phenomena and a non-representative sampling of soil macroporosity and the implications of these circumstances were particularly noticeable with reference to the estimate of the K_{fs} variability.

An investigation carried out at 34 Sicilian sites allowed Bagarello et al. (2012) to detect the existence of a statistically significant relationship between the field-saturated hydraulic conductivity measured with 0.30 m diam. (K_{fs30}, in mm h^{-1}) and 0.15 m diam. (K_{fs15}, in mm h^{-1}) cylinders (Fig. 2.43):

$$\ln\left(K_{fs30}\right) = 2.5309 + 0.6566\ \ln\left(K_{fs15}\right) \tag{2.118}$$

Equation (2.118) did not coincide with the identity line on the basis of the calculated 95 % confidence intervals for the intercept and the slope, and it suggested that $K_{fs30} > K_{fs15}$ should be expected for low K_{fs} values whereas more similar results occurs for high K_{fs} values. In particular, according to the fitted regression line, the discrepancy between K_{fs30} and K_{fs15} was substantial, i.e. by a factor of 10.4, for the lowest measured K_{fs15} value (~2 mm h^{-1}) but it was practically negligible, i.e. by a factor of 1.7, for the highest K_{fs15} values ($\geq$350 mm h^{-1}). The detected ring size effect was explained by considering that, in soils with a relatively low conductivity, macropores or other small zones with a locally high conductivity are rare. Therefore $K_{fs30} > K_{fs15}$ is the expected result because the wider surface sampled by the large cylinder implies a higher probability to intercept these zones. In soils with a relatively high conductivity, macropores or other high conductivity zones are more evenly distributed and even a small cylinder may yield a representative result, i.e. similar to the one obtained by a larger cylinder. Bagarello et al. (2012) also showed that the estimate of K_{fs30} can be improved if clay content, *cl* (%) and organic matter content, *OM* (%) are also available for the point sampled with the small cylinder. In particular, the following relationship was deduced on the same database of Sicilian soils:

$$\ln(K_{fs30}) = 0.793 + 0.450\ln(K_{fs15}) + 0.011cl + 0.076[OM \times \ln(K_{fs15}))] \tag{2.119}$$

According to Eq. (2.119), K_{fs30} increases with both *cl* and *OM* for a given K_{fs15} value, thus showing that the observed scale effects should be more noticeable in structured soils. Therefore, a large cylinder should be used with the SFH technique instead of a small cylinder, especially when the expected K_{fs} is relatively low. However, the developed equations establish that the measurements carried out with a small cylinder contain enough information to make an approximate prediction of the K_{fs} values that would be obtained at the same site with a larger cylinder.

Two adjacent plots of 4×11 m^2 established on the clay soil of Sparacia, in Sicily, were intensively sampled (sample size = 176 for a given plot) by the SFH technique with 0.15 m (Plot M, $K_{fs,15}$) and 0.30 m diam. (Plot N, $K_{fs,30}$) cylinders (Bagarello et al. 2013c). Plot N yielded 1.8 times higher and two times less variable K_{fs} values than plot M. The difference between $K_{fs,15}$ and $K_{fs,30}$ was not substantial according to the criteria defined by Elrick and Reynolds (1992a). The two datasets showed a similarity in terms of the highest measured conductivity, differing by a factor of 1.4, but also a noticeable difference with reference to the lowest value, differing by a factor of 55.6 (Fig. 2.44). Approximately one-fifth of the $K_{fs,15}$ data were lower than the lowest $K_{fs,30}$ value. Therefore, sampling a smaller portion of the plot (7.1 % and 28.3 % of the plot area with the small and the large cylinders, respectively) determined a wider range of K_{fs} results, since relatively low K_{fs} values were only measured with the small cylinder. Bagarello

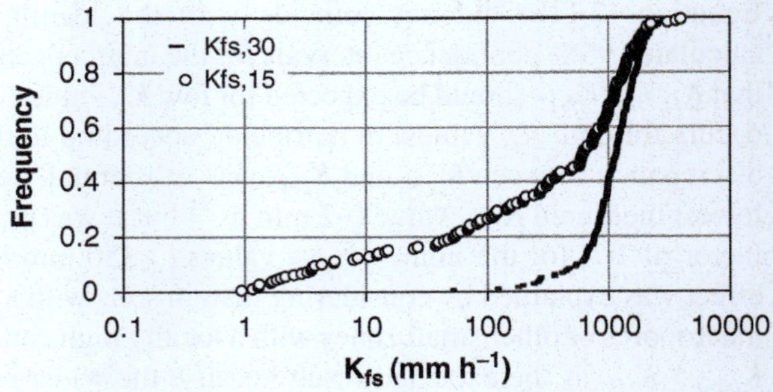

Fig. 2.44 Empirical cumulative frequency distribution of the field-saturated soil hydraulic conductivity data obtained with 0.15-m-diam. ($K_{fs,15}$) and 0.30-m-diam. ($K_{fs,30}$) cylinders (From Bagarello et al. 2013c, reprinted with permission)

et al. (2013c) suggested that the size effect was detected because the probability not to sample a representative elementary volume with an individual measurement was higher with the small cylinder than the large one. Taking into account that the upper end of the empirical cumulative frequency distribution of K_{fs} did not vary appreciably with the size of the cylinder, the investigation also suggested that a less noticeable size effect should be expected when the SFH technique is applied in highly conductive conditions. In conclusion, cylinders as large as practically possible should be used in general. This suggestion is particularly important to determine K_{fs} variability and for applications in slowly permeable soils.

Bagarello et al. (2012) investigated the usability of a literature estimate of the α* parameter by conducting two infiltration runs with the tension infiltrometer (TI) method and two SFH runs at each of 27 sites established within a Sicilian watershed. Equation (2.116) was used to calculate K_{fs} by using both a literature estimate of α* (K_{fs} denoted by the symbol K_{fsL}) and the α* value corresponding to the two highest pressure heads (−30 and −10 mm, i.e. close to saturation) established with the TI (K_{fsTI}). In particular, $\alpha^* = 4\ \text{m}^{-1}$ was used to calculate K_{fs} when the soil at the sampling point had a sand content, $sa < 20\ \%$. An α* value equal to $12\ \text{m}^{-1}$ was used for $20 \leq sa \leq 70\ \%$, and $\alpha^* = 36\ \text{m}^{-1}$ was chosen for soils with $sa > 70\ \%$. The K_{fsTI} values were moderately more variable than the K_{fsL} ones (coefficients of variation equal to 1.46 and 1.16, respectively) and the means ($100.5\ \text{mm h}^{-1}$ for K_{fsTI} and $92.3\ \text{mm h}^{-1}$ for K_{fsL}) differed by a not statistically significant, and practically negligible, 9 %. In addition, the correlation between the ln-transformed K_{fs} values was statistically significant and the corresponding regression line (Fig. 2.45) did not differ significantly from the identity line according to the calculated 95 % confidence intervals for the intercept and the slope. Therefore, the investigation suggested that using a field measurement of α* did not modify significantly the K_{fs} predictions as compared with the ones obtained by a simpler estimation of this parameter.

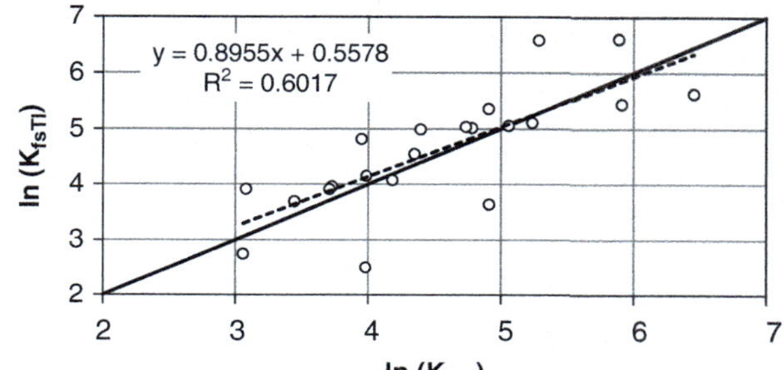

Fig. 2.45 Comparison between the field-saturated soil hydraulic conductivity values obtained by using a literature estimate of the α* parameter, K_{fsL} (mm h^{-1}), and a measurement of this parameter based on a tension infiltrometer experiment, K_{fsTI} (mm h^{-1}) (From Bagarello et al. 2012, reprinted with permission)

2.2.4.3 Comparison with Other Infiltration Techniques

An important point to be considered is what kind of information is contained in a measurement of K_{fs} carried out by the SFH technique. A practical means to go into this point is establishing comparisons with other K_{fs} measurement techniques, taking into account that such comparisons provide one of the few sources of information that practitioners can draw upon to select K_{fs} techniques that are appropriate for their circumstances (Reynolds et al. 2000). The single-ring pressure infiltrometer (PI) (Reynolds and Elrick 1990) and SFH techniques show similarities because both techniques are usable in initially unsaturated soil and sample similar soil volumes. With the PI, however, three-dimensional, steady, ponded flow out of the cylinder is used to estimate K_{fs} whereas a one-dimensional, transient infiltration process is considered by the SFH technique. Some comparisons between the SFH and PI techniques suggested a similarity of the K_{fs} values in relatively coarse textured soils (Bagarello et al. 2004; Bagarello and Sgroi 2007) and a tendency of the SFH technique to give higher results in finer soils, although a statistical similarity of the two techniques was also detected in this last case (Bagarello et al. 2012). The results obtained in the fine textured soil were explained by considering that the SFH runs were generally shorter than the PI ones and a longer run may promote short term swelling phenomena reducing macroporosity. This interpretation was indirectly supported by Bagarello et al. (2013c) because these authors obtained, in an intensively sampled clay soil by the SFH technique, a clear prevalence of high K_{fs} values, incompatible with runoff occurrence although runoff was frequently measured at the sampled field site. A more recent investigation (Bagarello et al. 2014c) confirmed that a short experiment with the SFH technique is expected to yield higher K_{fs} values than a long experiment with the PI technique. However, run duration did not explain all differences between the two techniques because relatively long SFH tests yielded higher K_{fs} values than shorter PI run, and differences between the two techniques were less substantial in soils with a low clay content. Soil disturbance at the surface of the sampled volume by the SFH run is expected to be minimal because a single, small water volume is gently poured on the infiltration surface from a small height (i.e., a few centimeters) but some PI devices, such as that used by Bagarello et al. (2014c) (Ciollaro and Lamaddalena 1998), may determine some turbulence close to the infiltration surface during the initial stages of the run, given that the water outlet diameter is relatively large (42 mm) in order to rapidly establish the ponding depth. Therefore, an additional factor determining the differences between the two techniques could be disturbance at the surface of the sampled soil (Arya et al. 1998; Dikinya et al. 2008; Bagarello et al. 2011b). In particular, Bagarello et al. (2014c) suggested that soil disturbance induced by the PI was more noticeable in the less stable soils with more clay, having greater opportunities for changing soil particle arrangement upon wetting, which can also mean development of an altered soil surface at the beginning of the run. To test this hypothesis, the relationship between the measured K_{fs} values and the soil structural index, *SSI*, by Pieri (1992) was examined. The PI results were found to be

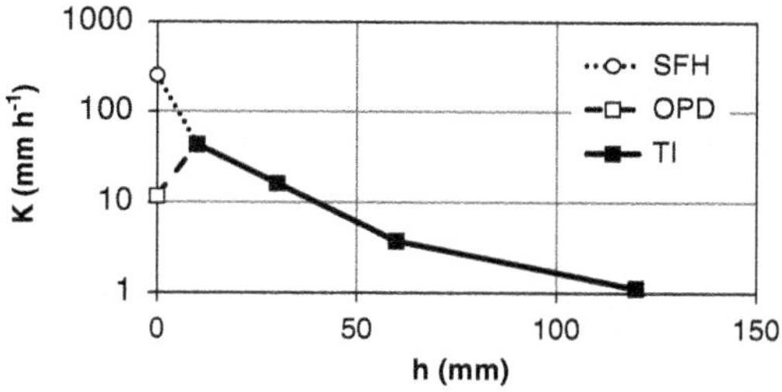

Fig. 2.46 Soil hydraulic conductivity, K, vs. pressure head, h, obtained experimentally at one of the sampled sites by Bagarello et al. (2014c) (SFH = SFH technique; OPD = single-ring pressure infiltrometer with the One-Ponding-Depth approach; TI = tension infiltrometer) (Reprinted with permission)

higher in more stable soils whereas a relationship between K_{fs} and SSI was not detected for the SFH technique. Therefore, this analysis supported the hypothesis that surface soil disturbance during the run was a factor determining the observed discrepancies between the two tested techniques. Another independent support to this interpretation was obtained by a comparison of K_{fs} with K_{-10}, i.e. the soil hydraulic conductivity corresponding to a pressure head of −10 mm, which was measured in the field with a TI. The K_{fs} values measured with the SFH technique were always higher than K_{-10} whereas in a few cases the PI technique yielded an unreliable K_{fs} result, i.e. lower than K_{-10} (Fig. 2.46). Taking into account that, with the TI, the soil surface disturbance is practically negligible, this comparison also was consistent with the suggested explanation. Therefore, the investigation by Bagarello et al. (2014c) showed that disturbance of the infiltration surface is more noticeable with the PI, and for this reason the two techniques should be expected to yield more similar results in more stable porous media. The investigation by Frey et al. (2012), using the double-ring infiltrometer method, was in line with this interpretation given that, according to these authors, the K_{fs} values obtained in the field are expected to be somewhat underestimated because several factors, such as structure deterioration, clay and organic matter swelling, slaking, eluviation and air entrapment, can potentially reduce the infiltration rates. Bagarello et al. (2014c) also suggested that using simultaneously the SFH and PI techniques might allow an improved interpretation and/or simulation of hydrological processes such as rainfall partition into infiltration and rainfall excess. The soil at the initial conditions, i.e. before occurrence of rainfall, can be better described with the SFH technique because the run is expected to alter only minimally the infiltration surface. A long and intense rainfall event can disturb appreciably the soil surface and this circumstance can be considered, in terms of measured K_{fs}, by carrying out a steady-state run with the PI device by Ciollaro and Lamaddalena (1998). It should also be noted that the tendency of early time transient flow data to yield higher K_{fs} values than steady flow data does not seem an uncommon result since it was also detected under constant head ponding conditions in a loam soil (Vauclin et al. 1994). In this case, however, the detected differences were attributed to a reduced accuracy of the early time estimates, due to the approximations involved in the calculation of K_{fs}.

The reliability of a measurement of K_{fs} obtained with the SFH technique was also investigated by establishing a comparison with the K_{fs} values determined by the BEST procedure of soil hydraulic characterization (Lassabatere et al. 2006) on a clay, a clay-loam and two sandy-loam soils (Bagarello et al. 2014a). With both techniques, water was applied from both a small (0.03 m; low, L, runs) and a large (1.5 m; high, H, runs) distance from the soil surface to test what happens with different levels of soil disturbance during the run. The height of water pouring did not affect significantly and/or appreciably the measured conductivities with the SFH technique since the means of K_{fs} obtained with the L and H runs differed at the most by a factor of 1.9. On the other hand, the height of water application influenced significantly and substantially the measured conductivities using BEST, with the L runs yielding higher means than the H ones by a factor of 11–35, depending on the soil. The SFH and BEST techniques showed similarities at most of the sampled sites when the height of water application was low since the means of K_{fs} differed by a factor of not more than 1.7 at three sites and of 9.3 in the soil with more silt. With a great height of pouring, the BEST technique yielded substantially lower K_{fs} values than the SFH technique, with differences between the two techniques by a factor varying with the site from 13 to 81. The different sensitivity of the two techniques to the water application height was attributed to the fact that water was applied once with the SFH technique and several times with the BEST procedure. Each water application generally contributed to alter the soil surface, and total energy of the applied water was found to be an appropriate predictor of the changes in K_{fs} when soil deterioration was not completed before concluding the infiltration run. On the basis of their results, Bagarello et al. (2014a) suggested that the choice of the methodology to be applied (SFH, BEST) and the height of water application (L, H) should depend on the intended use of the K_{fs} data. If the objective of the field campaign is to obtain data usable to explain surface runoff generation phenomena during intense rainfall events, the most appropriate choice among the tested ones should be BEST with a high water level pouring to mimic relatively prolonged rainfall effects at the soil surface. A low water application height is more appropriate to determine the saturated conductivity of a soil that is not directly impacted by rainfall, due for example to the presence of a mulch on the soil surface. In this case, the SFH and BEST techniques appear to yield relatively equivalent results in sandy-loam and clay soils, which suggests that the simpler SFH technique should be applied to determine K_{fs}. This technique should also be preferred in the more silty soils, where there were signs that a repeated water application promoted some soil disturbance decreasing K_{fs} also with a low application height.

2.2.4.4 Two-Level Analysis

Employing unskilled personnel in visual assessment of the soil textural and structural characteristics for an estimation of α* may introduce undetectable uncertainties in the estimated K_{fs} values. To reduce this risk, Bagarello et al. (2006b) modified the SFH run in an attempt to also obtain an estimate of α* usable for the

K_{fs} calculation by the so-called *Two-Level*, TL, analysis. In particular, the drop of the water level on the soil surface has to be measured twice during the transient, one-dimensional infiltration experiment. With the two (t, I) data pairs, i.e. (t_1, I_1) and (t_2, I_2), Eq. (2.115) is applied to obtain two estimates of K_{fs} corresponding to a pre-established value of α^*, i.e. K_{fs1} obtained by using the (t_1, I_1) data pair and K_{fs2} obtained by using the (t_2, I_2) data pair. In particular, the calculation of K_{fs1} and K_{fs2} is carried out for different positive values of α^*, ranging from 0.1 m^{-1} to 1000 m^{-1}. Because the correct value of α^* should produce the same value of K_{fs} from the two data pairs, the α^* value producing the lowest absolute value of the relative difference between K_{fs2} and K_{fs1} is assumed to be the estimate of α^* corresponding to the two selected (t, I) data pairs (α^*_{dp}, dp = data pair). Figure 2.47 shows an example of the relationship between $\left|\left[\left(K_{fs2} - K_{fs1}\right)/K_{fs1}\right]\right|$ and α^*, in which a clear minimum, suggesting a successful estimation of α^*, is detectable. Using a sequence of more than two (t, I) data pairs allows us to obtain an estimate of α^*_{dp} for each possible combination of two (t, I) data pairs measured within a falling head experiment. A representative value of α^* for the considered soil (α^*_s) is then obtained by averaging the α^*_{dp} results corresponding to each possible combination. The combinations producing an estimate of α^*_{dp} corresponding to an extreme value of the considered range (i.e., $\alpha^*_{dp} = 0.1$ m^{-1} or $\alpha^*_{dp} = 1000$ m^{-1}) are not included in the calculation of α^*_s. Extreme values of α^* falling outside the considered range ($0.1 \leq \alpha^* \leq 1000$ m^{-1}) have been reported (e.g., White and Sully 1992) but the literature suggests that only those values falling within the chosen range should be considered (Elrick and Reynolds 1992a; White and Sully 1992; Russo et al. 1997; Khaleel and Relyea 2001). Finally, K_{fs} can be calculated setting $\alpha^* = \alpha^*_s$ in Eq. (2.116) and using the measured values of $\Delta\theta$, t_a and H_0. A test of the TL analysis on repacked sandy-loam and loam soil cores (Bagarello et al. 2006b) showed that, for a given soil core, many calculations yielded an estimate of α^*_{dp}

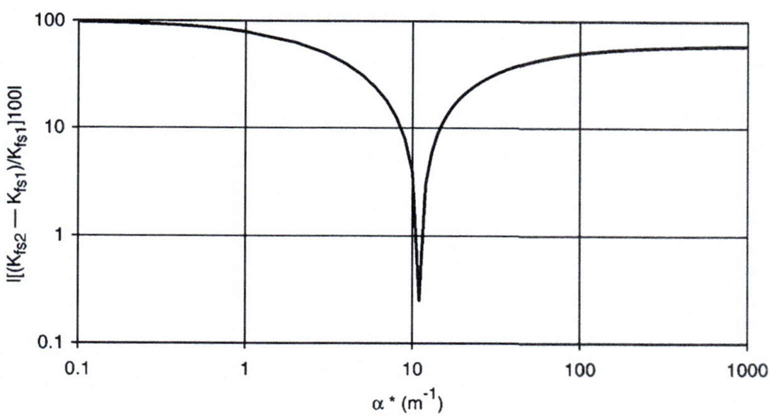

Fig. 2.47 Example of the plot of the relative difference between the field-saturated soil hydraulic conductivity values calculated for two infiltration data pairs (K_{fs1} and K_{fs2}) vs. the α^* parameter used in the calculations (From Bagarello et al. 2006b, reprinted with permission)

corresponding to an extreme value of the considered range. Therefore, a relatively low percentage of two-level calculations was generally usable for deducing α^*_s, and the valid estimates of α^* (α^*_{dp}) were highly variable. However, α^*_s was found to be relatively repeatable and reasonable because it did not vary appreciably with the soil core and it was close to the expected value on the basis of the physical characteristics of the considered soil. In addition, similar K_{fs} values (i.e., differences by a few percentage units) were obtained with the transient technique and a constant head permeameter steady-state procedure, supporting the applicability of the developed analysis. Finally, the original SFH approach yielded K_{fs} results that were practically equivalent to the K_{fs} values obtained by the more detailed analysis using many $I(t)$ data points, supporting the applicability of the simple SFH approach developed by Bagarello et al. (2004).

Example 2.7

The field-saturated soil hydraulic conductivity, K_{fs}, was measured at a sampling point with the Simplified Falling Head technique. A cylinder with an inner diameter of 0.15 m, inserted to a depth, $d = 0.12$ m was used. The initial volumetric water content of the sampled soil, θ_i, was of 0.156 m^3m^{-3}. The saturated soil water content, θ_{fs}, assumed to be equal to soil porosity, was estimated taking into account that the dry soil bulk density, ρ_b, was equal to 1364.7 kg m^{-3} and a soil particle density, ρ_s, of 2650 kg m^{-3} can be assumed:

$$\theta_{fs} = 1 - \frac{1364.7}{2650} = 0.485 \ \ m^3m^{-3}$$

Therefore, $\Delta\theta$ was equal to 0.485–0.156 = 0.329 m^3m^{-3}. The infiltration run was carried out by using a volume, V, of water equal to the volume of voids, V_p (L^3), within the soil volume confined by the ring, estimated by Eq. (2.117):

$$V_p = 0.12 \times \pi \frac{0.15^2}{4} \times 0.329 = 0.000698 \ \ m^3$$

In practice, $V = 0.7$ L was used, yielding an initial depth of water, H_0, of:

$$H_0 = \frac{0.0007}{\left(\pi \frac{0.15^2}{4}\right)} = 0.04 \ \ m$$

The run duration, t_a, was equal to 1912 s. Assuming an α^* parameter of 12 m^{-1}, the following K_{fs} value was obtained by Eq. (2.116):

$$K_{fs} = \frac{0.329}{(1-0.329)\times 1912}\left[\frac{0.04}{0.329} - \frac{0.04+\frac{1}{12}}{1-0.329}\ \ln\left(1+\frac{(1-0.329)\times 0.04}{0.329\times\left(0.04+\frac{1}{12}\right)}\right)\right] = 7.25\times 10^{-6} m\ s^{-1}.$$

2.3 Dripper or Point-Source Method

When a constant flow rate, Q (L^3T^{-1}), is applied to a point at the surface of an initially unsaturated soil, for example by a dripper for drip irrigation, water spreads over a small finite area of the surface forming a saturated disk that is easily detectable since it appears as a glistening surface. The rate of spreading is initially high and it decreases with time (Fig. 2.48) until the ponded zone approaches a constant circular or nearly circular area at some finite infiltration time. Once this saturated disk is formed, water continues spreading laterally on the soil surface forming an unsaturated zone (Yitayew et al. 1998; Shani et al. 1987; Al-Jabri et al. 2002a). The unsaturated front is identified by the change of the soil color from wet to dry conditions. The ultimate size of the saturated zone depends primarily on the field-saturated soil hydraulic conductivity, K_{fs} ($L\ T^{-1}$), and the application rate of the dripper. For a given application rate, a permeable soil shows a smaller saturated surface radius than a soil with a low permeability. For a given soil, the radius of the saturated zone increases with the applied flow rate (Yitayew et al. 1998).

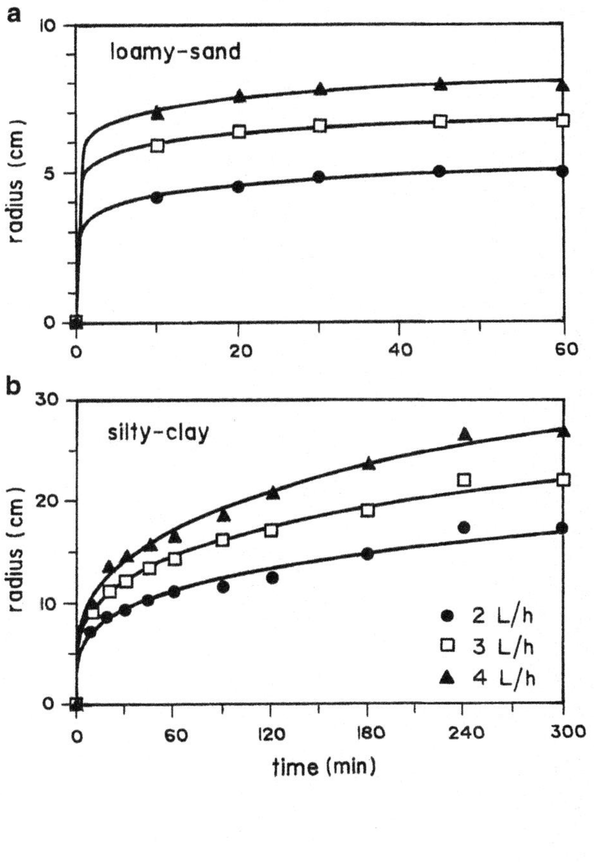

Fig. 2.48 Saturated water entry radius as a function of time with application rates of 2, 3 and 4 L h^{-1} for a (**a**) loamy-sand soil and (**b**) silty-clay soil (From Yitayew et al. 1998, reprinted with permission)

The *dripper* (Shani et al. 1987) or *point-source* (Al-Jabri et al. 2002a) method allows simultaneous determination of K_{fs} and the α (L^{-1}) parameter of the soil hydraulic conductivity function by Gardner (1958), or the macroscopic capillary length, λ_c (L), by a point application of a sequence of increasing, constant flow rates. For each flow rate, a steady-state condition can be established when the saturated zone does not vary with time. The radius, r_s (L), of this saturated zone can easily be measured before applying the next flow rate in the increasing sequence of Q values. If the initial soil water content is relatively low, the availability of (r_s, Q) data pairs for $N = 3$ or more flow rates allows estimation of K_{fs} and α or λ_c by the Wooding's (1968) equation:

$$i_s = \frac{Q}{\pi\ r_s^2} = K_{fs} + \frac{4\ K_{fs}}{\pi\ \alpha\ r_s} \tag{2.120}$$

where i_s ($L\,T^{-1}$) is the infiltration rate when the radius of the saturated zone is equal to r_s. Plotting i_s against $1/r_s$ yields a linear relationship with an intercept equal to K_{fs} and a slope, b_1, allowing estimation of α or $\lambda_c = \alpha^{-1}$:

$$\alpha = \frac{4\ K_{fs}}{\pi\ b_1} \tag{2.121}$$

The K_{fs} and b_1 values are simply obtained by a linear regression analysis of the i_s vs. $1/r_s$ data pairs.

Or (1996) developed the multiple-points permeameter for simultaneous in situ measurements at a large number of sampling points. The permeameter consists of four bundled polyethylene dripper lines, each 1.25 cm in diameter. Twenty-five equally spaced drippers (1.2 m spacing) were mounted on each dripper line, using a different dripper flow rate for each line, i.e. $Q = 2$, 4, 8 and 24 L h^{-1}. The four lines were connected to a PVC manifold with a separate control valve for each line. A typical measurement sequence begins with application of the lowest flow rate (2 L h^{-1}) while recording the saturated diameter of each pond. A particular flow rate (e.g., 12 L h^{-1}) can also be obtained by simultaneously turning on two different dripper lines (4 and 8 L h^{-1}).

Generally, commercially available drippers with flow rates varying between 2 and 24 L h^{-1} have been used. An accurate application of the method requires measurement of flow rates from each emitter. However, this measurement can be avoided if high-quality, pressure-compensated drippers are used. This was the case, for example, of the multiple-points permeameter by Or (1996), who reported coefficients of variation for flow rates along the permeameter lines varying between 0.04 and 0.06. The time to reach the constant saturated radius varies from soil to soil. In an investigation by Yitayew et al. (1998), steady-state conditions were established within 10–15 min for a loamy-sand soil but they were detected after 4–5 h from the starting of the run for a loam and a silty-clay soil. For other soils with loam and silt-loam texture, Or (1996) and Al-Jabri et al. (2002a, b, 2006) reported application times for a given flow rate of not more than one hour. According to

Shani et al. (1987), the time to approach steady-state increases with the discharge rate. In general, the ponded zone is expected to have a relatively small size. For example, with $Q \leq 4$ L h^{-1}, Yitayew et al. (1998) measured r_s values equal at the most to 8 cm in a loamy-sand soil and to 25–28 cm in loam and silty-clay soils.

Obviously, the method requires that the applied flow rates determine ponding conditions on the soil surface, i.e. they are higher than the infiltration rate of the soil (Al-Jabri et al. 2002a). For best results, the dripper method requires leveled soil surfaces (Or 1996), although Al-Jabri et al. (2002b) carried out an experiment on a gently sloped field, having a steepness of 1 %. Leveling and smoothing the soil surface allows a more symmetrical wetting geometry (Yitayew et al. 1998). Although small changes in microrelief can distort the shape of the ponded surface, measurements taken along several directions can be used to infer an equivalent pond radius (Or 1996).

Occasionally, a poor alignment of the ($1/r_s$, i_s) data pairs can be detected. A possible reason of this result is the presence of macropores in the sampled soil that hinder expansion of the saturated zone (Al-Jabri et al. 2002b). A less uncertain estimate of soil hydraulic properties can be obtained by using a relatively large number of ($1/r_s$, i_s) data pairs at a sampling point.

The dripper method can also be used for simultaneous determination of soil hydraulic and chemical transport properties with limited labor requirements (Al-Jabri et al. 2002a, b, 2006). Moreover, the characteristics of a large number of ponded areas can easily be retrieved from a photograph of the sampled field. Probably, the method could enable an accurate assessment of spatial-temporal variability in soil hydraulic properties at a field site. This can be explained by the fact that it is a non-destructive sampling of exactly the same points at each measurement campaign. However, only a limited number of applications of the dripper method can still be found in literature.

References

Agnese, C., Bagarello, V., Baiamonte, G., & Iovino, M. (2011). Comparing physical quality of forest and pasture soils in a Sicilian watershed. *Soil Science Society of America Journal, 75*, 1958–1970.

Al-Jabri, S. A., Horton, R., & Jaynes, D. B. (2002a). A point-source method for rapid simultaneous estimation of soil hydraulic and chemical transport properties. *Soil Science Society of America Journal, 66*, 12–18.

Al-Jabri, S. A., Horton, R., Jaynes, D. B., & Gaur, A. (2002b). Field determination of soil hydraulic and chemical transport properties. *Soil Science, 167*(6), 353–368.

Al-Jabri, S. A., Lee, J., Gaur, A., Horton, R., & Jaynes, D. B. (2006). A dripper-TDR method for in situ determination of hydraulic conductivity and chemical transport properties of surface soils. *Advances in Water Resources, 29*, 239–249.

Amoozegar, A. (1989). Comparison of the Glover solution with the simultaneous-equations approach for measuring hydraulic conductivity. *Soil Science Society of America Journal, 53*, 1362–1367.

Amoozegar, A. (1993). Comments on "methods for analyzing constant-head well permeameter data". *Soil Science Society of America Journal, 57*, 559–560.

Angulo-Jaramillo, R., Vandervaere, J.-P., Roulier, S., Thony, J. L., Gaudet, J. P., & Vauclin, M. (2000). Field measurement of soil surface hydraulic properties by disc and ring infiltrometers. A review and recent developments. *Soil & Tillage Research, 55*, 1–29.

Ankeny, M. D. (1992). Methods and theory for unconfined infiltration measurements. In G. C. Topp, W. D. Reynolds, & R. E. Green (Eds.), *Advances in measurement of soil physical properties: Bringing theory into practice* (SSSA special publication no.30, pp. 123–141). Madison: Soil Science Society of America, Inc.

Ankeny, M. D., Kaspar, T. C., & Horton, R. (1990). Characterization of tillage and traffic effects on unconfined infiltration measurements. *Soil Science Society of America Journal, 54*, 837–840.

Ankeny, M. D., Ahmed, M., Kaspar, T. C., & Horton, R. (1991). Simple field method for determining unsaturated hydraulic conductivity. *Soil Science Society of America Journal, 55*, 467–470.

Arya, L. M., Dierolf, T. S., Sofyan, A., Widjaja-Adhi, I. P. G., & van Genuchten, M. (1998). Field measurement of the saturated hydraulic conductivity of a macroporous soil with unstable subsoil structure. *Soil Science, 163*(11), 841–852.

Asare, S. N., Rudra, R. P., Dickinson, W. T., & Wall, G. J. (1993). Seasonal variability of hydraulic conductivity. *Transactions of the ASAE, 36*(2), 451–457.

Azam, M. G. (2005). Development of a location-specific soil hydraulic quality (SHQ) index: A case study from northeast Thailand. In M. Zoebisch, K. M. Cho, S. Hein, & R. Mowla (Eds.), *Integrated watershed management: Studies and experiences from Asia* (pp. 739–758). Bangkok: AIT.

Azam, M. G., Zoebisch, M. A., & Wickramarachchi, K. S. (2008). Effects of cropping systems on selected soil structural properties and crop yields in the Lam Phra Phloeng watershed northeast Thailand. *Journal of Agronomy, 7*(1), 56–62, ISSN 1812–5379.

Bagarello, V. (1993). Osservazioni sull'utilizzazione del Guelph permeameter per la misura in campo della conducibilità idrica satura. In *Atti del Convegno Il Ruolo dell'Ingegneria per l'Agricoltura del 2000*, V Convegno Nazionale A.I.G.R., Maratea (in Italian), pp. 151–161.

Bagarello, V. (1997). Influence of well preparation on field-saturated hydraulic conductivity measured with the Guelph permeameter. *Geoderma, 80*, 169–180.

Bagarello, V., & Provenzano, G. (1996). Factors affecting field and laboratory measurement of saturated hydraulic conductivity. *Transactions of the ASAE, 39*(1), 153–159.

Bagarello, V., & Giordano, G. (1999). Comparison of procedures to estimate steady flow rate in field measurement of saturated hydraulic conductivity with the Guelph permeameter method. *Journal of Agricultural Engineering Research, 74*, 63–71.

Bagarello, V., & Iovino, M. (1999). Spatial and temporal variability of saturated hydraulic conductivity in a central Sicily soil. In *Proceedings of the international workshop of EurAgEng's field of interest on soil and water "Modelling of Transport Processes in Soils"*, Leuven, Belgium, pp. 249–256.

Bagarello, V., & Sgroi, A. (2004). Using the single-ring infiltrometer method to detect temporal changes in surface soil field-saturated hydraulic conductivity. *Soil & Tillage Research, 76*, 13–24.

Bagarello, V., & Sgroi, A. (2007). Using the simplified falling head technique to detect temporal changes in field-saturated hydraulic conductivity at the surface of a sandy loam soil. *Soil & Tillage Research, 94*, 283–294.

Bagarello, V., & Sgroi, A. (2008). Testing soil encasing materials for measuring hydraulic conductivity of a sandy-loam soil by the cube methods. *Soil Science Society of America Journal, 72*, 1048–1057. doi:10.2136/sssaj2007.0022.

Bagarello, V., & Iovino, M. (2010). *Conducibilità idraulica del suolo – Metodi di misura nelle applicazioni idrologiche*. Milano: Ulrico Hoepli Editore S.p.A, 382 pp. (in Italian). ISBN 978-88-203-4411-5.

Bagarello, V., Iovino, M., & Reynolds, D. (1997). Impact of infiltration water sodicity on field-saturated soil hydraulic conductivity. In *Proceedings international conference on "Water Management, Salinity and Pollution Control towards Sustainable Irrigation in the Mediterranean Region", Volume IV, Wastewater re-use in irrigated agriculture*, Istituto Agronomico Mediterraneo, Valenzano (Bari), pp. 241–258.

Bagarello, V., Iovino, M., & Reynolds, W. D. (1999). Measuring hydraulic conductivity in a cracking clay soil using the Guelph permeameter. *Transactions of the ASAE, 42*(4), 957–964.

Bagarello, V., Iovino, M., & Tusa, G. (2000). Factors affecting measurement of the near saturated soil hydraulic conductivity. *Soil Science Society of America Journal, 64*, 1203–1210.

Bagarello, V., Iovino, M., & Elrick, D. (2004). A simplified falling-head technique for rapid determination of field-saturated hydraulic conductivity. *Soil Science Society of America Journal, 68*, 66–73.

Bagarello, V., Iovino, M., Palazzolo, E., Panno, M., & Reynolds, D. (2006a). Field and laboratory approaches for determining sodicity effects on soil hydraulic conductivity. *Geoderma, 130*, 1–13.

Bagarello, V., Elrick, D. E., Iovino, M., & Sgroi, A. (2006b). A laboratory analysis of falling head infiltration procedures for estimating the hydraulic conductivity of soils. *Geoderma, 135*, 322–334.

Bagarello, V., Sferlazza, S., & Sgroi, A. (2009a). Comparing two methods of analysis of single-ring infiltrometer data for a sandy-loam soil. *Geoderma, 149*, 415–420.

Bagarello, V., Giordano, G., Iovino, M., & Sgroi, A. (2009b). *Effetto della dimensione dell'anello sulla conducibilità idraulica del suolo saturo misurata con la tecnica SFH*. Atti del IX Convegno Nazionale dell'Associazione Italiana di Ingegneria Agraria, Ischia Porto, 12–16 Settembre, memoria n.3–4, in Italian.

Bagarello, V., Di Stefano, C., Ferro, V., Iovino, M., & Sgroi, A. (2010a). Physical and hydraulic characterization of a clay soil at the plot scale. *Journal of Hydrology, 387*, 54–64. doi:10.1016/j.hydrol.1010.03.029.

Bagarello, V., Giordano, G., Iovino, M., Sferlazza, S., & Sgroi, A. (2010b). *Approfondimenti su alcune tecniche infiltrometriche per la misura della conducibilità idraulica del suolo saturo*. Proceedings "XXXII Convegno Nazionale di Idraulica e Costruzioni Idrauliche", Palermo, 14–17 September, code D0433 (in Italian).

Bagarello, V., Di Prima, S., Iovino, M., Provenzano, G., & Sgroi, A. (2011a). Testing different approaches to characterize Burundian soils by the BEST procedure. *Geoderma, 162*, 141–150.

Bagarello, V., Giordano, G., Sferlazza, S., & Sgroi, A. (2011b). Effects of laboratory procedures on the integrity of a sandy-loam soil sample for bi-directional measurement of saturated hydraulic conductivity. *Applied Engineering in Agriculture, 27*(3), 351–358.

Bagarello, V., D'Asaro, F., & Iovino, M. (2012). A field assessment of the Simplified Falling Head technique to measure the saturated soil hydraulic conductivity. *Geoderma, 187–188*, 49–58. doi:10.1016/j.geoderma.2012.04.008.

Bagarello, V., Iovino, M., & Lai, J. (2013a). Field and numerical tests of the two-ponding depth procedure for analysis of single-ring pressure infiltrometer data. *Pedosphere, 23*, 779–789.

Bagarello, V., Castellini, M., Di Prima, S., Giordano, G., & Iovino, M. (2013b). Testing a simplified approach to determine field saturated soil hydraulic conductivity. *Procedia Environmental Sciences, 19*, 599–608. doi:10.1016/j.proenv.2013.06.068.

Bagarello, V., Di Stefano, C., Iovino, M., & Sgroi, A. (2013c). Using a transient infiltrometric technique for intensively sampling field-saturated hydraulic conductivity of a clay soil in two runoff plots. *Hydrological Processes, 27*, 3415–3423. doi:10.1002/hyp.9448.

Bagarello, V., Castellini, M., Di Prima, S., & Iovino, M. (2014a). Soil hydraulic properties determined by infiltration experiments and different heights of water pouring. *Geoderma, 213*, 492–501. doi:10.1016/j.geoderma.2013.08.032.

Bagarello, V., Di Prima, S., Iovino, M., & Provenzano, G. (2014b). Estimating field-saturated soil hydraulic conductivity by a simplified Beerkan infiltration experiment. *Hydrological Processes, 28*, 1095–1103. doi:10.1002/hyp.9649.

Bagarello, V., Baiamonte, G., Castellini, M., Di Prima, S., & Iovino, M. (2014c). A comparison between the single ring pressure infiltrometer and simplified falling head techniques. *Hydrological Processes, 28*, 4843–4853. doi:10.1002/hyp.9980.

Bell, R. W., & Schofield, N. J. (1990). Design and application of a constant head well permeameter for shallow high saturated hydraulic conductivity soil. *Hydrological Processes, 4*, 327–342.

Bird, T. L., Willis, T. M., & Melville, G. J. (1996). Subsoil hydraulic conductivity estimates for the Lower Macquaire Valley. *Australian Journal of Soil Research, 34*, 213–228.

Bodhinayake, W., Si, B. C., & Noborio, K. (2004). Determination of hydraulic properties in sloping landscapes from tension and double-ring infiltrometers. *Vadose Zone Journal, 3*, 964–970.

Bosch, D. D. (1997). Constant head permeameter formula dependence on alpha parameter. *Transactions of the ASAE, 40*(5), 1377–1379.

Bosch, D. D., & West, L. T. (1998). Hydraulic conductivity variability for two sandy soils. *Soil Science Society of America Journal, 62*, 90–98.

Bouma, J. (1983). Use of soil survey data to select measurement techniques for hydraulic conductivity. *Agricultural Water Management, 6*, 177–190.

Bouma, J. (1985). Soil spatial variability and soil survey. In J. Bouma & D. R. Nielsen (Eds.), *Soil spatial variability, proc. soil spatial variability workshop*, Las Vegas, NV, 30 Nov-1 Dec 1984, Int. Soil Sci. Soc. and Soil Sci. Soc. Am. PUDOC, Wageningen, pp. 130–149.

Bouma, J., & Dekker, L. W. (1981). A method for measuring the vertical and horizontal Ksat of clay soils with macropores. *Soil Science Society of America Journal, 45*, 662–663.

Bouwer, H. (1966). Rapid field measurement of air entry value and hydraulic conductivity as significant parameters in flow system analysis. *Water Resources Research, 1*, 729–738.

Bouwer, H. (1978). *Groundwater hydrology*. Toronto: McGraw-Hill Book Company.

Bouwer, H. (1986). Intake rate: Cylinder infiltrometer. In A. Klute (Ed.), *Methods of soil analysis, Part 1* (Agron. monogr. 9 2nd ed., pp. 825–844). Madison: ASA and SSSA.

Braud, I., De Condappa, D., Soria, J. M., Haverkamp, R., Angulo-Jaramillo, R., Galle, S., & Vauclin, M. (2005). Use of scaled forms of the infiltration equation for the estimation of unsaturated soil hydraulic properties (the Beerkan method). *European Journal of Soil Science, 56*, 361–374.

Brooks, R. H., & Corey, A. T. (1964). *Hydraulic properties of porous media* (Hydrology paper 3). Fort Collins: Colorado State University.

Brutsaert, W. (1977). Vertical infiltration in dry soil. *Water Resources Research, 13*, 363–368.

Burgy, R. H., & Luthin, J. N. (1956). A test of the single- and double-ring types of infiltrometers. *Transactions of the American Geophysical Union, 37*, 189–191.

Campbell, C. M., & Fritton, D. D. (1994). Factors affecting field-saturated hydraulic conductivity measured by the borehole permeameter technique. *Soil Science Society of America Journal, 58*, 1354–1357.

Carsel, R. F., & Parrish, R. S. (1988). Developing joint probability distributions of soil water retention characteristics. *Water Resources Research, 24*, 755–769.

Chappell, N. A., & Ternan, J. L. (1997). Ring permeametry: Design, operation and error analysis. *Earth Surface Processes and Landforms, 22*, 1197–1205.

Ciollaro, G., & Lamaddalena, N. (1998). Effect of tillage on the hydraulic properties of a vertic soil. *Journal of Agricultural Engineering Research, 71*, 147–155.

Clancy, K., & Alba, V. M. (2011). Temperature and time of day influence on double-ring infiltrometer steady-state infiltration rates. *Soil Science Society of America Journal, 75*(1), 241–245.

Clothier, B., & Scotter, D. (2002). 3.5 Unsaturated water transmission parameters obtained from infiltration. In J. H. Dane, & G. C. Topp (Co-Eds.), *Methods of soil analysis, Part 4 – Physical methods* (Number 5 in the Soil science society of America book series, pp. 879–889). Madison: Soil Science Society of America Inc.

Das, G. S., Mohanty, B. P., & Köhne, J. M. (2006). Soil hydraulic conductivities and their spatial and temporal variations in a vertisol. *Soil Science Society of America Journal, 70*, 1872–1881.

De Haro, J. M., Vanderlinden, K., Gómez, J. A., & Giráldez, J. V. (1998). Medida de la conductividad hidráulica saturada. In A. González et al. (Eds.), *Progresos en la Investigación de la Zona no Saturada* (Estudios de la Zona no Saturada 3, pp. 9–20). Huelva: Univ. de Huelva.

Dikinya, O., Hinz, C., & Aylmore, G. (2008). Decrease in hydraulic conductivity and particle release associated with self-filtration in saturated soil columns. *Geoderma, 146*, 192–200.

Dirk, S. M., Carlson, D. A., Cherkauer, D. S., & Malik, P. (1999). Scale dependency of hydraulic conductivity in heterogeneous media. *Ground Water, 37*, 904–919.

Dorsey, J. D., Ward, A. D., Fausey, N. R., & Bair, E. S. (1990). A comparison of four field methods for measuring saturated hydraulic conductivity. *Transactions of the ASAE, 33*(6), 1925–1931.

Dušek, J., Dohnal, M., & Vogel, T. (2009). Numerical analysis of ponded infiltration experiment under different experimental conditions. *Soil & Water Research, 4*, S22–S27.

Elrick, D. E., & Reynolds, W. D. (1992a). Methods for analyzing constant-head well permeameter data. *Soil Science Society of America Journal, 56*, 320–323.

Elrick, D. E., & Reynolds, W. D. (1992b). Infiltration from constant-head well permeameters and infiltrometers. In G. C. Topp, W. D. Reynolds, & R. E. Green (Eds.), *Advances in measurement of soil physical properties: Bringing theory into practice* (SSSA special publication no.30, pp. 1–24). Madison: Soil Science Society of America, Inc.

Elrick, D. E., Reynolds, W. D., Geering, H. R., & Tan, K. A. (1990). Estimating steady infiltration rate times for infiltrometers and permeameters. *Water Resources Research, 26*(4), 759–769.

Elrick, D. E., Parkin, G. W., Reynolds, W. D., & Fallow, D. J. (1995). Analysis of early-time and steady state single-ring infiltration under falling head conditions. *Water Resources Research, 31*, 1883–1893.

Elrick, D. E., Angulo-Jaramillo, R., Fallow, D. J., Reynolds, W. D., & Parkin, G. W. (2002). Analysis of infiltration under constant head and falling head conditions. In P. A. C. Raats, D. Smiles, & A. W. Warrick (Eds.), *Environmental mechanics: Water, mass and energy rransfer in the biosphere* (Geophysical monograph series, 129, pp. 47–53). Washington, DC: American Geophysical Union.

Fallow, D. J., & Elrick, D. E. (1996). Field measurement of air-entry and water-entry soil water pressure heads. *Soil Science Society of America Journal, 60*, 1036–1039.

Frey, S. K., Rudolph, D. L., & Parkin, G. W. (2012). Spatial and temporal influences on hydraulic properties in macroporous tile-drained soil. *Soil Science Society of America Journal, 76*, 350–360.

Gallichand, J., Madramootoo, C. A., Enright, P., & Barrington, S. F. (1990). An evaluation of the Guelph permeameter for measuring saturated hydraulic conductivity. *Transactions of the ASAE, 33*(4), 1179–1184.

Gardner, W. R. (1958). Some steady state solutions of the unsaturated moisture flow equation with application to evaporation from a water table. *Soil Science, 85*(4), 228–232.

Gómez, J. A., Giráldez, J. V., & Fereres, E. (2001). Analysis of infiltration and runoff in an olive orchard under no-till. *Soil Science Society of America Journal, 65*, 291–299.

Gómez, J. A., Vanderlinden, K., & Nearing, M. A. (2005). Spatial variability of surface roughness and hydraulic conductivity after disk tillage: Implications for runoff variability. *Journal of Hydrology, 311*, 143–156.

Green, W. H., & Ampt, G. A. (1911). Studies in soil physics. I. The flow of air and water through soils. *The Journal of Agricultural Science, 4*, 1–24.

Gupta, R. K., Rudra, R. P., Dickinson, W. T., Patni, N. K., & Wall, G. J. (1993). Comparison of saturated hydraulic conductivity measured by various field methods. *Transactions of the ASAE, 36*(1), 51–55.

Gwenzi, W., Hinz, C., Holmes, K., Phillips, I. R., & Mullins, I. J. (2011). Field-scale spatial variability of saturated hydraulic conductivity on a recently constructed artificial ecosystem. *Geoderma, 166*, 43–56.

Haverkamp, R., Kutilek, M., Parlange, J. Y., Rendon, L., & Krejca, M. (1988). Infiltration under ponded conditions: 2. Infiltration equations tested for parameter time-dependence and predictive use. *Soil Science, 145*(5), 317–329.

Haverkamp, R., Ross, P. J., Smettem, K. R. J., & Parlange, J. Y. (1994). Three-dimensional analysis of infiltration from the disc infiltrometer: 2. Physically based infiltration equation. *Water Resources Research, 30*(11), 2931–2935.

Haws, N. W., Liu, B., Boast, C. W., Rao, P. S. C., Kladivko, E. J., & Franzmeier, D. P. (2004). Spatial variability and measurement scale of infiltration rate on an agricultural landscape. *Soil Science Society of America Journal, 68*, 1818–1826.

Hayashi, M., & Quinton, W. L. (2004). A constant-head well permeameter method for measuring field-saturated hydraulic conductivity above an impermeable layer. *Canadian Journal of Soil Science, 84*, 255–264.

Hinnell, A. C., Lazarovitch, N., & Warrick, A. W. (2009). Explicit infiltration function for boreholes under constant head conditions. *Water Resources Research, 45*, W10429, 9 pages.

Hodnett, M. G., & Tomasella, J. (2002). Marked differences between van Genuchten soil water-retention parameters for temperate and tropical soils: A new water-retention pedo-transfer functions developed for tropical soils. *Geoderma, 108*, 155–180.

Horton, R. E. (1939). Analysis of infiltration in the hydrological cycle. *Transactions of the American Geophysical Union, 20*, 693–711.

Jabro, J. D. (1992). Estimation of saturated hydraulic conductivity of soils from particle size distribution and bulk density data. *Transactions of the ASAE, 35*(2), 557–560.

Jabro, J. D. (1996). Variability of field-saturated hydraulic conductivity in a Hagerstown soil as affected by initial water content. *Soil Science, 161*(11), 735–739.

Jabro, J. D., & Fritton, D. D. (1990). Simulation of water flow from a percolation test hole in layered soil. *Soil Science Society of America Journal, 54*, 1214–1218.

Jabro, J. D., & Evans, R. G. (2006). Discrepancies between analytical solutions of two borehole permeameters for estimating field-saturated hydraulic conductivity. *Applied Engineering in Agriculture, 22*(4), 549–554.

Jarvis, N. J., & Messing, I. (1995). Near-saturated hydraulic conductivity in soils of contrasting texture measured by tension infiltrometers. *Soil Science Society of America Journal, 59*, 27–34.

Kanwar, R. S., Rizvi, H. A., Ahmed, M., Horton, R., & Marley, S. J. (1989). Measurement of field-saturated hydraulic conductivity by using Guelph and velocity permeameters. *Transactions of the ASAE, 32*(6), 1885–1890.

Keller, T., Sutter, J. A., Nissen, K., & Rydberg, T. (2012). Using field measurement of saturated soil hydraulic conductivity to detect low-yielding zones in three Swedish fields. *Soil & Tillage Research, 124*, 68–77.

Khaleel, R., & Relyea, J. F. (2001). Variability of Gardner's α for coarse-textured sediments. *Water Resources Research, 37*, 1567–1575.

Kodešová, R., Šimůnek, J., Nikodem, A., & Jirků, V. (2010). Estimation of the dual-permeability model parameters using tension disk infiltrometer and Guelph permeameter. *Vadose Zone Journal, 9*, 213–225.

Koppi, A. J., & Geering, H. R. (1986). The preparation of unsmeared soil surfaces and an improved apparatus for infiltration measurements. *Journal of Soil Science, 37*, 177–181.

Lai, J., & Ren, L. (2007). Assessing the size dependency of measured hydraulic conductivity using double-ring infiltrometers and numerical simulation. *Soil Science Society of America Journal, 71*(6), 1667–1675.

Lai, J., Luo, Y., & Ren, L. (2010). Buffer index effects on hydraulic conductivity measurements using numerical simulations of double-ring infiltration. *Soil Science Society of America Journal, 74*, 1526–1536.

Lai, J., Luo, Y., & Ren, L. (2012). Numerical evaluation of depth effects of double-ring infiltrometers on soil saturated hydraulic conductivity measurements. *Soil Science Society of America Journal, 76*, 867–875.

Lassabatère, L., Angulo-Jaramillo, R., Soria Ugalde, J. M., Cuenca, R., Braud, I., & Haverkamp, R. (2006). Beerkan estimation of soil transfer parameters through infiltration experiments – BEST. *Soil Science Society of America Journal, 70*, 521–532.

Lauren, J. G., Wagenet, R. J., Bouma, J., & Wosten, J. H. M. (1988). Variability of saturated hydraulic conductivity in a glossaquic hapludalf with macropores. *Soil Science, 145*, 20–28.

Lee, D. M., Reynolds, W. D., Elrick, D. E., & Clothier, B. E. (1985). A comparison of three field methods for measuring saturated hydraulic conductivity. *Canadian Journal of Soil Science, 65*, 563–573.

Likhachev, E. R. (2003). Dependence of water viscosity on temperature and pressure. *Technical Physics, 48*, 514–515.

Lin, C., Greenwald, D., & Banin, A. (2003). Temperature dependence of infiltration rate during large scale water recharge into soils. *Soil Science Society of America Journal, 67*, 487–493.

Logsdon, S. D., & Jaynes, D. B. (1993). Methodology for determining hydraulic conductivity with tension infiltrometers. *Soil Science Society of America Journal, 57*, 1426–1431.

MacDonald, A. M., Maurice, L., Dobbs, M. R., Reeves, H. J., & Auton, C. A. (2012). Relating in situ hydraulic conductivity, particle size and relative density of superficial deposits in a heterogeneous catchment. *Journal of Hydrology, 434–435*, 130–141.

Mallants, D., Mohanty, B. P., Vervoort, A., & Feyen, J. (1997). Spatial analysis of saturated hydraulic conductivity in a soil with macropores. *Soil Technology, 10*, 115–131.

Mertens, J., Jacques, D., Vanderborght, J., & Feyen, J. (2002). Characterisation of the field-saturated hydraulic conductivity on a hillslope: In situ single ring pressure infiltrometer measurements. *Journal of Hydrology, 263*, 217–229.

Mirus, B. B., & Perkins, K. S. (2012). Practical estimates of field-saturated hydraulic conductivity of bedrock outcrops using a modified bottomless bucket method. *Water Resources Research, 48*, W09602, 6 pp.

Mohanty, B. P., Kanwar, R. S., & Everts, C. J. (1994a). Comparison of saturated hydraulic conductivity measurement methods for a glacial-till soil. *Soil Science Society of America Journal, 58*, 672–677.

Mohanty, B. P., Ankeny, M. D., Horton, R., & Kanwar, R. S. (1994b). Spatial analysis of hydraulic conductivity measured using disc infiltrometers. *Water Resources Research, 30*, 2489–2498.

Mohanty, B. P., Skaggs, T. H., & van Genuchten, M. T. (1998). Impact of saturated hydraulic conductivity on the prediction of tile flow. *Soil Science Society of America Journal, 62*, 1522–1529.

Mubarak, I., Angulo-Jaramillo, R., Mailhol, J. C., Ruelle, P., Khaledian, M., & Vauclin, M. (2010). Spatial analysis of soil surface hydraulic properties: Is infiltration method dependent? *Agricultural Water Management, 97*, 1517–1526.

Muñoz-Carpena, R., Regalado, C. M., Alvarez-Benedí, J., & Bartoli, F. (2002). Field evaluation of the new Philip-Dunne permeameter for measuring saturated hydraulic conductivity. *Soil Science, 167*(1), 9–24.

Nasta, P., Lassabatere, L., Kandelous, M. M., Šimůnek, J., & Angulo-Jaramillo, R. (2012). Analysis of the role of tortuosity and infiltration constants in the Beerkan method. *Soil Science Society of America Journal, 76*, 1999–2005. doi:10.2136/sssaj2012.0117n.

Nimmo, J. R., Schmidt, K. M., Perkins, K. S., & Stock, J. D. (2009). Rapid measurement of field-saturated hydraulic conductivity for areal characterization. *Vadose Zone Journal, 8*, 142–149.

Noshadi, M., Parvizi, H., & Sepaskhah, A. R. (2012). Evaluation of different methods for measuring field saturated hydraulic conductivity under high and low water table. *Vadose Zone Journal, 11*, 1–9.

Odell, B. P., Groenevelt, P. H., & Elrick, D. E. (1998). Rapid determination of hydraulic conductivity in clay liners by early-time analysis. *Soil Science Society of America Journal, 62*, 56–62.

Or, D. (1996). Drip irrigation in heterogeneous soils: Steady-state field experiments for stochastic model evaluation. *Soil Science Society of America Journal, 60*, 1339–1349.

Paige, G. B., & Hillel, D. (1993). Comparison of three methods for assessing soil hydraulic properties. *Soil Science, 155*(3), 175–189.

Philip, J. R. (1957a). The theory of infiltration, 1, The infiltration equation and its solution. *Soil Science, 83*, 345–357.

Philip, J. R. (1957b). The theory of infiltration: 4. Sorptivity and algebraic infiltration equations. *Soil Science, 84*, 257–264.

Philip, J. R. (1985). Approximate analysis of the borehole permeameter in unsaturated soil. *Water Resources Research, 21*(7), 1025–1033.

Philip, J. R. (1991). Hill slope infiltration: Planar slopes. *Water Resources Research, 27*, 109–117.

Philip, J. R. (1992). Falling head ponded infiltration. *Water Resources Research, 28*(8), 2147–2148.

Philip, J. R. (1993). Approximate analysis of falling-head lined borehole permeameter. *Water Resources Research, 29*(11), 3763–3768.

Pieri, C. J. M. G. (1992). *Fertility of soils: A future for farming in the West African Savannah.* Berlin: Springer.

Prieksat, M. A., Ankeny, M. D., & Kaspar, T. C. (1992). Design for an automated, self-regulating, single-ring infiltrometer. *Soil Science Society of America Journal, 56*, 1409–1411.

Ragab, R., & Cooper, J. D. (1993). Variability of unsaturated zone water transport parameters: implications for hydrological modelling. 1. In situ measurements. *Journal of Hydrology, 148*, 109–131.

Rawls, W. J., Nemes, A., & Pachepsky, Y. A. (2005). Effect of soil organic matter on soil hydraulic properties. In Y. A. Pachepsky & W. J. Rawls (Eds.), *Development of pedotransfer functions in soil hydrology* (pp. 95–114). Amsterdam/New York: Elsevier.

Regalado, C. M., Ritter, A., Alvarez-Benedí, J., & Muñoz-Carpena, R. (2005). Simplified method to estimate the Green-Ampt wetting front suction and soil sorptivity with the Philip-Dunne falling-head permeameter. *Vadose Zone Journal, 4*, 291–299.

Rex, J., & Dubé, S. (2006). Predicting the risk of wet ground areas in the Vanderhoof Forest District: Project description and progress report. *BC Journal of Ecosystems and Management, 7* (2), 57–71.

Reynolds, W. D. (1993). Chapter 56. Saturated hydraulic conductivity: Field measurement. In M. R. Carter (Ed.), *Soil sampling and methods of analysis, Canadian society of soil science* (pp. 599–613). Boca Raton: Lewis Publishers.

Reynolds, W. D. (1994). Letter to the editor. Comment on "Comparison of three methods for assessing soil hydraulic properties". *Soil Science, 157*(2), 116–119.

Reynolds, W. D. (2008). Chapter 76. Saturated hydraulic properties: Well permeameter. In M. R. Carter & E. G. Gregorich (Eds.), *Soil sampling and methods of analysis* (2nd ed., pp. 1025–1042). Boca Raton: Canadian Society of Soil Science.

Reynolds, W. D. (2010). Measuring soil hydraulic properties using a cased borehole permeameter: Steady flow analyses. *Vadose Zone Journal, 9*, 637–652.

Reynolds, W. D. (2011). Measuring soil hydraulic properties using a cased borehole permeameter: Falling-head analysis. *Vadose Zone Journal, 10*, 999–1015.

Reynolds, W. D., & Elrick, D. E. (1985a). *Measurement of field-saturated hydraulic conductivity, sorptivity and the conductivity – Pressure head relationship using the "Guelph permeameter"*. Proceedings of the National Water Well Association Conference on Characterization and Monitoring of the Vadose (Unsaturated) Zone, Denver, Colorado, November.

Reynolds, W. D., & Elrick, D. E. (1985b). In situ measurement of field-saturated hydraulic conductivity, sorptivity and the α-parameter using the Guelph permeameter. *Soil Science, 140*(4), 292–302.

Reynolds, W. D., & Elrick, D. E. (1986, Winter). *A method for simultaneous in situ measurement in the vadose zone of field-saturated hydraulic conductivity, sorptivity and the conductivity – Pressure head relationship*. GroundWater Monitoring Review, 84–95.

Reynolds, W. D., & Elrick, D. E. (1987). A laboratory and numerical assessment of the Guelph permeameter method. *Soil Science, 144*(4), 282–299.

Reynolds, W. D., & Elrick, D. E. (1990). Ponded infiltration from a single ring: I. Analysis of steady flow. *Soil Science Society of America Journal, 54*, 1233–1241.
Reynolds, W. D., & Elrick, D. E. (1991). Determination of hydraulic conductivity using a tension infiltrometer. *Soil Science Society of America Journal, 55*, 633–639.
Reynolds, W. D., & Zebchuk, W. D. (1996). Hydraulic conductivity in a clay soil: Two measurement techniques and spatial characterization. *Soil Science Society of America Journal, 60*, 1679–1685.
Reynolds, W. D., & Elrick, D. E. (2002a). 3.4.1.1 Principles and parameter definitions. In J. H. Dane, & G. C. Topp (Co-Eds.), *Methods of soil analysis, Part 4, Physical methods* (Number 5 in the Soil Science Society of America book series, pp.797–801). Madison: Soil Science Society of America, Inc.
Reynolds, W. D., & Elrick, D. E. (2002b). 3.4.3.2.b Pressure infiltrometer. In J. H. Dane, & G. C. Topp (Co-Eds.), *Methods of soil analysis, Part 4, Physical methods* (Number 5 in the Soil Science Society of America book series, pp. 826–836). Madison: Soil Science Society of America, Inc.
Reynolds, W. D., & Elrick, D. E. (2002c). 3.4.3.3 Constant head well permeameter (vadose zone). In J. H. Dane, & G. C. Topp (Co-Eds.), *Methods of soil analysis, Part 4, Physical methods* (Number 5 in the Soil Science Society of America book series, pp. 844–858). Madison: Soil Science Society of America, Inc.
Reynolds, W. D., & Elrick, D. E. (2005). Chapter 6 Measurement and characterization of soil hydraulic properties. In J. Álvarez-Benedí, & R. Muñoz-Carpena (Co-Eds.), *Soil-water-solute process characterization – An integrated approach*. Boca Raton: CRC Press.
Reynolds, W. D., & Lewis, J. K. (2012). A drive point application of the Guelph Permeameter method for coarse-textured soils. *Geoderma, 187–188*, 59–66.
Reynolds, W. D., Elrick, D. E., & Clothier, B. E. (1985). The constant head well permeameter: Effect of unsaturated flow. *Soil Science, 139*, 172–180.
Reynolds, W. D., Vieira, S. R., & Topp, G. C. (1992). An assessment of the single-head analysis for the constant head well permeameter. *Canadian Journal of Soil Science, 72*, 489–501.
Reynolds, W. D., Bowman, B. T., Brunke, R. R., Drury, C. F., & Tan, C. S. (2000). Comparison of tension infiltrometer, pressure infiltrometer, and soil core estimates of saturated hydraulic conductivity. *Soil Science Society of America Journal, 64*, 478–484.
Reynolds, W. D., Elrick, D. E., Youngs, E. G. (2002a). 3.4.3.2 Ring or cylinder infiltrometers (vadose zone). In J. H. Dane, & G. C. Topp (Co-Eds.), *Methods of soil analysis, Part 4, Physical methods* (Number 5 in the Soil Science Society of America book series, pp. 818–820). Madison: Soil Science Society of America, Inc.
Reynolds, W. D., Elrick, D. E., & Youngs, E. G. (2002b). 3.4.3.2.a Single-ring and double- or concentric-ring infiltrometers. In J. H. Dane, & G. C. Topp (Co-Eds.), *Methods of soil analysis, Part 4, Physical methods* (Number 5 in the Soil Science Society of America book series, pp. 821–826). Madison: Soil Science Society of America, Inc
Rienzner, M., & Gandolfi, C. (2014). Investigation of spatial and temporal variability of saturated soil hydraulic conductivity at the field-scale. *Soil & Tillage Research, 135*, 28–40.
Rojas, D. L. M., Rojas, P. H., & Menjívar Flórez, J. C. (2008). Estimación de la conductividad hidráulica saturada in situ en un suelo tratado con vinaza. *Acta Agronomica, 57*(2), 125–128.
Rovey, C. W., & Cherkauer, D. S. (1995). Scale dependency of hydraulic conductivity measurements. *Ground Water, 33*, 769–780.
Russo, D., Russo, I., & Laufer, A. (1997). On the spatial variability of parameters of unsaturated hydraulic conductivity. *Water Resources Research, 33*(5), 947–956.
Salverda, A. P., & Dane, J. H. (1993). An examination of the Guelph permeameter for measuring the soil's hydraulic properties. *Geoderma, 57*, 405–421.
Sgroi, A. (2005). *Approfondimenti metodologici e verifiche applicative di alcune tecni-che per la misura della conducibilità idraulica del suolo saturo*. Tesi di Dottorato di Ricerca in Idronomia Ambientale, XVII ciclo, Università di Palermo, 142 pp., in Italian.

Shani, U., Hanks, R. J., Bresler, E., & Oliveira, C. A. S. (1987). Field method for estimating hydraulic conductivity and matric potential – Water content relations. *Soil Science Society of America Journal, 51*, 298–302.

Shivakoti, B. R. (2005). Development of a location-specific soil resistance to erosion (SRE) index: A case study from northeast Thailand. In M. Zoebisch, K. M. Cho, S. Hein, & R. Mowla (Eds.), *Integrated watershed management: Studies and experiences from Asia* (pp. 719–736). Bangkok: AIT.

Shouse, P. J., Ellsworth, T. R., & Jobes, J. A. (1994). Steady-state infiltration as a function of measurement scale. *Soil Science, 157*(3), 129–136.

Šimůnek, J., Sejna, M., & van Genuchten, M. Th. (1999). *The HYDRUS-2D software package for simulating two-dimensional movement of water, heat, and multiple solutes in variably saturated media*. Version 2.0. IGWMC-TPS-53. Int. Ground Water Model. Ctr. Colorado School of Mines, Golden.

Smettem, K. R. J., Parlange, J. Y., Ross, J. P., & Haverkamp, R. (1994). Three-dimensional analysis of infiltration from disc infiltrometer. 1. A capillary-based theory. *Water Resources Research, 30*, 2925–2929.

Smettem, K. R. J., Ross, P. J., Haverkamp, R., & Parlange, J. Y. (1995). Three-dimensional analysis of infiltration from the disk infiltrometer. 3. Parameter estimation using a double-disk tension infiltrometer. *Water Resources Research, 31*(10), 2491–2495.

Smiles, D. E., & Knight, J. H. (1976). A note on the use of the Philip infiltration equation. *Australian Journal of Soil Research, 14*, 103–108.

Soilmoisture Equipment Corp. (1987). *2800KI operating instructions*. Santa Barbara: Soilmoisture Equipment Corp.

Soilmoisture Equipment Corp. (1992). *Operating instructions for the pressure infiltrometer attachments to the Guelph permeameter*. Santa Barbara: Soilmoisture Equipment Corp.

Soilmoisture Equipment Corp. (2005). *2805 Guelph pressure infiltrometer instructions*. Santa Barbara: Soilmoisture Equipment Corp.

Soilmoisture Equipment Corp. (2009). *2805D20 pressure infiltrometer attachment assembly*. Santa Barbara: Soilmoisture Equipment Corp.

Swartzendruber, D. (1987). A quasi-solution of Richards equation for the downward infiltration of water into soil. *Water Resources Research, 23*, 809–817.

Talsma, T., & Parlange, J. Y. (1972). One-dimensional vertical infiltration. *Australian Journal of Soil Research, 10*, 143–150.

Tomasella, J., & Hodnett, M. G. (2004). Chapter 21. Pedotransfer functions for tropical soils. In Y. Pachepsky & W. J. Rawls (Eds.), *Development of pedotransfer functions in soil hydrology* (Development in soil science, Vol. 30, pp. 415–429). New York: Elsevier.

Touma, J., Voltz, M., & Albergel, J. (2007). Determining soil saturated hydraulic conductivity and sorptivity from single ring infiltration tests. *European Journal of Soil Science, 58*, 229–238.

Vanderlinden, K., Gabriels, D., & Giráldez, J. V. (1998). Evaluation of infiltration measurements under olive trees in Córdoba. *Soil & Tillage Research, 48*, 303–315.

Vandervaere, J. P., Vauclin, M., & Elrick, D. E. (2000). Transient flow from tension infiltrometers: I. The two-parameter equation. *Soil Science Society of America Journal, 64*, 1263–1272.

van Es, H. M. (2002). 1.2 Soil variability. In J. H. Dane, & G. C. Topp (Co-Eds.), *Methods of soil analysis, Part 4, Physical methods* (Number 5 in the Soil Science Society of America book series, pp. 1–13). Madison: Soil Science Society of America, Inc.

van Genuchten, M. T. (1980). A closed-form equation for predicting the hydraulic conductivity of unsaturated soils. *Soil Science Society of America Journal, 44*(5), 892–898.

Vauclin, M., Elrick, D. E., Thony, J. L., Vachaud, G., Revol, P., & Ruelle, P. (1994). Hydraulic conductivity measurements of the spatial variability of a loamy soil. *Soil Technology, 7*, 181–195.

Vepraskas, M. J., & Williams, J. P. (1995). Hydraulic conductivity of saprolite as a function of sample dimensions and measurement technique. *Soil Science Society of America Journal, 59*, 975–981.

Verbist, K., Baetens, J., Cornelis, W. M., Gabriels, D., Torres, C., & Soto, G. (2009). Hydraulic conductivity as influenced by stoniness in degraded drylands of Chile. *Soil Science Society of America Journal, 73*, 471–484.

Verbist, K., Torfs, S., Cornelis, W. M., Oyarzún, R., Soto, G., & Gabriels, D. (2010). Comparison of single- and double-ring infiltrometer methods on stony soils. *Vadose Zone Journal, 9*, 462–475.

Verbist, K. M. J., Cornelis, W. M., Torfs, S., & Gabriels, D. (2013). Comparing methods to determine hydraulic conductivities on stony soils. *Soil Science Society of America Journal, 77*, 25–42.

Vieira, S. R., Reynolds, W. D., & Topp, G. C. (1988). *Spatial variability of hydraulic properties in a highly structured clay soil*. Proc. symp. validation of flow and transport models for the unsaturated zone, Ruidoso.

Wang, T., Zlotnik, V. A., Wedin, D., & Wally, K. D. (2008). Spatial trends in saturated hydraulic conductivity of vegetated dunes in the Nebraska Sand Hills: Effects of depth and topography. *Journal of Hydrology, 349*, 88–97.

Warrick, A. W. (2003). *Soil water dynamics*. New York: Oxford University Press.

Waugh, J., Smith, G., Danforth, B., Gee, G., Kothari, V., & Pauling, T. (2006). *Performance evaluation of the engineered cover at the Lakeview, Oregon, Uranium Mill Tailings Site*. WM'07 conference, February 25 – March 1, 2006, Tucson.

White, I., & Sully, M. J. (1987). Macroscopic and microscopic capillary length and time scales from field infiltration. *Water Resources Research, 23*(8), 1514–1522.

White, I., & Sully, M. J. (1992). On the variability and use of the hydraulic conductivity alpha parameter in stochastic treatments of unsaturated flow. *Water Resources Research, 28*(1), 209–213.

Wilson, G. V., Alfonsi, J. M., & Jardine, P. M. (1989). Spatial variability of saturated hydraulic conductivity of the subsoil of two forested watersheds. *Soil Science Society of America Journal, 53*, 679–685.

Wooding, R. A. (1968). Steady infiltration from a shallow circular pond. *Water Resources Research, 4*, 1259–1273.

Wu, L., & Pan, L. (1997). A generalized solution to infiltration from single-ring infiltrometers by scaling. *Soil Science Society of America Journal, 61*, 1318–1322.

Wu, L., Swan, J. B., Nieber, J. L., & Allmaras, R. R. (1993). Soil-macropore and layer influences on saturated hydraulic conductivity measured with borehole permeameters. *Soil Science Society of America Journal, 57*, 917–923.

Wu, L., Pan, L., Roberson, M. J., & Shouse, P. J. (1997). Numerical evaluation of ring-infiltrometers under various soil conditions. *Soil Science, 162*, 771–777.

Wu, L., Pan, L., Mitchell, J., & Sanden, B. (1999). Measuring saturated hydraulic conductivity using a generalized solution for single-ring infiltrometers. *Soil Science Society of America Journal, 63*, 788–792.

Wuest, S. B. (2005). Bias in ponded infiltration estimates due to sample volume and shape. *Vadose Zone Journal, 4*, 1183–1190.

Xu, D., & Mermoud, A. (2003). Modeling the soil water balance based on time-dependent hydraulic conductivity under different tillage practices. *Agricultural Water Management, 63*, 139–151.

Yitayew, M., Khan, A. A., & Warrick, A. W. (1998). In situ measurement of soil hydraulic conductivity using point application of water. *Applied Engineering in Agriculture, 14*(2), 115–120.

Youngs, E. G. (1987). Estimating hydraulic conductivity values from ring infiltrometer measurements. *Journal of Soil Science, 38*, 623–632.

Youngs, E. G., Leeds-Harrison, P. B., & Elrick, D. E. (1995). The hydraulic conductivity of low permeability wet soils used as landfill lining and capping material: analysis of pressure infiltrometer measurements. *Soil Technology, 8*, 153–160.

Zangar, C. N. (1953). *Theory and problems of water percolation* (Engineering monograph no.8). Denver: U.S. Department of the Interior, Bureau of Reclamation.
Zhang, R. D. (1997). Scale-dependent soil hydraulic conductivity. In M. M. Novak & T. G. Dewey (Eds.), *Fractal froniers* (pp. 383–392). London: World Scientific.
Zhang, Z. F., Groenevelt, P. H., & Parkin, G. W. (1998). The well-shape factor for the measurement of soil hydraulic properties using the Guelph permeameter. *Soil & Tillage Research, 49*, 219–221.

Chapter 3
Unsaturated Soil Hydraulic Properties

3.1 Tension or Disk Infiltrometer

3.1.1 Basis of the Technique

The tension infiltrometer (TI), also known as disk infiltrometer or disk permeameter (Perroux and White 1988; Reynolds 2008), primarily allows determination of unconfined axisymmetric infiltration from a surface circular source under a constant, negative pressure head. The TI is used primarily for field measurement of soil hydraulic conductivity, K (L T^{-1}), and sorptivity, S (L T$^{-1/2}$), corresponding to near-saturated conditions, that is for pore water pressure heads, h (L), in the range $-200 \leq h \leq 0$ mm. It can also be used, however, to determine near-saturated sorptive number, α^* (L^{-1}), flow-weighted mean pore radius, R_0 (L), and number of flow-weighted mean pores per unit area, N_0 (L^{-2}) (Reynolds 2008). Unconfined infiltration measurements made with devices that supply water to the soil under a slight tension reflect the impact of the fragile structural macropores that dominate flow at and close to saturation (Jarvis et al. 2013). Assessing this impact is important to properly interpret and simulate soil water fluxes and storages.

The device consists essentially of a 0.10–0.20 m diameter porous plate or disk connected to a water reservoir and a Mariotte-type bubble tower. The water reservoir supplies water to the porous disk and the bubble tower determines the pressure head, h_0 (L), at the bottom of this disk. A layer of fine sand or other contact material is frequently placed under the infiltrometer to ensure good hydraulic connection between the plate and the soil.

Water flow from the TI induces the development of an unsaturated wetting bulb in which h varies from h_0 at the soil surface under the disk to the antecedent soil pore pressure head, h_i (L), at the wetting front (Fig. 3.1). After a transient phase during which the infiltration rate, i (L T^{-1}), decreases, the flow approaches a steady-state condition characterized by a constant steady-state value, i_s (L T^{-1}). The rate of water flow or discharge out of the TI and into the porous medium can be monitored

R. Angulo-Jaramillo et al., *Infiltration Measurements for Soil Hydraulic Characterization*,
DOI 10.1007/978-3-319-31788-5_3

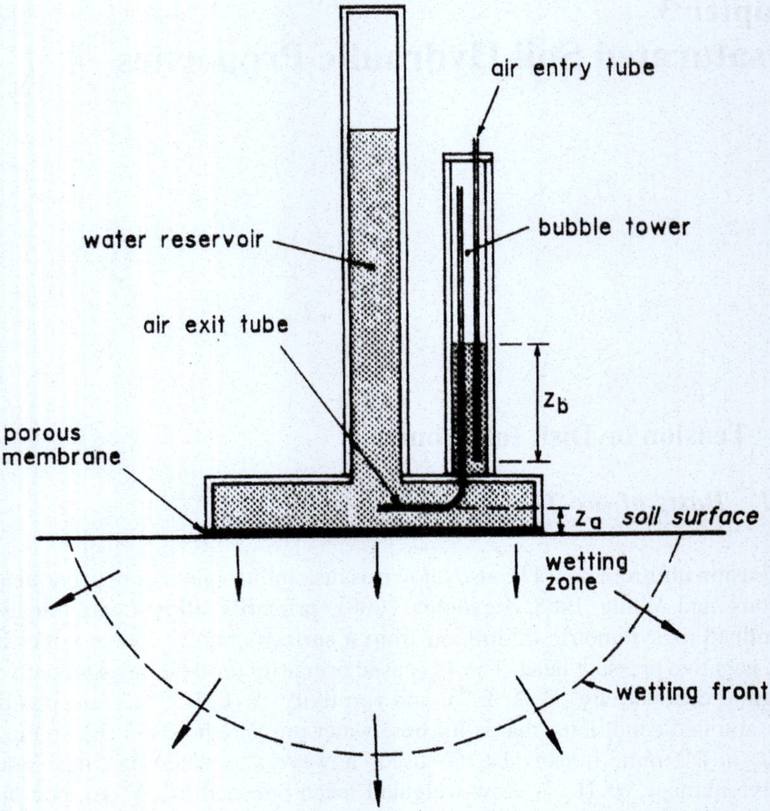

Fig. 3.1 Schematic view of the tension infiltrometer and three-dimensional wetting with axial symmetry below the porous disk

by manually reading the water level at the graduated scale of the reservoir at given time intervals or by automatized pressure transducer – data-logger systems (Ankeny et al. 1988; Ankeny 1992; Casey and Derby 2002). The duration of the TI experiment can vary between a few minutes and several hours, depending on the soil, the type of data analysis used and the number of pressure heads sequentially imposed at the infiltration surface. In particular, several steady-flow and transient-flow analysis procedures have been proposed which involve single, concentric, or multiple infiltrometers and single or multiple pressure heads (Elrick and Reynolds 1992a; White et al. 1992; Smettem and Smith 2002).

Water flux in the soil under the disk infiltrometer is due to capillarity and gravity. However, compared with one-dimensional infiltration, the importance of capillarity relative to gravity is greater. The time required to reach steady-state is reduced and the steady-state infiltration rate is greater for three-dimensional axisymmetric flow than for the corresponding one-dimensional infiltration (Clothier and Scotter 2002a).

An important strength of the TI method is that the apparatus is simple, inexpensive, portable, easily applied in both field and laboratory studies, and it requires only small volumes of water. Therefore, detailed investigations of spatial variability of soil hydraulic properties, field studies in areas with difficult access, and large-scale surveys are more feasible with this method than with many other methods (Reynolds 2008). Unlike borehole permeameters and some ring infiltrometers, the potential for the TI to disturb the sampled soil is minimized since augering a well or deeply inserting a ring are not required to apply the method. Therefore, the measured hydraulic properties of structured soils are expected to be highly plausible. Given that flow is three-dimensional, the measured $K(h)$ and capillary relationships are most relevant to three-dimensional unconfined flow in the near-saturated range. This characteristic is particularly important in a number of hydrologic, agronomic and environmental applications given that water flux and solute transport rates are highest close to saturation (Bagarello et al. 2007). According to Vandervaere (2002), using soil properties estimated with the TI to simulate hydrological processes different from infiltration is not recommended.

3.1.2 Apparatus and Procedures

The TI design and setup may be slightly different depending on the specific requirements by the soil scientists that developed the technique (Perroux and White 1988; Ankeny et al. 1988; Angulo-Jaramillo et al. 2000, among others). Commercial suppliers include Soilmoisture Equipment Corp., Goleta, CA, U.S.A. (www.soilmoisture.com), Soil Measurement Systems, Tucson, AZ, U.S.A. (www.soilmeasurement.com), Decagon Devices Inc., Pulmann, WA, U.S.A. (www.decagon.com), Eijkelkamp Agrisearch Equipment B.V., Giesbeek, The Netherlands (www.eijkelkamp.com), and SDEC, Reignac sur Erdre, France (www.sdec-france.com). Regardless of the specific design, all devices operate on the same physical principles and basically consist of three major components: (i) the bubble tower, which controls the tension (negative pressure) at the soil surface; (ii) the reservoir, which supplies water to the infiltration surface; and (iii) the porous disk, which establishes hydraulic continuity with the soil. The bubble tower contains an air-entry tube that can be moved vertically to control h_0 and an air-exit tube that establishes the negative pressure set into the bubble tower at the reservoir base. In particular, the pressure head imposed at the soil surface is the difference between the distance from the air-exit tube to the infiltration surface, z_a (L), and the height of water in the bubble tower above the bubble entry point, z_b (L) (Fig. 3.1). The distance z_a depends on the instrument design and can be considered as a constant if the disk is rigidly assembled to the water reservoir as in Fig. 3.1. For an infiltrometer with a separate base plate unit (Fig. 3.2), it is important that z_a in the field remains exactly equal to the value set during the laboratory calibration procedure. Pressure distribution in the water reservoir is hydrostatic with a maximum (i.e., less negative) value at the air-exit point, equal

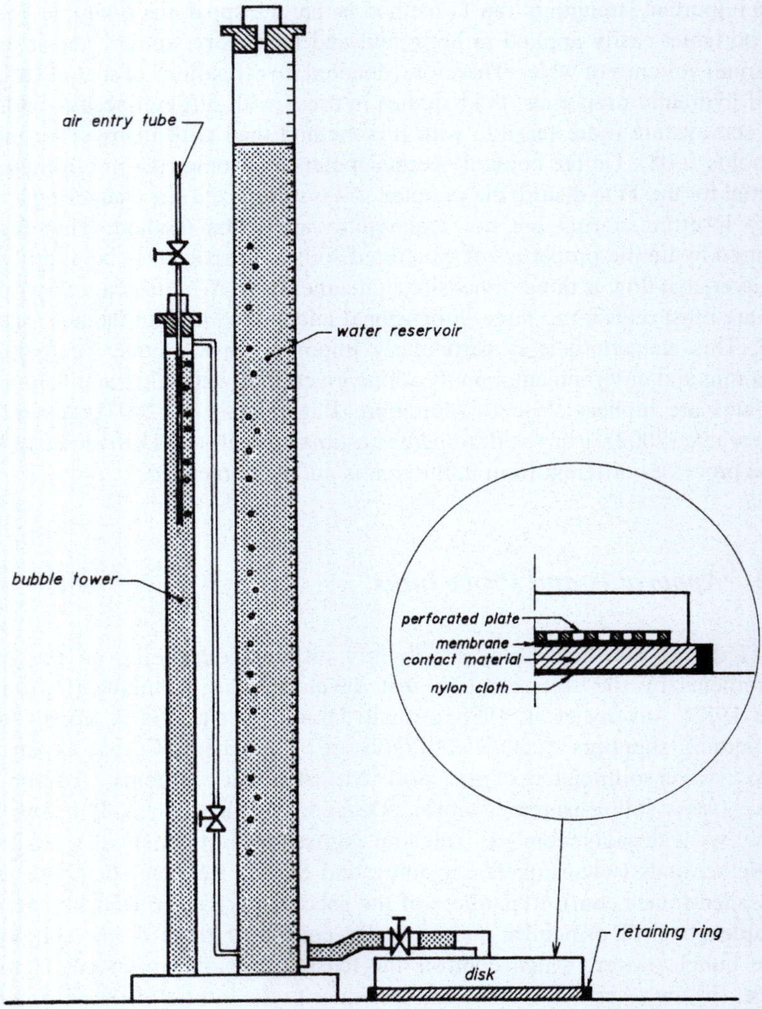

Fig. 3.2 Tension infiltrometer with separate baseplate unit (From Soilmoisture Equipment Corp. 2008a, reprinted with permission)

to $-z_b$, and a minimum value at the water surface, equal to $-z_b-L$, L (L) being the water level above the air-exit point.

The TI is generally fabricated from either transparent polycarbonate or polymethylmetacrylate (Plexiglas) that allows control of imposed pressure head in the bubble tower and visual reading of the water levels in the reservoir. The infiltrometer plate is constructed out of a rigid disk with a perforated bottom that supports a nylon membrane sealed to the plate by a tight-fitting O-ring. It is essential that the supply membrane be visible during infiltration to permit examination for air leaks (Perroux and White 1988).

The nylon membrane is available in a variety of pore sizes that enable h_0 to be selected in the range $-1 \leq h_0 \leq 0$ m. In any case, the air-entry pressure head of the porous membrane needs to be lower than the minimum pressure head to be applied to assure that, during infiltration, its pores remain saturated. Perroux and White (1988) used a 63 μm mesh size membrane with an air-entry pressure head of −0.2 m. A nylon membrane with similar characteristics was used by Bagarello et al. (2001). Castiglione et al. (2005) used a stainless-steel porous membrane having a mean pore size of 0.2 μm and a bubbling pressure of −0.60 m to characterize water flow in fractured rock formations where exposed rock surfaces are often jagged, with relatively sharp edges that may easily damage the porous nylon membrane used in commercially available infiltrometers. The conductance, that is the ratio between the saturated hydraulic conductivity and the thickness of the infiltrometer membrane, may dominate flow initially, resulting in a pressure head at the membrane/soil surface interface lower than the imposed h_0 value. However, for conductance values higher than 0.03 s^{-1}, the membrane conductance influences only the first 10 s of water adsorption (Perroux and White 1988). In the Guelph Tension Infiltrometer manufactured by Soilmoisture Equipment Corp. (Reynolds 2008), the infiltrometer plate consists of a hydrophilic polyethylene plastic plate connected to the same concentric water reservoir assembly used for the Guelph permeameter (Fig. 3.3).

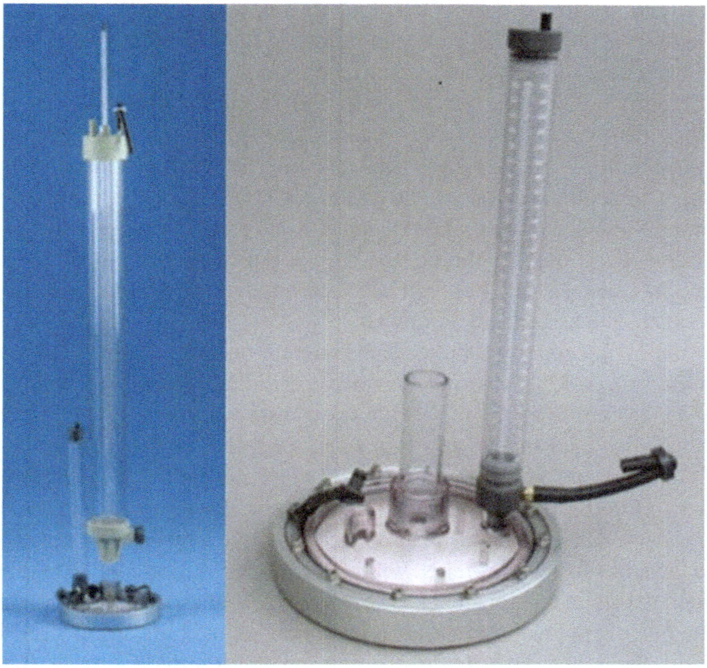

Fig. 3.3 Guelph Tension Infiltrometer with assembled baseplate unit (From Soilmoisture Equipment Corp. 2008b, reprinted with permission)

It is desirable for the porous disks to wet spontaneously when placed in water, as this produces a greater degree of saturation of the disk and consequently greater disk conductivity. To improve wettability, the porous disk can first be cleaned with a solvent (e.g., ethanol, dithionite-citrate solution, dilute bleach solution) to remove hydrophobic oils, oxides or organic matter, and then submerged in a surfactant solution (49 % by vol. isopropyl alcohol, 49 % distilled water, 2 % Triton × 100 surfactant) for 24–36 h. Hydraulic conductance of the porous plate should periodically be checked given that it may decrease with use as a consequence of pore blockage due to physical and biological clogging. To prevent damage and maintain saturation during transport, the infiltrometer plate should be placed on padding (e.g., sponge or rubber foam) in a flat-bottomed pail and maintained submerged below several centimetres of water. Porous plates with a radius, r (L), ranging from 27.5 to 125 mm are used under most common field and laboratory conditions (Angulo-Jaramillo et al. 2000). The approach to steady-state flow is more rapid with a small radius disk than with a large disk. However, the characteristic dimension of the source should be larger than the soil macroscopic capillary length, λ_c (L). In fact, as discussed in the next paragraph, the Wooding's (1968) solution, that is widely used for analysing steady-state TI data, is valid for $r/\lambda_c > 2$ (Cook and Broeren 1994).

In the TI originally designed by Perroux and White (1988), the porous plate is assembled to the water reservoir (Fig. 3.1). Similar devices were proposed, among others, by Ankeny (1992), Angulo-Jaramillo et al. (2000) and Reynolds (2008). With these devices, the weight of the water contained in the reservoir results in a small overload on the infiltration surface that can improve hydraulic contact with the soil. However, assembled TIs are more prone to wind induced oscillations with negative effects on the hydraulic continuity between the porous plate and the infiltration surface (Vandervaere et al. 2000a). Separate baseplate TIs, consisting of a porous disk connected to the water reservoir/bubble tower apparatus by a flexible plastic tube (Fig. 3.2), have been designed to overcome this drawback. This arrangement also allows a faster and easier substitution of the infiltration disk in case of membrane leakage or when disks with different diameters have to be used. Practical considerations for the device with a separate porous plate will now be discussed. In most cases, they can also be applied, if necessary with minimal adjustments, to the assembled TI device.

Water reservoir capacity influences the amount of water that can uninterruptedly infiltrate the soil. In highly permeable soils, periodic refilling may be necessary. It may best be accomplished at the end of a measurement for a given pressure head. Both the air-entry and the air-exit tubes and the valve of the flexible supply tube (Fig. 3.2) must be closed, then the stopper of the water reservoir is removed and the reservoir is refilled. This procedure establishes a positive pressure at the bottom of the reservoir. When infiltration restarts, this positive pressure is inevitably transmitted to the soil surface and hence will affect the infiltration rate. The time needed to restore the original tension depends on the volume of air pocket left upon refilling and on the rate of expansion of the air pocket (i.e., on the infiltration rate). This does not usually represent a problem for tension infiltration in soils (Castiglione et al. 2005). In any case, the refilling procedure should be completed as rapidly as

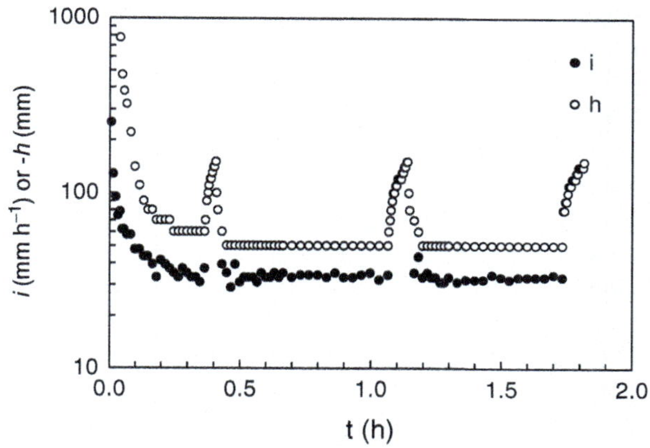

Fig. 3.4 Infiltration rate, *i*, and soil water pressure head, *h*, versus time, *t*, during an infiltration test that was interrupted for reservoir refilling at $t = 0.37$ h and $t = 1.07$ h and restarted after 0.05 h and 0.08 h, respectively (From Bagarello et al. 2001, reprinted with permission)

possible to limit its influence on the experiment. It is also recommended to provide the reservoir with an air vent valve to avoid overpressure in the air pocket during replacement of the stopper. In extremely low permeability media, like unsaturated fractured rock at relatively low pressure heads, the disturbance induced during refilling may last for a long time. Castiglione et al. (2005) developed an automated TI prototype, specifically designed for very low infiltration rate measurements, in which a system of solenoid valves controlled by a data-logger permits the automatic refilling of the reservoir in such a way that no volume of air is left above the water column. In this way, the negative pressure at the surface is instantaneously re-established as infiltration restarts. Bagarello et al. (2001) showed that removing the instrument from the infiltration surface for about 5 min did not affect the steady infiltration rates appreciably, and the removal/replacement procedure had a short term influence on the pressure head applied at the infiltration surface (Fig. 3.4). Nevertheless, to avoid refilling, it could be convenient to use large volume reservoirs or to connect two water reservoir units to the same porous plate, switching supply from one reservoir to the other during refilling. In this case, it is critically important that the two reservoirs connected in parallel are set to the same pressure head and that the reservoir bases and the disk are at the same height (Bagarello and Iovino 2010). Increasing the reservoir capacity may have adverse effects on measurement accuracy with both manual or automatic readings of the water level. Particularly in low permeability soils, water reservoirs with large cross-sectional areas may induce significant errors on infiltration rate estimation. For example, the reservoir's internal diameter of the TI manufactured by Soil Measurement Systems is 44.5 mm (cross-sectional area, $A_s = 1555$ mm^2). With a disk of radius $r = 120$ mm (infiltration area, $A_0 = 45239$ mm^2), each millimeter in water level drop corresponds to an infiltration height of 0.034 mm. Assuming that the precision of water level measurements by visual readings is 0.5 mm, it follows that water infiltration height can be estimated with a precision of approximately 0.017 mm. To increase accuracy at low infiltration rates, one or more plastic cylinders with known cross-sectional area can be placed inside the water reservoir in order to increase water level drop for a given infiltration volume (Jarvis and Messing 1995). Alternatively, a reservoir assembly constituted by two concentric cylindrical reservoirs, similar to that of the Guelph permeameter, can be used (Reynolds 2008). At low pressure heads or in low permeability soils, water is supplied from the inner reservoir only, whereas both the inner and the outer reservoirs are used at high pressure heads or in high permeability soils. A three-way valve at the bottom of the concentric reservoirs allows us to select the reservoir assembly configuration to be used for a run (inner reservoir alone, inner and outer reservoirs) on the basis of the sampled soil characteristics, to provide in any case adequate resolution of water level measurement.

Automatic reading of water levels can be conducted by pressure transducers connected to a data-logger for data acquisition and storage. Ankeny et al. (1991) used temperature-compensated, solid-state piezoelectric devices. These transducers give a linear voltage output as a function of tension (negative pressure). Head space tension in the air pocket at the top of the water reservoir is linearly related to the height of water in the column. A unit change in height causes a unit change in tension. Thus, infiltration can be calculated from the change in head space tension measured by the transducer mounted at the top of the water reservoir (Fig. 3.5a). In a Mariotte system, however, as bubbles expand and detach from the air-exit tube, the tension varies resulting in a characteristic fluctuation that can be cancelled out if a second transducer is mounted near the base of the water column (Fig. 3.5b). In this case, the difference between the top and the bottom transducers is dependent only upon the height of the water column (Ankeny 1992). For a TI device with separate plate, the bottom transducer can be mounted on the top of the disk, thus also allowing a control of the pressure head applied on the infiltration surface. The change in height registered by the transducers is multiplied by the cross-sectional area of the reservoir and divided by the infiltration surface area to obtain the infiltration rate. An improvement in measurement precision can be obtained by a differential pressure transducer with one port installed at the bottom position on the reservoir and the other port connected, using tubing, to the head space of the reservoir (Fig. 3.5c). The use of a differential transducer eliminates errors associated with the synchronization of the two transducers and also simplifies the extensive calibrations required by the two gauge transducer method (Casey and Derby 2002). Linearity of the relationship between output voltage and negative pressure has to be verified before use by the pressure transducer calibration procedure outlined in Sect. 2.2.1.1. An original automated design for TI data acquisition that is based on the use of time domain reflectometry (TDR), was proposed by Moret et al. (2004). In their device, a three-rod coaxial TDR probe is firmly placed in the centre of the water reservoir (Fig. 3.5d) and connected to a cable tester that, at predetermined intervals, measures the apparent dielectric constant of the two-phase system consisting of a probe partially immersed into water. The height of water in the reservoir is then calculated considering the probe length and the relative dielectric constants of air and water. The proposed technique allows for simultaneous TDR recording of both water flow and transient volumetric water content of soil below the infiltrometer disk, thus offering the possibility to integrate cumulative infiltration and soil moisture measurements into a single automated system.

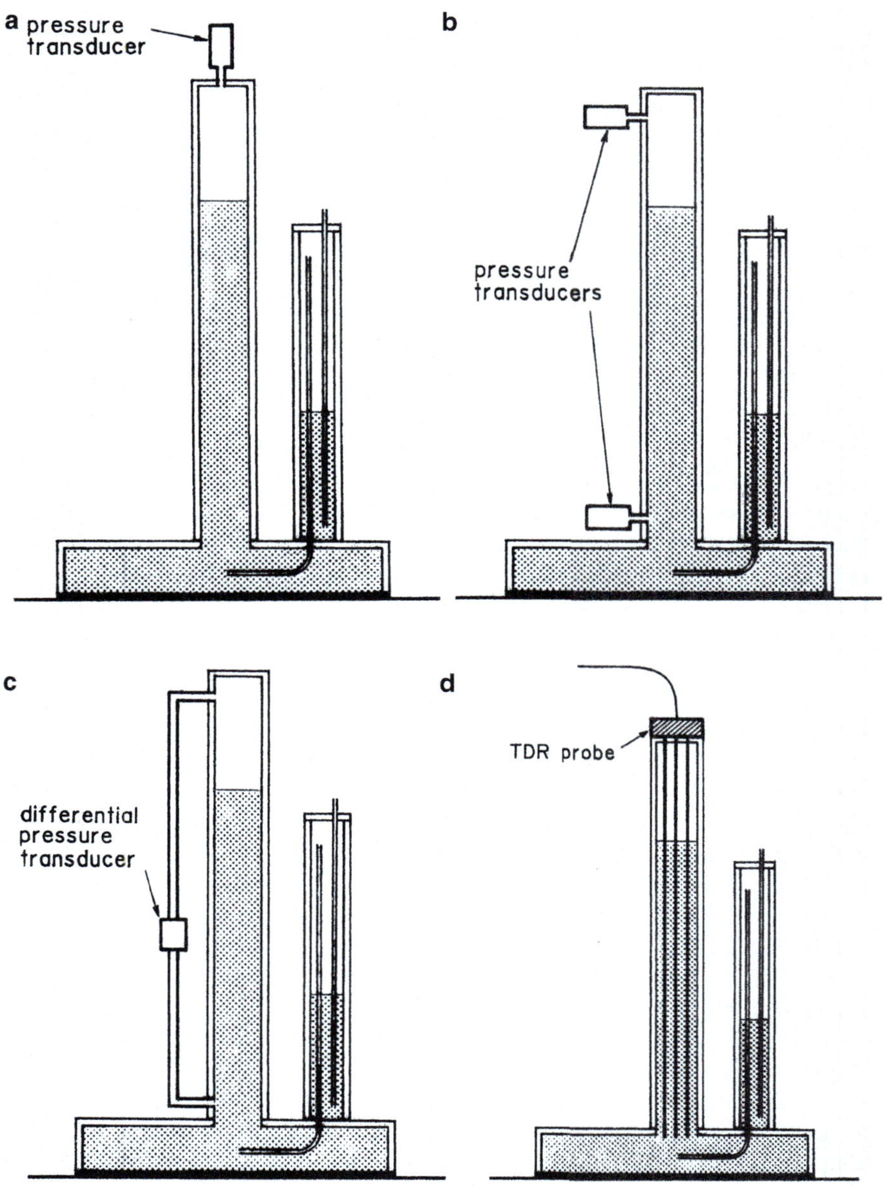

Fig. 3.5 Tension infiltrometer automation design with: (**a**) single pressure transducer, (**b**) double pressure transducer, (**c**) differential pressure transducer, (**d**) TDR probe

Pressure head at the soil surface can be regulated by moving the air-entry tube vertically so that the difference between the water level and the air-entry port in the bubble tower corresponds to the desired tension z_b (Fig. 3.1). A measuring tape on the wall of the bubble tower allows the setup of the desired pressure head. If

Fig. 3.6 Check for horizontality between the disk and the reservoir base

pressure head has to be changed during the TI application in the field, as with multi-potential experiments, the bubble-induced fluctuations in the bubble tower may induce erroneous assessments of the value of z_b. It is therefore recommended to preliminarily perform a laboratory calibration of the device, to set up accurately the differences z_b corresponding to the pressure heads, h_0, to be applied in the field (Eijkelkamp Agrisearch Equipment 2009). Calibration can be performed by connecting the reservoir outflow tube to a water manometer and applying a small vacuum in the head space of the reservoir in order to keep air slowly bubbling through the device (Ankeny 1992). With the base of the reservoir assumed as zero reference (i.e., soil surface), the distance to which the water is pulled out of the manometer represents the pressure head (in mm of water) that will be applied to the soil surface. The relative position of the water level and the air-entry tube in the bubble tower (i.e., z_b) must be noted and then accurately reproduced during the field experiment. To assure that the calibrated pressure head is the one really applied on the porous membrane of the device, a level and a sawhorse should be used to check that the disk and the reservoir base are always at the same height (Fig. 3.6).

Abrasion by contact material or soil particles may damage the porous membrane and air will leak through the disk. Air will appear as a stream of bubbles entering from the hole in the membrane. Such holes and break in the nylon weave can be plugged using a couple of drops of epoxy/resin glue, although membranes with several holes may need to be replaced. To check for porous membrane integrity, the previously saturated disk can be connected to an adjustable spillway (Fig. 3.7). With the bottom of the disk up, the spillway is lowered until a stream of bubbles flow into the disk. If the membrane is intact and fully saturated, the air should enter when the vertical distance between the disk surface and the spillway is equal to the air-entry potential of the membrane.

The porous membrane can be saturated by submerging the entire infiltrometer plate for several hours in a water bath containing several centimetres of de-aired, temperature-equilibrated water and periodically jarring the plate against the wall of the water bath to remove air that may be entrapped in the disk/membrane. The infiltrometer plate is connected to the water supply/bubble tower unit directly in the field, accurately

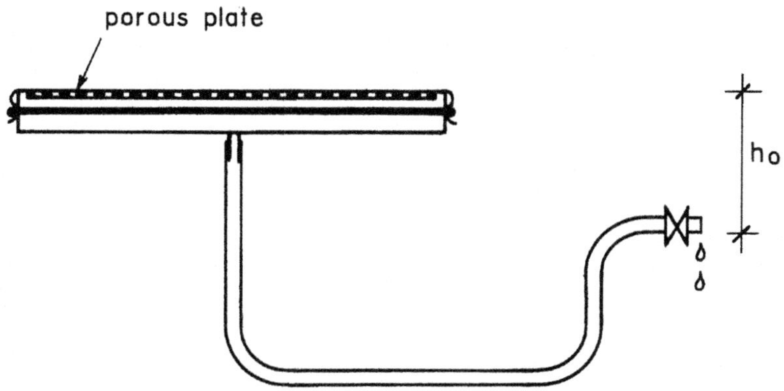

Fig. 3.7 Check for porous membrane leaks

checking that all air has been eliminated. For a TI with porous plate assembled with water reservoir, saturation is achieved by drawing the water up through the porous disk by a vacuum pump connected to the stopper of the water reservoir.

Use of tap water for conducting experiments is preferred to deionized water as the latter may cause clay soil dispersion with alteration of soil hydraulic properties (Bagarello et al. 2006b). However, use of a solution with electrolytic composition and concentration similar to the soil water is recommended in highly unstable soils (Reynolds et al. 2002). In any case, the water temperature should previously be equilibrated with that of soil water.

A layer of contact material should be placed under the tension infiltrometer to establish and maintain good hydraulic connection between porous disk and soil. This should be done regardless of whether the soil surface has been smoothed, levelled or left undisturbed (Perroux and White 1988; Bagarello et al. 2001; Vandervaere 2002; Smettem and Smith 2002) and regardless of whether steady-state or transient analyses are used (Vandervaere 2002), although investigations that do not make use of contact material can be found in the literature (e.g., Logsdon and Jaynes 1993). The contact material layer can introduce a difference between the pressure head set on the porous membrane and that applied on the soil surface. This difference must be accounted for in the analysis of the TI data (Reynolds and Zebchuk 1996; Bagarello et al. 2001; Reynolds 2006) (see Sect. 3.1.7). The air-dry contact material can be poured into a sharpened retaining ring with internal diameter equal to that of the infiltrometer plate. The primary purpose of the retaining ring is to ensure that the flow cross-section is circular and that flow within the contact material is vertical and rectilinear as required by theory (Reynolds 2008). Under the layer of contact material, it is advisable to lay a circle of flexible "guard" cloth, pre-cut to the same inside diameter of the retaining ring, with an air-entry potential lower than the minimum potential imposed by the TI (Bagarello et al. 2000). The purpose of the guard cloth is to prevent the contact material from infilling cracks, worm holes and other macropores present at the infiltration surface. Significant infilling of macropores can change the hydraulic properties of the porous medium, as well as greatly increase the amount of contact material required.

Sites including large attached debris, such as partially exposed roots, plant residues, stones and large clods, should be avoided as this material can cause poor hydraulic contact. Removal of attached debris or extraction of plant roots is not recommended as this operation can modify the soil structure and then impacts its hydraulic properties at the infiltration surface. Vegetation, if present, should be clipped flush with the soil surface. Subsurface measurements, or measurements on a hillslope, require the preliminary excavation of the site to expose a horizontal surface. Site preparation, in these cases, requires procedures that minimize smearing and compaction of soil surface. Use of the TI on sloping land is discussed in Sect. 3.1.8.

After preparing the infiltration site, the porous plate is extracted from the saturation bucket and it is carefully lowered on the contact material layer using a slight twist to ensure contact. At the same time, the valves of the air-entry and air-exit tubes of the bubble tower and the valve of the flexible plastic tube connecting the porous disk to the water reservoir/bubble tower apparatus (Fig. 3.2) are opened and the measurement of time begins. The infiltrometer is operating properly when air bubbles rise regularly up into the reservoir from the air-exit tube and into the bubble tower from the air-entry tube. Reading the water level at the graduated scale of the reservoir at given time intervals allows calculation of infiltration rates. Under a constant pressure head, h_0, the rate of fall in the water reservoir, R (L T^{-1}), normally decreases with time and approaches a constant value as the flow rate becomes quasi steady. Steady-state flow is assumed when the same R value is obtained over four or five consecutive measurements (i.e., R constant within about ±5 %, with no clear trend) (Reynolds 2008). Steady-state flow rate can also be obtained using the least squares regression slope of the visually selected linear portion of the cumulative infiltration, I (L), versus time, t (T), curve (Bagarello et al. 1999). The time required to achieve steady-state flow (time-to-steady-state) may then be estimated as the time corresponding to the first measured I value deviating from the corresponding regression-based cumulative infiltration by a fixed quantity (usually 2 %). Different factors, such as poor wettability due to water repellence, heterogeneous soil structure and/or changes in soil structure during the measurement, can influence time evolution of transient infiltration rate (Logsdon 1997).

For multi-potential experiments, once steady-state at a given h_0 value is attained, the position of the air-entry tube in the bubble tower is adjusted to set the successive values of desired water tensions. The pressure head values can be applied either in an ascending ($h_1 < h_2 < h_3 \ldots$) or descending ($h_1 > h_2 > h_3 \ldots$) order. A descending sequence is suggested to save time (Ankeny et al. 1991; Ankeny 1992). By conducting the highest pressure head step first, the wetting front advances as rapidly as possible and the unit gradient assumption is more likely to be valid (Mohanty et al. 1994). On the other hand, an ascending sequence is recommended in order to avoid hysteresis effects due to drainage occurring close to the disk while wetting continues at the infiltration front (Reynolds and Elrick 1991; Jarvis and Messing 1995). Actually, when an ascending sequence is applied, the initial soil water content of the sampled soil volume is not uniform starting from the second pressure head onwards, but a wetting process occurs for all pressure heads. Therefore, all the hydraulic conductivity values corresponding to the imposed pressure heads are obtained under wetting. When a descending sequence is applied, a wetting process is established for the first pressure head. Both drainage and wetting processes occur from the second pressure head onwards. Therefore, the measured conductivity derives from a combination of different processes. For these reasons, it is usually recommended that a sequence of 3–5 ascending pressure head values, with the last value at or close to zero, be used to provide adequate definition of the various near-saturated flow parameters (Reynolds 2008). However, both appreciable (Logsdon et al. 1993; McKenzie et al. 2001) and practically negligible (Bagarello et al. 2000; Mecke et al. 2000) effects of the pressure head sequence used for a multi-potential TI experiment on the measured conductivities have been reported. Bagarello et al. (2005) studied the influence of both the pressure head sequence and the highest h_0 value of the descending sequence on the hydraulic conductivity of a sandy-loam soil. Two increasing/decreasing sequences, only differing by the highest h_0 value, were considered. In particular, unsaturated conditions were maintained for the entire test in some sites (i.e., highest $h_0 = -10$ mm), whereas a slight positive pressure head ($h_0 = +5$ mm) was applied in other sites. Hydraulic conductivities obtained with the draining sequence were up to 3.3 times higher than the corresponding K_0 values measured with the wetting sequence and this result was in agreement with the hysteresis theory that predicts a higher conductivity with a drying sequence. Furthermore, the deviations between the two sequences were more noticeable in sites where saturated conditions were achieved at the end of the wetting sequence. Assuming as acceptable an uncertainty in the soil hydraulic conductivity estimates by a factor of 2–3 (Elrick and Reynolds 1992b; Reynolds and Zebchuk 1996; Elrick et al. 2002), Bagarello et al. (2005) concluded that both the pressure head sequence (ascending or descending) and the highest value of h_0 were minor factors for a rough hydraulic characterization of a study area. However, these two factors have to be considered to obtain truly comparable data from different experiments and, therefore, to reduce uncertainties in comparative studies.

3.1.3 Analysis of Steady-State Flow

Most analyses of steady-state infiltration from a TI set at a constant pressure head are based on the Wooding's (1968) "shallow pond" relationship, written here as (Warrick 1992):

$$Q_s = \pi r^2 K_0 + 4\frac{K_0}{K_0 - K_i}\phi_0 r \tag{3.1}$$

where Q_s (L^3T^{-1}) is the steady-state flow rate, r (L) is the disk radius, $K_0 = K(h_0)$ (L T^{-1}) is the soil hydraulic conductivity corresponding to the applied pressure head on the infiltration surface, h_0 (L), K_i (L T^{-1}) is the soil hydraulic conductivity corresponding to the antecedent pressure head in the soil before the infiltration takes place, h_i (L), and $\phi_0 = \phi(h_0)$ (L^2T^{-1}) is the matric flux potential (Gardner 1958):

$$\phi(h_0) = \int_{h_i}^{h_0} K(h)dh \tag{3.2}$$

h (L) being the soil water pressure head. The first term on the right hand side of Eq. (3.1) ($\pi r^2 K_0$) accounts for the gravity driven flow whereas the second term $[4 K_0 \phi_0 r/(K_0 - K_i)]$ expresses the capillary flow induced by the unsaturated soil matrix (Smettem et al. 1994). According to Angulo-Jaramillo et al. (2000), total flow is controlled by the soil hydraulic conductivity, K_0, which accounts for the effect of gravity, and the soil sorptivity, $S_0 = S(h_0)$ ($L\,T^{-1/2}$), that represents, in an integral sense, the soil's capillarity. The Wooding's solution of steady flow from a circular source in an unsaturated porous medium uses an exponential hydraulic conductivity function (Gardner 1958):

$$K(h) = K_{fs}\exp(\alpha h) \tag{3.3}$$

where K_{fs} ($L\,T^{-1}$) is the field-saturated soil hydraulic conductivity and α (L^{-1}) is a soil texture/structure parameter that expresses the relative importance of the gravity and capillary forces during water movement in unsaturated soil (Philip 1969). In the context of Eq. (3.1), a large α corresponds to tension infiltration dominated by gravity (i.e., the capillary forces are low) while a small α indicates infiltration dominated by capillarity (Reynolds and Elrick 1991). According to Philip (1969, 1985), Eq. (3.3) with α = constant represents the $K(h)$ relationship fairly well over the range of h which is of interest to most soil water studies. On the other hand, Parlange (1972) stated that, for real soils, $\alpha = \alpha(h)$ in Eq. (3.3).

As specified by Reynolds and Topp (2008), if the $K(h)$ relationship is ruled by Eq. (3.3), then Eq. (3.2) becomes:

$$\phi_0 = \phi(h_0) = \frac{K_0 - K_i}{\alpha(h_0)} \tag{3.4}$$

where $\alpha(h_0)$ (L^{-1}) gives the slope of ln K vs. h. Generally speaking, $\alpha(h_0)$ increases as h_0 increases, indicating an increase in the importance of the gravity component of infiltration relative to the capillary component as the soil gets wetter. The closer the $K(h)$ relationship is to a monotonic exponential function, i.e. Eq. (3.3), the closer the $\alpha(h_0)$ relationship is to a single constant value. For most natural porous materials at field capacity or drier, $h_i < -1$ m, $K_i \ll K_0$ and Eq. (3.4) can consequently be simplified to:

$$\phi_0 \approx \frac{K_0}{\alpha(h_0)} \tag{3.5}$$

Equation (3.5) is generally used rather than Eq. (3.4) because it avoids difficulties connected to the determination of hydraulic conductivity corresponding to very dry soil conditions. The assumption $K_i \ll K_0$ becomes progressively more incorrect as h_0 decreases, resulting in increasing error in Eq. (3.5) with smaller (more negative) h_0 values. The α parameter must consequently be used with caution when K_i is not substantially less than K_0, such as might occur in very wet porous materials or in fine-textured materials where $K(h)$ does not decrease rapidly with decreasing h.

If K_i is negligibly small relative to K_0 (Reynolds and Elrick 1991; Logsdon and Jaynes 1993), combining Eqs. (3.1) and (3.5) leads to :

$$i_s = K_0\left(1 + \frac{4\lambda_c}{\pi r}\right) \tag{3.6}$$

where $i_s = Q_s/\pi r^2$ (L T^{-1}) is the steady infiltration rate and λ_c (L) is the macroscopic capillary length that, under the assumptions of a hydraulic conductivity function expressed by Eq. (3.3) and $K_i \ll K_0$, is given by (Smettem and Smith 2002):

$$\lambda_c = \frac{\phi_0}{K_0} = \frac{1}{\alpha} \tag{3.7}$$

Equation (3.6) can be applied for both negative and null constant pressure head at the infiltration surface. Reynolds and Elrick (1991) extended the steady tension infiltration theory to a ring inserted at distance, d (L), into the soil. In their generalised solution, the term $(4\lambda_c/[\pi r])$ on the right hand side of Eq. (3.6) was substituted by $(\lambda_c/[G_r \pi r])$, in which G_r is a shape factor that was found to depend on soil hydraulic properties (namely on α) and on both d and r. For the special case of $d = 0$, corresponding to a surface disk, $G_r \approx 0.237$ was numerically obtained. This last value is approximately 5 % lower than the constant originally included in the Wooding's solution (i.e., 1/4 = 0.250). The conclusion by Reynolds and Elrick (1991) was that the Wooding's solution with $G_r = 0.237$ should be sufficiently accurate for practical use with most disk infiltrometers designed for field use (i.e., $r \geq 0.02$ m) and with the range of α values normally encountered in the field under conditions of near saturation (i.e., $1 \leq \alpha \leq 100$ m^{-1}).

Weir (1986) showed that Eq. (3.6) is an accurate approximation of steady flow from a circular source for large disk radii ($r > 2\lambda_c$). For $r \leq 2\lambda_c$, the following equation was shown to be more accurate than Eq. (3.6) (Weir 1987):

$$i_s = \frac{2K_0 \sin^2\tau}{\pi\tau^2 \sin\tau \cos\tau + 2\tau^2 \sin^2\tau \ln\tau - 1.073\tau^4} \tag{3.8}$$

where $\tau = r/(2\lambda_c)$. For $K_0 = K_s$, i.e., for saturated conditions at the soil surface ($h_0 = 0$), Clothier and Scotter (2002b) compared Eqs. (3.6) and (3.8) with numerically simulated flow data. For $r < 2\lambda_c$ ($\tau < 1$), the modified solution by Weir (1987) yielded more accurate i_s estimates than Eq. (3.6). The choice, for small disk sizes, of the refinement solution by Weir (1987) instead of the Wooding's (1968) method does not seem to be an absolute necessity on the basis of the results obtained by Wang et al. (1998b) on sandy-loam and sandy soils. In particular, these authors suggested that Wooding's approximate solution should be sufficient for most disk sizes used for TI measurements because soil spatial variability and macropores may play a larger role than the physical size of TI disk in determining soil hydraulic properties.

Three different approaches were proposed to solve Eq. (3.1), or Eq. (3.6), that contain two unknowns, i.e., K_0 and ϕ_0 or λ_c, if K_i can be neglected. In particular, an approach is based on the measurement of steady flow corresponding to the same imposed pressure head from two or more disks of different radius (Scotter et al. 1982; Smettem and Clothier 1989). Alternatively, multiple head procedures were proposed that apply a sequence of pressure heads on the same disk (Ankeny et al. 1991; Reynolds and Elrick 1991; Logsdon and Jaynes 1993; Jarvis and Messing 1995). The third approach, which is based on a single disk radius and a unique pressure head at the infiltration surface, makes use of both the steady-state and the early-time flow rates (Warrick 1992; White et al. 1992).

The main drawback of the multi-disk approaches is that the volume of soil sampled is different for each disk radius (Cook and Broeren 1994). Due to spatial variability of soil hydraulic conductivity, these methods may yield negative calculations and misleading determinations, particularly when only two ring sizes are used. Smettem and Clothier (1989) recommended the disk of the larger infiltrometer to be more than twice greater than the smaller one (i.e., $r_2 > 2r_1$) to obtain a reasonable sensitivity in the solution. Nevertheless, this means that the soil tested with the larger ring may differ from the soil tested with the smaller ring, which could lead to misestimating. Indeed, meaningless results were obtained by Logsdon and Jaynes (1993) for 30 out of 40 infiltration tests conducted using two disks of 76 and 230 mm in diameter, respectively. To reduce the spatial variability problem, the method should be used on averages of several determinations for each infiltrometer base size (Logsdon and Jaynes 1993). Alternatively, more than two disk sizes can be used to find the best fitting values of K_0 and ϕ_0 or λ_c by fitting Eq. (3.6) to the data (Thony et al. 1991).

Cook and Broeren (1994) compared different methods based on the Wooding's equation and they concluded that all methods gave similar estimate of K_0. A similar comparison was conducted by Hussen and Warrick (1993b) who concluded that methods based on a single disk of large radius and multiple heads (more than three points) should be expected to give stable, accurate and repeatable results. Therefore, considering the above mentioned limitations of the double- (or multi-) disk approach, only the single disk approach will be described in detail. In particular, different multi-potential methods and a single-potential method will be illustrated.

With the Ankeny et al. (1991) approach, the steady-state flow rate is measured with the same source for at least two different surface pressure head values (h_x and h_y). Assuming that the macroscopic capillary length, λ_c, is constant for the potential range $\Delta h = h_x - h_y$, i.e., between the two successive values of the imposed pressure head, the soil hydraulic conductivity values, K_x and K_y, corresponding to h_x and h_y, respectively, as well as λ_c can be obtained from (Cook and Broeren 1994):

$$K_x = \frac{Q_{sx}}{\pi r^2 + 2\Delta h r \frac{Q_{sx}+Q_{sy}}{Q_{sx}-Q_{sy}}} \tag{3.9a}$$

$$K_y = K_x \frac{Q_{sy}}{Q_{sx}} \tag{3.9b}$$

$$\lambda_c = \Delta h \frac{K_x + K_y}{2(K_x - K_y)} \tag{3.9c}$$

in which Q_{sx} and Q_{sy} are the steady flow rates corresponding to h_x and h_y, respectively. If three or more pressure heads are imposed on the infiltration surface, two estimates of K_0 can be obtained for a given h_0 value. In this case, the best estimate of K_0 can be assumed as the arithmetic average of the available estimates.

The Reynolds and Elrick (1991) approach relies on the hypothesis that the soil hydraulic conductivity function over the range $-\infty < h \leq 0$ is adequately described by a piecewise exponential relationship of the form:

$$K(h) = \overline{K}_{x,y} \exp(\overline{\alpha}_{x,y} h) \tag{3.10}$$

where $\overline{K}_{x,y}$ (L T^{-1}) and $\overline{\alpha}_{x,y}$ (L^{-1}) are the saturated hydraulic conductivity and the α parameter, respectively, for each segment of the $K(h)$ relationship between two adjacent pressure heads, h_x and h_y, with $x = 1, 2, \ldots$ and $y = x + 1$. Since $K(h)$ is a piecewise exponential, then successive $\overline{\alpha}_{x,y}$ and $\overline{K}_{x,y}$ values (i.e., $\overline{\alpha}_{1,2}, \overline{\alpha}_{2,3}, \ldots; \overline{K}_{1,2}, \overline{K}_{2,3}, \ldots$) will generally have different magnitudes. For $K_i \ll K_0$, substituting Eq. (3.10) into Eq. (3.1) and applying the logarithmic transformation produces:

$$\ln Q_s = \overline{\alpha}_{x,y} h_0 + \ln\left[\left(\frac{r}{G_d \overline{\alpha}_{x,y}} + \pi r^2\right)\overline{K}_{x,y}\right] \tag{3.11}$$

where $G_d = 0.237$ is the shape factor for a circular surface source with depth of insertion $d = 0$. Equation (3.11) describes a piecewise linear plot of $\ln(Q_s)$ versus h_0, from which $\overline{\alpha}_{x,y}$ can be calculated as the piecewise slope:

$$\overline{\alpha}_{x,y} = \frac{\ln(Q_{sx}/Q_{sy})}{(h_x - h_y)} \tag{3.12}$$

and $\overline{K}_{x,y}$ is determined from the piecewise intercept:

$$\overline{K}_{x,y} = \frac{G_d \overline{\alpha}_{x,y} Q_{sx}}{r(1 + G_d \overline{\alpha}_{x,y} \pi r)(Q_{sx}/Q_{sy})^P} \tag{3.13}$$

where Q_{sx} and Q_{sy} are the steady-state flow rates corresponding to h_x and h_y, respectively, and $P = h_x/(h_x - h_y)$. Estimation of $K(h)$ within the pressure head range (h_x, h_y) can be obtained by substituting Eqs.(3.12) and (3.13) into Eq. (3.10). If more than two potentials are used, two estimates of K_0 can be obtained for each intermediate h_0 value, one for $\overline{\alpha}_{x,y}$ and another for $\overline{\alpha}_{x+1,y+1}$. As for the Ankeny et al. (1991) method, these individual $K(h_0)$ estimates can then be averaged to yield one K_0 estimate for each h_0.

The Ankeny et al. (1991) and Reynolds and Elrick (1991) methods are expected to yield similar results for a wide range of soil properties and steady-state infiltration rates. However, the former approach, that is based on simultaneous-equation solution, is more susceptible than the latter to errors resulting from soil heterogeneity (Reynolds and Elrick 1991). According to these authors, small pressure head increments ($\Delta h_0 \approx 0.01$ m) are required to maintain a high degree of accuracy (i.e., $\approx \pm 7$ %) in estimation of soil hydraulic conductivity when the $K(h)$ relationship is very steep. Such a small Δh_0 appreciably increases the number of (h_0, Q_s) data pairs needed to cover the wet-end range of h_0 (-0.15 m $\leq h_0 \leq 0$) (Reynolds and Elrick 1991). However, most practical applications do not require such accuracy and, therefore, the number of pressure heads needed to estimate the wet end of the $K(h)$ relationship with sufficient accuracy is not excessive under most circumstances. Limits in the use of small Δh_0 values were highlighted by Logsdon and Jaynes (1993), who found that the simultaneous equation approach of Ankeny et al. (1991) yielded very small values of K_0 when the two consecutive pressure heads were very close in value.

Logsdon and Jaynes (1993) proposed an alternative method for calculating the hydraulic conductivity from multi-head experiments that involves a simultaneous fitting of all the data. Substituting Eq. (3.7) into Eq. (3.6) and assuming an exponential relationship for $K(h)$ in the form of Eq. (3.3) gives:

$$i_s = K_{fs}\exp(\alpha h)\left(1 + \frac{4}{\alpha \pi r}\right) \tag{3.14}$$

which contains two unknowns (K_{fs} and α). With infiltrometer measurements made at two or more pressure heads, a nonlinear least squares regression of Eq. (3.14) can be used to estimate K_{fs} and α. Values for $K(h)$ can then be found by substituting the resulting values of α and K_{fs} into Eq. (3.3). If measurements are made at only two pressure heads, the regression procedure becomes similar to the approach by Reynolds and Elrick (1991). The nonlinear regression method produced mean K_0 values close to the one-dimensional conductivity measured in the field on gypsum-coated soil columns (Logsdon and Jaynes 1993). Moreover, the regression method was found to be practically equivalent to the Ankeny et al. (1991) approach for the estimation of the unsaturated hydraulic conductivity of a sandy-loam soil in the range of h_0 values from -120 to -30 mm (Fig. 3.8) (Bagarello and Iovino 2010). According to Hussen and Warrick (1993b), the results of the nonlinear regression method are averaged over more measurements and therefore they are more stable than those based on the solution of simultaneous equations. On the other hand, the main drawback of the approach relies on the assumption that Eq. (3.3) can provide an accurate description of the soil hydraulic conductivity function. Mostly in structured soils, the choice of a constant α value over the range of pressure heads close to zero could not be realistic, since α varies with the considered pressure head (Jarvis and Messing 1995; Ventrella et al. 2005; Bagarello et al. 2006a; Reynolds and Topp 2008). In these cases, piecewise accommodation of the $K(h)$ relationship, like the ones embedded in the Ankeny et al. (1991) and Reynolds and Elrick (1991)

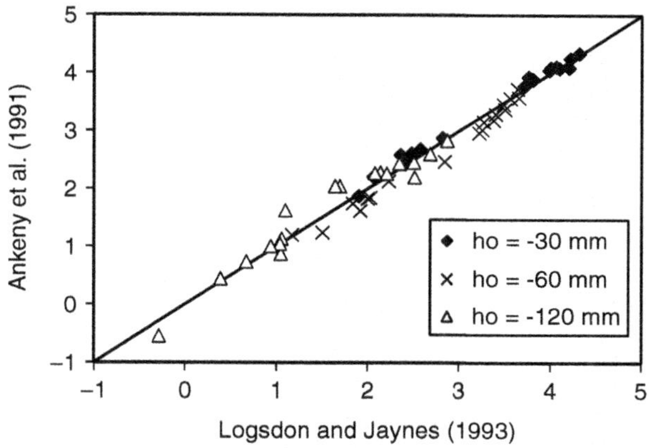

Fig. 3.8 Comparison between log-transformed hydraulic conductivity values, K_0 (mm h^{-1}), estimated for a sandy-loam soil by the Ankeny et al. (1991) and Logsdon and Jaynes (1993) approaches for different applied pressure head, h_0, values with the tension infiltrometer (From Bagarello and Iovino 2010)

approaches, appears more adequate to describe the soil hydraulic conductivity in the pressure head range close to saturation. An advantage of the multiple head approaches is that they do not require soil moisture measurements, so avoiding the problem to collect separate samples to be oven-dried or the use of a more cumbersome apparatus for in-situ measurement of soil water content (e.g., TDR).

For a run with a single pressure head, h_0, White et al. (1992) proposed to determine K_0 using Eq. (3.6) and the soil sorptivity, estimated by the slope of cumulative infiltration, I (L), vs. $\sqrt{t}$, t (T) being the time, from the early-time infiltration data. The approximate relationship between λ_c and $S_0 = S(h_0)$ (L T$^{-0.5}$) can be written as (White and Sully 1987; Hussen and Warrick 1993a; Angulo-Jaramillo et al. 2000):

$$\lambda_c = \frac{\phi_0}{K_0} = \frac{bS_0^2}{(\theta_0 - \theta_i)K_0} \tag{3.15}$$

where b is a constant ($1/2 \le b \le \pi/4$) depending on the shape of the soil water diffusivity function, θ_0 (L^{3}L^{-3}) is the volumetric soil water content corresponding to h_0 and θ_i (L^{3}L^{-3}) is the initial volumetric soil water content. White and Sully (1987) suggested that a reasonable value of b for most soil is 0.55. Substituting Eq. (3.15) into Eq. (3.6) gives:

$$i_s = K_0 + \frac{2.2S_0^2}{(\theta_0 - \theta_i)\pi r} \tag{3.16}$$

Estimation of K_0 by Eq. (3.16) requires data collected at both the early-time and steady-state stages of the infiltration process from a disk as well as measurement of θ_0 and θ_i. Implicit in this method is the assumption that the early-time and the steady-state capillary forces are the same. However, if there is a change in either the soil water content or the soil structure with depth, the values of K_0 calculated with this method may be incorrect (Cook and Broeren 1994). Reliable determinations of S_0 require that the soil be reasonably dry before infiltration and accurate infiltration rate measurements be made during early time (Logsdon and Jaynes 1993).

According to White and Sully (1987), the representative mean pore radius, R_0 (L), is (Reynolds and Topp 2008):

$$R_0 = \frac{\sigma K_0}{\rho_w g \phi_0} = \frac{\sigma \alpha(h_0)}{\rho_w g} \tag{3.17}$$

where σ (M T^{-2}) is the surface tension of water, ρ_w (M L^{-3}) is the density of water, and g (L T^{-2}) is the acceleration due to gravity. The pore index R_0 represents an effective equivalent mean pore radius that is conducting water when infiltration occurs at $h = h_0$. The concentration of hydraulically active pores at the soil surface, N_0 (L^{-2}), may be derived from Poiseuille's law for flow in smooth, cylindrical capillary tubes (Reynolds et al. 1995):

$$N_0 = \frac{8\mu K_0}{\rho_w g \pi R_0^4} \tag{3.18}$$

where μ (M L^{-1}T^{-1}) is the dynamic viscosity of water. Equations (3.17) and (3.18) have been used in investigations on soil water conducting macro- and meso-porosity (e.g., Cameira et al. 2003; Moret and Arrúe 2007; Soracco et al. 2011) and soil physical quality assessment (Iovino et al. 2013).

Deciding when steady flow, Q_s, has been attained is somewhat arbitrary and it will depend to some extent on the experience of the operator. For multi-dimensional infiltration, Philip (1986) suggested that the time scale for steady flow rate corresponds to the time when flow rate from the source reaches 1.05 times the real steady-state flow rate. When the characteristic dimension of the supply source equals or exceeds λ_c, the time to steady flow can be assumed equal to $t_{\mathrm{grav}} = (S_0/K_0)^2$ (Cook and Broeren 1994). Physically, t_{grav} is the time "at which the effect of gravity on the process can be expected to be as great as that of capillarity" (Philip 1969). Using numerical simulation, Warrick (1992) confirmed that the time to approach $1.05Q_s$ was of the same order of magnitude as t_{grav}. However, for field experiments it would be extremely difficult to ascertain when true steady-state was reached without knowing the true value of the characteristic time t_{grav}. Furthermore, true steady flow will only occur in completely homogeneous soil, which probably does not exist in natural environments (Reynolds and Elrick 2002). Hence, an estimate of steady flow, i.e., a measurement of quasi steady-state flow, is generally considered acceptable in practice. Attainment of quasi steady-state flow in medium to fine textured soils may be particularly long, which enhances the risk that the simplifying assumptions of the analysis (i.e., homogeneous, non-swelling soil with uniform initial water content) are violated. In general, gradients in water content and soil bulk density, soil layering, and changes in soil texture that occur near the soil surface can be reflected in negative, and hence meaningless, values of hydraulic conductivity (Hussen and Warrick 1993b).

The requirement $K_i \ll K_0$ theoretically limits the analysis to soils with relatively low initial water content. However, this could not be a great problem since most applications of practical interest are conducted in relatively dry soil (i.e., soils at field capacity or drier, $h_i \leq -1$ m).

Near-saturated soil hydraulic properties are often extremely sensitive to small changes in pore water pressure head. Particularly in structured soils, $K(h_0)$ often changes by several orders of magnitude in the range $-150 \leq h_0 \leq 0$ mm. Therefore, it is important that the infiltrometer be accurately calibrated (see procedure outlined before) and levelled. When working on sloped surfaces, cutting a level bench could become necessary, so that the applied pressure head is the same at all points on the infiltration surface.

According to Reynolds (2008), important strengths of the TI approaches based on the Wooding's solution of steady flow include well-established experimental methods and theory, robustness, provision of measurements at several pressure heads on a single infiltration surface, relatively large (and thereby more representative) sample volumes. Weaknesses of the steady flow analysis include potentially long equilibration time and potentially great susceptibility to soil heterogeneity and non-uniform water content due to the relatively large sample volume.

Example 3.1

A multi-potential TI experiment was carried out by establishing an ascending sequence of pressure heads on the infiltration surface, i.e., $h_0 = -150$, -75, -30 and -10 mm. The TI reservoir had an inner cross-sectional area, $A_s = 1555$ mm^2, and the radius of the porous disk was $r = 120$ mm, corresponding to an infiltration surface, $A = 45239$ mm^2. Table 3.1 lists the water level readings at the reservoir of the instrument, L (mm), at different times, t (min), during the infiltration process for the four pressure head values. The soil hydraulic conductivity, K_0, has to be calculated with the approaches by Ankeny et al. (1991), Reynolds and Elrick (1991) and Logsdon and Jaynes (1993).

The rate of fall of the water level in the reservoir, R (mm h^{-1}), can be calculated as $R = \Delta L/\Delta t$ (Table 3.1). The corresponding infiltration rate, i (mm h^{-1}), is given by:

$$i = R\frac{A_s}{A} \tag{3.19}$$

A steady-state condition, indicated by five or more consecutive constant values of R, and hence i, was obtained for each of the four applied pressure heads. The corresponding steady-state infiltration rates were $i_{s1} = 6.2$ mm h^{-1}, $i_{s2} = 15.5$ mm h^{-1}, $i_{s3} = 32.0$ mm h^{-1} and $i_{s4} = 41.3$ mm h^{-1}, where the subscripts 1–4 denote the four applied pressure heads in the sequence.

When detection of steady-state flow is problematic due, for example, to a certain scattering in the R measurements, an alternative approach consists of plotting

Table 3.1 Readings at the reservoir of the tension infiltrometer device, L (mm), at different times, t (min), and corresponding values of cumulative infiltration, I (mm), rate of fall of water level in the reservoir, R (mm h^{-1}), and infiltration rate, i (mm h^{-1}), for the multi-potential experiment of the Example 3.1 (h_0 = imposed pressure head on the soil surface)

t	L	I	R	i	t	L	I	R	i
$h_0 = -150$ mm					$h_0 = -75$ mm				
0	0	0.0			0	0	0.0		
0.5	12	0.4	1440	49.5	2	17	0.6	510	17.5
1	23	0.8	1320	45.4	4	34	1.2	510	17.5
2	44	1.5	1260	43.3	6	50	1.7	480	16.5
3	64	2.2	1200	41.3	8	67	2.3	510	17.5
4	82	2.8	1080	37.1	10	84	2.9	510	17.5
6	112	3.9	900	30.9	12	100	3.4	480	16.5
8	130	4.5	540	18.6	14	117	4.0	510	17.5
10	143	4.9	390	13.4	16	132	4.5	450	15.5
15	166	5.7	276	9.5	18	148	5.1	480	16.5
20	185	6.4	228	7.8	20	163	5.6	450	15.5
25	203	7.0	216	7.4	22	178	6.1	450	15.5
30	219	7.5	192	6.6	24	193	6.6	450	15.5
35	234	8.0	180	6.2	26	208	7.2	450	15.5
40	249	8.6	180	6.2	28	223	7.7	450	15.5
45	264	9.1	180	6.2	30	238	8.2	450	15.5
50	279	9.6	180	6.2					
55	294	10.1	180	6.2					
60	309	10.6	180	6.2					
$h_0 = -30$ mm					$h_0 = -10$ mm				
0	0	0.0			0	0	0.0		
1	16	0.6	960	33.0	2	36	1.2	1080	37.1
3	47	1.6	930	32.0	4	73	2.5	1110	38.2
5	78	2.7	930	32.0	6	111	3.8	1140	39.2
7	108	3.7	900	30.9	8	149	5.1	1140	39.2
9	138	4.7	900	30.9	10	189	6.5	1200	41.3
11	169	5.8	930	32.0	12	229	7.9	1200	41.3
13	200	6.9	930	32.0	14	269	9.2	1200	41.3
15	231	7.9	930	32.0	16	309	10.6	1200	41.3
17	262	9.0	930	32.0	18	349	12.0	1200	41.3
19	293	10.1	930	32.0	20	389	13.4	1200	41.3

cumulative infiltration, I (T), versus time, t (T), data and determining the slope of the least squares regression line corresponding to the linear part of the $I(t)$ curve (Fig. 3.9). Steady-state flow is assumed when the relative error, E, between estimated and measured cumulative infiltration is lower than 2 % (Bagarello et al. 1999):

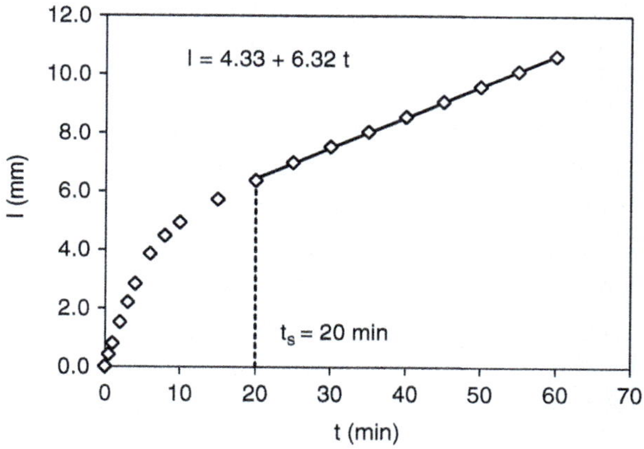

Fig. 3.9 Application of the regression approach to estimate steady-state infiltration rate and time-to-steady-state from the cumulative infiltration data of the Example 3.1 (applied pressure head, $h_0 = -150$ mm)

Table 3.2 Steady-state infiltration rate, i_s (mm h^{-1}), steady-state flow rate, Q_s (mm^{3}h^{-1}), and time-to-steady-state, t_s (min), obtained by the regression approach for each established pressure head, h_0 (mm), with the multi-potential experiment of the Example 3.1

h_0	i_s	Q_s	t_s
−150	6.3	285,861	20
−75	15.6	705,322	14
−30	31.7	1,433,690	1
−10	41.1	1,858,565	6

$$E = \left| \frac{I_i - I_i^*}{I_i} \right| \le 2\% \qquad (3.20)$$

in which I_i is the measured cumulative infiltration at time t_i and I_i^* is the corresponding cumulative infiltration value estimated by the linear regression line. The equilibration time or time-to-steady-state, t_s (T), is then estimated as the shortest time for which the condition expressed by Eq. (3.20) is fulfilled. For each applied pressure head value, Table 3.2 lists the steady-state infiltration rates estimated by the regression approach, the steady-state flow rates, $Q_s = i_s \pi r^2$, and the equilibration time, t_s. For this example, the i_s values estimated by the regression approach are close to those estimated by five or more consecutive equal values of R. However, this is not a general result and different i_s estimates may be obtained by the two approaches. In these cases, the use of the regression approach is recommended as it is based on a more robust estimation of cumulative quantities. In the following, the i_s and Q_s data obtained by the regression approach were used (Table 3.2).

The Ankeny et al. (1991) approach applied to the first two pressure head values ($h_1 = -150$ mm and $h_2 = -75$ mm) gives:

$$K_1 = \frac{285861}{\pi 120^2 + 2(-150 + 75)120\frac{285861+705322}{285861-705322}} = 3.3 \text{ mm h}^{-1}$$

$$K_2 = 3.3\frac{705322}{285861} = 8.0 \text{ mm h}^{-1}$$

whereas, for the successive pressure head pairs, it gives:

$h_2 = -75$ mm and $h_3 = -30$ mm: $K_2 = 9.2$ mm h^{-1}, $K_3 = 18.6$ mm h^{-1}
$h_3 = -30$ mm and $h_4 = -10$ mm: $K_3 = 17.4$ mm h^{-1}, $K_4 = 22.5$ mm h^{-1}

For a given h_0 value within the sequence, i.e., excluding the extreme h_0 values (-150 and -10 mm), an estimate of K_0 can be obtained as the arithmetic average of the estimates corresponding to the two pressure head pairs (h_{i-1}, h_i) and (h_i, h_{i+1}):

$$K_2 = \frac{8.0 + 9.2}{2} = 8.6 \quad \text{mm h}^{-1}$$

$$K_3 = \frac{18.6 + 17.4}{2} = 18.0 \text{ mm h}^{-1}$$

For the first pressure head data pair ($h_0 = -150$ and -75 mm), the Reynolds and Elrick (1991) approach gives:

$$\overline{\alpha}_{1,2} = \frac{\ln(285861/705322)}{(-150 + 75)} = 0.0120 \quad \text{mm}^{-1}$$

$$\overline{K}_{1,2} = \frac{0.237 \times 0.0120 \times 285861}{120(1 + 0.237 \times 0.0120 \times 120\pi)(285861/705322)^2} = 19.9 \text{ mm h}^{-1}$$

Equation (3.10) with these estimates of $\overline{\alpha}_{x,y}$ and $\overline{K}_{x,y}$ allows calculation of the soil hydraulic conductivities corresponding to the considered pressure heads:

$$K_1 = 19.9\exp(-0.0120 \times 150) = 3.3 \quad \text{mm h}^{-1}$$

$$K_2 = 19.9\exp(-0.0120 \times 75) = 8.1 \quad \text{mm h}^{-1}$$

For the successive pressure head pairs, the Reynolds and Elrick (1991) approach gives:

$h_2 = -75$ mm and $h_3 = -30$ mm : $\overline{\alpha}_{2,3} = 0.0158$ mm^{-1}, $\overline{K}_{2,3} = 29.7$ mm h^{-1}
$h_3 = -30$ mm and $h_4 = -10$ mm : $\overline{\alpha}_{3,4} = 0.0130$ mm^{-1}, $\overline{K}_{3,4} = 25.1$ mm h^{-1}

and the corresponding soil hydraulic conductivities are:

$h_2 = -75$ mm and $h_3 = -30$ mm : $K_2 = 9.1$ mm h^{-1}, $K_3 = 18.5$ mm h^{-1}
$h_3 = -30$ mm and $h_4 = -10$ mm : $K_3 = 17.0$ mm h^{-1}, $K_4 = 22.1$ mm h^{-1}

Table 3.3 Application of the approach by Logsdon and Jaynes (1993) to the data of the Example 3.1

Initial values of parameters				
$K_{fs} = 20$ mm h^{-1}				
$\alpha = 0.015$ mm^{-1}				
h_0	i_s	i_s^*	$(i_s - i_s^*)^2$	
−150	6.3	3.6	7.398	
−75	15.6	11.1	20.296	
−30	31.7	21.8	98.374	
−10	41.1	29.4	136.717	
		SSD	*262.784*	
Final values of parameters				
$K_{fs} = 27.22$ mm h^{-1}				
$\alpha = 0.0142$ mm^{-1}				
h_0	i_s	i_s^*	$(i_s - i_s^*)^2$	$K(h_0)$
−150	6.3	5.7	0.423	3.2
−75	15.6	16.4	0.693	9.4
−30	31.7	31.1	0.358	17.8
−10	41.1	41.3	0.043	23.6
		SSD	*1.518*	

K_{fs} = field-saturated soil hydraulic conductivity; α = soil texture/structure parameter; h_0 (mm) = established pressure head on the soil surface; i_s (mm h^{-1}) = measured value of steady-state infiltration rate; i_s^* (mm h^{-1}) = estimated value of steady-state infiltration rate; *SSD* = sum of squared deviations; *K* (mm h^{-1}) = soil hydraulic conductivity

The K_0 values corresponding to the intermediate potentials can be estimated as the arithmetic average of the available estimates, which yields $K_2 = 8.6$ mm h^{-1} and $K_3 = 17.8$ mm h^{-1}.

Finally, $K(h_0)$ values can also be estimated by the Logsdon and Jaynes (1993) approach which applies a nonlinear least squares regression method to estimate the parameters K_{fs} and α of Eq. (3.3). At this aim, the Microsoft Excel Solver can be used according to the following procedure. First, initial values of K_{fs} and α ($K_{fs} = 20$ mm h^{-1} and $\alpha = 0.015$ mm^{-1} in this example) are arbitrarily chosen and a first attempt value of i_s is estimated for each applied pressure head by Eq. (3.14). Then, the sum of the squared deviations, *SSD*, between the measured and the estimated i_s values is calculated since *SSD* represents the quantity to be minimized by the Microsoft Excel Solver. Starting from the first attempt values, the iterative procedure finds the "best fit" values of K_{fs} and α that minimize *SSD*. Finally, the hydraulic conductivity values corresponding to the different h_0 values can be obtained from the estimated K_{fs} and α values by the Gardner's (1958) equation, i.e., by Eq. (3.3). The results of the optimization procedure are listed in Table 3.3.

3.1.4 Analysis of Transient Flow

In the last two decades, several researchers oriented their work on determining the soil hydraulic properties by the transient phase alone of a TI infiltration process. Analysis of transient, three-dimensional flow from the device has several potential advantages as compared with the more classic methods of analysis based on the Wooding's steady flow equation (Angulo-Jaramillo et al. 2000). First of all, analysis of transient flow means shorter experiments and smaller sampled volumes of soil, which is obviously in better agreement with the hypotheses of homogeneity and initial uniform water content. Uncertainties about the time at which steady infiltration flux is attained are overcome (Hussen and Warrick 1993a, b; Quadri et al. 1994; Vandervaere et al. 1997, 2000a) and much useful information that is lost by only considering steady flow can be accounted for (Logsdon 1997). Furthermore, shorter experiments allow to increase the number of sampling points over a field, which is of particular interest for spatial variability studies, and to assess the vertical variations of soil hydraulic properties with high resolution by conducting infiltration tests at closely spaced depths (Vandervaere et al. 2000a). In particular, some of the proposed transient approaches have been adapted to determine hydraulic conductivity and sorptivity of crusted soils (Vandervaere et al. 1997) (see Chap. 4).

Basically, two approaches have been proposed to analyze transient infiltration data collected by a TI: (i) analytical models, that will be discussed in this section, and (ii) inverse methods, that will be discussed in Sect. 3.1.5. Note that a TI run performed with a single water pressure head fixed at zero can also be analyzed with the BEST method (Lassabatere et al. 2006) that is described in Sect. 3.2.

The most common analytical models (Warrick 1992; Haverkamp et al. 1994; Zhang 1997a, b) have in common the following two-term cumulative infiltration equation, formally identical to the Philip's (1957) one-dimensional infiltration model:

$$I = C_1\sqrt{t} + C_2 t \tag{3.21}$$

where I (L) is the cumulative infiltration, t (T) is the time and C_1 (L T$^{-1/2}$) and C_2 (L T^{-1}) are coefficients that differ according to the considered model. In particular, neglecting the effect of gravity, Warrick (1992) proposed expressions for C_1 and C_2 that are essentially of empirical nature (Vandervaere et al. 2000a). Zhang (1997a, b) related C_1 to capillary forces and C_2 to gravity forces. This assumption seems questionable, as Smettem et al. (1994) showed that the term accounting for lateral capillary flow from a circular source is linear in time and hence it should appear in coefficient C_2 (Vandervaere et al. 2000a). Notwithstanding this, Zhang's model has been recommended for analysing the data with the Mini Disk Infiltrometer (Decagon Device Inc. 2012) and it will be described later in the chapter (Sect. 3.1.6).

The model proposed by Haverkamp et al. (1994), that will be discussed in more detail below, is based on previous work by Turner and Parlange (1978) and it accounts for the lateral flux at the edge of the circular infiltration surface by the approximate expression proposed by Smettem et al. (1994). The following expressions for coefficients C_1 and C_2 of Eq. (3.21) were proposed to define a physically based, three-dimensional infiltration equation for disk infiltrometers valid for short to medium times, and considering that initial hydraulic conductivity is negligible with regards to K_0:

$$C_1 = S_0 \tag{3.22}$$

$$C_2 = \frac{2-\beta}{3}K_0 + \frac{\gamma S_0^2}{r(\theta_0 - \theta_i)} \tag{3.23}$$

where S_0 (L $T^{-1/2}$) is the soil sorptivity corresponding to the imposed pressure head, h_0 (L), β is a parameter depending on the capillary diffusivity function that lies in the interval [0, 1], K_0 (L T^{-1}) is the soil hydraulic conductivity corresponding to h_0, γ is a constant approximately equal to 0.75, θ_0 (L^3L^{-3}) is the volumetric soil water content corresponding to h_0 and θ_i (L^3L^{-3}) is the initial soil water content. For most soils, provided the initial water content is relatively low (i.e., relative saturation < 0.25), an average value of β = 0.60 can be assumed (Smettem et al. 1994; Haverkamp et al. 1994). Substituting Eqs. (3.22) and (3.23) into Eq. (3.21) yields:

$$I = S_0\sqrt{t} + \left[\frac{2-\beta}{3}K_0 + \frac{\gamma S_0^2}{r(\theta_0 - \theta_i)}\right]t \tag{3.24}$$

The first term of the right-hand side of Eq. (3.24) corresponds to the vertical capillary flow and dominates infiltration during its early stage. The second term corresponds to the gravity-driven vertical flow and the third one represents the lateral capillary component flow that, as shown by Smettem et al. (1994), is linear with time. According to Haverkamp et al. (1994), Eq. (3.24) can adequately describe the axisymmetric flow from the disk of a TI for times smaller than the characteristic time scale, t_{grav} (T) (Philip 1969):

$$t_{grav} = \left(\frac{S_0}{K_0}\right)^2 \tag{3.25}$$

Vandervaere et al. (2000b) proposed four methods to determine K_0 and S_0 that are based on estimation of coefficients C_1 and C_2 from a transient infiltration experiment and application of Eqs. (3.22) and (3.23). The Transient Single-Test (TST) method consists of conducting a single-radius infiltration experiment for a given pressure head value. Estimates of sorptivity and hydraulic conductivity are then obtained from:

$$S_0 = C_1 \tag{3.26}$$

$$K_0 = \frac{3}{2-\beta}\left[C_2 - \frac{\gamma C_1^2}{r(\theta_0 - \theta_i)}\right] \tag{3.27}$$

Determining K_0 also requires knowledge of the difference between the soil water content corresponding to h_0 (θ_0) and the initial soil water content (θ_i). The TST method is conceptually similar to the steady-state single-test (SST) method (i.e., Eq. 3.16) proposed by White et al. (1992).

When the early-time estimation of sorptivity (i.e., coefficient C_1) proves difficult or doubtful, as when the contact layer masks the initial infiltration regime, a possible alternative is offered by the Transient Multi-Radii (TMR) method. When two infiltration experiments are conducted with disks of different radius and a common h_0 value, determination of coefficient C_2 for each of them allows estimation of S_0 and K_0 from the two-equation system obtained by Eq. (3.23) written for the two experiments. The TMR method is conceptually analogous to the Scotter et al. (1982) method for analysis of steady-state ponded infiltration. The TMR method has two advantages over the TST method: (i) it is the squared value of sorptivity that is calculated by the former method, which minimizes the error in S_0; (ii) the K_0 estimation does not require volumetric soil water content measurements. On the other hand, the drawback of the TMR method lies in the fact that two or more disks have to be used at different locations, which introduces complications due to the short-distance spatial variability of soil properties (Vandervaere et al. 2000b).

Estimation of K_0 can also be obtained from sorptivity measurements at different values of h_0 (White and Perroux 1987). Assuming that the initial soil water pressure head, h_i, does not vary much between experiments performed at different h_0 values, Eqs. (3.2) and (3.15) yield:

$$K_0 = \frac{\partial \phi_0}{\partial h_0} = \frac{\partial}{\partial h_0}\left[\frac{bS_0^2}{(\theta_0 - \theta_i)}\right] \tag{3.28}$$

With the Transient Multi-Sorptivity (TMS) method, two experiments are performed at supply pressure heads h_1 and h_2 ($h_1 < h_2$) and K_0 is estimated with Eq. (3.28) written in a finite difference form:

$$K\left(\frac{h_1 + h_2}{2}\right) = \frac{b}{h_2 - h_1}\left[\frac{S_2^2}{(\theta_2 - \theta_i)} - \frac{S_1^2}{(\theta_1 - \theta_i)}\right] \tag{3.29}$$

where S_1 and S_2 are the sorptivities and θ_1 and θ_2 are the volumetric water contents corresponding to h_1 and h_2, respectively. Equation (3.29) assumes that variation of b with h_0 can be neglected. The TMS method seems to be an interesting alternative to the TST and TMR methods when K (gravity forces) plays a minor role in the axisymmetric infiltration process relative to S (capillary forces), i.e., in situations where the TST and TMR methods are likely to be inaccurate. The sorptivity estimation at any given pressure head can be obtained with a single-disk experiment (method TMS1) or with a multi-radii experiment (method TMS2). In this last case, four experimental locations need to be sampled to obtain a single value of K_0.

In conclusion, the TST method, which allows the estimation of both S_0 and K_0 from a unique transient experiment, appears to be the most attractive method from a practical point of view. The TMR and TMS methods require a more complex experimental approach and they are more prone to the effects of small scale spatial variability of soil hydraulic properties.

Applicability of the different methods depends on the dominant term in the infiltration process (Vandervaere et al. 2000b). In particular, the relative importance of gravity flow as compared to capillary flow can be identified by comparing the soil sorptivity to the so-called S_{opt} (L $T^{-1/2}$) parameter:

$$S_{opt} = \sqrt{\frac{r\Delta\theta(2-\beta)K_0}{3\gamma}} \quad (3.30)$$

For given r, $\Delta\theta = (\theta_0 - \theta_i)$, β, K_0 and γ, S_{opt} is the sorptivity value for which gravity and lateral capillary terms have equivalent weights in the flow process. If $S_0 < S_{opt}$, gravity prevails over lateral capillarity. Consequently, a precise estimation of S_0 (coefficient C_1) is unlikely, but the conditions are very good for estimating K_0 with the transient methods TST and TMR, the SST method (White et al. 1992) and the steady-state multi-radii method (Smettem and Clothier 1989). The multi-potential steady-state methods (Ankeny et al. 1991; Reynolds and Elrick 1991) and the transient methods TMS1 and TMS2 are not recommended (Vandervaere et al. 2000b). In order to improve the S_0^2 estimation, the disk radius chosen must be small enough to enhance the lateral capillary term. The estimation of S_0 will also be improved when the initial conditions are driest. The condition $S_0 \approx S_{opt}$ is the most desirable if both S_0 and K_0 are to be estimated by the TST or SST (Steady State Test) methods. The TMR method and the steady-state multi-radii method also provide acceptable estimates of both variables. To estimate K_0 accurately, the disk radius must be as large as possible. If $S_0 > S_{opt}$, then the estimation of K_0 is very sensitive to measurement errors. In this case, Vandervaere et al. (2000b) recommended the multi-potential steady-state method (Ankeny et al. 1991; Reynolds and Elrick 1991) or the TMS methods. An interesting strategy for improving the K_0 estimation consists of working in relatively wet initial conditions that reduce sorptivity, taking however into account that the condition $K_i \ll K_0$ must be fulfilled. Ultimately, the choice of the most appropriate methods requires a few preliminary experiments to estimate S_0 for the site of interest.

Estimation of coefficients C_1 and C_2 can be obtained by fitting Eq. (3.21) to the experimental I vs. t data once the adequacy of the model with the data has been established. According to Vandervaere et al. (2000a), the commonly applied non-linear techniques (Cumulative Infiltration, CI, method, Fig. 3.10) offer no check for the adequacy of the form of the two-term equation with the data. Indeed, as a consequence of the intercompensation between t and $\sqrt{t}$, any monothonically increasing data set convex-upwards will provide fitted values of C_1 and C_2 even if

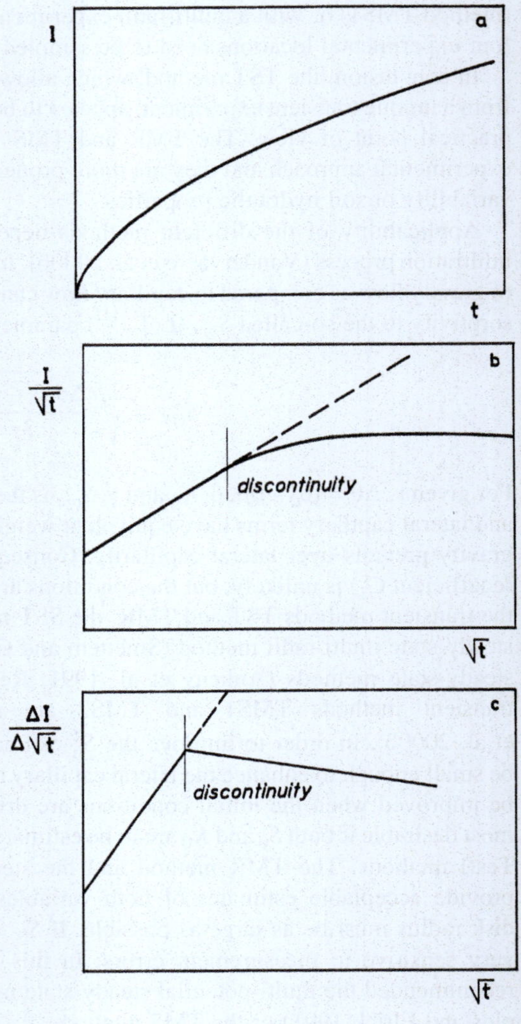

Fig. 3.10 Qualitative representation of an infiltration process with a discontinuity at a certain instant on a (**a**) cumulative infiltration, I (L), vs. time, t (T), plot (Cumulative Infiltration, CI, method); (**b**) $I/\sqrt{t}$ vs. $\sqrt{t}$ plot (Cumulative Linearization, CL, method); and (**c**) $\Delta I/\Delta\sqrt{t}$ vs. $\sqrt{t}$ plot (Differentiated Linearization, DL, method)

these values have no physical meaning. Vandervaere et al. (2000a) effectively summarized this circumstance stating that best fit does not necessarily mean good fit. Visual observation of the quality of the fit cannot stand as a test for the adequacy of Eq. (3.21) because of the difficulty of distinguishing the possible inadequacy of the equation and the scatter of the data points. To account for this, Smiles and Knight (1976) proposed linearizing Eq. (3.21) by dividing both sides by $\sqrt{t}$ (Cumulative Linearization, CL, method), giving:

$$\frac{I}{\sqrt{t}} = C_1 + C_2\sqrt{t} \tag{3.31}$$

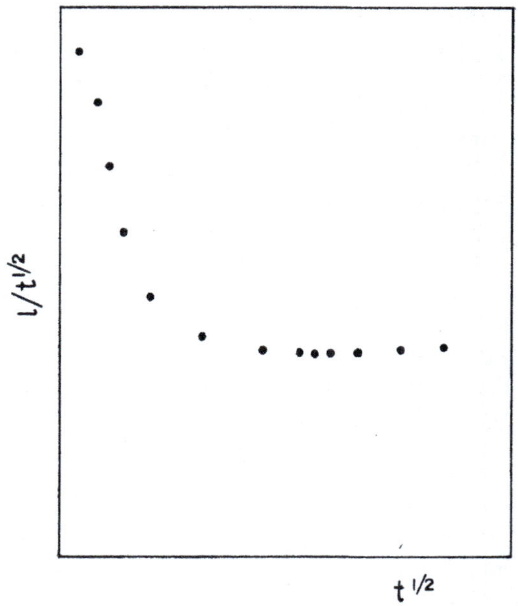

Fig. 3.11 Qualitative effect of contact material on infiltration with the Cumulative Linearization method of representation of the cumulative infiltration, I (L), vs. time, t (T), data pairs

In this form, it is easy to determine C_1 as the intercept and C_2 as the slope of the regression line between $I/\sqrt{t}$ and $\sqrt{t}$, and to test the adequacy of Eq. (3.21) by checking the linearity of the data set. In particular, discontinuities in the infiltration process can easily be detected with the CL method but not with the CI method (Fig. 3.10). When a layer of contact material is used to ensure the hydraulic contact between the disk and the soil, application of Eq. (3.31) is compromised by the water initially stored in the contact material during the early stages of infiltration (Fig. 3.11). In this case, applicability of Eq. (3.21) and estimation of coefficients C_1 and C_2 can be performed by differentiating the cumulative infiltration data with respect to the square root of time (Differentiated Linearization, DL, method, Fig. 3.10):

$$\frac{dI}{d\sqrt{t}} = C_1 + 2C_2\sqrt{t} \tag{3.32}$$

If Eq. (3.21) adequately describes the experimental data, plotting of $dI/d\sqrt{t}$ vs. $\sqrt{t}$ should be linear, with C_1 equal to the intercept and C_2 equal to half the slope of the regression line. The influence of the contact material is easy to detect since it corresponds to the initially sharply decreasing part of the experimental curve, deviating from the monotonically increasing linear behaviour, as shown in the Example 3.2. Linear regression can then be restricted to the rest of the data set, thus providing values of C_1 and C_2 without bias (Vandervaere et al. 2000a). Obviously Eq. (3.32) is also applicable in absence of a contact material layer on the soil surface. In this case, the linear trend should also include the first $dI/d\sqrt{t}$ vs. $\sqrt{t}$ points if Eq. (3.21) correctly describes the infiltration process (Fig. 3.10). Practical procedures to calculate $dI/d\sqrt{t}$ and $\sqrt{t}$ of Eq. (3.32) are described in the Example 3.2.

Use of Eq. (3.32) yields estimates of C_1 and, consequently, of soil sorptivity potentially more accurate than those determined from the slope of the Philip's (1957) one-dimensional horizontal infiltration equation (Vandervaere et al. 2000b):

$$I = S_0\sqrt{t} \tag{3.33}$$

In fact, this last approach neglects the gravity and the lateral diffusion effects (Smettem et al. 1995), thus resulting in an overestimation of S_0 that depends on the chosen time interval for which Eq. (3.33) is considered valid (Bonnell and Williams 1986). If the following limit is used:

$$S_0 = \lim_{t \to 0} \left(\frac{dI}{d\sqrt{t}} \right) \tag{3.34}$$

it is not necessary to formulate any hypothesis on the time validity of Eq. (3.33) and S_0 can be deduced by the DL method using the complete experimental information. In particular, it is not necessary to assume that gravity and/or geometry have a negligible impact at the beginning of the experiment because they influence coefficient C_2 and do not need to be specifically considered in S_0 estimation (Vandervaere et al. 2000b). If no contact material is used, Eq. (3.31) can also be used to estimate S_0.

Uncertainties in sorptivity estimation can yield large errors in determining hydraulic conductivity by the SST (Steady State Test) method, given that the squared value of S_0 is introduced in Eq. (3.16). Bagarello and Iovino (2004) applied both Eqs. (3.31) and (3.33) to calculate K_0 by the SST method. The usually applied approach, based on the early-time linear regression of the I vs. $\sqrt{t}$ data, overestimated S_0 thus resulting in a frequent occurrence of negative, and meaningless, K_0 values. The CL method appeared appropriate to reduce the risk of overestimating S_0, and hydraulic conductivity values were always positive. Therefore, linearization of the transient stage of infiltration process to estimate soil sorptivity seems also promising for applying the SST method by White et al. (1992).

Due to its practical interest, the TST method has been tested with specific reference to: (i) the applicability of the Haverkamp et al. (1994) model (Ferraris et al. 2003); (ii) the applicability of the DL method to estimate coefficients C_1 and C_2 (Bagarello and Iovino 2003, 2004); and (iii) the reliability of the hydraulic conductivity predictions in soils situated in the lateral capillarity domain, i.e., when $S_0 > S_{opt}$ (Bagarello et al. 2004).

The explicit two-term equation (Eq. 3.21) is a simplified time expansion of a more complete implicit three-dimensional infiltration equation (Haverkamp et al. 1994). Accuracy of Eq. (3.21) in reproducing a transient infiltration experiment could be one of the sources of errors when the TST method is applied. Ferraris et al. (2003) compared explicit and implicit models showing that, under relatively wet conditions, the simplified model was more accurate than the implicit model to reproduce numerically simulated infiltration data. An opposite result (i.e., implicit equation yielding estimates of I closer to the simulated values than the simplified equation) was generally obtained in initially dry conditions. The observed discrepancies between the analytically and numerically simulated data were attributed to the assumption of fixed values for the constants β and γ regardless of the soil type and the initial and boundary conditions of the infiltration process (Haverkamp et al. 1994). Uncertainty in the estimation of parameter β was also considered a possible source of inaccuracy of transient methods based on the Haverkamp et al. (1994) infiltration model by Vandervaere (2002). The possibility that an improved estimate of β and γ enables a more accurate prediction of soil hydraulic properties was recently suggested by Nasta et al. (2012) with reference to the so-called BEST procedure of soil hydraulic characterization (Lassabatere et al. 2006), making use of the Haverkamp et al. (1994) infiltration model (see Sect. 3.2).

In some cases, estimating the two coefficients by the DL method can be difficult or even impossible. This happens when transition time from infiltration into the contact material to infiltration into the soil is difficult to detect because of overlap between the two phenomena (Vandervaere et al. 1997). According to Jacques et al. (2002), establishing the starting time of infiltration into the soil by the DL method can involve a rather subjective evaluation of the data. Bagarello and Iovino (2003) showed that a clear discontinuity between the sharply decreasing, early portion of the $dI/d\sqrt{t}$ vs. $\sqrt{t}$ data and the linearly increasing behavior for the rest of the dataset was detectable when the capillary forces of the contact material were much higher than the soil capillary forces, i.e., when the sorptivity of the contact material was 10–12 times higher than soil sorptivity. A possible strategy to increase this ratio consists of using initially dry contact material on relatively wet soil. Furthermore, in soils situated in the lateral capillarity domain, relatively wet conditions reduce sorptivity and, hence, increase the accuracy of the K_0 estimates (Vandervaere et al. 2000b).

For given experimental conditions, the CL and DL methods should yield identical estimates of coefficients C_1 and C_2. A test of the expected equivalence of the two methods was conducted by Bagarello and Iovino (2004). To overcome the influence of contact material that prevents application of the CL method, the investigation was conducted on laboratory repacked samples without using contact material (Haverkamp et al. 1994; Quadri et al. 1994; Vandervaere et al. 2000a). The CL and DL methods were also compared with a few field tests conducted without contact material on a sandy-loam soil by Hussen and Warrick (1993a). The two linearization methods were not perfectly equivalent. In particular, in eight out of the 26 experiments considered, selection of data points to be used for estimating C_1 and C_2 differed between the two methods (Fig. 3.12). Furthermore, even when the same experimental information was used, the linearization technique (CL or DL) influenced the estimates of coefficients C_1 and C_2 (Fig. 3.13).

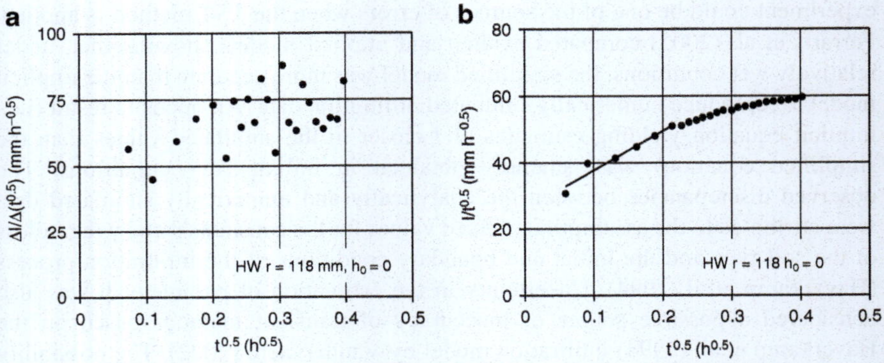

Fig. 3.12 Application of the (**a**) Differentiated Linearization method and (**b**) Cumulative Linearization method to the experiment conducted by Hussen and Warrick (1993a) with a disk of radius $r = 118$ mm and an imposed pressure head $h_0 = 0$ (I = cumulative infiltration; t = time; From Bagarello and Iovino 2004)

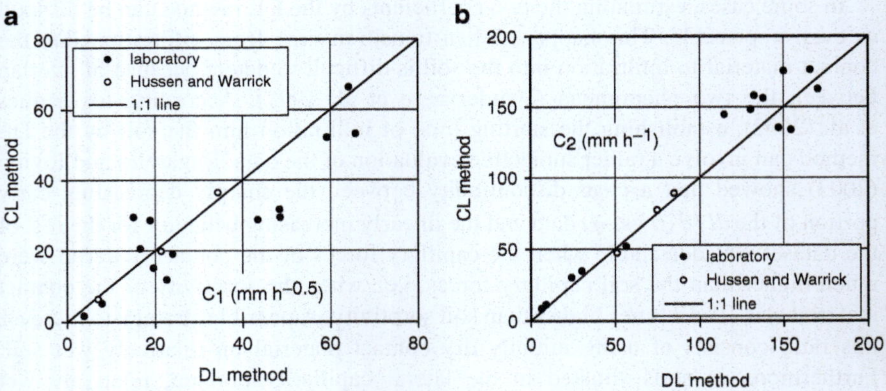

Fig. 3.13 Comparison between the (**a**) C_1 and (**b**) C_2 values of the two-parameter infiltration equation obtained with the Cumulative Linearization, CL, and Differentiated Linearization, DL, methods (From Bagarello and Iovino 2004)

Estimation of coefficients of the two-term cumulative infiltration equation (Eq. 3.21) could also be influenced by the duration of the experiment, as suggested by different authors (Haverkamp et al. 1994; Clausnitzer et al. 1998; Vandervaere et al. 2000a). For a sandy-loam soil and a clay soil situated in the domain of gravity, Bagarello and Iovino (2003) recognized that the experiment duration had a much more pronounced influence on the estimates of C_1 than of those of C_2. In particular, differences between the maximum and the minimum values of C_1 obtained with different durations of the experiment reached 15000 %. The same difference calculated for C_2 was 446 % at the most. However, variability of K_0 was controlled primarily by variability of C_2, while variability of C_1 did not have a substantial

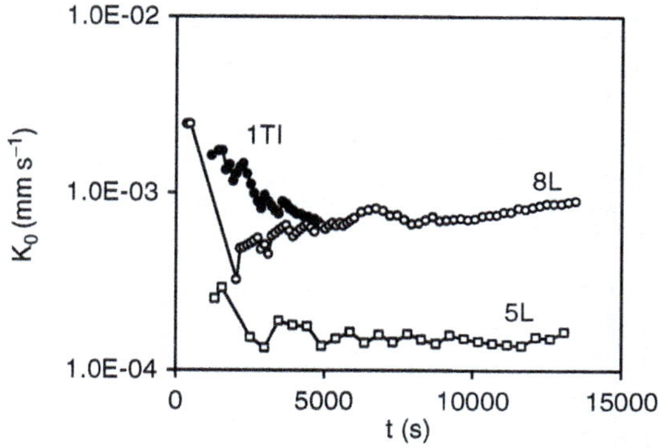

Fig. 3.14 Effect of experiment duration on soil hydraulic conductivity, K_0, estimated by the Transient Single-Test method for three infiltration tests (1TI, 5 L, 8 L; t = time) (From Bagarello and Iovino 2003, modified, reprinted with permission)

effect on the estimates of K_0. Therefore, choosing different durations of the experiment produced values of K_0 that, in most cases, varied with the duration of the experiment by less than 40 %. This result was attributed to the fact that both soils were situated in the domain of gravity, where the estimation of C_1 is made difficult by the importance of the gravity term but it has little importance on the calculation of K_0. The time-dependence of the K_0 calculations was particularly noticeable during the initial part of the experiment (Fig. 3.14). Therefore, conducting TST experiments of relatively long duration (i.e., approximately a hour or more) should be recommended to reduce the risk of obtaining predictions of K_0 that depend strongly on the duration of the experiment.

Both TST and SST methods are expected to provide poor K_0 predictions in soils situated in the lateral capillarity domain, particularly when experiments are conducted under initially dry soil conditions (Vandervaere et al. 2000b). This situation can cause significant problems if using the TI method given that the lateral capillarity domain, that seems to prevail in repacked soils (Vandervaere et al. 2000b), has also been detected in the field (Jacques et al. 2002). Bagarello et al. (2004) simulated numerical infiltration experiments to evaluate the best strategy for conducting a single-test experiment in these soils. The steady-state (SST) and transient (TST) methods were applied to estimate K_0 for a sandy-loam and a clay soil using different values of the initial pressure head, h_i, and the applied pressure head, h_0, and different durations of the experiment (4 h for the sandy loam soil and 12 h for the clay soil). The estimates of K_0 obtained by the SST method were always greater than the corresponding theoretical values, particularly for the clay soil. The values of K_0 calculated by the TST method varied with the duration of the experiment and the most accurate predictions of K_0 were obtained in initially wet soil. In particular, the ratio, K_R, between the estimated and theoretical values of K_0 ranged between 0.4 and 2.0. Such accuracy is generally considered adequate for many applications (Elrick and Reynolds 1992b), and even could be further improved by choosing an appropriate duration for the experiment. As a matter of fact, a duration of about 6 h in the clay soil and about 1 h in the sandy-loam soil produced values of K_R close to unity. For these durations, the SST method gave K_R

values of 3.5–4.0 in the clay soil and of 1.0–3.0 in the sandy-loam soil. The TST method, even when the experiment was prolonged over the optimal duration, yielded K_R values not higher than 1.5. Therefore, within the range of durations of practical interest, a long experiment should be preferred to a short experiment to reduce the time-dependence of the K_0 estimates and to improve their accuracy. However, the time-dependent validity of the transient solution (Eq. 3.21) should be considered to avoid the risk to include the steady phase of the infiltration process into the analysed dataset. As a matter of fact, analysis of the existing literature confirms that the transient single-test method could be threatened by several problems (validity of the infiltration model, experiment duration, initial soil water content). This needs to be considered in order to select the best experimental strategy.

Jarvis et al. (2013) recently developed a global dataset for hydraulic conductivity obtained in the field with the TI by pooling together data obtained with different approaches, including the steady-state, multi-potential, single (Logsdon and Jaynes 1993) or piece-wise (Ankeny et al. 1991; Reynolds and Elrick 1991) log-linear approach, the steady-state, multiple disk radii method (Smettem and Clothier 1989), transient methods requiring early-time transient infiltration measurements (Vandervaere et al. 2000b), and also other methods. This means that differences in the conductivity estimates due to the use of different experimental approaches were not considered to be substantial enough to prevent development of a single set of data.

Example 3.2

Table 3.4 lists the cumulative infiltration data, *I*, measured at different times, *t*, during an infiltration test conducted with a TI having a porous disk of radius

Table 3.4 Data for the Example 3.2

t	I	$t_m^{0.5}$	$\Delta I/\Delta t^{0.5}$
0	0		
0.017	2.51	0.154	14.202
0.033	3.27	0.202	8.360
0.050	3.61	0.240	3.967
0.067	3.75	0.286	0.593
0.100	3.78	0.359	0.747
0.167	3.85	0.444	0.691
0.233	3.90	0.514	0.796
0.300	3.95	0.576	0.892
0.367	4.01	0.631	0.979
0.433	4.06	0.682	0.702
0.500	4.09	0.730	0.753
0.567	4.13	0.794	1.025
0.700	4.21	0.874	0.902
0.833	4.28		

t (h) = time; I (mm) = cumulative infiltration; t_m (h) = geometric mean time

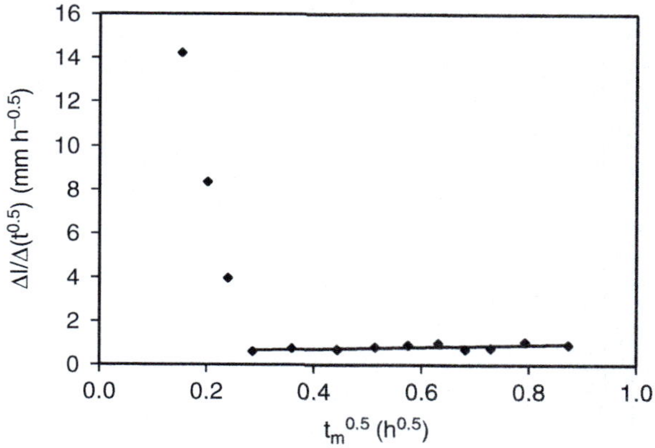

Fig. 3.15 Application of the Differentiated Linearization, DL, method to the data of the Example 3.2 (I = cumulative infiltration; t = time; t_m = geometric mean time)

r = 120 mm and applying a single pressure head at the soil surface, i.e. $h_0 = -50$ mm. To assure good hydraulic contact between the porous plate and the soil surface, a layer of contact material 10 mm thick was used. The initial volumetric soil water content, measured close to the sampling point before beginning the infiltration test, was $\theta_i = 0.175$ m^{3}m^{-3}. Final volumetric water content, measured below the porous plate soon after the end of the infiltration test was $\theta_0 = 0.422$ m^{3}m^{-3}. Estimation of coefficients C_1 and C_2 of Eq. (3.21) was conducted by the DL method (Eq. 3.32) where $dI/d\sqrt{t}$ was approximated by (Vandervaere et al. 2000a):

$$\frac{\Delta I}{\Delta\sqrt{t}} = \frac{I_{i+1} - I_i}{\sqrt{t_{i+1}} - \sqrt{t_i}} = C_1 + 2C_2\sqrt{t_m} \quad (i = 1,\ 2,\ \ldots n-1) \tag{3.35}$$

where n is the number of data points, I_i and I_{i+1} are cumulative infiltrations at times t_i and t_{i+1}, respectively, and t_m is the associated geometric mean time, given by:

$$t_m = \sqrt{t_i\ t_{i+1}} \tag{3.36}$$

The $\Delta I/\Delta\sqrt{t}$ data were plotted against $\sqrt{t_m}$ in Fig. 3.15. The contact material effect on the infiltration process was clearly indicated by the first three data points that showed a decreasing trend, deviating from the remaining data points defining a monotonically increasing linear relationship. Once applicability of the DL method was assessed, the coefficients C_1 and C_2 were determined as the intercept and half the slope, respectively, of the linear regression line fitted to the $\Delta I/\Delta\sqrt{t}$ vs. $\sqrt{t_m}$ data showing an increasing linear behaviour: $C_1 = 0.524$ mm h$^{-1/2}$ and $C_2 = 0.262$ mm h^{-1}. Then, the soil sorptivity, S_0, and hydraulic conductivity, K_0, values corresponding to the applied pressure head were calculated from Eqs.(3.26) to (3.27), respectively, i.e., by the TST method:

$$S_0 = 0.524 \quad \text{mm} \quad \text{h}^{-1/2}$$

$$K_0 = \frac{3}{2 - 0.6}\left[0.262 - \frac{0.75 \times 0.524^2}{120(0.422 - 0.175)}\right] = 0.547 \text{ mm h}^{-1}$$

Finally, the validity of Eq. (3.21) was tested by checking that the experimental duration, equal to 0.83 h, was less than the characteristic time, t_{grav}:

$$t_{grav} = \left(\frac{0.524}{0.547}\right)^2 = 0.92 \text{ h}$$

3.1.5 Numerical Inversion of Tension Disk Infiltrometer Data

Parameter optimization methods based on numerical inversion of the unsaturated water flow equation allow simultaneous estimation of the soil hydraulic properties, i.e., the soil water retention, $\theta(h)$, and hydraulic conductivity, $K(h)$, functions, from a single transient experiment (Hopmans et al. 2002). Provided the unsaturated soil hydraulic properties are described by a given set of parametric relationships, the unknown parameters are estimated by minimizing a suitable objective function, which expresses the discrepancy between the observed values of some selected variables and the system response predicted by numerical solution of the Richards' equation. While initially applied to laboratory experiments, inverse methods are equally applicable to field data or some appropriate combination of laboratory and field data. An important advantage of parameter optimization is that an error analysis of the estimated parameters can be developed. However, inverse methods experience a number of problems related to computational efficiency, convergence and parameter uniqueness, especially when many hydraulic parameters must be estimated simultaneously (van Genuchten and Leij 1992). Application of the inverse methodology to transient tension infiltrometer data was first explored in the pioneering papers by Šimůnek and van Genuchten (1996, 1997) and Šimůnek et al. (1998a, b, 1999a).

The governing flow equation for axisymmetric, isothermal Darcian flow in a variably saturated, isotropic, rigid porous medium is given by the following modified form of the Richards' equation (Warrick 1992):

$$\frac{\partial \theta}{\partial t} = \frac{1}{R}\frac{\partial}{\partial R}\left(RK(h)\frac{\partial h}{\partial R}\right) + \frac{\partial}{\partial Z}\left(K(h)\frac{\partial h}{\partial Z}\right) - \frac{\partial K(h)}{\partial Z} \quad (3.37)$$

where θ (L^3L^{-3}) is the volumetric soil water content, h (L) is the pressure head, K ($L\,T^{-1}$) is the soil hydraulic conductivity, R (L) is the radial coordinate, Z (L) is the vertical coordinate, positive downward, and t (T) is the time. For a given initial condition, generally expressed in terms of water content distribution, Eq. (3.37) is solved numerically by quasi-three-dimensional finite element codes, like HYDRUS-2D (Šimůnek et al. 2006), for the following boundary conditions:

$$h(R,Z,t) = h_0(t) \quad 0 < R \leq r \quad Z = 0 \tag{3.38a}$$

$$\partial h(R,Z,t)/\partial Z = 1 \quad R > r \quad Z = 0 \tag{3.38b}$$

$$h(R,Z,t) = h_i \quad R^2 + Z^2 \rightarrow \infty \tag{3.38c}$$

where h_0 (L) is the time variable surface pressure head imposed by the TI, r (L) is the disk radius and h_i (L) is the initial pressure head corresponding to the initial soil water content, θ_i (L^3L^{-3}). Equation (3.38a) specifies the boundary condition under the disk, variable with time in case of multi-potential experiments. Equation (3.38b) assumes a zero flux at the remainder of the soil surface while Eq. (3.38c) states that the other boundaries are sufficiently distant from the infiltration source so that they do not influence the flow process (Šimůnek et al. 1998a). The boundary condition at the axis of symmetry ($R = 0$) is a no flow condition.

Inverse solution of Eq. (3.37) requires an a-priori assumption of unsaturated soil hydraulic properties, that are generally expressed by the van Genuchten-Mualem (VGM) functions (van Genuchten 1980):

$$S_e = \frac{\theta - \theta_r}{\theta_s - \theta_r} = \frac{1}{[1 + (\alpha_{vG}|h|^n)]^m} \tag{3.39}$$

$$K(S_e) = K_s S_e^{0.5} \left[1 - \left(1 - S_e^{1/m}\right)^m\right]^2 \tag{3.40}$$

where S_e is the effective saturation, θ_r (L^3L^{-3}) and θ_s (L^3L^{-3}) denote the residual and saturated soil water contents, respectively, K_s (LT^{-1}) is the saturated soil hydraulic conductivity, and α_{vG} (L^{-1}), n and m ($=1 - 1/n$) are empirical parameters.

The objective function Φ to be minimized during the parameter estimation process can be expressed in terms of differences between the observed and the simulated values of an arbitrary combination of selected variables, q, as (Šimůnek and van Genuchten 1996):

$$\Phi(\mathbf{b}, q) = \sum_{j=1}^{M} \left(v_j \sum_{i=1}^{N_j} w_{ij} \left[q_j(t_i) - q_j^*(t_i, \mathbf{b})\right]^2 \right) \tag{3.41}$$

where $\mathbf{b} = \{\theta_s, \theta_r, \alpha_{vG}, n, K_s\}$ is the vector of optimized parameters, M is the number of different sets of measurements (e.g., infiltration data, pressure heads and/or water contents at specific points in the flow space, unsaturated hydraulic conductivities obtained by the Wooding's analysis), N_j is the number of measurements in a particular set of data, $q_j(t_i)$ is the specific measurement at time t_i for the jth measurement set, q_j^* (t_i, $\mathbf{b}$) represents the corresponding model predictions for parameter vector $\mathbf{b}$, and v_j and w_{ij} are weights associated with a particular measurement set j or a measurement i within set j, respectively. The weighting coefficients v_j are used to minimize the influence of the discrepancy between weights related to data of different types in the objective function because of different absolute values and numbers of data involved. Usually, weighting coefficients v_j are calculated by (Ramos et al. 2006):

$$v_j = \frac{1}{N_j \sigma_j^2} \tag{3.42}$$

This approach defines the objective function as an average weighted squared deviation normalized by measurement variances σ_j^2 (Angulo-Jaramillo et al. 2000). The weighting coefficients for a given measurement set, w_{ij}, are generally assumed equal to 1, that is, all measurement errors inside a particular set are assumed equal (Šimůnek and van Genuchten 1996). Minimization of the objective function Φ in HYDRUS-2D is accomplished using the Levenberg–Marquardt nonlinear minimization method (Marquardt 1963).

According to Šimůnek and van Genuchten (1997), the best strategy for applying the inverse method consists of measuring cumulative infiltration, I (L), vs. time, t (T), during a sequence of pressure heads consecutively imposed on the infiltration surface and the final soil water content, θ_f (L^3L^{-3}). In this case, the objective function assumes the following form:

$$\Phi(\mathbf{b}, I, \theta_f) = w_I \sum_{i=1}^{N} w_i \left[I(t_i) - I^*(t_i, \mathbf{b})\right]^2 + w_\theta \left[\theta_f - \theta_f^*(\mathbf{b})\right]^2 \tag{3.43}$$

that allows for estimation of up to four unknown parameters (θ_s, α_{vG}, n, K_s) whereas θ_r is fixed to zero. Numerical investigations showed that measurement of cumulative infiltration or instantaneous infiltration rate alone did not provide a unique solution in the three-dimensional parameter space $\alpha_{vG} - n - K_s$ given that the same infiltration curve can be calculated using an infinite number of combinations of these three parameters (Šimůnek and van Genuchten 1996; Schwartz and Evett 2002).

Despite the advantages of using the inverse method in conjunction with TI measurements, these methods are hampered by a number of practical problems that must be overcome so that they can successfully be applied in the field. For instance, a possible source of error when applying the experimental procedure proposed by Šimůnek and van Genuchten (1997) is associated with determination of volumetric soil water content in equilibrium with the last imposed pressure head. Errors can arise when the soil surface is sampled after the removal of the disk infiltrometer due to the small sampling depth required and the fact that the bulk density must be estimated for this thin layer (Angulo-Jaramillo et al. 2000). Furthermore, the most pertinent soil volume of interest directly beneath the disk is typically inaccessible to soil moisture sensors (Schwartz and Evett 2002).

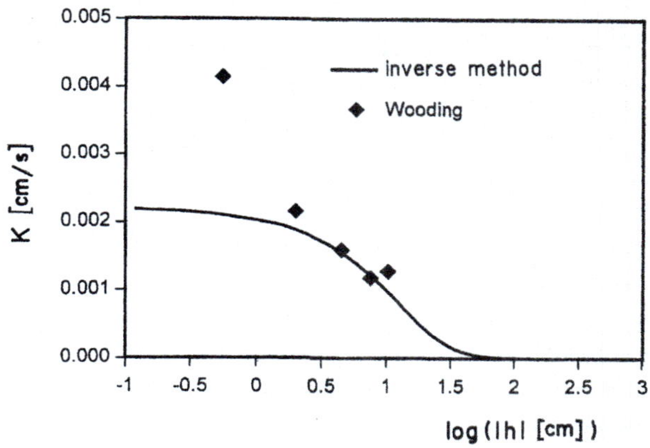

Fig. 3.16 Unsaturated hydraulic conductivities calculated using Wooding's analytical solution for particular pressure heads and the complete function obtained with numerical inversion (Adapted from Šimůnek et al. 1998a, reprinted with permission)

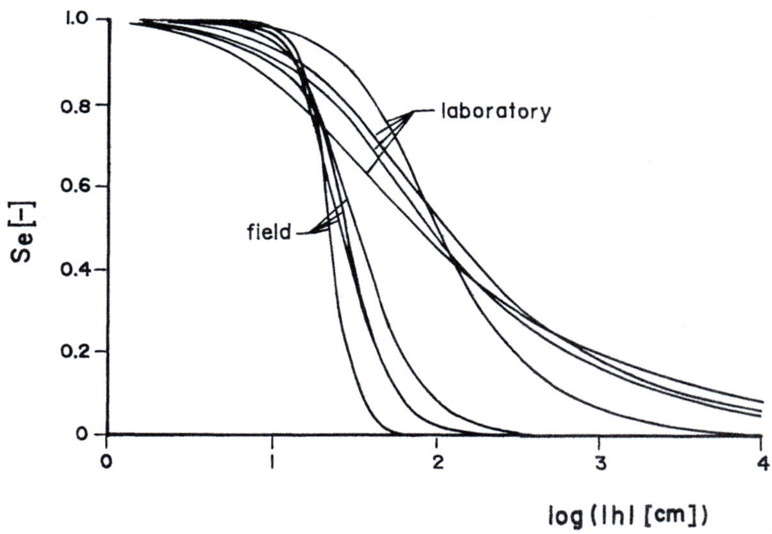

Fig. 3.17 Soil water retention curves obtained in the laboratory and by numerical inversion of the field infiltration data (Adapted from Šimůnek et al. 1998c, reprinted with permission)

Unsaturated hydraulic conductivities obtained with parameter estimation corresponded well with the results of Wooding (1968) analysis (Fig. 3.16) (Šimůnek et al. 1998a, c; Ventrella et al. 2005) as well as with soil hydraulic conductivities obtained in the drier range using the evaporation method (Šimůnek et al. 1999a). However, soil water retention curves estimated by the inverse method were generally in poor agreement with those determined by classical laboratory methods (Fig. 3.17) (Šimůnek et al. 1998c). The observed differences were attributed to the simplifying assumptions for the numerical experiment, including homogeneity, isotropy and uniform initial conditions and the inappropriate description of the unsaturated hydraulic functions. In particular, parameters in the retention curve are closely coupled with those in the hydraulic conductivity function by the VGM model. Therefore, the numerical inversion of a TI experiment is expected to yield a more reliable estimation of parameters expressive of dynamic rather than static properties of soil.

Castellini and Iovino (2005) also showed that the inverse method underestimated θ for a given pressure head compared with equilibrium laboratory methods. Similar results were also obtained by Šimůnek et al. (1999a) with the transient evaporation method for a loamy-sand soil. These last authors attributed the discrepancies to hysteresis and non-equilibrium behaviour and concluded that retention curves estimated from infiltration experiments may be more useful in describing infiltration and contaminant transport in the vadose zone than the retention curves determined by steady-state methods or from transient processes of a different nature. However, water retention was predicted more accurately than hydraulic conductivity by Ramos et al. (2006). These authors suggested that one of the major limitations of the inverse method is its extreme dependence on the initial water content which, unlike the final water content, cannot be determined at exactly the same location where the TI measurements are performed.

Schwartz and Evett (2002) developed a specific optimization strategy for estimating only the $K(h)$ relationship of fine-textured soils. Simultaneous identification of the three parameters, α_{vG}, n and K_s, is very difficult, if not impractical, in these soils partly due to the enhancement of capillarity over gravity. Therefore, they recommended that α_{vG} be estimated from independent water retention data and thereafter be fixed at this value for inverse fit to cumulative infiltration data. In order to estimate hydraulic conductivity function, they proposed using both one-potential TI experiments with final water content measurement and multi-potential TI experiments. In particular, a single TI experiment at $h_0 = -15$ cm made it possible to estimate parameters n and K_s using Eq. (3.43), the VGM model being used for $h \leq -15$ cm. Close to saturation ($h > -15$ cm), inverse optimization made use of multi-potential TI data with $K(h)$ defined using a piecewise log-linear expression instead of the VGM model and an objective function in the form of $\Phi[I(t)]$. According to Schwartz and Evett (2002), if a single fit of the VGM model is used over the entire pressure head range, then large values of K_s and small values of n are required to adequately describe the conductivity and the water retention relationship of fine-textured soils for $h \leq -15$ cm, which in turn poorly represents $K(h)$ near saturation. The method, that may not necessarily be appropriate for coarser-textured soils, probably deserves further attention given that it aims at overcoming the inadequacy of the VGM model, which assumes a unimodal pore size distribution, in describing the soil hydraulic conductivity of structured soils close to saturation.

Alternative optimization strategies were proposed to improve soil water retention curve estimations, which include, in the objective function, soil water content data collected in the wetted zone during the transient infiltration experiment. Indeed, auxiliary $\theta(t)$ data proved to be more efficient than the corresponding $h(t)$ data in the optimization procedure (Šimůnek et al. 1999a). Use of TDR to measure water content during infiltration probably has the greatest potential for field application because the response time of the instrument is fast, measurements are valid for essentially all water contents, and the rigidity and small diameter of probes permit their insertion into the soil with little disturbance to the flow field (Wang et al. 1998a).

Schwartz and Evett (2003) tested an inverse optimization procedure that used cumulative infiltration and water content from TDR probes inserted at a 30° angle into the soil close to the disk's edge and oriented towards the vertical axis of the infiltrometer. To better describe water retention in the dry region, the objective function should anyway include at least one independently measured $\theta(h)$ value at a pressure head sufficiently lower than the lowest applied h_0 value. Experiments conducted on a repacked loamy-sand soil column highlighted some limitations on the use of the TDR in conjunction with the TI (Šimůnek et al. 1999a). In fact, the simulated water contents reached the final value corresponding to each applied pressure head faster than the measured ones that tended to increase more gradually with time. In other terms, the simulated steep increase in water content corresponding to the arrival of the wetting front was not experimentally observed. The detected lag in time was partly attributed to space averaging by the TDR probes as compared with the use of point values in numerical simulations. An even more probable explanation is the occurrence of non-equilibrium flow (Wang et al. 1998a; Šimůnek et al. 1999a). Water initially may move primarily through the larger pores that provide the main paths for water infiltration and hence may bypass the smaller pores, which are vulnerable to air entrapment. This situation would cause the pressure head and the infiltration rates to reach apparent steady-state conditions much faster than the water content (Wang et al. 1998a). In consequence, the water retention curve estimated by the inverse method underestimates the equilibrium water content determined by steady laboratory techniques (Fig. 3.17).

In conclusion, the inverse method applied to TI experiments can not yet be considered a standard field technique given that reliability of results depends on several factors including: (i) experimental strategy; (ii) type and precision of auxiliary data to be included in the objective function; (iii) choice of the parametric model for the unsaturated soil hydraulic properties; and (iv) choice of parameters to be optimized. Therefore, further field investigations are necessary, also because most studies were performed on laboratory repacked soil columns. A point needing specific assessment is the practical advantage of performing TDR measurements during a tension infiltration experiment to improve soil hydraulic characterization with numerical inversion procedures.

3.1.6 Particular Devices

The mini disk infiltrometer (MDI, Decagon Devices Inc. 2012) is a miniaturized tension infiltrometer allowing simple and rapid determination of soil hydrodynamic parameters. The instrument manufactured by Decagon Devices Inc. (Fig. 3.18)

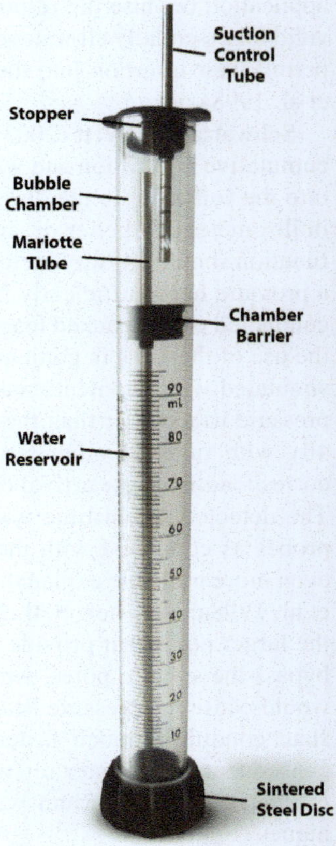

Fig. 3.18 Scheme of the mini disk infiltrometer (From Decagon Devices Inc. 2014, reprinted with permission)

consists of a Plexiglas tube, 31 mm in diameter and about 300 mm long, partitioned into two chambers by a rubber septum. The upper, or bubble, chamber controls the suction whereas the lower chamber, or reservoir, contains about 135 mL of water that infiltrates into the soil through a porous sintered stainless steel disk (45 mm diameter and 3 mm thick) that does not allow water to leak in open air. The adjustable suction control tube allows the operator to apply pressure heads from −0.5 to −7 cm on the soil surface. The infiltrometer should be secured to a ring stand that is lowered onto the soil surface when timing begins (Kirkham and Clothier 2000). The MDI is becoming popular due to its compact size and the small amount of water needed for its operation. These features make the device particularly suitable for field research in remote sites where several constraints limit the use of larger infiltrometers because of complicated experimental configuration, difficult manual transportation of equipment, limited access to water, sloping soil surfaces with locally extreme microtopography, low and dense tree branches (Madsen and Chandler 2007).

Different methods can be applied to analyse infiltration data collected with the MDI for a given pressure head, h_0 (L), imposed on the soil surface. The method proposed by Zhang (1997a), describing transient infiltration by a two-term infiltration model (Philip 1957), was recommended by Decagon Device Inc. (2012) to analyse the MDI data since it was considered quite simple and appropriate for infiltration measurements in dry soil. In the case of a homogeneous, isotropic soil and a uniform initial soil water content profile, the infiltration process under the MDI disk can be described by Eq. (3.21), where the parameters C_1 (L T$^{-1/2}$) and C_2 (L T^{-1}) can be described as functions of soil sorptivity, S_0 (L T$^{-0.5}$), and hydraulic conductivity, K_0 (L T^{-1}), at the applied pressure head, h_0 (L):

$$C_1 = A_1 S_0 \tag{3.44}$$

$$C_2 = A_2 K_0 \tag{3.45}$$

where A_1 and A_2 are dimensionless coefficients that depend on soil water content, soil water retention and infiltrometer parameters. For soils with the van Genuchten (1980) type retention function, the expressions of these coefficients were determined using numerically simulated infiltration data (Zhang 1997a):

$$A_1 = \frac{1.4b^{0.5}(\theta_0 - \theta_i)^{0.25}\exp[3(n - 1.9)\alpha_{vG}h_0]}{(\alpha_{vG}r)^{0.15}} \tag{3.46}$$

$$A_2 = \frac{11.65(n^{0.1} - 1)\exp[2.92(n - 1.9)\alpha_{vG}h_0]}{(\alpha_{vG}r)^{0.91}} \quad \text{for} \quad n \geq 1.9 \tag{3.47a}$$

$$A_2 = \frac{11.65(n^{0.1} - 1)\exp[7.5(n - 1.9)\alpha_{vG}h_0]}{(\alpha_{vG}r)^{0.91}} \quad \text{for} \quad n < 1.9 \tag{3.47b}$$

where h_0 (cm) is the applied pressure head during the infiltration process (h_0 is negative for unsaturated conditions), n and α_{vG} (cm^{-1}) are the water retention parameters, r (cm) is the radius of the infiltrometer, θ_0 (cm^{3}cm^{-3}) is the soil water content at h_0, θ_i (cm^{3}cm^{-3}) is the initial soil water content, and b is a parameter set at 0.55 (Warrick and Broadbridge 1992). Determination of S_0 and K_0 consists of using Eqs. (3.44) and (3.45) once the parameters C_1 and C_2 have been estimated by fitting Eq. (3.21) to the I vs. t data using a maximum neighbourhood method (Marquardt 1963) or a linear fitting technique, such as the so-called differentiated linearization (DL) technique (Vandervaere et al. 1997). The van Genuchten parameters, required by the procedure, can be estimated from soil texture (e.g., Carsel and Parrish 1988) or they can be obtained by fitting laboratory determined soil water retention data, although in this last case the procedure of soil hydraulic characterization becomes more complicated from an experimental point of view.

Using numerically simulated MDI data for 12 soils with different textures, Dohnal et al. (2010) showed that the unsaturated hydraulic conductivity obtained by Zhang's (1997a) method was generally overestimated, with absolute relative errors in the K_0 predictions varying from zero to approximately 70 %. The errors were larger than those reported by Zhang (1997a) probably because the two investigations differed in both the pressure head range (−0.5 to −6 cm in Dohnal et al. (2010) and −5 to −25 cm in Zhang (1997a)), and the size of the disk source (15.2 cm^2 in the former investigation and 38.5–726 cm^2 in the latter one). The largest errors were detected for soils with $n \leq 1.35$. Therefore, the following alternative expression, specifically optimized for a small disk size (15.2 cm^2) and a limited range of disk pressures ($-6 \leq h_0 \leq -0.5$ cm), was developed to estimate A_2 for these soils (Dohnal et al. 2010):

$$A_2 = \frac{11.65(n^{0.82} - 1)\exp[34.65(n - 1.19)\alpha_{vG}h_0]}{(\alpha_{vG}r)^{0.6}} \quad \text{for} \quad n \leq 1.35 \qquad (3.48)$$

The average relative error in the determination of K_0 using this equation was <10 % and the maximum error was of 31 %. According to Table 1 by Dohnal et al. (2010), listing the parameters α_{vG} and n of the van Genuchten water retention curve for soils with different texture, Eq. (3.48) should be used for sandy-clay-loam, sandy-clay, silty-clay and clay soils.

Zhang's (1997a, b) idea to relate the coefficient C_2 to gravity forces alone was questioned by Angulo-Jaramillo et al. (2000) and Vandervaere et al. (2000a) since this coefficient, accounting for three-dimensional effects on the flow process from the disk source, also depends on soil sorptivity (Haverkamp et al. 1994). For this reason, Dohnal et al. (2010) also tested alternative methods for the determination of near-saturated hydraulic conductivity from MDI measurements, including the physically based method by Haverkamp et al. (1994) and the so-called White and Sully (1987) method, i.e., Eq. (3.16), using the Wooding's (1968) equation, the estimated sorptivity and apparent steady-state infiltration rate. In general, both alternative methods did not perform well, probably because flow in the numerically tested soils was dominantly driven by lateral capillary forces as opposed to gravity forces. A means to move the process from the lateral capillarity domain to the gravity domain is to increase the disk size (Dohnal et al. 2010) abandoning, in practice, the MDI. This choice does not seem to be necessary due to the availability of an alternative data analysis procedure, based on the work by Zhang (1997a, b) and Dohnal et al. (2010). This last procedure has some physical weaknesses but it appears to yield reasonably reliable estimates of K_0.

The MDI has been used for several purposes, including determining infiltration rates of soil crusts (Li et al. 2005), estimating the hydraulic conductivity of a sandy soil under different plant covers (Homolák et al. 2009), testing method dependency of saturated soil hydraulic conductivity measurements (Fodor et al. 2011), and characterizing the saturated hydraulic conductivity of an alpine glacial till (Ronayne et al. 2012). Gonzalez-Sosa et al. (2010) combined the MDI with the BEST procedure of soil hydraulic characterization (Lassabatere et al. 2006) to estimate the topsoil hydraulic conductivity and sorptivity in a small French suburban catchment located near Lyon, France. The MDI has also been used to evaluate soil water repellency induced by fire (Lewis et al. 2006) or vegetation (Lichner et al. 2007), and to evaluate the effects of a low-intensity prescribed fire on soil hydraulic properties in a wooded scrubland ecosystem (Chief et al. 2012). To determine the water repellency index, soil sorptivity has to be measured using both ethanol, i.e., a non-polar liquid, and water. In both cases, sorptivity can be estimated as the slope of the cumulative infiltration vs. square root of time measurements at a pressure head $h_0 = -2$ cm (Decagon Device Inc. 2012). Only the infiltrometers with polycarbonate reservoirs, produced after 2005, should be filled with ethanol. Moreover, care must be taken to avoid spillage of ethanol since it can damage the numbering on the reservoir. The bubble chamber of the device has to be filled with fresh or tap water for measurements with both water and ethanol.

An automation of the MDI was developed by Madsen and Chandler (2007). The reservoir was connected to a differential pressure transducer and the transducer output was recorded by a commercial data-logger. Repeated use of the device may determine clogging of the sintered steel disk. To reduce this risk, Kirkham and Clothier (2000) drilled twenty holes with a diameter of 0.25–0.30 mm in the porous plate on the bottom of the infiltrometer. A slight scoring was made in the plastic around the perimeter of the infiltrometer just above the porous plate into which a rubber O-ring was secured to hold a 5.5×5.5 cm^2 piece of nylon mesh. With this modified infiltrometer, clogging was never observed.

The concentric disk tension infiltrometer (CDTI) is a modification of the classical TI by Perroux and White (1988). In particular, the CDTI allows simultaneous one- (1D) and three-dimensional (3D) infiltration measurements that are subject to identical initial and boundary conditions and hence it yields more information than the TI, which only allows 3D infiltration measurements. The basal unit of the CDTI is composed of two hydraulically independent portions, i.e., an inner disk surrounded by an outer buffer disk (Fig. 3.19) (Smettem et al. 1995, 1998; Zhang et al. 1999). Water is supplied independently to both disks from two calibrated water reservoirs but a single bubbling tower is used to control and balance the negative pressure head, h_0 (L), over the two disks. By the CDTI, that is the conceptual offspring of the double ring infiltrometer, a region of 1D flow is delineated under the inner disk and a region of 3D flow is established under the basal unit of the device, comprising the inner and outer portions of the disk.

The two parts of the basal unit must have appropriate sizes to ensure consistency between the experimental data and the theoretical assumptions about flow dimensionality. Zhang et al. (1999) developed design criteria for the inner disk of a CDTI by distinguishing between a zone of 1D flow, i.e., a zone under the basal unit where the assumption of vertical streamlines is acceptable, and an outside zone, where capillary forces distort these streamlines horizontally. Using numerically simulated infiltration data, these authors suggested that a small basal unit, dry initial conditions, fine soil texture and small pressure heads established on the infiltration surface should lead to the use of small inner disks since these factors, which enhance capillary flow relative to gravity flow, reduce the radius of the 1D flow zone under the basal unit of the device. Zhang et al. (1999) also suggested that, for an outer disk radius of 150 mm, which probably represents a practical maximum, an

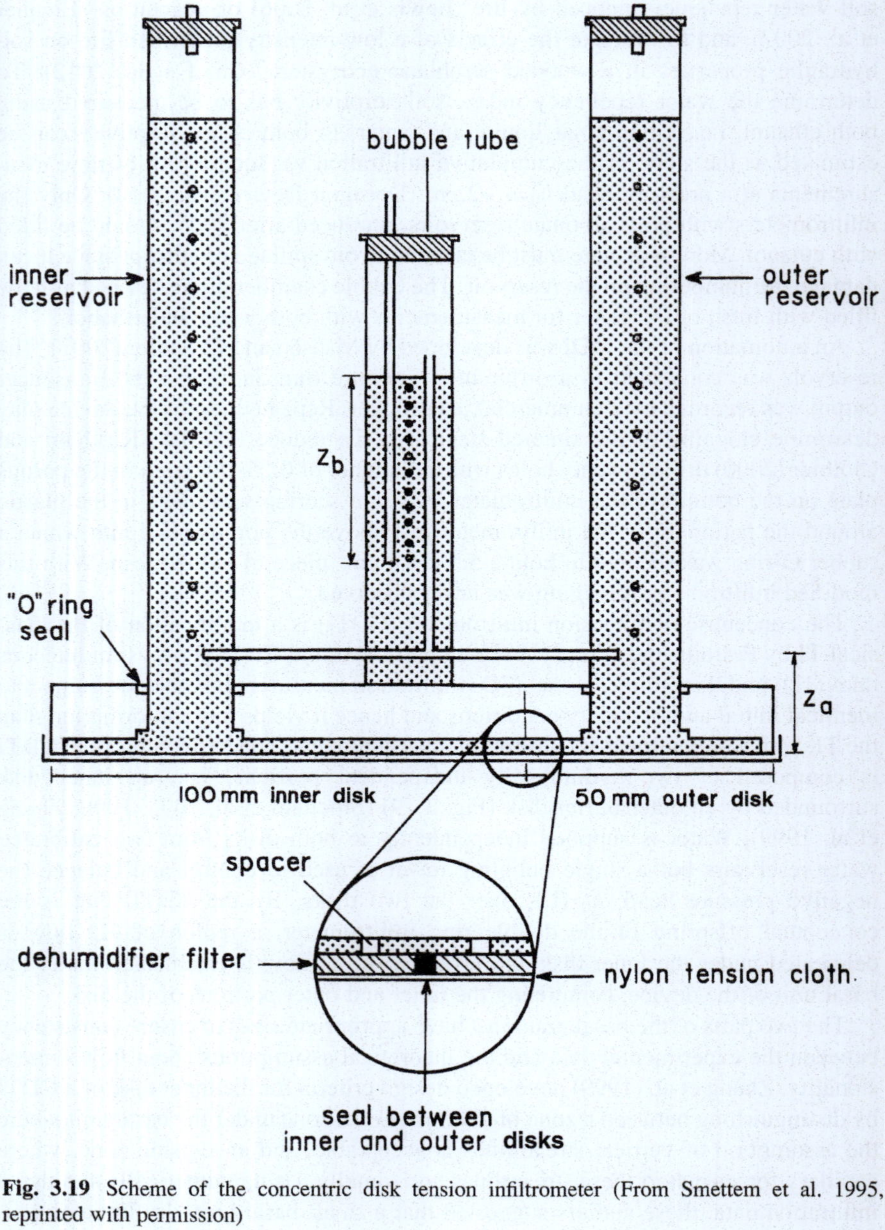

Fig. 3.19 Scheme of the concentric disk tension infiltrometer (From Smettem et al. 1995, reprinted with permission)

inner disk with a radius of no more than 50 mm would provide 1D flow in homogeneous soil under most conditions. In the device used by Smettem et al. (1995), the radii of the basal unit and the buffered inner disk were 150 and 100 mm, respectively. The thickness of the seal between the two disks was 2 mm, which should have a negligible effect on the measured infiltration process since the surface area of the seal was only 4 % of the basal area of the inner disk. The radius of the inner disk used by Smettem et al. (1995) was twice that recommended by Zhang et al. (1999) but, according to these last authors, errors should be expected to be small even in this case if the device is used in coarse textured soils with high, i.e., close to zero, pressure heads. Smettem et al. (1998) used a field device having an inner disk radius of 62.5 mm and a combined inner and outer disk radius of 126 mm. The basal unit and the inner disk of the CDTI realized by Bagarello et al. (2010) had a radius of 107.5 and 50 mm, respectively. In this last device, water was supplied to the two portions of the disk from two reservoirs with a different internal diameter, i.e., 43.5 mm for the inner portion and 64.1 mm for the outer zone, to account for both the different infiltration surfaces (7854 mm^2 for the inner disk and 28451 mm^2 for the outer zone) and flow fields (1D, 3D).

If contact material is not used for the CDTI run, establishing a given pressure head at the soil surface and simultaneously monitoring 1D and 3D infiltration allows estimation of both the soil sorptivity, S_0 (L $T^{-1/2}$), corresponding to h_0 and the so-called γ parameter of Eq. (3.24) (Smettem et al. 1995). The S_0 and γ values obtained from a set of experimental data can then be used to calculate the contribution of gravitational flow during 3D infiltration. This information allows predictions on the reliability of the hydraulic conductivity estimates (Smettem et al. 1995).

In particular, sorptivity can be estimated by using both the 1D (I_{1D}, L) and 3D (I_{3D}, L) cumulative infiltration data, which is advantageous because a comparison between two different estimates of the same variable can be made. At the initial stage, S_0 is assumed to be given by Eq. (3.33) (Philip 1957, 1969) where I_{1D}, measured by the inner disk of the device, is considered. Therefore, an estimate of S_0 is obtained by the slope of the linear regression line fitted to the early time I_{1D} vs. $t^{1/2}$ data. Use of the I_{3D} data is based on the infiltration model by Haverkamp et al. (1994) given by Eq. (3.24), that is expected to be valid over most of the experimental time range. Dividing both sides of Eq. (3.24) by $t^{1/2}$ yields:

$$\frac{I_{3D}}{\sqrt{t}} = S_0 + \left[\frac{2-\beta}{3}K_0 + \frac{\gamma S_0^2}{r(\theta_0 - \theta_i)}\right]\sqrt{t} \tag{3.49}$$

Therefore, S_0 can be estimated from the intercept of the regression line fitted to the $I_{3D}/t^{1/2}$ vs. $t^{1/2}$ data. The parameters describing gravity and edge effects are lumped into the slope term and do not require evaluation in order to estimate sorptivity. In a laboratory experiment performed on an initially dry ($\theta_i = 0.04$ m^3m^{-3}) sandy-loam soil with $h_0 = -30$ mm ($\theta_0 = 0.32$ m^3m^{-3}), Eqs. (3.33) and (3.49) yielded very similar S_0 values, i.e., 51.1–52.6 mm $h^{-1/2}$ (Smettem et al. 1995). It should be noted that, in principle, a third estimate of S_0 could be obtained with a CDTI run because Eq. (3.33) could be used with the early-time 3D infiltration data. By early-time we mean the data collected before $t = t_{geom}$ (T), i.e., the time at which geometry can be expected to swamp the initially one-dimensional character of the process (Philip 1969):

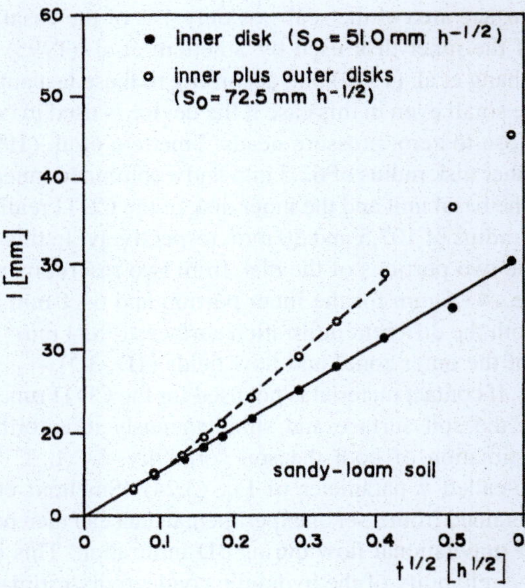

Fig. 3.20 Cumulative infiltration, I, plotted as a function of the square root of time, t, for a laboratory experiment with the concentric disk tension infiltrometer (Adapted from Smettem et al. 1995, reprinted with permission)

$$t_{geom} = \left[\frac{r(\theta_0 - \theta_i)}{S_0}\right]^2 \tag{3.50}$$

However, infiltration data depart from linearity much earlier than given by t_{geom} (Cook and Broeren 1994; Warrick 1992) and, in any case, t_{geom} cannot be evaluated a priori so that any error in estimating sorptivity from Eq. (3.33) and 3D infiltration data will lead to errors in t_{geom} (Smettem et al. 1998). In the experiment by Smettem et al. (1995), this last approach yielded $S_0 = 72.5$ mm h$^{-1/2}$, with a high coefficient of determination ($R^2 = 0.997$, Fig. 3.20). This value overestimated S_0 but it could appear acceptable without recourse to comparison with 1D flow. Therefore, two alternatives are possible to estimate S_0 with the CDTI, i.e., 1D flow and Eq. (3.33) or 3D flow and Eq. (3.49). Only this last approach should be applied for sorptivity estimation by a classical TI run since Eq. (3.33) with 3D infiltration data tends to overestimate S_0 even if a good linearity of the experimental I_{3D} vs. $t^{1/2}$ relationship data is detected.

The simplified, explicit, one-dimensional infiltration model by Smettem et al. (1994) and Haverkamp et al. (1994) is:

$$I_{1D} = S_0\sqrt{t} + \frac{2-\beta}{3}K_0 t \tag{3.51}$$

Therefore, an estimate of the γ coefficient of Eq. (3.24) can simply be obtained using the estimated sorptivity and the slope of the linear regression line between $I_{3D} - I_{1D}$ and t (Fig. 3.21):

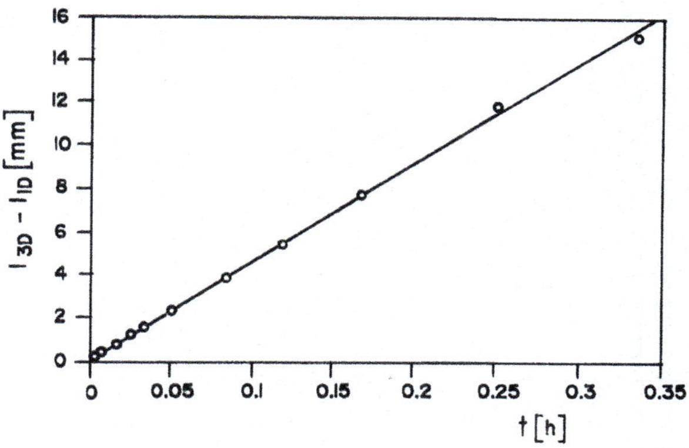

Fig. 3.21 Difference between three- (I_{3D}) and one-dimensional (I_{1D}) cumulative infiltration plotted as a function of time, t, for a laboratory experiment carried out with the concentric disk tension infiltrometer on a sandy-loam soil (From Smettem et al. 1995, reprinted with permission)

$$I_{3D} - I_{1D} = \frac{\gamma S_0^2}{r(\theta_0 - \theta_i)} t \tag{3.52}$$

Plotting $I_{3D} - I_{1D}$ against t allows verification of the suggestion by Haverkamp et al. (1994) that, from an experimental viewpoint, γ could be taken as a constant.

The gravity term can be expressed as:

$$\frac{2-\beta}{3} K_0 t = \varepsilon I_{3D} \tag{3.53}$$

where ε is the fractional influence of gravity on cumulative 3D infiltration. Introducing Eq. (3.53) into (3.24) gives:

$$I_{3D}(1-\varepsilon) = S_0 t^{1/2} + \frac{\gamma S_0^2}{r(\theta_0 - \theta_i)} t \tag{3.54}$$

As I_{3D} is known from the experiment together with an estimate of S_0 and γ, it is easy to obtain ε as a function of time. At the end of the laboratory experiment by Smettem et al. (1995), the value of ε obtained by Eq. (3.54) was very close to zero, suggesting a negligible gravity effect and, hence, uncertainties in the estimation of K_0 because 3D infiltration was dominated by capillary flow effects. On the other hand, gravity substantially influenced infiltration in a field application of the CDTI carried out on a loam soil with $h_0 = -20$ mm (Smettem et al. 1998).

The method developed by Zhang et al. (1999) to analyze the CDTI data makes use of the Wooding (1968) equation that can be written as:

$$i_0 = K_0 + \frac{4\phi_0}{\pi r} \tag{3.55}$$

where, for an established h_0 value, i_0 (L T^{-1}) is the steady-state 3D infiltration rate from the source and ϕ_0 (L^2T^{-1}) is the matric flux potential. This method avoids the need to estimate S_0, and this circumstance was considered advantageous by Zhang et al. (1999) because field infiltration runs generally require use of contact material, complicating the interpretation of the initial infiltration data. Moreover, soil water content data are not required and hydraulic conductivity values corresponding to different pressure heads can be obtained at a single sampling point. The method by Zhang et al. (1999) assumes that, when the CDTI is properly designed, the water flux from the inner disk is essentially 1D and asymptotic to K_0 because unit hydraulic gradient is approached in a vertically homogeneous porous medium. Moreover, Eq. (3.55) can be used to determine ϕ_0 which depends on the difference between the total 3D flux (i_0) and the inner disk 1D flux (K_0) at steady-state:

$$\phi_0 = \frac{\pi r}{4}(i_0 - K_0) \tag{3.56}$$

According to Zhang et al. (1999), determination of ϕ_0 allows parameter estimation of selected soil hydraulic conductivity functions and hence estimation of the hydraulic conductivity at potentials below the measured range.

The most accurate estimates of K_0 should be obtained in coarse-textured soils, where a unit hydraulic gradient beneath the central disk is most likely, and at high (i.e., less negative) pressure heads, since steady-state is approached more rapidly. In many cases, however, the approach of unit hydraulic gradient requires too long a waiting period and is incompatible with the normal duration of a field run (approximately 0.5–1.5 h), implying that field measured steady-state infiltration rates are expected to overestimate K_0. To avoid or at least reduce this possibility, Zhang et al. (1999) suggested extrapolating quasi steady-state infiltration data to long time for estimation of K_0 from the inner disk flux. In particular, measured 1D infiltration rates, i_{1D} (L T^{-1}), have to be plotted against $t^{-1/2}$. Using the linearly increasing portion of i_{1D} vs. $t^{-1/2}$ corresponding to the longest experimental times, an estimation of K_0 can be obtained by extrapolation of the fitted regression line to $t^{-1/2}=0$. The extrapolation technique improved the reliability of K_0 estimates, particularly in fine-textured soil conditions and for low h_0 values (Zhang et al. 1999).

The CDTI allows determination of both confined (inner disk, K_c) and unconfined (inner and outer disk, K_u) soil hydraulic conductivity. Establishing the differences between these two conductivity estimates at a site of interest may have practical importance for properly characterizing the field site. For example, an appreciable difference between these two estimates of K_0 could suggest that K_c should be used for simulating vertical infiltration or redistribution processes whereas K_u could be more appropriate in a study of the infiltration process from a point source. The estimates of K_c and K_u for a clay soil and two different sandy-loam soils were compared by Bagarello et al. (2010). Using multi-potential CDTI runs ($h_0=-120$, -60 and -30 mm), K_u was calculated according to Ankeny et al. (1991) and K_c was estimated from the measured steady-state infiltration rate under the inner disk of the device. The ratio between the means of K_c and K_u varied from 0.96 to 1.45, depending on the considered soil/pressure head combination, and the ratio between the associated coefficients of variation ranged between 0.97 and 2.36. In most cases, K_c was higher and more variable than K_u and a statistical significance of differences was detected for some of the established comparisons. Possible factors determining the observed differences included overestimation of steady-state 1D flow rate and reduced ability of the relatively small inner disk to sample a site-representative surface macroporosity compared to the relatively large disk used for 3D measurement. The conclusion by Bagarello et al. (2010) was that differences between K_u and K_c were small and probably negligible for many practical applications (e.g., Elrick and Reynolds 1992b) but K_c was less reliable than K_u. An improved quality of the K_c data can be expected with large inner disks, since a more representative area is sampled by an individual measurement, and a longer infiltration run, since it improves estimation of steady-state conditions. In any case, there are practical limits to possible improvements. For example, a very large inner disk cannot be used since a small inner disk and a large basal unit are required to improve the 1D character of flow under the disk's central zone (Zhang et al. 1999). In addition, the basal unit cannot be excessively large since, in this case, a good contact between the disk and the soil could be difficult or even impossible to achieve.

3.1.7 *Hydraulic Contact at the Infiltration Surface*

Soil hydraulic conductivity measurements conducted by the TI can be invalidated by incomplete hydraulic contact between the infiltration disk and the soil surface (Perroux and White 1988). According to some studies, establishing a good hydraulic contact is not a critical problem when the soil surface is relatively smoothed and levelled given that possible small hollows at the soil surface can be filled by sieved soil collected at the sampling site (Logsdon and Jaynes 1993; Wang et al. 1998b). According to other studies, however, a layer of contact material (usually natural sand, uniform glass beads, or some other fine particulate material) is always required to promote hydraulic continuity between the disk and the soil (Perroux and White 1988; Bagarello et al. 2001; Vandervaere 2002; Reynolds 2008). For example, Bagarello et al. (2001) showed that convincing, apparent steady-state infiltration rates were measured when a layer of contact material was not used over the smoothed and levelled surface of a sandy-loam soil. However, these infiltration rates were about 30 % lower than those obtained with a layer of contact material and the pressure head at the soil surface was also lower (Fig. 3.22). Similar results were obtained by Minasny and McBratney (2000) for a clay soil, and the authors recommended using contact material to prevent the risk of obtaining unrepresentatively low infiltration rates due to a poor contact between the disk base and the soil surface.

The contact material layer on the soil surface should not influence infiltration rates but Everts and Kanwar (1993) observed that a ponded infiltrometer yielded infiltration rates of at least one order of magnitude higher than those obtained at the same sampling point with TI set at the same positive pressure head and placed on a 20-mm layer of sand. More recently, Schwärzel and Punzel (2007) observed that

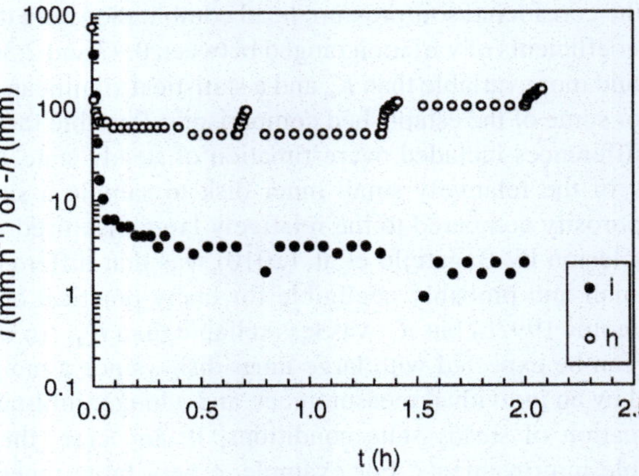

Fig. 3.22 Infiltration rate, *i*, and pressure head at the soil surface, *h*, vs. time, *t*, during an infiltration test. The TI baseplate was removed at $t = 0.71$ h and put again in place after 15 s. The TI baseplate was then removed at $t = 1.41$ h and put again in place after 15 s once the contact material had been eliminated (From Bagarello et al. 2001, reprinted with permission)

saturated hydraulic conductivities measured with a disk infiltrometer placed on a sand contact layer were ten times smaller than the corresponding values measured by a hood infiltrometer. According to Reynolds (2008), however, a contact sand layer having the correct hydraulic properties and placed correctly cannot decrease the infiltration rates.

Non-uniform wetting of the contact material layer may be a cause of error in TI measurements. As a consequence of air pockets below the disk membrane, part of the contact material surface can be excluded from infiltration. Tracer experiments conducted by Close et al. (1998) showed that non-uniform wetting occurred even under controlled laboratory conditions on repacked soil columns (Fig. 3.23) and this occurrence was erratic. The membrane/sand contact was particularly problematic when high tension (more negative pressure head) was applied. In this case, fingering of the wetting front occurred, denoting an incomplete contact between the porous disk and the sand. The lateral redistribution within the sand layer was not effective in compensating for imperfect membrane contact at high tensions.

Use of contact material has been the subject of several studies specifically focused on: (i) determining the influence of contact material hydraulic properties on infiltration; (ii) finding the best commercial contact material for TI application; (iii) allowing analysis of transient flow from the TI when contact material is used; and (iv) developing alternative disk designs to omit contact material use.

It is generally assumed that contact material has no effect on TI operation or results. In many applications, it is simply stated that a thin layer of sand or a layer of sieved soil was used with no mention of the hydraulic properties or thickness of this layer. However, the hydraulic properties and the thickness of the contact material layer can have a large influence on the water pressure head at the soil surface and must be accounted for in TI analyses (Reynolds and Zebchuk 1996; Reynolds 2006; Vandervaere et al. 2000a).

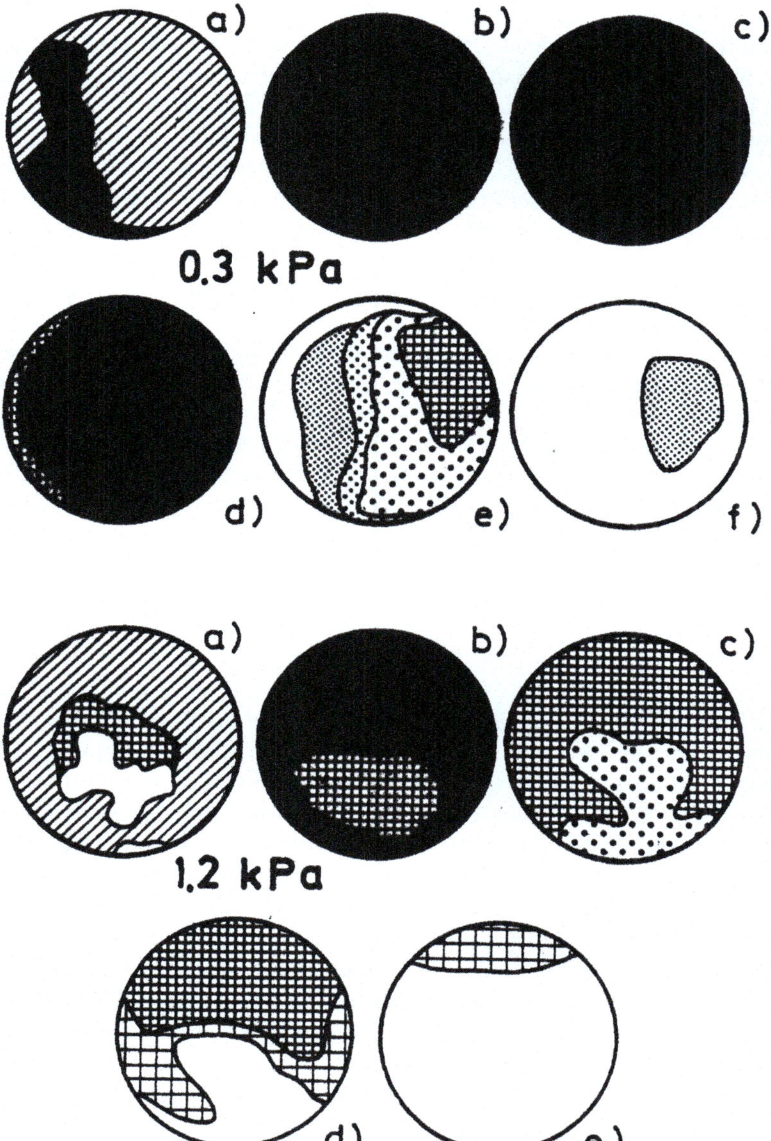

Fig. 3.23 Spatial distribution of soil water content at depths increasing from (**a**) to (**f**) below the disk of a tension infiltrometer for two values of the applied tension. *Darker grey* tones indicate higher soil water content (From Close et al. 1998, modified, reprinted with permission)

According to Perroux and White (1988), the characteristics of the contact material for steady infiltrometer flow should include a saturated hydraulic conductivity, K_{cm} (L T^{-1}), higher than that of the soil, a water entry pressure head, h_w (L), lower (more negative) than the minimum pressure head applied to the infiltrometer membrane, and as small a thickness as possible to minimize the material effect on flow. A thickness, T_{cm} (L), of 3–5 mm is considered adequate for most conditions. However, many intact soil surfaces are often sloping or undulating and $T_{cm} \geq 10$ mm can be required (Thony et al. 1991). Reynolds and Zebchuk (1996) recommended that the required characteristics of contact material should also include low variability of hydraulic properties, chemical inertness and, preferably, the possibility to retrieve and re-use the material.

The presence of a layer of contact material introduces an offset, Δh (L), between the pressure head at the soil surface, h_s (L), and the pressure head set on the TI membrane, h_0 (L), which consists of a constant component caused by the finite thickness of the layer of contact material, T_{cm}, and a variable component caused by flow-induced head loss within the contact material (Reynolds and Zebchuk 1996). Providing that flow in the contact material layer is reasonably steady, saturated and rectilinear, the pressure head offset can be determined via the Darcy's law (Reynolds 2006):

$$\Delta h = (h_s - h_0) = \left[T_{cm} - \frac{i(h_0)T_{cm}}{K_{cm}}\right] \tag{3.57}$$

where $i(h_0)$ (L) is the steady flux density out of the infiltrometer and into the soil at membrane pressure head $h = h_0$. The condition $\Delta h = T_{cm}$, which corresponds to a pressure head offset equal to the contact material thickness, occurs for the special case of $i(h_0) = 0$ (i.e, no flow out of the TI). As the flux density increases, the pressure head offset due to the disk elevation is compensated by the frictional head loss. In particular, if $i(h_0) = K_{cm}$, then $\Delta h = 0$ and, consequently, the pressure head at the soil surface equals the pressure head set at the TI membrane, i.e., $h_s = h_0$. This indicates an exact counterbalance of the elevation and the head-loss effects as well as a pressure head gradient in the contact material layer equal to unity. Therefore, the contact material layer does not influence h_s provided that steady flux density is equal to K_{cm}. For $i(h_0) \neq K_{cm}$, the offset changes linearly with increasing contact material thickness, with the slope of the relationship dependent on $i(h_0)$ and K_{cm}. In particular, if $i(h_0) > K_{cm}$, then $\Delta h < 0$ thus resulting in a pressure head at the soil surface lower (more negative) than the one imposed at the TI membrane. Otherwise, when $i(h_0) < K_{cm}$, the pressure head offset is positive (i.e., $h_s > h_0$). Equation (3.57) shows a linear relationship between Δh and the inverse of K_{cm}. When the hydraulic conductivity of the contact material is very high (i.e., $1/K_{cm}$ approaches zero), Δh tends to assume a constant value equal to T_{cm}. Given that T_{cm} can also be thicker than 10 mm in many field applications and that $i(h_0)$ is controlled by soil characteristics, it is advantageous to have K_{cm} as large as possible, as this reduces the dependence of Δh on the flux density (Reynolds 2006). There is, however, a practical limit on the magnitude of K_{cm} given that the water entry potential, h_w, increases (i.e., becomes less negative) with K_{cm}. Consequently, contact materials having $K_{cm} > 0.1 - 0.5$ mm s^{-1} have a water-entry value that is too large to maintain saturation (and thereby constant hydraulic conductivity) across the required range of h_0 values, which is usually $-200 \leq h_0 \leq 0$ mm. It is worth noting that only under saturated conditions the hydraulic conductivity of the contact material is constant and independent of the applied pressure head.

If the TI flux densities are low enough to avoid appreciable head-loss effect, as with the more negative pressure heads, the offset in pressure head is close to the thickness of the contact material. Such an offset can readily be accounted for by simply adjusting the bubble tower pressure head by an amount equal to T_{cm}. For example, if $T_{cm} = 10$ mm and $h_s = -120$ mm is desired on the soil surface, then the bubble tower should be set to yield $h_0 = -130$ mm on the TI membrane. When the flux density is relatively high, as for the less negative pressure heads, it is advisable to use Eq. (3.57) to determine the soil surface pressure head taking into account elevation and head-loss effect (Reynolds 2006).

In addition to the K_{cm} and h_w requirements described above, there are certain physical attributes associated with the contact material that are desirable for convenient field use. First, the hydraulic properties of the contact material should be stable and characterized by a low variability. Furthermore, the contact material should be chemically inert so that it neither leaches nor sorbs dissolved constituents and is thus usable for TI-based solute transport studies. Finally, the contact material should be easily obtained, inexpensive and reusable, if necessary (Reynolds 2006). A contact material that matches the above attributes relatively well is the Spheriglass no.2227 (Potters Industries, LaPraire, Canada), that was successfully used by Reynolds and Zebchuk (1996) and Bagarello et al. (2001). This contact material is made by uniform spheres of amorphous and relatively inert soda-lime glass. The 99.2 % (w/w) of the glass beads have fine sand texture (50–250 μm). The spherical shape of the particles facilitates a highly stable, close packing arrangement of the contact material layer thus promoting constant and repeatable K_{cm} and h_w values. For the Spheriglass no.2227 spheres, Bagarello et al. (2001) obtained a mean K_{cm} value of 265 mm h^{-1} (coefficient of variation = 1.5 %), and this value was practically constant after repeated use both in the laboratory and in the field. Reusing a natural sand, usable as an alternative contact material in TI experiments, produced a 27 % increase of the mean K_{cm} due to breakdown and progressive loss of fine particles. The main water retention curves obtained from new and field-used Spheriglass no. 2227 spheres were highly reproducible. In particular, the water entry pressure head, h_w, varied between −640 and −650 mm. Therefore, the hydraulic properties of the Spheriglass no. 2227 spheres appeared adequate given that the measured h_w values were lower than the minimum pressure head value imposed by the TI ($h_0 \approx -200$ mm) and the K_{cm} values were sufficiently high to allow use in most agricultural soils (Bagarello et al. 2001).

The contact material layer influences transient analysis of TI experiments (Angulo-Jaramillo et al. 2000; Reynolds 2008). In this case, separating the wetting phase of the contact material from the early-time infiltration into the soil is necessary in order to consider exclusively the latter phase in data analysis. Due to a more or less prolonged transition zone between the two stages of the TI run, estimating the instant at which infiltration into the soil starts may be difficult from plots of cumulative infiltration or infiltration rate vs. time (Vandervaere et al. 1997).

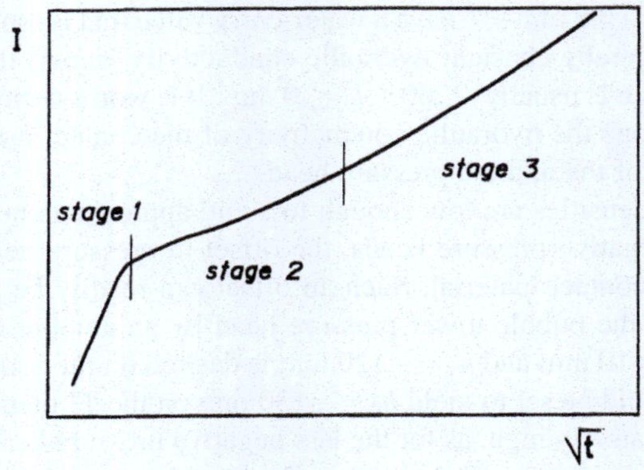

Fig. 3.24 Qualitative representation of cumulative infiltration, I, versus the square root of time, t, for a tension infiltrometer experiment. Stage 1 represents wetting of the contact material. Stage 2 is when soil capillarity controls the infiltration process. In stage 3, the process is influenced by the geometry of the source

Separation of the contact material wetting phase from the phase where soil sorptivity dominates flow can be made visually on the cumulative infiltration plot, I (L), versus the square root of time, t (T), measured from $t = 0$ (Fig. 3.24) (Cook and Broeren 1994). Transition from the stage in which wetting of the contact material occurs to that in which infiltration is dominated by soil sorptivity is highlighted by an abrupt change in the slope of the I vs. $\sqrt{t}$ relationship.

One of the advantages of the differentiated linearization, DL, method (see Sect. 3.1.4) is that it easily allows a visual assessment of the wetting phase of the contact material (Vandervaere et al. 2000a) since, in the $\Delta I/\Delta\sqrt{t}$ vs. $\sqrt{t}$ plot, the influence of the contact material corresponds to the initially sharply decreasing part of the curve, deviating from the monotonically increasing linear behaviour (Fig. 3.15). However, application of the DL method can be difficult or even impossible when the transition time from infiltration into the contact material to infiltration into the soil is difficult to detect because of overlap between the two phenomena (Vandervaere et al. 1997). Minasny and McBratney (2000) applied the DL method to numerically simulated infiltration data and observed that an early-time perturbation, which should be indicative of the wetting phase of the contact material, was also detected when a layer of contact material was not used. Jacques et al. (2002) showed that establishing the starting time of infiltration into the soil by the DL method can involve a rather subjective evaluation of the data.

Bagarello and Iovino (2003) observed an abrupt change in slope between the sharply decreasing, early portion of the $\Delta I/\Delta\sqrt{t}$ vs. $\sqrt{t}$ dataset and the linearly increasing part of the dataset for about the 40 % of the experiments conducted in a clay soil and a sandy-clay soil. For the remaining experiments, a transition zone between the first and the second part of the dataset was observed, and this zone complicated or even impeded estimation of the initial time of the infiltration process into the soil. Separation between the wetting phase of the contact material and the early-time infiltration into the soil was always possible on an I vs. $\sqrt{t}$ plot. Estimates of the sorptivity of both the contact material (S_{cm}) and the soil (S_0) were obtained

Fig. 3.25 View of the malleable disk base (From Moret-Fernández et al. 2013, reprinted with permission)

from linear interpolation of the data corresponding to stages 1 and 2 of Fig. 3.24. Calculation of the S_{cm}/S_0 ratios allowed the authors to conclude that the applicability of the DL method depended on differences in capillary forces between the contact material and the soil. In particular, the DL method should be applicable when the ratio S_{cm}/S_0 is relatively high (i.e., greater than 10–12 according to Bagarello and Iovino 2003). A possible strategy to increase this ratio consists of using initially dry contact material on relatively wet soil.

Alternative approaches to overcome the drawbacks due to an incomplete hydraulic contact rely on alternative designs of the infiltration disk.

The malleable disk base (MDB) for TI was proposed by Moret-Fernández et al. (2013) as an alternative to the conventional rigid base in order to improve contact with the soil surface without using a contact material layer. The MDB consists of 100 mm diameter disk covered with a loosened, malleable nylon cloth of 20-μm mesh (air-entry value of about −750 mm) filled with 100 g of coarse sand (1.0–1.5 mm grain size) that forms a deformable layer, approximately 0.5 cm thick, capable of adapting to a relatively smooth area when the infiltrometer is placed on the soil surface (Fig. 3.25). The MDB was tested in structured soils with different forms of tillage management. The DL and CL methods were applied to estimate the soil hydraulic properties and the results were compared with those obtained by applying the DL method to data collected with a conventional disk base and a layer of contact sand. The direct contact of the MDB with the soil surface eliminated the jump in the cumulative infiltration curve at the beginning of the experiment, which is commonly observed with conventional devices as a consequence of the additional water stored in the contact material. The DL method was easily applied with the MDB data but not with the data collected with the conventional device (Fig. 3.26a, b). An excellent linear behaviour of the data was also found when the CL method was applied (Fig. 3.26c). Furthermore, the CL method applied to the MDB data resulted in higher coefficients of determination and significance as regards the regression models and lower standard errors of both S_0 and K_0 estimates. In general, the DL method applied to data collected by the conventional disk was

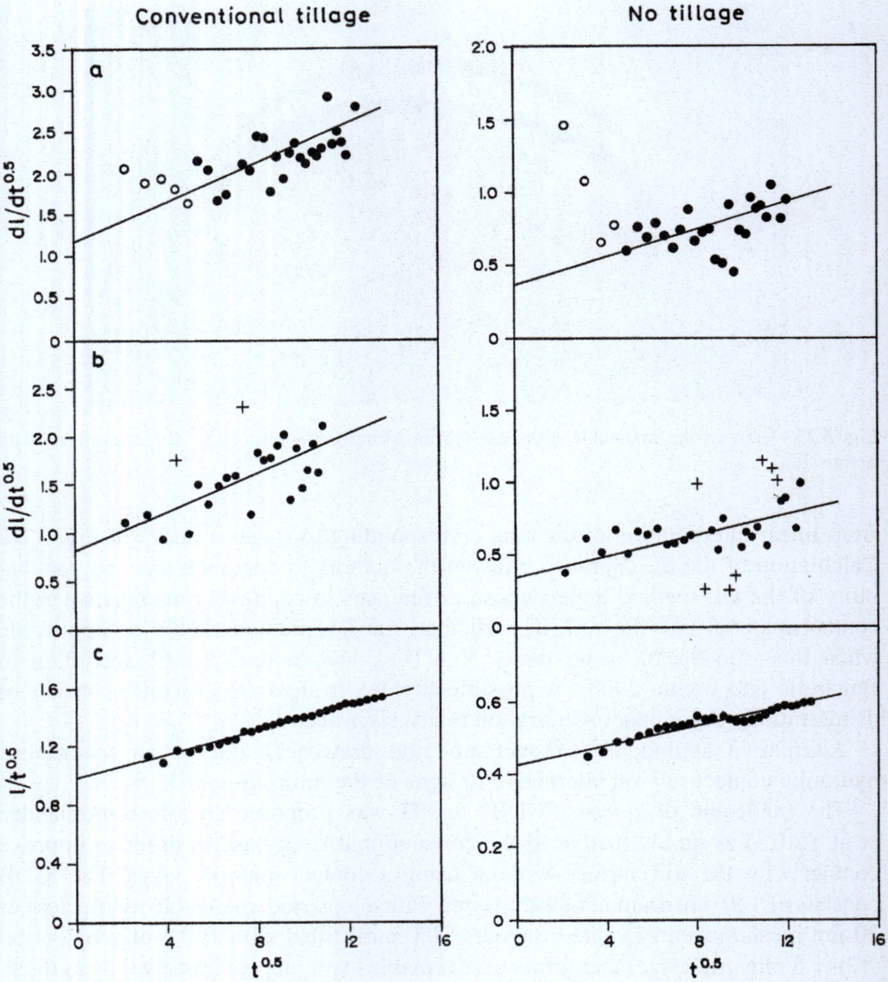

Fig. 3.26 Application of the Differentiated Linearization method to infiltration data measured with the (**a**) conventional and (**b**) malleable disk bases. *White* and *grey circles* in (**a**) denote the section of the linear fitting curve corresponding to the contact sand layer and surface soil, respectively. *Crosses* indicate removed points. Figure (**c**) shows application of the Cumulative Linearization method to data collected by the malleable *disk* base (From Moret-Fernández et al. 2013, reprinted with permission)

considered a more inaccurate approach, also because of the subjective procedure required to remove the first points of the infiltration curve corresponding to the wetting of contact sand.

The Hood infiltrometer (HI) (Schwärzel and Punzel 2007) is a relatively new type of infiltrometer, specifically developed to establish a good hydraulic contact between the soil and the device. Instead of requiring a disk and contact material, it

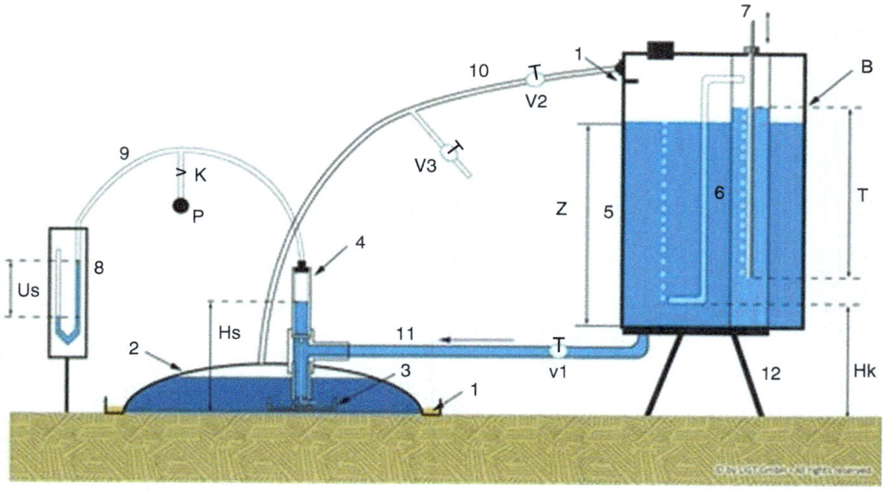

Fig. 3.27 Schematic of the Hood Infiltrometer (From UGT GmbH 2012, reprinted with permission)

places a water-filled hood, open side down, onto the soil surface. The HI consists of three components: a hood, a Mariotte water supply, and a U-tube manometer (Fig. 3.27). The hood is made of acrylic material and is placed open side down onto the soil with a retaining ring. Water is supplied by a conventional Mariotte water reservoir/bubble tower system. In contrast to the disk of the TI device, an additional air outlet tube connects the head space of the water reservoir with the head space of the hood. The hood also contains a standpipe that is joined to the U-tube manometer. The purpose of this manometer is to measure the effective pressure head on the soil surface. Water infiltration takes place from the hood and no perforated plate, nylon membrane, or contact material is required on the infiltration surface. Starting of the infiltration run requires a two-step procedure: (i) first, the connection tube between the hood and the water reservoir is opened and the water moves into a buffer cup inside the hood; (ii) then, the air outlet tube is slowly opened to cause the air inside the hood to go below the sub-pressure. At this moment, the buffer cup overflows, thus filling the space below the hood with water and infiltration starts. According to Schwärzel and Punzel (2007), the HI allows the hydraulic properties to be measured from saturation up to the bubble point of the soil, defined as the pressure head required to force air through the pores of a water-saturated soil. The field tests performed by these authors on a sandy-loam soil are, to our knowledge, the only experimental validation of the new device. They performed a sequence of three experiments conducted at the same place: (i) HI test; (ii) TI test with a 10-mm-thick contact material layer, and (iii) HI test with the contact material left on the infiltration surface. The HI measurements without a contact layer were almost one order of magnitude greater than the corresponding values of the disk and hood experiments with a contact layer. However, the contact material effect (Eq. 3.57) explained only 10 % of the observed differences between the hood and disk

Table 3.5 Data for the Example 3.3

t (s)	I (mm)	$t_m^{0.5}$ ($s^{0.5}$)	$\Delta I/\Delta t^{0.5}$ (mm $s^{-0.5}$)
0	0.000		
15	1.272	4.606	0.536
30	2.132	6.062	0.503
45	2.750	7.208	0.331
60	3.094	9.212	0.129
120	3.507	12.123	0.056
180	3.644	14.417	0.033
240	3.713	16.381	0.038
300	3.782	18.128	0.042
360	3.851	19.719	0.045
420	3.919	21.190	0.049
480	3.988	22.564	0.026
540	4.022	23.858	0.055
600	4.091	25.086	0.029
660	4.126	26.255	0.060
720	4.194	27.375	0.031
780	4.229	28.451	0.033
840	4.263	29.967	0.034
960	4.332	31.910	0.046
1080	4.418	33.740	0.048
1200	4.504	35.476	0.031
1320	4.555	37.131	0.053
1440	4.641	38.714	0.056
1560	4.727	40.235	0.046
1680	4.796	41.701	0.048
1800	4.865		

t = time; I = cumulative infiltration; t_m = geometric mean time

infiltrometer results. It was therefore supposed that preparation of the soil surface for the TI measurements led to sealing and smearing of the pores at the soil surface, and applying pressure heads near saturation caused mobile fine-textured particles of the contact material to clog the macropores. The proposed thesis was tested with numerical simulation, which showed how additional flow impedance of the seal layer prevented the full range of available pore sizes from conducting water. The authors concluded that the classical TI measurements were expected to provide unrealistic near-saturated hydraulic conductivity values, at least for the sandy-loam soil investigated in their study.

Example 3.3
Table 3.5 lists the cumulative infiltration data collected during a Transient Single-Test experiment (Sect. 3.1.4). The pressure head established at the soil surface was $h_0 = -100$ mm and a layer of Spheriglass no.2227 spheres with a thickness $T_{cm} = 10$ mm was used to assure good hydraulic contact with the soil. To detect the contact material wetting phase, the infiltration data were plotted on a I vs. $\sqrt{t}$ plot (Fig. 3.28). The abrupt change in the slope is indicative of the time, t_0, when

Fig. 3.28 Determination of the starting time of infiltration into the soil for the data of the Example 3.3 from the I vs. $t^{0.5}$ plot (I = cumulative infiltration; t = time)

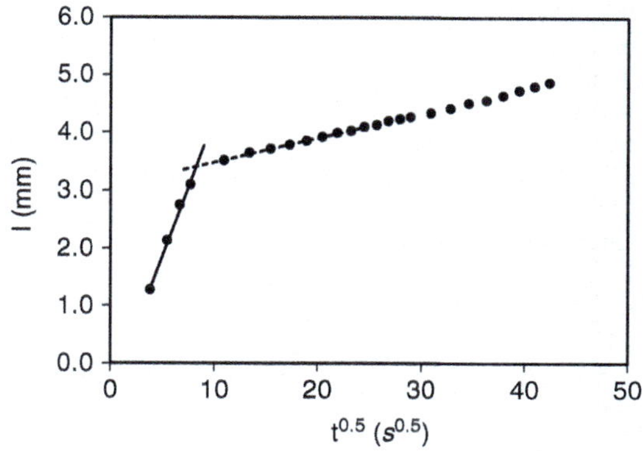

infiltration into the soil starts. The sorptivity of contact material was estimated as the slope of the linear regression of the first part of the data, corresponding to the contact material wetting. Using the first four cumulative infiltration data ($0 \leq t \leq 60$ s), $S_{mc} = 0.478$ mm s$^{-1/2}$ was calculated. Similarly, linear regression of the I vs. $\sqrt{t}$ data collected between 120 and 600 s allowed estimation of soil sorptivity, $S_0 = 0.042$ mm s$^{-1/2}$. The intersection between the regression line corresponding to stage 1:

$$I = 0.478\sqrt{t} - 0.530$$

and the line interpolating the experimental data collected for stage 2:

$$I = 0.042\sqrt{t} + 3.063$$

gave:

$$\sqrt{t_0} = -\frac{-0.530 - 3.063}{0.478 - 0.042} = 8.24 \text{ s}^{1/2}$$

Therefore, the time after which water flow was dominated by soil sorptivity was estimated to be $t_0 \approx 68$ s. Figure 3.29 shows the same infiltration data plotted on a $\Delta I/\Delta\sqrt{t}$ vs. $\sqrt{t_m}$ plot (see also Example 3.2). In this case, transition from infiltration into the contact material to infiltration into the soil, which should be highlighted by the abrupt change in slope between the sharply decreasing, early portion of the $\Delta I/\Delta\sqrt{t}$ vs. $\sqrt{t}$ data and the linearly increasing part of the dataset, was more uncertain. In fact, the S_{cm}/S_0 ratio was 11.4, that is very close to the lower limit of the S_{cm}/S_0 range ($S_{mc}/S_0 \leq 12$) preventing or at least complicating the applicability of the DL method (Bagarello and Iovino 2003). Assuming that the beginning of infiltration into the soil corresponds to the first point of the monotonically increasing $\Delta I/\Delta\sqrt{t}$ vs. $\sqrt{t_m}$ data, t_0 was estimated at 180 s.

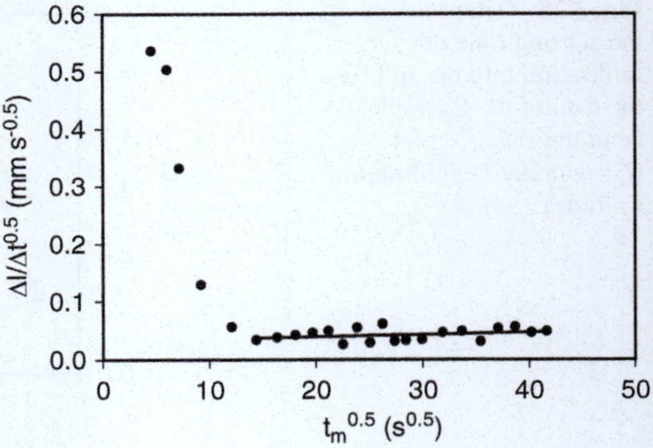

Fig. 3.29 Determination of the starting time of infiltration into the soil for the data of the Example 3.3 from the $\Delta I/\Delta\sqrt{t}$ vs. $\sqrt{t_m}$ plot (Differentiated Linearization method; I = cumulative infiltration; t = time; t_m = geometric mean time)

3.1.8 Applying the Tension Infiltrometer on Sloping Land

In many parts of the world, most of the landscapes under crop cultivation and watersheds are not levelled (slope steepness > 1 %) but very few measurement techniques are available for determining soil hydraulic characteristics in situ on hillslopes (Bodhinayake et al. 2004). However, some contributions on the use of the TI on sloping land can be found in literature (Casanova et al. 2000; Joel and Messing 2000; Bodhinayake et al. 2004; Meshgi and Chui 2014).

When a TI test is carried out on a sloping surface, the supply pressure head will vary from the upslope to the downslope side of the disk/soil interface. In particular, the pre-determined pressure head, h_0 (L), is valid for the midpoint of the disk. In any other point of the infiltration surface, the applied pressure head will be equal to $h_0 + \Delta z$, Δz (L) being the difference in soil surface height between the midpoint of the disk and the considered point. The Δz term is positive for the points lower than the midpoint of the disk and negative for the higher points. Therefore, the applied pressure head will be highest at the lower edge of the disk and lowest at its upper edge. For example, if a disk having a diameter of 200 mm is used on a surface forming an angle of 5°71′ with the horizontal plane (steepness = 10 %), the applied pressure head on the infiltration surface will vary by a quantity, $\Delta h_0 = 2 \times \Delta z = 19.9$ mm in all. If h_0 at the midpoint of the disk is −30 mm, the applied pressure head on the infiltration surface will vary linearly from −20.05 mm to −39.95 mm moving from the lower edge to the upper edge of the disk. A slightly negative pressure (i.e., near saturation) applied at the centre of the disk may impose a concomitant positive pressure head at the lower edge (Casanova et al. 2000).

Casanova et al. (2000) and Joel and Messing (2000) tested steepness effects on the hydraulic conductivity data obtained in medium textured soils by a porous plate of 200 mm in diameter. The investigations were carried out applying h_0 values ranging from −100 to −10 mm on surfaces having steepness values of 11–29 %. Increasing steepness imposed some problems with the device stability to maintain full contact with the soil surface. This problem was overcome with stabilizing supports. In the investigation by Joel and Messing (2000), superficial water flow was occasionally observed with the highest applied pressure head at the midpoint of the disk ($h_0 = -10$ mm). In both investigations, soil hydraulic conductivity, K, increased with the slope steepness, and the increase was by even more than an order of magnitude in the experiment by Casanova et al. (2000). The hypothesis that these results were influenced by variation of pressure head along the infiltration surface was considered plausible. In particular, the reasoning was that the downslope part, where the applied pressure head was higher than at the middle of the disk, had a larger influence on K because of the exponential relation between K and h, giving rise to larger effective K values than would be obtained on level ground. However, Joel and Messing (2000) also concluded that determining the effective applied pressure as a result of variation in the pressure between upslope and downslope sides of the disk is difficult.

Bodhinayake et al. (2004) tested the dependence of K on surface steepness for a silt-loam soil both experimentally and using computer simulations. A disk diameter of 200 mm, nominal pressure heads of −220 to −30 mm and steepness values of 0–20 % were considered. Disks of different diameters (from 40 to 1000 mm) were also included in the computer simulations. No significant differences were detected among slope treatments for the field measured infiltration rates, and the experimental results were generally confirmed by the simulation study. Bodhinayake et al. (2004) explained this result suggesting that the infiltration rate was lower at the upslope side than at the downslope side. However, for a relatively homogeneous soil such as the sampled one, the lower infiltration rate at the upslope side was offset by the higher infiltration rate at the downslope side, resulting in no substantial differences among slopes. It was therefore concluded that the TI was suitable for characterizing soil hydraulic properties in lands with slopes reaching up to 20 %.

The starting point of a recent investigation by Meshgi and Chui (2014) was that, for horizontal surfaces, soil hydraulic properties can be obtained from numerical inversion of TI data with an axisymmetric two-dimensional (2D) domain using the HYDRUS-2D code (Šimůnek et al. 2006) because in this case the three-dimensional (3D) infiltration process can be simplified to an axisymmetric 2D (i.e., depth and radius) process. However, when the TI is placed on a slope, it is no longer an axisymmetric 2D problem but there is not yet any software package developed to analyse TI data on the basis of 3D inverse modeling. Therefore, Meshgi and Chui (2014) attempted to assess the impact of analysing TI data from sloped surfaces using the 2D approximation. Having carried out extensive experiments in the field, the authors were able to conclude the following; (i) the effect of the 2D approximation on the 3D problem should be expected to be more noticeable as the soil hydraulic conductivity increases. In particular, for accurate estimations of K, TI measurements can be carried out on slopes as steep as 25° in clayey soils but on almost horizontal surfaces (i.e., angles less than 3°) in sandy soils, and (ii) the maximum allowable slope for the application of the 2D approximation to estimate K from TI data by avoiding any significant difference with horizontal estimation decreases with decreasing initial soil water content since more infiltration occurs in dry soils.

According to Joel and Messing (2000), one way of diminishing the impact of the larger applied pressure at the lower portion of the interface between the disk and the soil could be to use devices with smaller disks. However, this would at the same time make the infiltrating area less representative of the variation of macropores at the soil surface. On steep slopes, it may be necessary to create a bench (Sullivan et al. 1996), but this may disturb and remove surface soil, which may be of relevance in studies of the impact of crusting or sealing on infiltration.

3.2 BEST – Beerkan Estimation of Soil Transfer Parameters Procedure

3.2.1 Soil Hydraulic Characteristic Curves

BEST focuses specifically on the van Genuchten (1980) relationship for the water retention curve with the Burdine (1953) condition and the Brooks and Corey (1964) relationship for hydraulic conductivity, since they were found to be accurate and relevant for describing the hydraulic behavior of most soils (Fuentes et al. 1992):

$$\frac{\theta - \theta_r}{\theta_s - \theta_r} = \left[1 + \left(\frac{h}{h_g}\right)^n\right]^{-m} \tag{3.58a}$$

$$m = 1 - \frac{k_m}{n} \tag{3.58b}$$

$$\frac{K(\theta)}{K_s} = \left(\frac{\theta - \theta_r}{\theta_s - \theta_r}\right)^\eta \tag{3.59a}$$

$$\eta = \frac{2}{mn} + 2 + p \tag{3.59b}$$

where θ (L^3L^{-3}) is the volumetric soil water content, h (L) is the soil water pressure head, K ($L\,T^{-1}$) is the soil hydraulic conductivity, n, m and η are shape parameters, k_m is a user index (Haverkamp et al. 2005), p is a tortuosity parameter, and h_g (L), representing the inflection point of the water retention curve, θ_s (L^3L^{-3}, field-saturated soil water content), θ_r (L^3L^{-3}, residual soil water content) and K_s ($L\,T^{-1}$, field-saturated soil hydraulic conductivity) are scale parameters. In BEST, θ_r is assumed to be zero. The Burdine's (1953) model is considered for the water retention curve, meaning that $k_m = 2$, $n > 2$ and $p = 1$. A small value of n refers to a water retention curve with a gentle slope, implying a wide range of pore size distribution (Minasny and McBratney 2007). A large value of n refers to sandy soils where the water retention curve has a prominent "leg" and a narrow range of pore sizes.

To estimate all hydraulic parameters, BEST methods require two kinds of data: (i) particle-size distribution (PSD) of the soil and bulk density, and (ii) cumulative infiltration along with the initial and final soil water contents (Fig. 3.30). The experimental data needed as input for BEST methods are presented successively

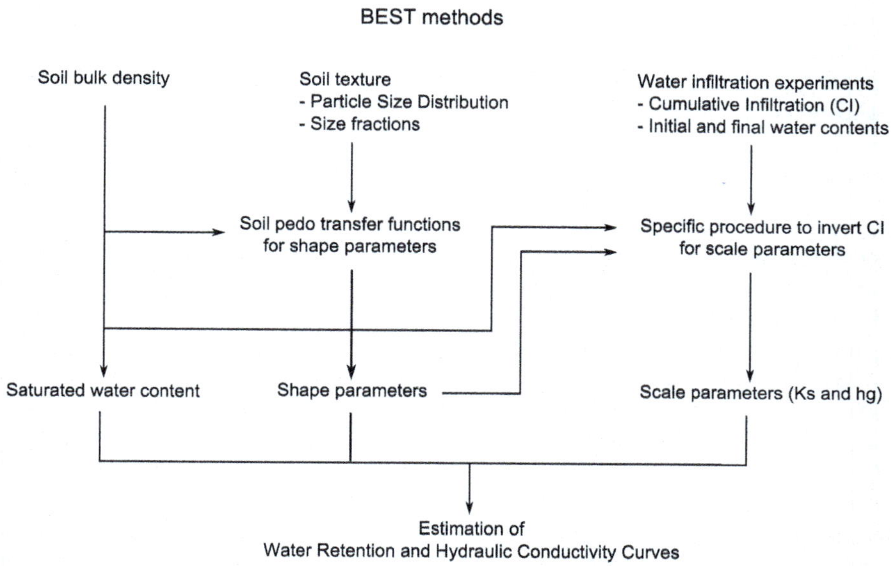

Fig. 3.30 BEST flow chart

below. Then, the use of the data by BEST methods and related algorithms are described in detail.

3.2.2 *Soil Particle-Size Distribution, Bulk Density and Water Content*

On the field, practically, a single sample is used for the determination of PSD, bulk density and initial water content. A ring of known volume, V_{ring} (L^3), is driven into the soil near the infiltration location to sample undisturbed in situ material. The ring size can be adapted in function to the maximum particle size. The weight of the sampled soil is recorded before and after oven-drying to determine the initial gravimetric soil water content, w_i (M M^{-1}). Soil bulk density, ρ_b (M L^{-3}), is derived from the dry mass of soil in the ring, m_s (M):

$$\rho_b = \frac{m_s}{V_{ring}} \tag{3.60}$$

and the initial volumetric soil water content, θ_i (L^3L^{-3}), is derived from w_i and ρ_b using the general relation between volumetric and gravimetric water contents:

$$\theta_i = w_i \rho_b \tag{3.61}$$

Finally, the same sample is used to determine the PSD, combining mechanical sieving, laser diffractometry and/or sedimentation. Soil porosity, f, has also to be calculated assuming either usual (2.65 g cm^{-3}) or measured values for mineral density, ρ_s:

$$f = 1 - \frac{\rho_b}{\rho_s} \tag{3.62}$$

In several applications of the method, the soil porosity was also used to estimate the volumetric saturated soil water content, θ_s (L^{3}L^{-3}) (Mubarak et al. 2009a, 2010; Xu et al. 2009; Yilmaz et al. 2010; Bagarello et al. 2011). Such a procedure was questioned by Gonzalez-Sosa et al. (2010) who estimated θ_s from field final water content and compared it with the corresponding values of porosity over 57 samples. They obtained an average θ_s/f ratio of 0.7 ± 0.18, which is much lower than the values commonly reported in literature (i.e., 0.8–1; Rogowski 1971). Therefore, Gonzalez-Sosa et al. (2010) concluded that accurate determination of saturated water content was a key issue for characterization of soil hydraulic properties.

In some instances, reduced experimental equipment may be available, but this condition does not preclude necessarily the BEST experiment. For example, only a small number of stainless steel cylinders and a battery-operated balance were available for an investigation carried out in Burundi by Bagarello et al. (2011). Therefore, an undisturbed soil core was weighed in the field immediately after sampling. The soil was then extracted from the cylinder of known weight and it was stored in a small plastic bag that was labeled and closed. The disturbed soil sample corresponding to a known bulk (undisturbed) soil volume was then oven-dried in Italy to determine both the soil bulk density and the soil water content at the time of sampling.

3.2.3 *Water Infiltration Experiments*

Water infiltration experiments are performed according to the Beerkan method that was pioneered by Braud et al. (2005). A ring of a diameter in the order of several centimeters is inserted a short distance (e.g., 0.5–1 cm) into the soil to pour water volumes at soil surface and to avoid lateral loss of ponded water (Fig. 3.31). The surface vegetation is removed before ring insertion while the roots remain in situ. Then, a fixed, small volume of water, V_w (L^3), is poured into the cylinder at time, t (T), zero, and the time elapsed during the infiltration of the known volume of water is measured. When the first volume has completely infiltrated, a second volume of water equal to V_w is added to the cylinder, and the time needed for it to infiltrate is measured. The procedure is repeated for about 8–15 known volumes and cumulative infiltration, I (L), is recorded. Finally, the dataset is made up of a number of N_{tot} discrete points (t_i, I_i) describing an experimental cumulative infiltration curve. For example, for the third water application, I_3 (L) is equal to $3 \times V_w/\sigma_{inf}$, σ_{inf} (L^2) being

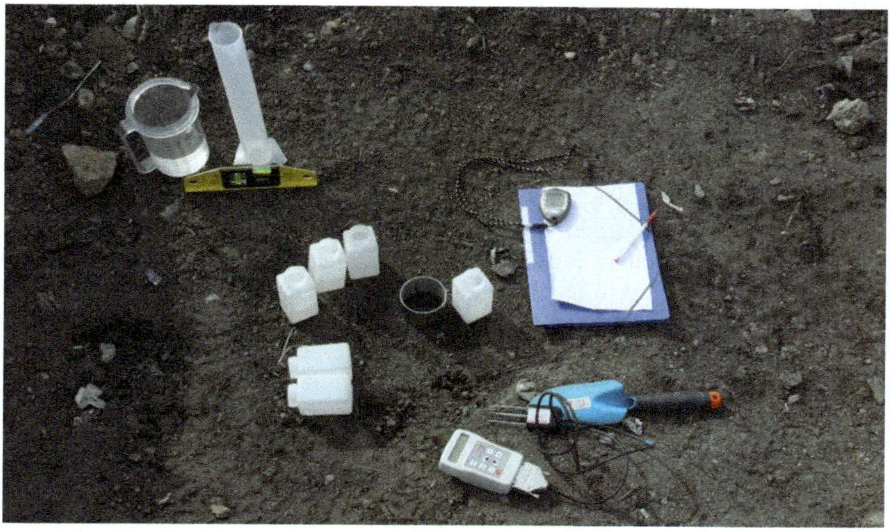

Fig. 3.31 Beerkan experiment apparatus: spirit level to prepare the soil, ring inserted into the ground, prepared volumes of water to be infiltrated, timer and notebook, water content probe

the infiltration surface, and the associated t_3 (T) value is the time elapsed from beginning of the experiment to the instant when the third water volume has completely infiltrated. At the end of the infiltration run, the saturated soil within the ring is sampled to determine the gravimetric saturated water content, w_s (M M^{-1}). The volumetric saturated soil water content can be obtained by $\theta_s = w_s \cdot \rho_b$, as with the determination of the initial water content, i.e. Eq. (3.61).

If the cumulative infiltration curve is obtained directly from the N_{tot} discrete points (t_i, I_i), the infiltration rate needs to be calculated from the derivation of the cumulative infiltration with regards to time. Lassabatere et al. (2006) suggested calculating the infiltration rate, i^{exp} (L T^{-1}), and the associated time, t_i* (T), as follows:

$$i^{\exp}(t_i*) = \frac{I_{i+1} - I_i}{t_{i+1} - t_i} \quad (i = 1 \,..N_{tot} - 1) \tag{3.63a}$$

$$t_i* = \left(\frac{\sqrt{t_i} + \sqrt{t_{i+1}}}{2} \right)^2 \quad (i = 1 \,..N_{tot} - 1) \tag{3.63b}$$

The experimental steady-state infiltration rate, i_s^{exp} (L T^{-1}), is obtained from the last points of the infiltration curve, when water infiltration stabilizes:

$$i_s^{\exp} = \underset{i=(N_{tot}-N_{end})\to N_{tot}}{slope} (t_i, I_i) \tag{3.64}$$

where N_{end}, that is the number of points considered for the linear regression, must be defined to include only the last points of the $I(t)$ curve that define the steady-state condition.

Ring radius and volume of water must be adapted to the experimental conditions and the kind of soils. Most of the time, the cylinder's radii are in the order of a few centimeters and infiltrated volumes in the order of 100–200 mL (Lassabatere et al. 2010b; Yilmaz et al. 2010; Xu et al. 2012). For instance, Lassabatere et al. (2006) used volumes of water for each pouring of 75–140 mL with a ring radius of 7.5 cm, resulting in initial ponded heads in the ring of 0.25–0.80 cm. In another site exhibiting contrasting kinds of soils including coarse soils, the same authors considered ring radii between 4.75 and 10 cm and volumes of water of 100–500 cm^3 (Lassabatere et al. 2010b). The number of volumes to be infiltrated was fixed large enough to reach steady-state. For instance, Yilmaz et al. (2010) applied 8–20 known volumes of water, depending on the run. Apparent steady-state was assumed to be reached when two (Mubarak et al. 2009a) or three (Mubarak et al. 2010) consecutive infiltration times were identical.

Infiltration runs with large cylinders, i.e., having a radius of 20 cm, were carried out by Gonzalez-Sosa et al. (2010) in an attempt to average the soil spatial heterogeneity for obtaining measurements representative of the field. This choice implied a change in the infiltration protocol. The cylinders were inserted 2–5 cm deep into the soil to ensure water tightness and avoid leaks, without disturbing too much the three-dimensional water flow. A total of 12 L of water were poured within a plastic sheet, sealed to the cylinder, to minimize the disturbance in topsoil that frequently occurs when water is added directly. The plastic sheet was then removed and the chronometer was started. The initial height was measured by a ruler. Then, the infiltration height as a function of elapsed time was followed by reading the ruler. During the first minutes, small time intervals of a few seconds were used and the time interval was increased after 3 or 5 min. The operation was terminated when all the water had infiltrated the soil. Therefore, a falling head test with an initial height of 9.5 cm was considered to be analyzable with the BEST relationships and procedures by Gonzalez-Sosa et al. (2010).

Theoretically, the water pressure head on the infiltration surface should be zero. In fact, the applied experimental procedure triggers a small oscillation of water pressure head at the soil surface with a maximum value just after pouring a new volume of water and a minimum value just before. From a practical point of view, however, this circumstance can be ignored since the established ponded head of water is generally very small (<1 cm according to Lassabatere et al. 2006) and small changes in the head of water do not have appreciable impact on experimental data (Touma et al. 2007; Xu et al. 2009, 2012).

3.2.4 *Estimating Shape Parameters by Inverting Particle-Size Distribution Data*

BEST estimation of shape parameters relies on the approach by Haverkamp et al. (2005). These authors introduced a water retention shape index, p_m, which constitutes an integral measure of the slope of the water retention curve and entirely characterizes the retention behavior of a particular soil. For the case of Eq. (3.58a) with a null value for residual soil water content, the shape index can be defined as (Leji et al. 2005):

$$p_m = \frac{mn}{1+m} = \frac{mn(k_m + mn)}{k_m + 2mn} \tag{3.65}$$

A more complex definition of this index with non-null θ_r values can be found in Leji et al. (2005). The water retention shape index was proved to be independent of the parameterization of the water retention equation and in particular of the choice of k_m. On the same basis, Haverkamp et al. (2006) considered the shape index of the cumulative PSD, referred to as the particle-size shape index, p_M,:

$$p_M = \frac{MN}{1+M} = \frac{MN(k_M + MN)}{k_M + 2MN} \tag{3.66}$$

where k_M is the user index for the PSD and M and N are the shape parameters of the following theoretical PSD:

$$P(D) = \left[1 + \left(\frac{D_g}{D}\right)^N\right]^{-M} \tag{3.67a}$$

$$M = 1 - \frac{k_M}{N} \tag{3.67b}$$

where *P(D)* is the fraction by mass of particles having a diameter below the specific value D (L), and D_g (L) is a scale parameter. Haverkamp et al. (2005, 2006) suggested a set of pedotransfer functions to relate the water retention and particle-size shape indexes, these being used in the BEST algorithm to derive parameters n, m and η from particle-size distribution and porosity data.

For BEST methods, we consider the specific case of $k_m = k_M = 2$. Firstly Eq. (3.67a) is fitted to the experimental particle-size distribution data to derive M and N and to calculate the particle-size shape index p_M in agreement with Eq. (3.66). The water retention shape index, p_m, is then derived on the basis of the relations proposed by Haverkamp et al. (2006):

$$p_m = p_M(1 + \kappa)^{-1} \tag{3.68}$$

where κ is a coefficient defined as:

$$\kappa = \frac{2s - 1}{2s(1 - s)} \tag{3.69}$$

where the fractal dimension of the media, s, varying from 0.5 to 1 (Minasny and McBratney 2007), is defined as the root of the following equation:

$$(1-f)^s + f^{2s} = 1 \tag{3.70}$$

in which f (L^3L^{-3}) is the soil porosity, that can be obtained from the ρ_b measurement. The m parameter is derived from the value of the water retention shape index and considering the positive root of Eq. (3.65), which leads to:

$$m = \frac{1}{p_m}\left(\sqrt{1+p_m^2} - 1\right) \tag{3.71}$$

The n and η parameters are then obtained by using Eqs.(3.58b) and (3.59b), respectively, which completes estimation of hydraulic shape parameters.

Several alternatives have been proposed. For instance, Minasny and McBratney (2007) suggested simplifying the estimation of the $(1+\kappa)^{-1}$ term of Eq. (3.68) by using the following relationship:

$$(1+\kappa)^{-1} = 0.7467 - 0.6266\ f + 0.5456\ f^2 - 0.3813\ f^3 \tag{3.72}$$

Equation (3.72), having an average absolute error of 6×10^{-5} in the range $0.3 \leq f \leq 0.7$, was obtained by fitting a cubic polynomial to the empirical observations between $(1+\kappa)^{-1}$ and f.

In many conventional laboratory analyses, particle-size is only provided for three fractions, i.e., clay, silt and sand. Furthermore, in many instances, particle-size analysis is not available and field texture (hand texture) may be the only data that was collected. In these cases, the lack of an appropriate number of particle-size fractions is expected to generally preclude use of Eq. (3.67a), impeding estimation of shape parameters. Using three soil databases with hydraulic properties and PSD (UNSODA, GRIZZLY, *Australian soil database*, total sample size = 2262), Minasny and McBratney (2007) derived the following empirical pedotransfer function for predicting the shape parameter, n, from sand, sa (%, USDA classification system), and clay, cl (%), content, that can be used in the Beerkan method, i.e., when $\theta_r = 0$ and $m = 1 - 2/n$:

$$n = 2.18 + 0.11[48.087 - 44.954\ S(x_1) - 1.023\ S(x_2) - 3.896\ S(x_3)] \tag{3.73a}$$

where

$$x_1 = 24.547 - 0.238\ sa - 0.082\ cl \tag{3.73b}$$

$$x_2 = -3.569 + 0.081\ sa \tag{3.73c}$$

$$x_3 = 0.694 - 0.024\ sa + 0.048\ cl \tag{3.73d}$$

$$S(x) = \frac{1}{1+\exp(-x)} \tag{3.73e}$$

On the basis of these equations, Minasny and McBratney (2007) also calculated average values for different soil texture classes.

3.2.5 Estimating Scale Parameters by Inverting Water Infiltration Data

BEST estimates sorptivity, S (L T$^{-0.5}$), and saturated soil hydraulic conductivity, Ks (L T^{-1}), through fitting experimental cumulative infiltration on the analytical model for steady and transient states proposed by Haverkamp et al. (1994). Then, the hydraulic scale parameters that complete soil hydraulic characterization are derived from the previous estimates of shape parameters, sorptivity and saturated hydraulic conductivity. In fact, the sole parameter to be estimated is the scale parameter for water pressure head, h_g.

Soil sorptivity can be expressed as a function of hydraulic parameters and initial and final water contents using the approximation by Parlange (1975). For the specific case of water retention and hydraulic conductivity curves defined by Eqs. (3.58a and 3.58b) and (3.59a and 3.59b), a zero initial soil water content and a saturated final water content, sorptivity can be defined as follows (Haverkamp et al. 1999):

$$S^2(0,\theta_s) = -c_p(n,m,\eta)\theta_s K_s h_g \tag{3.74a}$$

$$c_p(n,m,\eta) = \Gamma\left(1+\frac{1}{n}\right)\left\{\frac{\Gamma\left(m\eta-\frac{1}{n}\right)}{\Gamma(m\eta)} + \frac{\Gamma\left(m\eta+m-\frac{1}{n}\right)}{\Gamma(m\eta+m)}\right\} \tag{3.74b}$$

where Γ stands for the gamma function. Usually, the initial water content, θ_i, differs from zero, and sorptivity $S(\theta_i, \theta_s)$ can be linked to the case of null initial water content, $S(0, \theta_s)$, using the approximation proposed by Haverkamp et al. (1999):

$$S^2(\theta_i,\theta_s) \approx S^2(0,\theta_s)\frac{K_s - K_i}{K_s}\frac{\theta_s-\theta_i}{\theta_s} \tag{3.75a}$$

$$S^2(\theta_i,\theta_s) = -c_p(n,m,\eta)\theta_s K_s h_g\left(1-\frac{\theta_i}{\theta_s}\right)\left[1-\left(\frac{\theta_i}{\theta_s}\right)^\eta\right] \tag{3.75b}$$

where $K_i = K(\theta_i)$ is the soil hydraulic conductivity corresponding to the initial soil water content. Then, the scale parameter for water pressure head, h_g, can easily be estimated from knowledge of soil sorptivity, saturated hydraulic conductivity and shape parameters (Lassabatere et al. 2006):

$$h_g = -\frac{S^2(\theta_i,\theta_s)}{c_p(n,m,\eta)(\theta_s-\theta_i)\left[1-\left(\frac{\theta_i}{\theta_s}\right)^\eta\right]K_s} \tag{3.76}$$

Hereafter, the sorptivity, S, refers to the case of initial soil water content different from zero, i.e., $S = S(\theta_i, \theta_s)$.

To estimate sorptivity and saturated soil hydraulic conductivity, BEST inverts the experimental data obtained with the Beerkan method using the approximations of the quasi-exact implicit formulation proposed by Haverkamp et al. (1994). These authors consider that, for an infiltration experiment with zero water pressure on a circular surface of radius, r (L), above a uniform soil with a uniform water content, θ_i, the three-dimensional cumulative infiltration, I (L), and the infiltration rate, i (L T^{-1}), can be approached by the following explicit transient [Eqs. (3.77a) and (3.77b)] and steady-state [Eqs. (3.77c) and (3.77d)] expressions (Haverkamp et al. 1994):

$$I(t) = S\sqrt{t} + \left(AS^2 + BK_s\right)\, t \tag{3.77a}$$

$$i(t) = \frac{S}{2\sqrt{t}} + \left(AS^2 + BK_s\right) \tag{3.77b}$$

$$I_s(t) = \left(AS^2 + K_s\right)t + C\frac{S^2}{K_s} \tag{3.77c}$$

$$i_s(t) = AS^2 + \; K_s \tag{3.77d}$$

where t (T) is the time and A (L^{-1}), B and C are constants that can be defined for the specific case of a Brooks and Corey (1964) relationship as:

$$A = \frac{\gamma}{r(\theta_s - \theta_i)} \tag{3.78a}$$

$$B = \frac{2-\beta}{3}\left[1 - \left(\frac{\theta_i}{\theta_s}\right)^{\eta}\right] + \left(\frac{\theta_i}{\theta_s}\right)^{\eta} \tag{3.78b}$$

$$C = \frac{1}{2(1-\beta)\left[1 - \left(\frac{\theta_i}{\theta_s}\right)^{\eta}\right]}\ln\left(\frac{1}{\beta}\right) \tag{3.78c}$$

where β and γ are coefficients that are commonly set at 0.6 and 0.75, respectively, for $\theta_i < 0.25\ \theta_s$ (Smettem et al. 1994; Haverkamp et al. 1994). More recent investigations suggested that the values of β and γ scarcely depend on the initial degree of saturation (Xu et al. 2012) but can be soil dependent (Lassabatere et al. 2009).

Two main BEST methods have been proposed, that is (i) the version originally developed by Lassabatere et al. (2006) and (ii) an alternative method that fixes the shortcomings and errors of the first version when used for coarse materials (Yilmaz et al. 2010). In fact, the two methods differ only by the way experimental data are fitted to Eq. (3.77a). The estimation of shape parameters and the derivation of parameter h_g from previous estimations of saturated hydraulic conductivity and sorptivity are similar.

Inverting the data on Eq. (3.77a) through the simultaneous optimization of sorptivity, S, and saturated hydraulic conductivity, K_s, leads to several problems of non-uniqueness. In practical terms, the simultaneous optimization of S and K_s is impossible and the problem must be defined in terms of optimization of one sole parameter to avoid the problem of non-uniqueness. For both BEST methods, K_s is defined as a function of sorptivity and the fit of experimental data on the model is

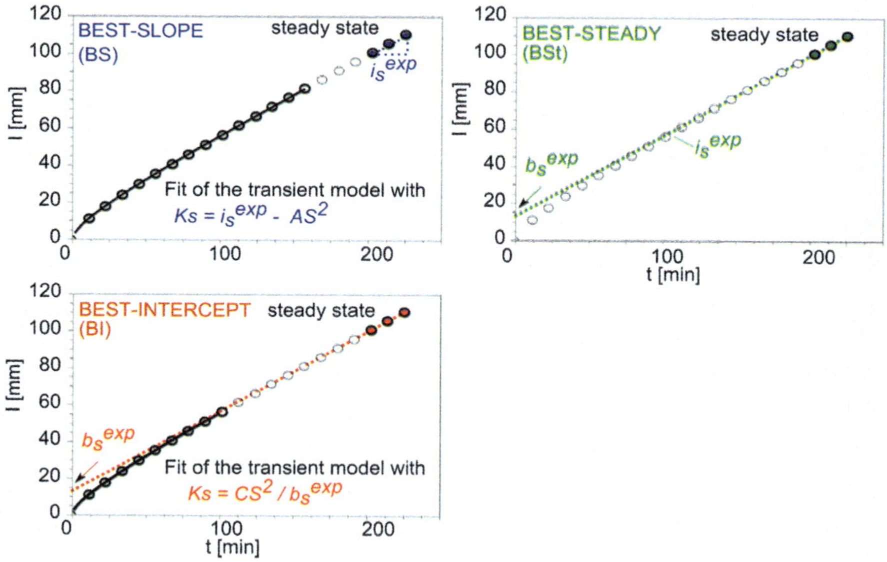

Fig. 3.32 Description of the fitting procedure for BEST-Slope, BEST-Intercept and BEST-Steady algorithms

performed through the optimization of sorptivity. Both methods consider that steady-state is reached at the end of the experiment (Fig. 3.32). The straight line fitted to the cumulative infiltration data describing steady-state conditions can be modelled by Eq. (3.77c). Consequently, its slope, i_s^{exp}, and intercept, b_s^{exp}, can be defined as follows:

$$i_s^{\exp} \approx AS^2 + K_s \tag{3.79a}$$

$$b_s^{\exp} \approx C\frac{S^2}{K_s} \tag{3.79b}$$

BEST-Slope, the original version proposed by Lassabatere et al. (2006), and BEST-Intercept, the alternative version proposed by Yilmaz et al. (2010), use the slope and the intercept of the steady-state expansion, respectively, i.e., of Eq. (3.77c), to define K_s as a function of S, leading to:

$$K_s = i_s^{\exp} - AS^2 \tag{3.80a}$$

$$K_s = C\frac{S^2}{b_s^{\exp}} \tag{3.80b}$$

The estimation of S is carried out by fitting experimental data to the new model for transient state infiltration as defined by the combination of Eqs. (3.77a) and (3.80a or 3.80b), leading to the following expansions for BEST-Slope and BEST-Intercept, respectively:

$$I(t) = S\sqrt{t} + \left[A(1-B)S^2 + Bi_s^{\exp}\right]\ t \tag{3.81a}$$

$$I(t) = S\sqrt{t} + \left[A + \frac{BC}{b_s^{\exp}}\right]\ S^2 t \tag{3.81b}$$

One of the shortcomings of Eqs. (3.81a or 3.81b) is that they apply only to the transient state, i.e., not all the experimental dataset. However, for a proper use of BEST methods, steady-state should be reached (Fig. 3.32). Thus, Eqs. (3.81a or 3.81b) must be fitted to a part of the whole data set. As the range of the transient state is not known *a priori*, the fit is performed successively for all the sets composed by the first k points, k varying from a minimum of five points to a maximum of the total number of data points (N_{tot}) (Lassabatere et al. 2006). The fit is performed by minimizing the classical objective function for cumulative infiltration:

$$f(S, K_s, k) = \sum_{i=1}^{k} \left[I_i^{\exp} - I(t_i)\right]^2 \tag{3.82}$$

where I^{exp} and I are the measured and the estimated cumulative infiltration, respectively. The estimate for sorptivity obtained with k points, S_k, is used to derive an estimate of the saturated hydraulic conductivity, $K_{s,k}$ through Eqs. (3.80a or 3.80b), depending on the method used (BEST-Slope or BEST-Intercept). Finally, the estimates are used to quantify the time, t_{max} (T), defined as the maximum time for which the transient expression can be considered valid, using the equation proposed by Lassabatere et al. (2006):

$$t_{\max} = \frac{1}{4(1-B)^2}\left(\frac{S}{K_s}\right)^2 \tag{3.83}$$

where $(S/K_s)^2$ corresponds to the gravity time defined by Philip (1969). Then, the maximum time $t_{max,k}$ is compared with the maximum time of the experimental dataset used for the fit, i.e., t_k. The values of S and K_s are not considered valid unless $t_k \leq t_{max,k}$. Among all values of S and K_s that fulfill this condition, the S and K_s values corresponding to the largest k (k_{max}) are retained since they are considered more precise (Fig. 3.33). In practice, it is also possible to move backwards, starting from N_{tot} and stopping the procedure when the prescribed condition ($t_k \leq t_{max,k}$) is satisfied.

The optimization function requires a first estimate for sorptivity to initialize the objective function defined by Eq. (3.82). Lassabatere et al. (2006) suggested applying the Cumulative Linearization method proposed by Vandervaere et al. (2000a) to the whole dataset. In addition, the same authors suggested optimizing between zero and the maximum value of sorptivity, S_{max}, to ensure positive estimates for K_s. A maximum value for sorptivity is estimated through the same methodology while considering that gravity had no effect, i.e., $B = 0$ in Eqs. (3.81a or 3.81b). In order to obtain positive values for K_s, it is necessary to apply the additional condition that ${S_{max}}^2$ must be less than the steady-state infiltration rate (${i_s}^{exp}$) divided

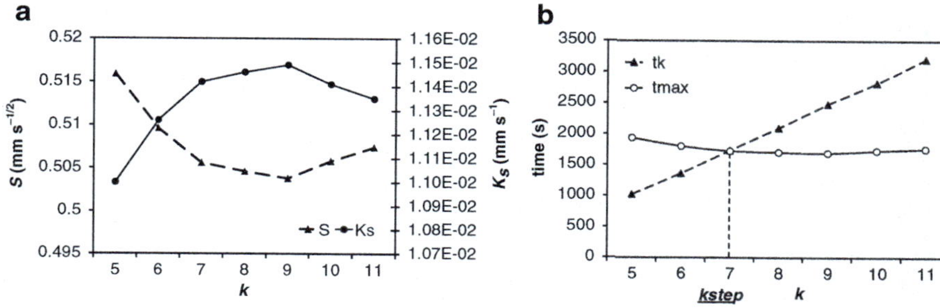

Fig. 3.33 Example of BEST-Slope estimation of the (**a**) sorptivity (S) and saturated soil hydraulic conductivity (K_s), and (**b**) longest time of the data subset (t_k) and maximum time ($t_{max,k}$) versus the number of points used for the fit (k) for the case of a sandy soil from Chernobyl (From Lassabatere et al. 2006, reprinted with permission)

by coefficient A. Thus, maximum sorptivity is defined as the minimum value between the estimate provided by the fit with $B = 0$ and the steady-state infiltration rate (i_s^{exp}) divided by coefficient A.

Finally BEST methods offer a complete characterization of water retention and hydraulic conductivity curves, as depicted by Fig. 3.34 for the cases studied by Lassabatere et al. (2006). Of course, the S and K_s values must provide both accurate modelling of experimental data (Fig. 3.35) and agreement with basic criteria related to the physics of water infiltration in the soils. To quantify the accuracy of the fit, Lassabatere et al. (2006) also suggested calculating the relative error, E_r, between the model and the experimental data:

$$E_r = \sqrt{\frac{\sum_{i=1}^{k} \left[I^{\exp}(t_i) - I(t_i)\right]^2}{\sum_{i=1}^{k} \left[I^{\exp}(t_i)\right]^2}} \tag{3.84}$$

In their applications, Lassabatere et al. (2006) obtained values for E_r not exceeding 3.5 %. However, on the basis of the check of alternative fitting methods carried out by these authors, $E_r \leq 5.5$ % can be considered to denote an acceptable error for transient cumulative infiltration. Finally, the modelled infiltration at steady-state must be in agreement with the experimental data. With reference to this last point, Lassabatere et al. (2006) provided an example suggesting in particular that the modelled steady-state infiltration rate, i_s, should not be higher than the experimental infiltration rate at steady-state, i_s^{exp}.

Recently, Bagarello et al. (2014) proposed an alternative approach, named BEST-Steady, that makes use of the intercept and the slope of the straight line fitted to the data describing steady-state conditions on the plot of I versus

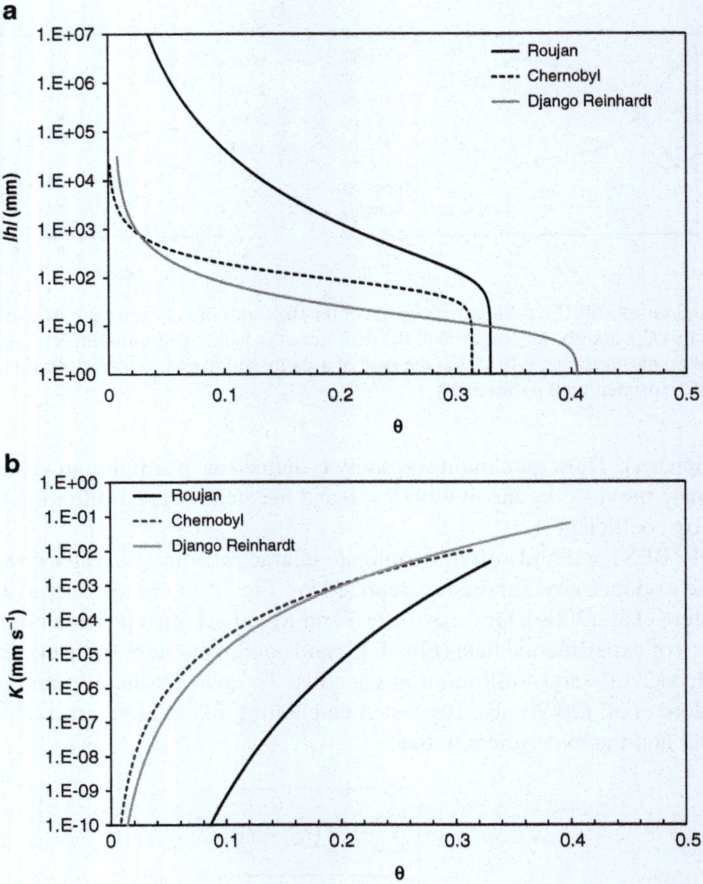

Fig. 3.34 Examples of soil hydraulic characteristic curves obtained with the BEST-Slope algorithm: (**a**) water retention curve, $h(\theta)$, and (**b**) soil hydraulic conductivity function, $K(\theta)$ (From Lassabatere et al. 2006, reprinted with permission)

t (Fig. 3.32). The combination of Eqs. (3.79a) and (3.79b) leads to the following estimate for sorptivity:

$$S = \sqrt{\frac{i_s^{\exp}}{A + \frac{C}{b_s^{\exp}}}} \tag{3.85}$$

Then, saturated hydraulic conductivity can be estimated using either Eq. (3.80a) or (3.80b). BEST-Steady does not make any direct use of the data describing the transient state. However, monitoring this stage is necessary to establish when a steady-state condition begins (estimating i_s^{exp}) and how much water infiltrates the soil before reaching this condition (estimating b_s^{exp}).

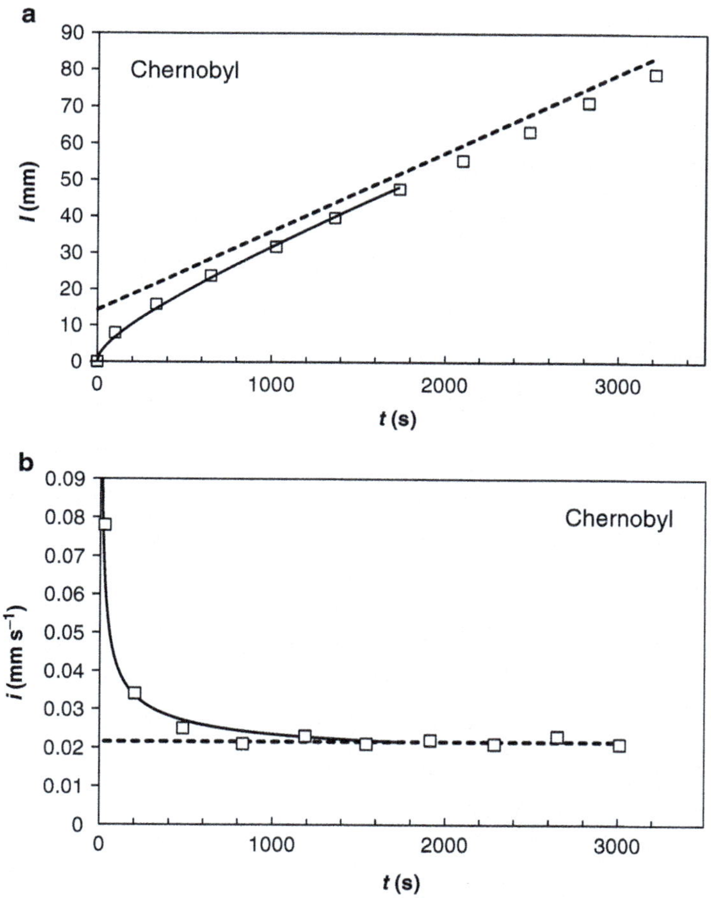

Fig. 3.35 Modeling with BEST-Slope of (**a**) cumulative infiltration, I, and (**b**) infiltration rate, i, for the Chernobyl soil (From Lassabatere et al. 2006, reprinted with permission)

3.2.6 Issues on Estimation of Shape Parameters

Several uses and tests of the BEST methods have been reported in the literature. A point that was considered was treatment of particle-size distribution. The capability of Eqs. (3.67a and 3.67b) to accurately reproduce experimental data was questioned by several authors (Minasny and McBratney 2007; Xu et al. 2012). Bagarello et al. (2009) tested the fitting ability of the BEST PSD model for 114 soil samples collected in Burundi. For each sample, sieving analysis was carried out using six sieves with mesh sizes of 2, 0.85, 0.425, 0.25, 0.106 and 0.075 mm. Eight fine fraction data points were obtained by the hydrometer method, measuring the suspension density at times, $t_s = 2$, 5, 15, 30, 60, 180, 1440 and 2880 min (mean measured

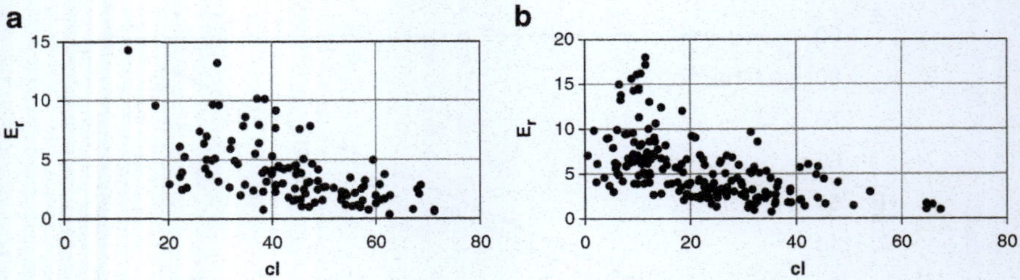

Fig. 3.36 Relationship between the relative error, E_r (%), of the BEST particle-size distribution model and the clay content, *cl* (%), for (**a**) a dataset of Burundian soils (From Bagarello et al. 2009, reprinted with permission), and (**b**) a dataset of Sicilian soils (From Bagarello and Iovino 2012, reprinted with permission)

diameters = 0.030, 0.019, 0.011, 0.0081, 0.0058, 0.0034, 0.0012, and 0.00089 mm, respectively). Two fractions finer than 0.002 mm were determined (i.e., a total of 14 particle-size limits) because using a large range of measured diameters was considered advisable to evaluate the ability of a particular model to reproduce the complete PSD, although the suggested standard procedure consists of measuring the suspension density for $t_s \leq 1440$ min (Gee and Or 2002). The fit of the data on Eqs. (3.67a and 3.67b) proved to be quite accurate most of the time. The relative errors, E_r [Eq. (3.84)], between measured and predicted soil particle fractions varied from 0.3 to 14.3 % with a mean of 3.9 %, and $E_r < 5$ % was obtained for 76 % of the considered soil samples. A significant effect of the clay content on the BEST PSD model fitting performances was found since E_r decreased as *cl* increased (Fig. 3.36a). In particular, both high (>5 %) and low (<5 %) E_r values were obtained in soils with a relatively low clay content whereas $E_r < 5$ % was always obtained for high *cl* values. A possible reason of this result was that the PSD of soils with a high clay content has a simpler form (Hwang et al. 2002) and the range of values to cover by the model is smaller. A similar result was obtained by Bagarello and Iovino (2012) with reference to a dataset of approximately 200 Sicilian soils (Fig. 3.36b). Encouraging results were also obtained for the Burundian dataset by considering a shorter measurement time of suspension density, since the means of E_r were equal to 4.3 % and 4.8 % for $t_s \leq 180$ min (12 measured points) and 60 min (11 points), respectively (Bagarello et al. 2009). Simply fitting Eqs. (3.67a and 3.67b) to the percentages by mass of particles lower than 0.002, 0.05 and 2 mm (i.e., considering in practice the three soil textural fractions) yielded higher E_r values, as would be expected, with distances from the corresponding E_r values obtained with the complete PSD that however decreased as *cl* increased (Fig. 3.37). The conclusions of these studies show that the mean performance of the PSD model can be considered satisfactory and predictions are expected to be particularly accurate for soils with high clay content. As a practical alternative, particularly in fine textured soils, using experimental information reduced to only three points is proposed.

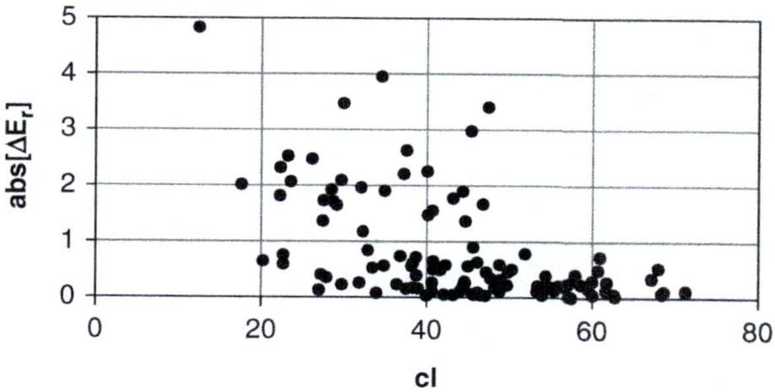

Fig. 3.37 Absolute difference, $abs[\Delta E_r]$ (%), between the relative errors obtained, for a given soil sample, by fitting the BEST particle-size distribution model with the reduced (three data points) and the complete (14 data points) experimental information as a function of the clay content of the sample, *cl* (%), for a dataset of Burundian soils (From Bagarello et al. 2009, reprinted with permission)

Bagarello et al. (2011) investigated the possibility to use BEST methods with few data for particle-size distribution. On several fields, the original BEST-Slope procedure, applied with a PSD including 14 measured data points, was tested against different alternative procedures making use of a less detailed PSD information and the methodology proposed by Minasny and McBratney (2007). Bagarello et al. (2011) defined several scenarios: (i) the BEST/M1 scenario, that estimates the parameter n from soil *sa* and *cl* contents using Eqs. (3.73a, 3.73b, 3.73c and 3.73d), (ii) the BEST/M2 scenario, that estimates n from soil textural class according to Minasny and McBratney (2007), (iii) the BEST/B1 scenario, that fits the first 11 data points of the measured PSD to Eqs. (3.67a and 3.67b) to estimate n, and finally (iv) the BEST/B2 scenario, that only uses Eqs. (3.67a and 3.67b) with the three determined soil textural fractions, i.e. sand, silt and clay contents. In comparison with the original BEST procedure, both BEST/M1 and BEST/M2 yielded significantly different results in terms of $\ln(m)$, $\ln(n)$, η and abs (h_g), whereas the $\ln(m)$, $\ln(n)$, η, abs(h_g) and $\ln(K_s)$ values obtained with both BEST/B1 and BEST/B2 did not show any statistically significant difference (Bagarello et al. 2011). Therefore, these last two procedures were found to be reliable practical alternatives to the original one. The poorer performances of the BEST/M1 and BEST/M2 procedures were probably influenced by the kind of soils studied, i.e. tropical soils, that contrast with the soils used for the establishment of Eqs. (3.73a, 3.73b, 3.73c and 3.73d). Indeed, the procedure developed by Minasny and McBratney (2007) was found to work well with soils from a temperate climate environment (Bagarello and Iovino 2012).

Regarding the analysis of particle-size distribution and the derivation of shape parameters, Bagarello and Iovino (2012) investigated the change of k_M in Eq. (3.67a). In BEST, $k_M = 2$ is used, but $k_M = 1$ and a fitted k_M to the experimental PSD were also considered by Bagarello and Iovino (2012). Differences among alternative estimates of p_M decreased as the *cl* content of the soil increased (Fig. 3.38), probably because the fitting accuracy of all considered PSD models increased with the *cl* content of the soil. In other words, the calculation procedure of

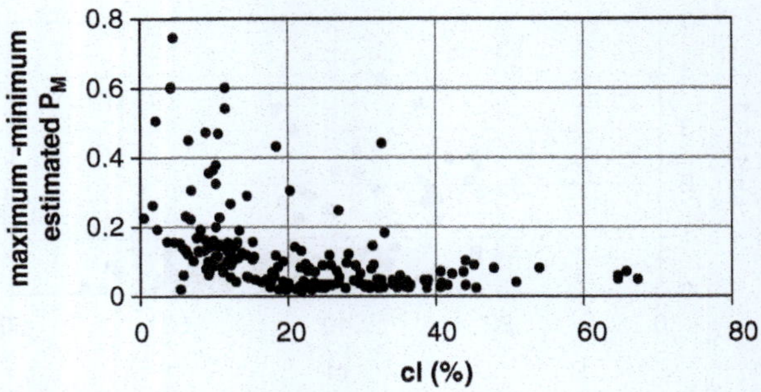

Fig. 3.38 Difference between the maximum and the minimum estimate of the particle-size shape index, p_M, calculated by alternative procedures against the clay content, *cl*, for a set of Sicilian soils (From Bagarello and Iovino 2012, reprinted with permission)

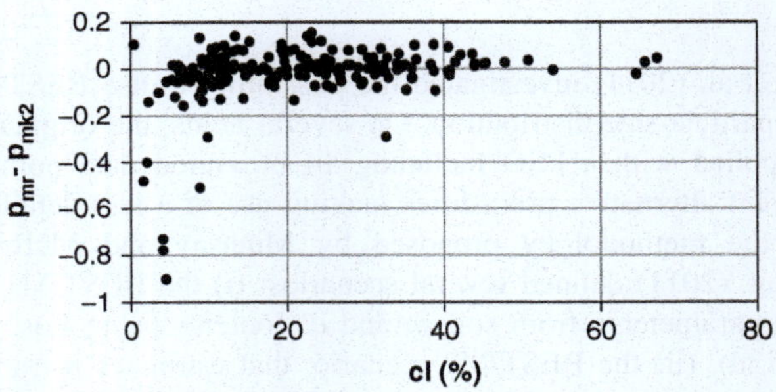

Fig. 3.39 Difference between p_{mr} (water retention shape index obtained by the measured water retention curve) and p_{mk2} (water retention shape index estimated according to the BEST procedure) against the clay content, *cl*, for a set of Sicilian soils (From Bagarello and Iovino 2012, reprinted with permission)

p_M should have a limited importance if the alternative models describe accurately the data. Bagarello and Iovino (2012) also compared BEST estimates of soil water retention with laboratory data. In particular, the van Genuchten's (1980) equation was fitted to the water retention data obtained in the laboratory and p_m was calculated using the fitted parameters and Eq. (3.65). The laboratory determined water retention shape index, p_m, did not coincide with the estimated one according to BEST (mean factor of difference = 1.33, maximum = 4.6), particularly in soils with a low *cl* content (<10 %, Fig. 3.39). However, alternative procedures, using different PSD models to estimate p_M, did not improve the quality of the p_m predictions. Taking into account that alternative PSD models performed better than the BEST one in terms of fitting accuracy, it was suggested that a common user index for the PSD and water retention models is more important than an accurate description of the PSD data in the indirect prediction of p_m included in the BEST procedure. The pedotransfer function developed by Minasny and McBratney (2007), allowing an estimate of p_m on the basis of the soil's clay and sand percentages, was an improvement in the estimation of p_m as compared with the original BEST procedure for the Sicilian database, since the maximum factor of difference with the laboratory determined shape index was 2.3 and the two datasets differed, on average, by a factor of 1.29. A possible reason of this result was that Minasny and McBratney (2007) used a very large number of laboratory determined water retention curves to deduce Eq. (3.73a, 3.73b, 3.73c and 3.73d) and the Sicilian database showed similarities with the databases considered by Minasny and McBratney (2007).

3.2.7 Issues on Estimation of Scale Parameters

BEST methods have also been evaluated with regards to the treatment of infiltration data. The success of the two methods (BEST-Slope, BEST-Intercept) or one of the two methods depends upon the kind of soils and experimental conditions. BEST-Slope may fail when the lateral infiltration of water due to the geometric effect, AS^2, is large in comparison to the contribution of gravity flow, K_s. In such a case, the steady-state infiltration is much more conditioned by the term AS^2, and the subtraction in Eq. (3.80a) becomes too sensitive, leading to negative or wrong estimates for K_s. In such a case, BEST-Slope may fail whereas BEST-Intercept provides the whole set of hydrodynamic parameters (Yilmaz et al. 2010). If BEST-Slope does not fail and provides values, the BEST-Slope estimates may differ significantly from BEST-Intercept (Fig. 3.40).

The success of the BEST methods also depends on the quality of the infiltration data, including the precision of description of the transient state and the attainment of steady-state. BEST methods may be successful if it is possible to get enough experimental points for both transient state and steady-state. Several large databases attest that the BEST method usually provides good fits of experimental data. For an experiment carried out in Burundi on mostly clay soils with a ring of $r = 0.075$ m and $V_w = 150$ mL, Bagarello et al. (2011) obtained a satisfactory agreement between the infiltration model and the data over a large dataset (85 infiltration runs, mean of $E_r = 2.3$ %) and the probability of obtaining low E_r values was found to increase with the cumulative infiltration volume that was modeled by the transient infiltration relationship. In an investigation on some Irish soils (Xu et al. 2012), fits were generally acceptable with BEST-Slope and BEST-Intercept, even if E_r was slightly higher than for some other models like the so-called method 1 by Wu et al. (1999) (Sect. 2.2).

Unfortunately, under certain circumstances, two opposite challenging situations occur: (i) the steady-state is unreachable and all the experimental points describe the transient state, or (ii) steady-state is reached too quickly, with only few points for the transient state, meaning a poor description of the transient state. These two contrasting cases were depicted experimentally by Gonzalez-Sosa et al. (2010). When the influence of capillary forces on the infiltration process is very short and the gravitational time is reached soon after the beginning of the run, the sorptivity is estimated with only a small number of points which can be detrimental to the

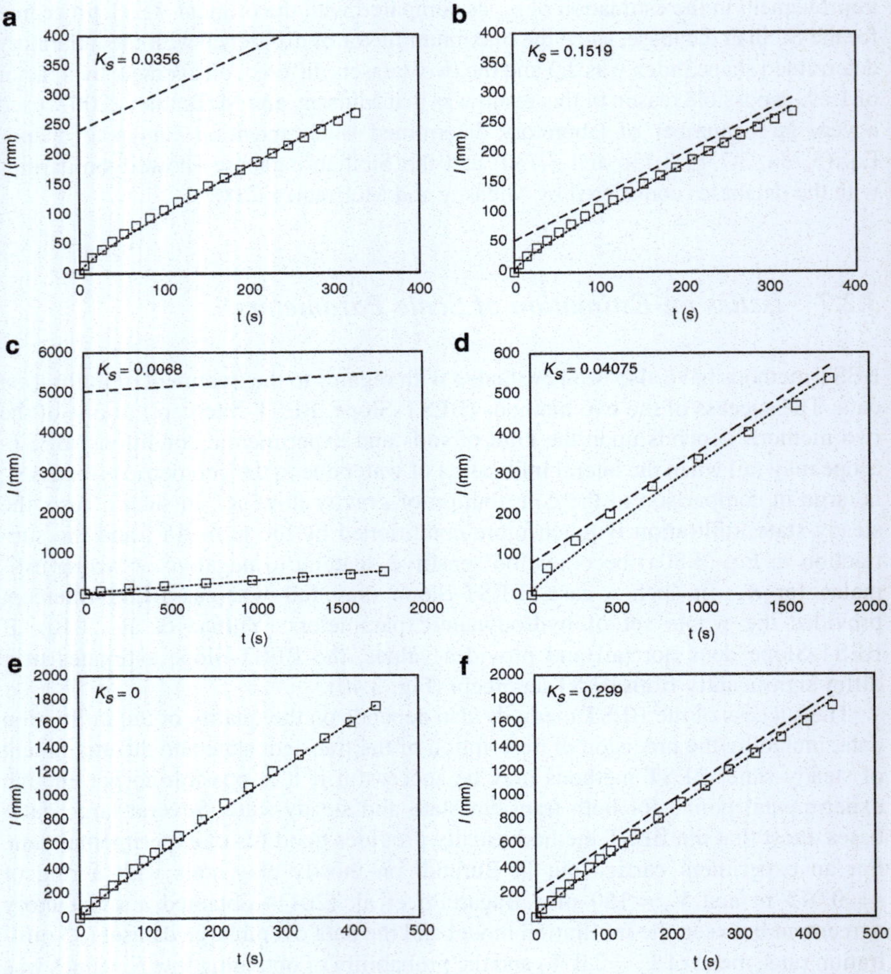

Fig. 3.40 Comparison of (**a**, **c** and **e**) BEST-Slope and (**b**, **d** and **f**) BEST-Intercept for several levels of differences between the two algorithms: (**a** and **b**) significant; (**c** and **d**) very significant; and (**e** and **f**) no result for BEST-Slope but results for BEST-Intercept. Experimental data (*points*) and BEST model (*lines*); I is infiltration, t is time and K_s is saturated soil hydraulic conductivity (From Yilmaz et al. 2010, reprinted with permission)

robustness of the optimization but is theoretically adequate (Fig. 3.41a). On the other hand, when the transient infiltration equation is valid for the whole infiltration test duration, water infiltration data indicate transient state, with a robust estimate of sorptivity, but the estimation of the asymptotic regime, and thus K_s, is less easy (Fig. 3.41b).

When steady-state is unreachable, most of the points describe the transient state. In this case, the straight line fitted to the data supposed to describe steady-state

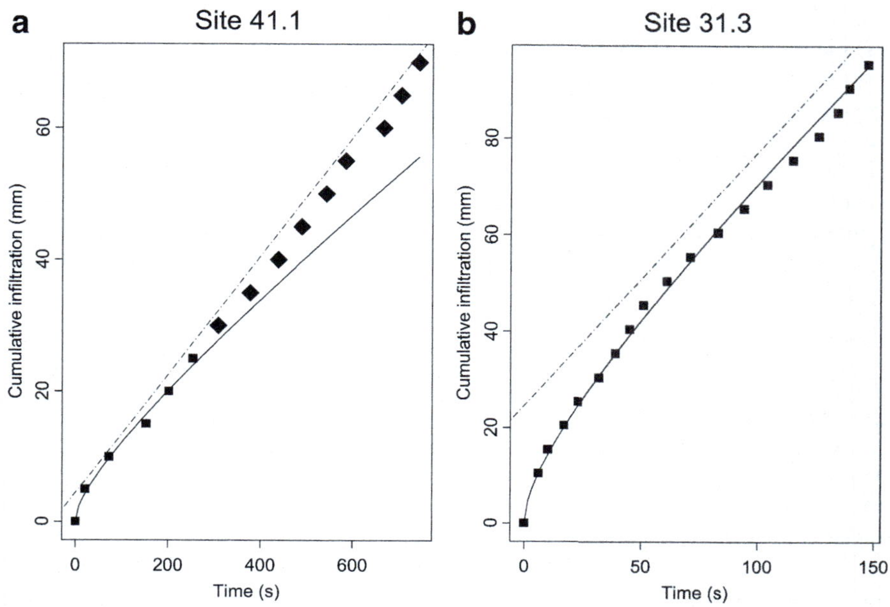

Fig. 3.41 Observed (points) and fitted cumulative infiltration data for two sites (site 41.1 and site 31.3). The full line is the fit of the transient infiltration model for short time steps and the dashed line is the fit of the asymptotic infiltration model for large time steps. The transient infiltration model was fitted to the points plotted with squares (From Gonzalez-Sosa et al. 2010, reprinted with permission)

conditions does not properly represent steady-state. The related slope, i_s^{exp}, and intercept, b_s^{exp}, may be poor estimates for the slope and the intercept of the real steady-state. Then, Eqs. (3.79a, b) may become erroneous, rendering both sets of equations, Eqs. (3.80a or 3.80b) and (3.81a or 3.81b), erroneous as well. In other words, sorptivity and saturated hydraulic conductivity may be inaccurately estimated. Lassabatere et al. (2010a) investigated the repercussions of the non-fulfillment of steady-state condition on estimates for S and K_s. They used BEST methods to treat analytically generated data, considering complete data sets including both transient state and steady-state and truncated datasets with only the transient state. The analytical data were generated using the implicit quasi-exact infiltration model by Haverkamp et al. (1994) for different reference values of S and K_s. The ratios between the estimated and target values did not depend on the reference S and K_s values, provided that the data subsets were truncated at the same scaled time, t_{sc}, defined as follows:

$$t_{sc} = \frac{2K_s^2}{S^2} t \tag{3.86}$$

Their results indicate that both BEST-Slope and BEST-Intercept provide accurate estimates when steady-state is reached. The relative errors between target and estimated values never exceed a few percent, with slight under or overestimations depending on the method and the ring radius. In general, BEST-Intercept performed better than BEST-Slope for a rather wide range of ring radii. On the contrary, if the datasets are truncated, i.e. mimicking the end of the experiment before reaching steady-state, sorptivity is always underestimated and thus saturated hydraulic conductivity is always overestimated. In agreement with Eq. (3.76), these trends are expected to yield higher values of h_g than the real ones (Lassabatere et al. 2010a).

When infiltration reaches steady-state early, only a few data points under transient flow state are available and BEST methods fail in estimating sorptivity. For this reason, Xu et al. (2009) suggested that the field experiments should be conducted under a wide range of soil water conditions from initial to saturated, to have enough data points for the transient flow state. Indeed, dry initial conditions favor the development of the transient state and may ease the acquisition of enough data points for the transient state. In particular, the constraint $\theta_i < 0.25 \times \theta_s$ should be respected (Haverkamp et al. 1994), even though this condition may be difficult for soils with strong water retention by capillarity (e.g. clays). In addition to this precaution, experimental devices and protocols should be designed to allow small increments in terms of infiltrated volumes and thus allow a better description of the transient state. To deal with this problem, Lassabatere et al. (2010a) suggested estimating K_s from the infiltration rate at the end of the experiment, assuming that the term related to sorptivity, AS^2, is negligible in Eq. (3.80a), leading to:

$$K_s = i_s^{\text{exp}} - AS^2 \approx i_s^{\text{exp}} \tag{3.87}$$

Bagarello et al. (2011) suggested some modifications to render BEST methods more robust with regards to the inaccurate description of transient state. When the transient state is not accurately described, the condition $t \leq t_{max}$ may not be fulfilled, leading to BEST algorithm failure. When the condition is fulfilled but with a reduced number of data points, the fit is performed on a small data set leading to misestimations and poor fits. Indeed, Bagarello et al. (2011) noticed, for soils studied in Burundi, that worst fits, i.e. larger relative errors between the model and the experimental data, were obtained when the number of data points considered for the fit, k_{max}, was small (Fig. 3.42). In fact, the need to determine the maximum number of data points that can be described by the transient infiltration model (k_{max}) complicates the data analysis and renders the BEST algorithm too sensitive. With reference to the same experiment carried out in Burundi, Bagarello et al. (2011) established a comparison between the scale parameters deduced according to the original BEST-Slope procedure and the corresponding ones obtained by considering the complete experimental infiltration curve, i.e. by assuming $k_{max} = N_{tot}$. For both variables, the differences between the two datasets were not statistically significant (Fig. 3.43), suggesting that using the complete dataset of measured cumulative infiltration values instead of the reduced transient series did not affect soil hydraulic characterization obtained by BEST. In other terms, for the sampled soils, an infiltration run carried out

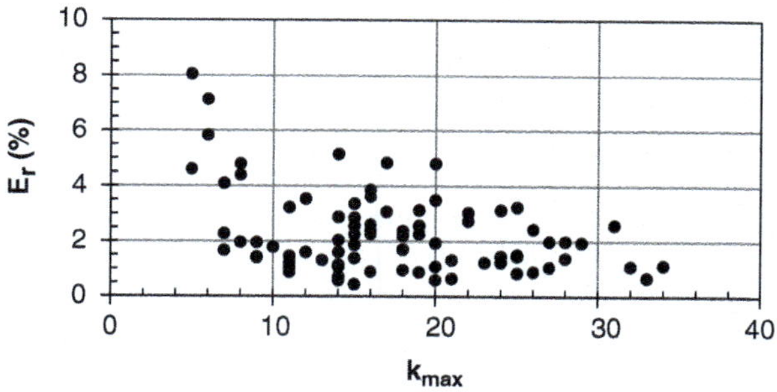

Fig. 3.42 Relationship between the relative error, E_r (%), of the infiltration model calculated for each individual infiltration run carried out on Burundian soils and the number of data points describing a transient process, k_{max} (From Bagarello et al. 2011, reprinted with permission)

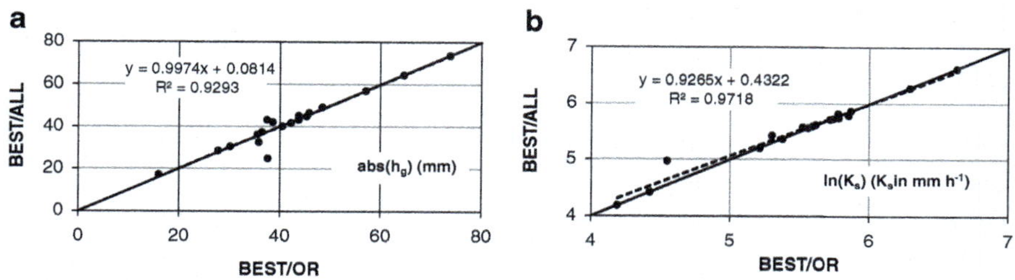

Fig. 3.43 Comparison between the BEST/OR (original BEST-Slope algorithm) and BEST/ALL (BEST-Slope algorithm applied by ignoring the constraint on the time validity of the transient infiltration model) in terms of (**a**) abs(h_g) results and (**b**) ln(K_s) results (From Bagarello et al. 2011, reprinted with permission)

until three consecutive infiltration rates were nearly constant and a simplified analysis using the entire measured cumulative infiltration curve could be an interesting alternative to simplify calculation procedures.

BEST methods have also been questioned with regards to the chosen values for the infiltration constants, γ and β, that are implemented into the analytical model, i.e. Eqs. (3.77a, 3.77b, 3.77c and 3.77d and 3.78a, 3.78b, 3.78c and 3.78d), and the tortuosity parameter, p, introduced in Eq. (3.59b). Nasta et al. (2012) assessed the sensitivity of the BEST algorithm regarding the choice of these parameters in their feasible ranges: $0 \leq p \leq 1.33$; $0.6 < \gamma < 0.8$; $0 < \beta < 2$. The originally considered values of the infiltration constants ($\gamma = 0.75$ and $\beta = 0.6$) were not optimal for the three experimental soils by Lassabatere et al. (2006). The analysis showed that tortuosity (p) plays only a secondary role compared with constants β and γ, which are more important for the estimation of the scale parameters. The proper calibration of β and γ as a function of the soil type could significantly improve the estimates of the soil hydraulic parameters (Nasta et al. 2012). Yet, a proper definition of these constants could be obtained through fitting numerical data onto the quasi-exact implicit formulation as already performed by Lassabatere et al. (2009) for only four kinds of soils. Such identification of infiltration constants β and γ requires further investigations.

Despite these shortcomings and questions, BEST methods appear quite a useful and efficient tool for the hydraulic characterization of soils. The BEST algorithms have been found to be usable for many kinds of soils, even hydrophobic sandy soils (Lassabatere et al. 2012). In this last case, a compromise must be found for initial water content between the two following constraints: (i) the initial water content must remain lower than 1/4 of the saturated water content but (ii) it cannot be very low, to prevent or at least reduce the hydrophobic behavior of the soil and the failure of the method.

On the basis of approximately 400 runs carried out in Italy and in Burundi, Bagarello et al. (2014) suggested that BEST-Slope and BEST-Intercept should be expected to differ by the number of successful runs (positive S and K_s values), with BEST-Intercept yielding a higher success percentage than BEST-Slope at the expense of a poorer performance in terms of data representation by the infiltration model. The BEST-Steady algorithm, allowing a success percentage of 100 % in the investigation by Bagarello et al. (2014), yields S and K_s estimates that are closer to those obtained with BEST-Intercept than BEST-Slope.

3.2.8 Tests of the Soil Hydraulic Characteristic Curves

BEST methods have also been tested against real measures of water retention and hydraulic conductivity. Minasny and McBratney (2007) suggested that the hydraulic conductivity curve can be predicted reasonably well, assuming that it conforms to the Brooks and Corey model, and water retention can only be predicted near saturation, due to the large error in estimating the air-entry parameter, h_g.

Bagarello and Iovino (2012) tested the BEST procedures to predict water retention characteristics of approximately 200 Sicilian soils. The van Genuchten (vG) model with the Burdine condition used in BEST showed, on average, a satisfactory fitting accuracy when it was fitted to laboratory measured water retention values (mean of $E_r = 3.6\,\%$), and it performed well ($E_r \leq 5\,\%$) for approximately 80 % of the soil samples. The highest differences between the retention data and the fitted curve were observed for the lowest (i.e. more negative) pressure heads (Fig. 3.44a). Moreover, the E_r values of the vG model were always low in soils with a high clay content (i.e. $cl \geq 44\,\%$, Fig. 3.44b). In the other cases, the mean performance of the vG equation was still satisfactory (mean of $E_r = 3.6\,\%$), but it was unpredictable using the cl content for a given sampling point since both high and low E_r results were obtained.

3.2.9 Using the Soil Hydraulic Parameters Obtained with BEST

BEST methods provide the whole set of hydraulic parameters considering Eqs. (3.58a and 3.58b) and (3.59a and 3.59b). However, these equations cannot

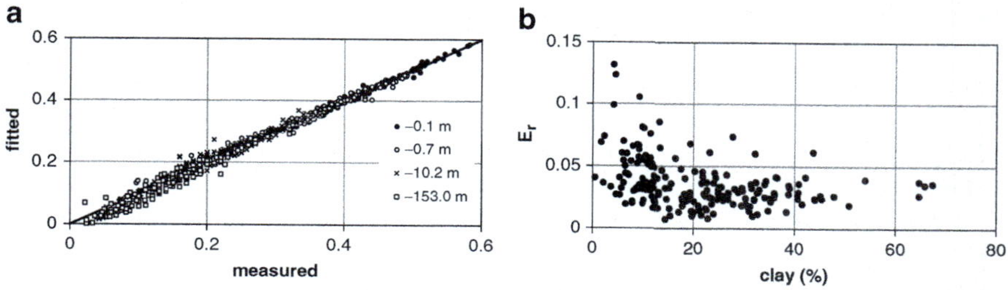

Fig. 3.44 Comparison between the measured volumetric soil water content, θ (L^3L^{-3}), and the one estimated by fitting the van Genuchten (1980) soil water retention model used in BEST for four selected pressure heads (**a**) and relationship between the relative error, E_r, of the van Genuchten (1980) soil water retention model used in BEST and the clay content of the soil (**b**) for a dataset of water retention curves measured in Sicily (From Bagarello and Iovino 2012, reprinted with permission)

be used with the HYDRUS-2D model (Šimůnek et al. 1999b) since it requires that soil hydraulic properties are described by either the van Genuchten/Mualem or the Brooks and Corey relations. In particular, the van Genuchten relationships for both θ(*h*) and *K*(θ) with Mualem (1976) condition are:

$$\frac{\theta-\theta_r}{\theta_s-\theta_r}=\left(1+|\alpha_{vG}h|^n\right)^{-m} \tag{3.88a}$$

$$m=1-\frac{1}{n} \tag{3.88b}$$

$$\frac{K(\theta)}{K_s}=\left(\frac{\theta-\theta_r}{\theta_s-\theta_r}\right)^l\left[1-\left(1-\left(\frac{\theta-\theta_r}{\theta_s-\theta_r}\right)^{1/m}\right)^m\right]^2 \tag{3.89}$$

where α_{vG} (L^{-1}) is simply called the alpha parameter of van Genucthen model, and *n*, *m* and *l* are shape parameters. Therefore, the soil hydraulic parameters obtained with BEST have to be converted into a set of six hydraulic parameters (θ_r, θ_s, *n*, α_{vG}, K_s and *l*) of Eqs. (3.88) and (3.89) for applying HYDRUS-2D.

The conversion procedure applied by Mubarak et al. (2009b) includes the following steps: (i) θ_r is considered to remain equal to 0 as in the calculations made by the BEST method; (ii) because of the difficulty of eliminating all the air bubbles in the soil in order to saturate it completely, the natural saturated water content, θ_s, is set at 85 % of the total soil porosity; (iii) according to Leji et al. (2005), the water retention shape index corresponding to $\theta_r=0$, p_m, is constant for different parameterizations [see Eq. (3.65)]. Therefore, p_m does not change between Eq. (3.58a and 3.58b) with $k_m=2$ and Eq. (3.88). Taking into account that the relationship between *m* and *n* is known, Eq. (3.65) can easily be applied to convert the shape parameters, *n* and *m*, in the Burdine condition to the Mualem condition; (iv) to estimate the alpha parameter, the RETC program (van Genuchten et al. 1991) is used to fit Eq. (3.88a) on soil water retention data generated by

Eq. (3.58a) for a range from near saturation to very dry; and (v) Eq. (3.89) is fitted to the soil hydraulic conductivity data generated by Eq. (3.59a) using the RETC code with a value of $l = 0.5$. Lassabatere et al. (2010b) also adapted the BEST algorithm to the description of $\theta(h)$ and $K(\theta)$ with the van Genuchten relationships and the Mualem (1976) condition. In this case, however, the l shape parameter was considered a fitting parameter. Recently, Nasta et al. (2012) reported that the Look-up Table option of HYDRUS-2D/3D was adapted to provide the definition of the soil hydraulic properties using an external text input file, *Mater.in*. The conversion procedure by Mubarak et al. (2009b) needs testing and maybe developments, also considering that the same authors suggested that the direct development of the soil parameters for the van Genuchten/Mualem relations into the BEST algorithm is possible but it requires supplementary research, especially on the shape parameters estimation of these relations from particle-size analysis. A point that probably should be considered is that both θ_s and K_s, that are representative of a saturated soil condition, should be independent of the model describing unsaturated soil hydraulic properties. In other terms, an alternative approach to convert the soil hydraulic parameters could consider θ_s and K_s values that do not vary with the applied water retention and hydraulic conductivity model.

Mubarak et al. (2010) developed as well the comparison between estimated hydraulic parameters with different kinds of experiments and methods. They dealt with the practical problem to establish a comparison between a complete soil hydraulic characterization carried out with BEST on a given year (2007) and a characterization performed in the past (1990) by the single-ring pressure infiltrometer method (Reynolds and Elrick 1990; Vauclin et al. 1994), yielding an estimate of the $K(h)$ relationship according to Eq. (3.3).

The capillary length obtained by Eq. (3.7) was used to obtain the following estimate of the mean characteristic microscopic pore radius, i.e. the mean characteristic dimension of hydraulically functioning pores, ξ_m (L):

$$\xi_m = \frac{\sigma}{g\ \rho_w\ \lambda_c} \tag{3.90}$$

where σ (M T^{-2}) is the surface tension, g (L T^{-2}) is the acceleration due to gravity and ρ_w (M L^{-3}) is the density of water. According to Mubarak et al. (2010), when describing unsaturated water flow subject to a given set of initial and boundary conditions, the water flow behavior of the soil should be independent of the choice of the soil hydraulic functional relationships. These authors used the soil hydraulic parameters obtained with BEST to obtain another estimate of the capillary length, denoted by the symbol α_h (L), as:

$$\alpha_h = -h_g \tag{3.91}$$

with h_g calculated by Eq. (3.76). The characteristic microscopic pore radius, ξ_m, was also calculated by the following relationship:

$$\xi_m = \frac{\sigma}{g\ \rho_w\ \alpha_h} \tag{3.92}$$

For example, for pure water at 20 °C and both α_h and ξ_m expressed in mm, Eq. (3.92) reduces to:

$$\xi_m = \frac{7.44}{\alpha_h} \tag{3.93}$$

The mean characteristic pore size and the capillary length are actual estimates of the capillary component of water transfer into the soil. Therefore, the higher the ξ_m and the lower the α_h the greater the effect of gravity compared to capillarity as the infiltration driving force. Assessing temporal variation of these parameters allows us to establish the changes in structure of the fine soil fraction (Mubarak et al. 2009a). The concept of capillary length led the authors to conclude that the estimates of ξ_m obtained by Eqs. (3.90) and (3.92) were directly comparable, allowing a comparison between the two sampling dates also in terms of characteristic microscopic pore radius. The analysis by Mubarak et al. (2010) was based on the hypothesis that a soil hydraulic characterization carried out with the single-ring pressure infiltrometer is comparable with the one performed with the BEST procedure. This hypothesis needs experimental testing since different methods often yield dissimilar results, especially in terms of K_s values (Reynolds et al. 2000), also considering that the pressure infiltrometer is a steady-state method while BEST also considers the transient stage of the infiltration process.

Recently, Souza et al. (2014) suggested that the estimates of K_s and ξ_m can also be used to estimate the maximum number of effective straight-cylindrical pores per unit area, N_0 (L^{-2}), by the following relationship (Watson and Luxmoore 1986):

$$N_0 = \frac{8\mu K_s}{\rho_w g \pi \xi_m^4} \tag{3.94}$$

where μ ($M\,L^{-1}T^{-1}$) is the dynamic viscosity of water.

According to Gonzalez-Sosa et al. (2010), the BEST protocol can also be used when a negative water pressure is maintained at the soil surface by a tension infiltrometer. In particular, these authors used a mini-disk infiltrometer (Decagon Devices Inc., Pullman, WA) with a surface pressure set at −20 mm and a thin layer (<1 mm) of fine sand to ensure good contact between the infiltrometer disk and the soil. In this case, and assuming that the contact layer has no significant influence on early stages of infiltration, the soil hydraulic conductivity corresponding to the established, negative pressure head is obtained by analyzing the data. However, using the mini-disk infiltrometer does not allow estimation of the scale parameter, h_g, of the water retention curve. Scalenghe and Ferraris (2009) also analyzed tension infiltration tests with the BEST method.

The BEST procedure of soil hydraulic characterization also proved to be an effective tool to experimentally assess, in stormwater retention/infiltration basins, the impact of the heterogeneity of sediment distribution on soil water transfer properties (Cannavo et al. 2010), and the role of spontaneous vegetation on the hydraulic performances of the basin (Gonzalez-Merchan et al. 2014).

Therefore, BEST is very attractive since it substantially facilitates the hydraulic characterization of unsaturated soils, and it is becoming an applied method in soil science. Knowledge about factors affecting the BEST results is also increasing but there are several points still needing checking and, maybe, developments. Probably, the first and most obvious assessment concerns the ability of the procedure to yield a reliable soil hydraulic characterization. The estimated water retention and soil hydraulic conductivity curves should be compared with independent measurements of these properties carried out in different soils and soil conditions to establish if the simplified method is a practical alternative to more cumbersome and time consuming methods and also to detect points in the indirect procedure needing specific adjustments or developments. Another point needing consideration is probably the estimation of S_{max}, also because $B=0$ has to be assumed for this calculation but this constant has a positive value in the BEST procedure. Taking into account that determining the saturated soil water content may be difficult in the field, also for practical reasons, it would be advisable to develop estimating procedures of θ_s on the basis of the measured variables with the BEST experiment. Pouring water when the previously applied amount had completely infiltrated may promote air entrapment phenomena in the sampled soil volume and may also favor soil structure alteration phenomena at infiltration surface. Therefore, the impact of the suggested procedure, that has the obvious advantage of being very simple, on the soil hydraulic characterization should specifically be taken into account. With this aim, a comparison with infiltration runs carried out by steadily maintaining a small (i.e., very close to zero) depth of water on the infiltration surface should be developed. An automated infiltrometer, potentially usable for this kind of comparison was recently developed by Di Prima (2015). Procedures should also be developed for predicting the parameters of alternative models for the soil hydraulic characteristic curves, also including macropore effects.

Example 3.4

An application of the BEST procedure of soil hydraulic characterization was carried out at a Burundian site, on a clay soil having sand, *sa* (0.05–2.0 mm), silt, *si*, and clay, *cl*, percentages equal to 19.9 %, 20.1 % and 60.0 %, respectively. The experimentally determined particle size distribution (PSD) is reported in Table 3.6. The dry soil bulk density, ρ_b, was equal to 916 kg m^{-3} and the corresponding porosity, f, calculated by assuming a density of the solid particles, ρ_s, of 2650 kg m^{-3}, was equal to $f=1-916/2650=0.654$.

The experimentally determined PSD was fitted on Eq. (3.67a), taking into account Eq. (3.67b), by using the Solver routine of the Microsoft Excel software (Fig. 3.45a), and $N=2.091$, $M=0.0436$, and $D_g=0.535$ mm was obtained. The relative error, E_r [Eq. (3.84)], between estimated and measured values of $P(D)$, was equal to 1.3 %, denoting a good correspondence between the model and the data. The shape index for PSD, p_M, was equal to:

$$p_M=\frac{0.0436\times 2.091}{1+0.0436}=0.0873$$

Table 3.6 Measured particle size data pairs (diameter, *d* – frequency by weight, *F*) for the soil of the Example 3.4

d (mm)	*F*
2	1
0.850	0.9934
0.425	0.9752
0.250	0.9280
0.106	0.8438
0.074	0.8148
0.02816	0.7816
0.01814	0.7416
0.01070	0.7032
0.00764	0.6836
0.00547	0.6540
0.00319	0.6228
0.00115	0.5732
0.00082	0.5520

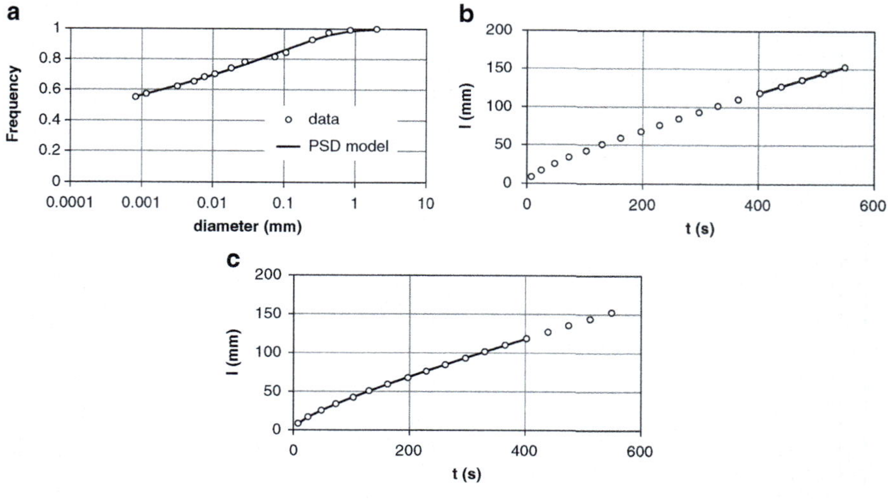

Fig. 3.45 Plots of the data of the Example 3.4: (**a**) fitting of the particle-size data on the BEST PSD model; (**b**) estimation of the steady-state infiltration rate, and (**c**) fitting of the infiltration model to the transient phase of the infiltration process (I = cumulative infiltration; t = time)

Using Eq. (3.70) and the Solver routine, an estimate of $s = 0.7294$ was obtained, and κ was calculated by Eq. (3.69):

$$\kappa = \frac{2 \times 0.7294 - 1}{2 \times 0.7294 \times (1 - 0.7294)} = 1.1622$$

The value of $(1+\kappa)^{-1}$ was 0.4625 which is very close (difference, $\Delta = 0.2\ \%$) to the $(1+\kappa)^{-1}$ estimate obtained by Eq. (3.72), equal to 0.4634. The water retention shape index, p_m, was estimated according to Eq. (3.68):

$$p_m = 0.0873 \times (1 + 1.1622)^{-1} = 0.0404$$

The shape parameters m, n and η were then derived on the basis of Eqs. (3.71), (3.58b) and (3.59b), respectively:

$$m = \frac{1}{0.0404}\left(\sqrt{1 + 0.0404^2} - 1\right) = 0.0202$$

$$n = \frac{2}{1 - 0.0202} = 2.0412$$

$$\eta = \frac{2}{0.0202 \times 2.0412} + 2 + 1 = 51.56$$

Using Eqs. (3.73a, 3.73b, 3.73c and 3.73d) by Minasny and McBratney (2007) yielded $n = 2.0844$. This estimate of n is close ($\Delta = 2.1\ \%$) to the one obtained from the PSD, that was used for the example calculations. The coefficient c_p, calculated by Eq. (3.74b), was equal to:

$$c_p = 0.886\left\{\frac{1.600}{0.978} + \frac{1.549}{0.968}\right\} = 2.867$$

The volumetric soil water content at the time of the field infiltration experiment, θ_i, was equal to 0.142 m^3m^{-3}. A ring with a radius of 75 mm was used and water was applied on the infiltration surface 18 times. A volume of water $V_w = 150$ mL was poured each time. The cumulative infiltration data are reported in Table 3.7. Fig. 3.45b, showing the experimental I vs. t relationship, suggested to carry out a linear regression analysis of the last five data pairs to obtain an estimate of the steady-state infiltration rate, i_s^{exp}, equal to 0.231 mm s^{-1} (coefficient of determination, $R^2 = 0.99998$). Due to practical limitations, measuring the soil water content at the end of the infiltration experiment was not possible. Therefore, the saturated volumetric soil water content, θ_s, was assumed to be equal to the soil porosity, f.

For comparative purposes, the original BEST-Slope algorithm and the alternative BEST-Intercept and BEST-Steady algorithms were applied.

For $\theta_s = f$, $\theta_i/\theta_s = 0.217$ was obtained, showing that $\beta = 0.6$ and $\gamma = 0.75$ were usable to analyze the data. The following estimates of the A and B constants were obtained by Eqs. (3.78):

$$A = \frac{0.75}{75 \times (0.654 - 0.142)} = 0.0195\ \text{mm}^{-1}$$

$$B = \frac{2 - 0.6}{3}\left[1 - \left(\frac{0.142}{0.654}\right)^{51.56}\right] + \left(\frac{0.142}{0.654}\right)^{51.56} = 0.4667$$

Table 3.7 Cumulative infiltration, I, and associated time, t, defining the measured infiltration curve for the Example 3.4

No.	I (mm)	t (s)
1	8.5	8
2	17.0	25
3	25.5	48
4	33.9	73
5	42.4	103
6	50.9	130
7	59.4	162
8	67.9	197
9	76.4	229
10	84.9	262
11	93.4	297
12	101.9	330
13	110.3	365
14	118.8	402
15	127.3	439
16	135.8	475
17	144.3	512
18	152.8	549

Table 3.8 Maximum sorptivity, S_{max}, and sum of the squared differences between measured and estimated cumulative infiltration, SSD, corresponding to different numbers of infiltration data, k, for the Example 3.4

k	S_{max} (mm $s^{-1/2}$)	SSD (mm^2)
5	2.715	0.287
6	2.736	1.254
7	2.747	1.822
8	2.750	1.900
9	2.758	2.570
10	2.766	3.944
11	2.773	5.348
12	2.781	8.368
13	2.789	11.901
14	2.795	14.800
15	2.800	17.571
16	2.805	21.275
17	2.810	25.361
18	2.814	30.101

Equation (3.81a) with $B = 0$ was applied to calculate the maximum sorptivity, S_{max}. For a number of infiltration data, k, varying between 5 and 18, Table 3.8 gives the estimated S_{max} value and the sum of the squared differences between the measured, $I^{exp}(t_i)$, and the estimated, $I(t_i)$, cumulative infiltration, SSD. The highest calculated S_{max} was equal to 2.814 mm $s^{-1/2}$ and the associated $S_{max}{}^2$ was 7.921 mm^2s^{-1}. This last value was lower than $i_s^{exp}/A = 0.231/0.0195 = 11.853$ mm^2s^{-1}. Therefore, the maximum sorptivity for the considered experiment was of 2.814 mm $s^{-1/2}$.

Table 3.9 Application of the BEST-Slope algorithm to the data of the Example 3.4 for determining soil sorptivity, S, and saturated soil hydraulic conductivity, K_s, for different numbers of infiltration data, k, and hence different durations of the transient phase of the infiltration run, t_k, and associated maximum time, t_{max}

k	t_k (s)	S (mm s$^{-1/2}$)	K_s (mm s^{-1})	t_{max} (s)
18	549	2.479	0.1114	435
17	512	2.478	0.1115	434
16	475	2.477	0.1116	433
15	439	2.476	0.1117	432
14	402	2.475	0.1117	431

For a given k value, corresponding to a duration of the experiment equal to t_k, the experimental data were then fitted on Eq. (3.81a) written, for the particular application, as:

$$I(t) = S\sqrt{t} + \left[0.0195 \times (1 - 0.4667)S^2 + 0.4667 \times 0.231\right]t$$

The fit, that was performed by minimizing the objective function defined by Eq. (3.82), yielded an estimate of S. This estimate and the experimental steady-state infiltration rate were then used to obtain an estimate of K_s on the basis of Eq. (3.80a) written as:

$$K_s = 0.231 - 0.0195 \times S^2$$

The time, t_{max}, was calculated by Eq. (3.83). In this example, we moved backwards, i.e. we started from N_{tot} and deleted an experimental (t_i, I_i) data pair each time. The procedure was stopped when the prescribed condition ($t_k < t_{max}$) was satisfied. The results of these calculations, summarized in Table 3.9, showed that 14 experimental points yielded a run duration ($t_k = 402$ s) lower than the maximum time ($t_{max} = 431$ s). The infiltration model described well the data (Fig. 3.45c) and the relative error, E_r, calculated by Eq. (3.84) was equal to 0.44 %. The S (=2.475 mm s$^{-1/2}$) and K_s (=0.1117 mm s^{-1}) values corresponding to $t_k = 402$ s were considered representative for the sampled site. The scale parameter for water pressure, h_g, was then estimated by Eq. (3.76):

$$h_g = -\frac{2.475^2}{2.867 \times (0.654 - 0.142)\left[1 - \left(\frac{0.142}{0.654}\right)^{51.56}\right]0.1117} = -37.3 \text{ mm}$$

The S and K_s values were positive and they defined modelled infiltration rates higher at transient than at steady-state. In particular, $i(t)$, calculated by Eq. (3.77b) for $10 \le t \le 431$ s, was equal to 1.0001–2.43 times the i_s value calculated by Eq. (3.77d), equal to 0.231 mm s^{-1}. This last value coincided with the experimentally determined steady-state infiltration rate, i_s^{exp}.

Table 3.10 Parameters of the water retention curve and the soil hydraulic conductivity function, i.e. Eqs. (3.58a and 3.58b) and (3.59a and 3.59b), obtained with different BEST algorithms and procedures for the data of the Example 3.4

Algorithm	m	n	η	θ_s (m^{3}m^{-3})	K_s (mm s^{-1})	h_g (mm)
BEST-Slope	0.0202	2.0412	51.56	0.654	0.1117	−37.3
BEST-Intercept					0.1412	−27.6
BEST-Steady					0.1291	−27.3

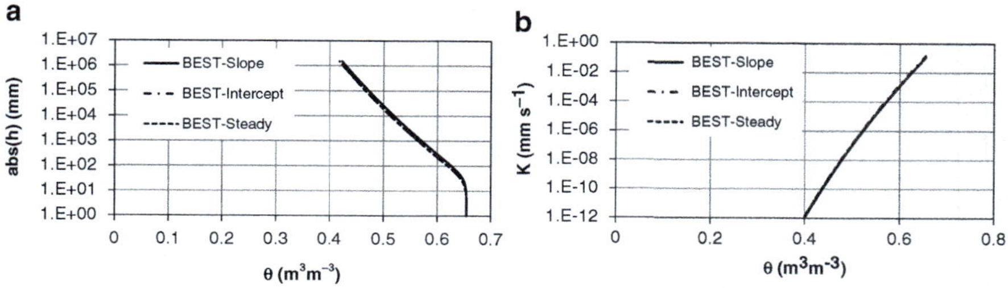

Fig. 3.46 Soil hydraulic characteristic curves obtained for the Example 3.4 with different BEST algorithms and assumptions: (**a**) water retention curve, and (**b**) soil hydraulic conductivity function

Table 3.10 summarizes the parameters of the soil hydraulic characteristic curves estimated by the BEST-Slope algorithm. The water retention curve is plotted in Fig. 3.46a whereas the soil hydraulic conductivity function is shown in Fig. 3.46b.

Calculation of the scale parameters was repeated using the BEST-Intercept algorithm. The linear regression analysis of the last five (t_i, I_i) data pairs, yielding an estimate of i_s^{exp} for the BEST-Slope application, also gave an intercept (b_s^{end}) of the asymptotic expansion, $I_s(t)$, equal to 25.86 mm. The C coefficient was equal to:

$$C = \frac{1}{2\left[1 - \left(\frac{0.142}{0.654}\right)^{51.56}\right](1 - 0.6)} \ln\left(\frac{1}{0.6}\right) = 0.6385$$

For a given k value, the experimental data were then fitted on Eq. (3.81b) written, for the particular application, as:

$$I(t) = S\sqrt{t} + \left(0.0195 + \frac{0.4667 \times 0.6385}{25.86}\right) S^2 t$$

The fit, that was performed by minimizing the objective function defined by Eq. (3.82), yielded an estimate of S. This estimate and the experimentally determined intercept of the asymptotic expansion were then used to obtain an estimate of K_s by Eq. (3.80b) written as:

$$K_s = 0.6385 \frac{S^2}{25.86}$$

Table 3.11 Application of the BEST-Intercept algorithm to the data of the Example 3.4 for determining soil sorptivity, S, and saturated hydraulic conductivity, K_s, for different numbers of infiltration data, k, and hence different durations of the transient phase of the infiltration run, t_k, and associated maximum time, t_{max}

k	t_k (s)	S (mm s$^{-1/2}$)	K_s (mm s^{-1})	t_{max} (s)
18	549	2.386	0.1406	253
17	512	2.386	0.1406	253
16	475	2.387	0.1406	253
15	439	2.387	0.1407	253
14	402	2.388	0.1408	253
13	365	2.388	0.1409	253
12	330	2.388	0.1408	253
11	297	2.388	0.1408	253
10	262	2.390	0.1410	252
9	229	2.391	0.1412	252

The time, t_{max}, was calculated by Eq. (3.83). Also in this case, the procedure was applied by initially considering N_{tot} data pairs and deleting an experimental (t_i, I_i) data pair each time. Calculations were stopped when the prescribed condition ($t_k < t_{max}$) was satisfied. The results of these calculations, summarized in Table 3.11, showed that nine experimental points yielded a run duration ($t_k = 229$ s) lower than the maximum time ($t_{max} = 252$ s). The transient infiltration model described well the data, given that a relative error $E_r = 0.96\,\%$ was obtained by Eq. (3.84). The S (=2.391 mm s$^{-1/2}$) and K_s (=0.1412 mm s^{-1}) values corresponding to $t_k = 229$ s were considered representative for the sampled site and $h_g = -27.6$ mm was obtained by Eq. (3.76).

With BEST-Steady, the following estimates of S and K_s were obtained by Eq. (3.85) and Eq. (3.80a), respectively:

$$S = \sqrt{\frac{0.231}{0.0195 + \frac{0.6385}{25.86}}} = 2.286 \text{ mm s}^{-1/2}$$

$$K_s = 0.231 - 0.0195 \times 2.286^2 = 0.1291 \text{ mm s}^{-1}$$

Equation (3.76) yielded the corresponding h_g value that was equal to −27.3 mm. Table 3.10 also summarizes the scale parameters estimated by the BEST-Intercept and BEST-Steady algorithms. In this application, the considered algorithm did not influence appreciably the estimated water retention curve (Fig. 3.46a) and soil hydraulic conductivity function (Fig. 3.46b).

References

Angulo-Jaramillo, R., Vandervaere, J. P., Roulier, S., Thony, J. L., Gaudet, J. P., & Vauclin, M. (2000). Field measurement of soil surface hydraulic properties by disc and ring infiltrometers. A review and recent developments. *Soil & Tillage Research, 55*, 1–29.

Ankeny, M. D. (1992). Methods and theory for unconfined infiltration measurements. In G. C. Topp, W. D. Reynolds, & R. E. Green (Eds.), *Advances in measurement of soil physical properties: Bringing theory into practice* (SSSA special publication no. 30, pp. 123–141). Madison: Soil Science Society of America, Inc.

Ankeny, M. D., Kaspar, T. C., & Horton, R. (1988). Design for an automated tension infiltrometer. *Soil Science Society of America Journal, 52*, 893–896.

Ankeny, M. D., Ahmed, M., Kaspar, T. C., & Horton, R. (1991). Simple field method for determining unsaturated hydraulic conductivity. *Soil Science Society of America Journal, 55*, 467–470.

Bagarello, V., & Iovino, M. (2003). Field testing parameter sensitivity of the two-term infiltration equation using differentiated linearization. *Vadose Zone Journal, 2*, 358–367.

Bagarello, V., & Iovino, M. (2004). Indagine di laboratorio su una metodologia innovativa per la determinazione della conducibilità idraulica del suolo con l'infiltrometro a depressione. *Rivista di Ingegneria Agraria, 2*, 81–92 (in Italian).

Bagarello, V., & Iovino, M. (2010). *Conducibilità idraulica del suolo. Metodi di misura nelle applicazioni idrologiche*. Milano: Hoepli. 382 pp. (in Italian).

Bagarello, V., & Iovino, M. (2012). Testing the BEST procedure to estimate the soil water retention curve. *Geoderma, 187–188*, 67–76.

Bagarello, V., Iovino, M., & Reynolds, W. D. (1999). Measuring hydraulic conductivity in a cracking clay soil using the Guelph permeameter. *Transactions of the ASAE, 42*(4), 957–964.

Bagarello, V., Iovino, M., & Tusa, G. (2000). Factors affecting measurement of the near saturated soil hydraulic conductivity. *Soil Science Society of America Journal, 64*, 1203–1210.

Bagarello, V., Iovino, M., & Tusa, G. (2001). Effect of contact material on tension infiltrometer measurements. *Transactions of the ASAE, 44*(4), 911–916.

Bagarello, V., Ferraris, S., & Iovino, M. (2004). An evaluation of the single-test tension infiltrometer method for determining the hydraulic conductivity of lateral capillary domain soils. *Biosystems Engineering, 87*(2), 247–255.

Bagarello, V., Castellini, M., & Iovino, M. (2005). Influence of the pressure head sequence on the soil hydraulic conductivity determined with tension infiltrometer. *Applied Engineering in Agriculture, 21*(3), 383–391.

Bagarello, V., Castellini, M., Iovino, M., & Ventrella, D. (2006a). *Variability of soil hydraulic conductivity and Gardner's α for different soils of south Italy*. Proc. of int. conference "SPATIAL", 14–15 September 2006, Foggia, Italy, 4 pp.

Bagarello, V., Iovino, M., Palazzolo, E., Panno, M., & Reynolds, W. D. (2006b). *Field and laboratory approaches for determining sodicity effects on soil hydraulic conductivity. Geoderma, 130*, 1–13.

Bagarello, V., Castellini, M., & Iovino, M. (2007). Comparison of unconfined and confined unsaturated hydraulic conductivity. *Geoderma, 137*, 394–400.

Bagarello, V., Provenzano, G., & Sgroi, A. (2009). Fitting particle size distribution models to data from Burundian soils for the BEST procedure and other purposes. *Biosystems Engineering, 104*, 435–441.

Bagarello, V., Castellini, M., Iovino, M., & Sgroi, A. (2010). Testing the concentric-disk tension infiltrometer for field measurement of soil hydraulic conductivity. *Geoderma, 158*, 427–435. doi:10.1016/j.geoderma.2010.06.018.

Bagarello, V., Di Prima, S., Iovino, M., Provenzano, G., & Sgroi, A. (2011). Testing different approaches to characterize Burundian soils by the BEST procedure. *Geoderma, 162*, 141–150.

Bagarello, V., Di Prima, S., & Iovino, M. (2014). Comparing alternative algorithms to analyze the beerkan infiltration experiment. *Soil Science Society of America Journal, 78*(3), 724–736. doi:10.2136/sssaj2013.06.0231.

Bodhinayake, W., Si, B. C., & Noborio, K. (2004). Determination of hydraulic properties in sloping landscapes from tension and double-ring infiltrometers. *Vadose Zone Journal, 3*, 964–970.

Bonnell, M., & Williams, J. (1986). The two parameters of the Philip infiltration equation. Their properties and spatial and temporal heterogeneity in a red earth of tropical semi-arid Queensland. *Journal of Hydrology, 87*, 9–31.

Braud, I., De Condappa, D., Soria, J. M., Haverkamp, R., Angulo-Jaramillo, R., Galle, S., & Vauclin, M. (2005). Use of scaled forms of the infiltration equation for the estimation of unsaturated soil hydraulic properties (the Beerkan method). *European Journal of Soil Science, 56*, 361–374.

Brooks, R. H., & Corey, C. T. (1964). *Hydraulic properties of porous media* (Hydrology paper, 3). Fort Collins: Colorado State University.

Burdine, N. T. (1953). Relative permeability calculation from pore size distribution data. *Petr. Trans. Am. Inst. Min. Metall. Eng., 198*, 71–77.

Cameira, M. R., Fernando, R. M., & Pereira, L. S. (2003). Soil macropore dynamics affected by tillage and irrigation for a silty loam alluvial soil in southern Portugal. *Soil & Tillage Research, 70*, 131–140.

Cannavo, P., Vidal-Beaudet, L., Béchet, B., Lassabatere, L., & Charpentier, S. (2010). Spatial distribution of sediments and transfer properties in soils in a stormwater infiltration basin. *Journal of Soils and Sediments, 10*, 1499–1509.

Carsel, R. F., & Parrish, R. S. (1988). Developing joint probability distributions of soil water retention characteristics. *Water Resources Research, 24*, 755–769.

Casanova, M., Messing, I., & Joel, A. (2000). Influence of aspect and slope gradient on hydraulic conductivity measured by tension infiltrometer. *Hydrological Processes, 14*, 155–164.

Casey, F. X. M., & Derby, N. (2002). Improved design for an automated tension infiltrometer. *Soil Science Society of America Journal, 66*, 64–67.

Castellini, M., Iovino, M. (2005). *Applicazione del metodo inverso per la determinazione delle proprietà idrauliche del suolo con l'infiltrometro a depressione*. Proc. VIII Convegno Nazionale AIIA, Catania, 27–30 June (in Italian with English abstract).

Castiglione, P., Shouse, P. J., Mohanty, B., Hudson, D., & van Genuchten, M. T. (2005). Improved tension infiltrometer for measuring low fluid flow rates in unsaturated fractured rock. *Vadose Zone Journal, 4*, 885–890.

Chief, K., Young, M. H., & Shafer, D. S. (2012). Changes in soil structure and hydraulic properties in a wooded-shrubland ecosystem following a prescribed fire. *Soil Science Society of America Journal, 76*, 1965–1977.

Clausnitzer, V., Hopmans, J. W., & Starr, J. L. (1998). Parameter uncertainty analysis of common infiltration models. *Soil Science Society of America Journal, 62*, 1477–1487.

Close, K. R., Frasier, G., Dunn, G. H., & Loftis, J. C. (1998). Tension infiltrometer contact interface evaluation by use of a potassium iodide tracer. *Transactions of the ASAE, 41*, 995–1004.

Clothier, B., Scotter, D. (2002a). 3.5. Unsaturated water transmission parameters obtained from infiltration. In J. H. Dane & G. C. Topp (Co-Eds.), *Methods of soil analysis, Part 4 – Physical methods* (Number 5 in the Soil Science Society of America book series, pp. 879–889). Madison: Soil Science Society of America Inc.

Clothier, B., Scotter, D. (2002b). 3.5.4.2. Steady state observations. In J. H. Dane & G. C. Topp (Co-Eds.), *Methods of soil analysis, Part 4 – Physical methods* (Number 5 in the Soil Science Society of America book series, pp. 894–898). Madison: Soil Science Society of America Inc.

Cook, F. J., & Broeren, A. (1994). Six methods for determining sorptivity and hydraulic conductivity with disc permeameters. *Soil Science, 157*(1), 2–11.

Decagon Devices Inc. (2012). *Mini disk infiltrometer user's manual, version 10*. 2365 NE Hopkins Court, Pullman, WA 99163.

Decagon Devices Inc. (2014). *Mini disk infiltrometer user's manual, version September 10, 2014*. 2365 NE Hopkins Court, Pullman, WA 99163.

Di Prima, S. (2015). *Automated single ring infiltrometer with a low-cost microcontroller circuit. Computer and Electronics in Agriculture, 218*, 390–395.

Dohnal, M., Dusek, J., & Vogel, T. (2010). Improving hydraulic conductivity estimates from minidisk infiltrometer measurements for soils with wide pore-size distributions. *Soil Science Society of America Journal, 74*, 804–811.

Eijkelkamp Agrisearch Equipment. (2009). *09.09 Tension infiltrometer. User manual*. Giesbeek: Eijkelkamp Agrisearch Equipment B.V., 16 pp.

Elrick, D. E., & Reynolds, W. D. (1992a). Infiltration from constant-head well permeameters and infiltrometers. In G. C. Topp, W. D. Reynolds, & R. E. Green (Eds.), *Advances in measurement of soil physical properties: Bringing theory into practice* (SSSA special publication, no. 30, pp. 1–24). Madison: Soil Science Society of America, Inc.

Elrick, D. E., & Reynolds, W. D. (1992b). Methods for analyzing constant-head well permeameter data. *Soil Science Society of America Journal, 56*, 320–323.

Elrick, D. E., Angulo-Jaramillo, R., Fallow, D. J., Reynolds, W. D., & Parkin, G. W. (2002). Analysis of infiltration under constant head and falling head conditions. In P. A. C. Raats, D. Smiles, & A. W. Warrick (Eds.), *Environmental mechanics: Water, mass and energy transfer in the biosphere* (Geophysical monograph series, 129, pp. 47–53). Washington, DC: American Geophysical Union.

Everts, C. J., & Kanwar, R. S. (1993). Interpreting tension-infiltrometer data for quantifying soil macropores: Some practical considerations. *Transaction of the ASAE, 36*, 423–428.

Ferraris, S., Iovino, M., Bagarello, V. (2003). A comparison between analytically and numerically simulated tension infiltrometer data. In P. Piccarolo (Ed.), *Management and technology application to empower agriculture and agro-food system*. XXX CIOSTA-CIGR V Congress proceedings, Torino, pp. 1192–1201.

Fodor, N., Sándor, R., Orfanus, T., Lichner, L., & Rajkai, K. (2011). Evaluation method dependency of measured saturated hydraulic conductivity. *Geoderma, 165*, 60–68.

Fuentes, C., Haverkamp, R., & Parlange, J.-Y. (1992). Parameter constraints on closed-form soil-water relationships. *Journal of Hydrology, 134*, 117–142.

Gardner, W. R. (1958). Some steady-state solutions of the unsaturated moisture flow equation with application to evaporation from a water table. *Soil Science, 85*(4), 228–232.

Gee, G. W., & Or, D. (2002). 2.4 Particle-size analysis. In Dane J.H., & Topp G.C., (Co-Eds.), *Methods of soil analysis. Part 4, Physical methods* (SSSA book series 5, pp. 255–293). Madison: Soil Science Society of America, Inc.

Gonzalez-Merchan, C., Barraud, S., & Bedell, J.-P. (2014). Influence of spontaneous vegetation in stormwater infiltration system clogging. *Environmental Science and Pollution Research, 21*, 5419–5426.

Gonzalez-Sosa, E., Braud, I., Dehotin, J., Lassabatere, L., Angulo-Jaramillo, R., Lagouy, M., Branger, F., Jacqueminet, C., Kermadi, S., & Michel, M. (2010). Impact of land use on the hydraulic properties of the topsoil in a small French catchment. *Hydrological Processes, 24*, 2382–2399.

Haverkamp, R., Ross, P. J., Smettem, K. R. J., & Parlange, J. Y. (1994). Three-dimensional analysis of infiltration from the disc infiltrometer. 2. Physically based infiltration equation. *Water Resources Research, 30*, 2931–2935.

Haverkamp, R., Bouraoui, F., Zammit, C., Angulo-Jaramillo, R., & Delleur, J. W. (1999). Soil properties and moisture movement in the unsaturated zone. In J. W. Delleur (Ed.), *The handbook of groundwater engineering* (pp. 2931–2935). Boca Raton: CRC.

Haverkamp, R., Leij, F. J., Fuentes, C., Sciortino, A., & Ross, P. J. (2005). Soil water retention: I. Introduction of a shape index. *Soil Science Society of America Journal, 69*, 1881–1890.

Haverkamp, R., Debionne, S., Vialet, P., Angulo-Jaramillo, R., & de Condappa, D. (2006). Soil properties and moisture movement in the unsaturated zone. In J. W. Delleur (Ed.), *The handbook of groundwater engineering*. Boca Raton: CRC.

Homolák, M., Capuliak, J., Pichler, V., & Lichner, L. (2009). Estimating hydraulic conductivity of a sandy soil under different plant covers using minidisk infiltrometer and a dye tracer experiment. *Biologia, 64*, 600–604.

Hopmans, J. W., Šimůnek, J., Romano, N., & Durner, W. (2002). 3.6.2. Simultaneous determination of water transmission parameters and retention properties: Inverse methods. In J. H. Dane & G. C. Topp (Co-Eds.), *Methods of soil analysis, Part 4, Physical methods* (Number 5 in the Soil Science Society of America book series, pp. 963–1008). Madison: Soil Science Society of America, Inc.

Hussen, A. A., & Warrick, A. W. (1993a). Algebraic models for disc tension permeameters. *Water Resources Research, 29*(8), 2779–2786.

Hussen, A. A., & Warrick, A. W. (1993b). Alternative analyses of hydraulic data from disc tension infiltrometers. *Water Resources Research, 29*(12), 4103–4108.

Hwang, S. I., Lee, K. P., Lee, D. S., & Powers, S. E. (2002). Models for estimating soil particle size distribution. *Soil Science Society of America Journal, 66*, 1143–1150.

Iovino, M., Castellini, M., Bagarello, V., & Giordano, G. (2013). Using static and dynamic indicators to evaluate soil physical quality in a Sicilian area. *Land Degradation & Development*. Published online on Wiley Online Library, doi:10.1002/ldr.2263.

Jacques, D., Mohanty, B. P., & Feyen, J. (2002). Comparison of alternative methods for deriving hydraulic properties and scaling factors from single-disc tension infiltrometer measurements. *Water Resources Research, 38*(7), 1120, 25–14.

Jarvis, N. J., & Messing, I. (1995). Near-saturated hydraulic conductivity in soils of contrasting texture measured by tension infiltrometers. *Soil Science Society of America Journal, 59*, 27–34.

Jarvis, N., Koestel, J., Messing, I., Moeys, J., & Lindahl, A. (2013). Influence of soil, land use and climatic factors on the hydraulic conductivity of soil. *Hydrology and Earth System Sciences, 17*, 5185–5195.

Joel, A., & Messing, I. (2000). Application of two methods to determine hydraulic conductivity using disc permeameters on sloping land. *European Journal of Soil Science, 51*, 93–98.

Kirkham, M. B., & Clothier, B. E. (2000). Infiltration into a New Zealand native forest soil. In C. Rosenzwerg, S. Russo & D. Hillel (Eds.), *A spectrum of achievement in agronomy: Women fellows of the tri-societies* (ASA special publication, no. 62, pp. 13–26). Madison: American Society of Agronomy.

Lassabatere, L., Angulo-Jaramillo, R., Soria Ugalde, J. M., Cuenca, R., Braud, I., & Haverkamp, R. (2006). Beerkan estimation of soil transfer parameters through infiltration experiments – BEST. *Soil Science Society of America Journal, 70*, 521–532.

Lassabatere, L., Angulo-Jaramillo, R., Soria Ugalde, J. M., Šimůnek, J., & Haverkamp, R. (2009). Analytical and numerical modeling of water infiltration experiments. *Water Resources Research, 45*, W12415.

Lassabatere, L., Angulo-Jaramillo, R., Goutaland, D., Letellier, L., Gaudet, J. P., Winiarski, T., & Delolme, C. (2010a). Effect of settlement of sediments on water infiltration in two urban infiltration basins. *Geoderma, 156*, 316–325.

Lassabatere, L., Yilmaz, D., Angulo-Jaramillo, R., Soria Ugalde, J.,M., Braud, I., & Šimůnek, J. (2010a). *Numerical evaluation of inverse modeling methods for 1D and 3D water infiltration experiments in homogeneous soils*. Proceedings 19th World Congress of Soil Science, Soil Solution for a Changing World, 1–6 August, Brisbane, Australia.

Lassabatere, L., Loizeau, S., Angulo-Jaramillo, R., Winiarski, T., Rossier, Y., Delolme, C., & Gaudet, J. P. (2012). *Influence of the initial soil water content on Beerkan water infiltration experiments*. Geophysical Research Abstracts, Vol. 14, EGU2012-2278-2, EGU General Assembly.

Leji, F. J., Haverkamp, R., Fuentes, C., Zatarain, F., & Ross, P. J. (2005). Soil water retention: II. Derivation and application of shape index. *Soil Science Society of America Journal, 69*, 1891–1991.

Lewis, S. A., Wu, J. Q., & Robichaud, P. R. (2006). Assessing burn severity and comparing soil water repellency, Hayman Fire, Colorado. *Hydrological Processes, 20*(1), 1–16.

Li, X.-Y., González, A., & Solé-Benet, A. (2005). Laboratory methods for the estimation of infiltration rate of soil crusts in the Tabernas Desert badlands. *Catena, 60*, 255–266.

Lichner, L., Orfánus, T., Novákova, K., Šír, M., & Tesař, M. (2007). The impact of vegetation on hydraulic conductivity of sandy soil. *Soil & Water Research, 2*, 59–66.

Logsdon, S. D. (1997). Transient variation in the infiltration rate during measurement with tension infiltrometers. *Soil Science, 162*(4), 233–241.

Logsdon, S. D., & Jaynes, D. B. (1993). Methodology for determining hydraulic conductivity with tension infiltrometers. *Soil Science Society of America Journal, 57*, 1426–1431.

Logsdon, S. D., McCoy, E. L., Allmaras, R. R., & Linden, D. R. (1993). Macropore characterization by indirect methods. *Soil Science, 155*, 316–324.

Madsen, M. D., & Chandler, D. G. (2007). Automation and use of mini disk infiltrometers. *Soil Science Society of America Journal, 71*, 1469–1472.

Marquardt, D. W. (1963). An algorithm for least-squares estimation of nonlinear parameters. *SIAM Journal on Applied Mathematics, 11*, 413–441.

McKenzie, N. J., Cresswell, H. P., Rath, H., & Jacquier, D. (2001). Measurement of unsaturated hydraulic conductivity using tension and drip infiltrometers. *Australian Journal of Soil Research, 39*, 823–836.

Mecke, M., Westman, C. J., & Ilvesniemi, H. (2000). Prediction of near-saturated hydraulic conductivity in three podzolic boreal forest soils. *Soil Science Society of America Journal, 64*, 485–492.

Meshgi, A., & Chui, T. F. M. (2014). Analysing tension infiltrometer data from sloped surface using two-dimensional approximation. *Hydrological Processes, 28*(3), 744–752.

Minasny, B., & McBratney, A. B. (2000). Estimation of sorptivity from disc-permeameter measurements. *Geoderma, 95*, 305–324.

Minasny, B., & McBratney, A. B. (2007). Estimating the water retention shape parameter from sand and clay content. *Soil Science Society of America Journal, 71*, 1105–1110.

Mohanty, B. P., Ankeny, M. D., Horton, R., & Kanwar, R. S. (1994). Spatial analysis of hydraulic conductivity measured using disc infiltrometers. *Water Resources Research, 30*(9), 2489–2498.

Moret, D., & Arrúe, J. L. (2007). Characterizing soil water-conducting macro- and mesoporosity as influenced by tillage using tension infiltrometry. *Soil Science Society of America Journal, 71*, 500–506.

Moret, D., López, M. V., & Arrúe, J. L. (2004). TDR application for automated water level measurement from Mariotte reservoirs in tension disc infiltrometers. *Journal of Hydrology, 297*, 229–235.

Moret-Fernández, D., Blanco, N., Martínez-Chueca, V., & Bielsa, A. (2013). Malleable disc base for direct infiltration measurements using the tension infiltrometry technique. *Hydrological Processes, 27*(2), 275–283.

Mualem, Y. (1976). A new model for predicting the hydraulic conductivity of unsaturated porous media. *Water Resources Research, 12*, 513–522.

Mubarak, I., Mailhol, J. C., Angulo-Jaramillo, R., Ruelle, P., Boivin, P., & Khaledian, M. (2009a). Temporal variability in soil hydraulic properties under drip irrigation. *Geoderma, 150*, 158–165.

Mubarak, I., Mailhol, J. C., Angulo-Jaramillo, R., Bouarfa, S., & Ruelle, P. (2009b). Effect of temporal variability in soil hydraulic properties on simulated water transfer under high-frequency drip irrigation. *Agricultural Water Management, 96*, 1547–1559.

Mubarak, I., Angulo-Jaramillo, R., Mailhol, J. C., Ruelle, P., Khaledian, M., & Vauclin, M. (2010). Spatial analysis of soil surface hydraulic properties: Is infiltration method dependent? *Agricultural Water Management, 97*, 1517–1526.

Nasta, P., Lassabatere, L., Kandelous, M. M., Šimůnek, J., & Angulo-Jaramillo, R. (2012). Analysis of the role of tortuosity and infiltration constants in the Beerkan method. *Soil Science Society of America Journal, 76*, 1999–2005.

Parlange, J.-Y. (1972). Theory of water movement in soils: 4. Two and three dimensional steady infiltration. *Soil Science, 113*, 96–101.

Parlange, J.-Y. (1975). On solving the flow equation in unsaturated soils by optimization: Horizontal infiltration. *Soil Science Society of America Journal, 39*, 415–418.

Perroux, K. M., & White, I. (1988). Designs for disc permeameters. *Soil Science Society of America Journal, 52*, 1205–1215.

Philip, J. R. (1957). The theory of infiltration: 4. Sorptivity and algebraic infiltration equations. *Soil Science, 84*, 257–264.

Philip, J. R. (1969). Theory of infiltration. *Advances in Hydroscience, 5*, 215–296.

Philip, J. R. (1985). Approximate analysis of the borehole permeameter in unsaturated soil. *Water Resources Research, 21*, 1025–1033.

Philip, J. R. (1986). Linearized unsteady multidimensional infiltration. *Water Resources Research, 22*, 1717–1727.

Quadri, M. B., Clothier, B. E., Angulo-Jaramillo, R., Vauclin, M., & Green, S. R. (1994). Axisymmetric transport of water and solute underneath a disk permeameter: Experiments and numerical model. *Soil Science Society of America Journal, 58*, 696–703.

Ramos, T., Gonçalves, M. C., Martins, J. C., van Genuchten, M. T., & Pires, F. P. (2006). Estimation of soil hydraulic properties from numerical inversion of tension disk infiltrometer data. *Vadose Zone Journal, 5*, 684–696.

Reynolds, W. D. (2006). Tension infiltrometer measurements: Implications of pressure head offset due to contact sand. *Vadose Zone Journal, 5*, 1287–1292.

Reynolds, W. D. (2008). Chapter 82: Unsaturated hydraulic properties: Field tension infiltrometer. In M. R. Carter & E. G. Gregorich (Eds.), *Soil sampling and methods of analysis* (Canadian Society of Soil Science 2nd ed., pp. 1107–1127). Boca Raton: CRC Press.

Reynolds, W. D., & Elrick, D. E. (1990). Ponded infiltration from a single ring: I. Analysis of steady flow. *Soil Science Society of America Journal, 54*, 1233–1241.

Reynolds, W. D., & Elrick, D. E. (1991). Determination of hydraulic conductivity using a tension infiltrometer. *Soil Science Society of America Journal, 55*, 633–639.

Reynolds, W. D., & Elrick, D. E. (2002). 3.4.3.3 Constant head well permeameter (vadose zone). In J. H. Dane & G. C. Topp (Co-Eds.), *Methods of soil analysis, Part 4, Physical methods* (Number 5 in the Soil Science Society of America book series, pp. 844–858). Madison: Soil Science Society of America, Inc.

Reynolds, W. D., & Topp, G. C. (2008). Chapter 69: Soil water. Analyses, principles and parameters. In M. R. Carter & E. G. Gregorich (Eds.), *Soil sampling and methods of analysis* (Canadian Society of Soil Science 2nd ed.). Boca Raton: CRC Press.

Reynolds, W. D., & Zebchuk, W. D. (1996). Use of contact material in tension infiltrometer measurements. *Soil Technology, 9*, 141–159.

Reynolds, W. D., Gregorich, E. G., & Curnoe, W. E. (1995). Characterisation of water transmission properties in tilled and untilled soils using tension infiltrometers. *Soil & Tillage Research, 33*, 117–131.

Reynolds, W. D., Bowman, B. T., Brunke, R. R., Drury, C. F., & Tan, C. S. (2000). Comparison of tension infiltrometer, pressure infiltrometer, and soil core estimates of saturated hydraulic conductivity. *Soil Science Society of America Journal, 64*, 478–484.

Reynolds, W. D., Elrick, D. E., Youngs, E. G., Booltink, H. W. G., & Bouma, J. (2002). 3.4.2. Laboratory methods. In J. H. Dane & G. C. Topp (Co-Eds.), *Methods of soil analysis, Part 4 – Physical methods* (Number 5 in the Soil Science Society of America book series, pp. 802–817). Madison: Soil Science Society of America Inc.

Rogowski, A. S. (1971). Watershed physics: Model of soil characteristic. *Water Resources Research, 12*, 513–522.

Ronayne, M. J., Houghton, T. B., & Stednick, J. D. (2012). Field characterization of hydraulic conductivity in a heterogeneous alpine glacial till. *Journal of Hydrology, 458–459*, 103–109.

Scalenghe, R., & Ferraris, S. (2009). The first forty years of a Technosol. *Pedosphere, 19*(1), 40–52.

Schwartz, R. C., & Evett, S. R. (2002). Estimating hydraulic properties of a fine-textured soil using a disc infiltrometer. *Soil Science Society of America Journal, 66*, 1409–1423.

Schwartz, R. C., & Evett, S. R. (2003). Conjunctive use of tension infiltrometry and time-domain reflectometry for inverse estimation of soil hydraulic properties. *Vadose Zone Journal, 2*, 530–538.

Schwärzel, K., & Punzel, J. (2007). Hood infiltrometer—A new type of tension infiltrometer. *Soil Science Society of America Journal, 71*(5), 1438–1447.

Scotter, D. R., Clothier, B. E., & Harper, E. R. (1982). Measuring saturated hydraulic conductivity and sorptivity using twin rings. *Australian Journal of Soil Research, 20*, 295–304.

Šimůnek, J., & van Genuchten, M. T. (1996). Estimating soil hydraulic properties from tension-disc infiltrometer data by numerical inversion. *Water Resources Research, 32*, 2683–2696.

Šimůnek, J., & van Genuchten, M. T. (1997). Estimating unsaturated soil hydraulic properties from multiple tension disc infiltrometer data. *Soil Science, 162*(6), 383–398.

Šimůnek, J., Angulo-Jaramillo, R., Schaap, M. G., Vandervaere, J. P., & van Genuchten, M. T. (1998a). Using an inverse method to estimate the hydraulic properties of crusted soils from tension disc infiltrometer data. *Geoderma, 86*, 61–81.

Šimůnek, J., van Genuchten, M. T., Gribb, M. M., & Hopmans, J. W. (1998b). Parameter estimation of unsaturated soil hydraulic properties from transient flow processes. *Soil & Tillage Research, 47*, 27–36.

Šimůnek, J., Wang, D., Shouse, P. J., & van Genuchten, M. T. (1998c). Analysis of field tension disc infiltrometer data by parameter estimation. *International Agrophysics, 12*, 167–180.

Šimůnek, J., Wendroth, O., & van Genuchten, M. T. (1999a). Estimating unsaturated soil hydraulic properties from laboratory tension disc infiltrometer experiments. *Water Resources Research, 35*(10), 2965–2979.

Šimůnek, J., Sejna, M., & van Genuchten, M. Th. (1999b). *HYDRUS-2D. Simulating water flow and solute transport in two-dimensional variably saturated media*. Version 2.0. Golden: International Ground Water Modeling Center, 227 pp.

Šimůnek, J, van Genuchten, M. T., & Šejna, M. (2006). *The HYDRUS software package for simulating two- and three-dimensional movement of water, heat, and multiple solutes in variably-saturated media*. Technical manual, Version 1.0. Prague: PC Progress, 241 pp.

Smettem, K. R. J., & Clothier, B. E. (1989). Measuring unsaturated sorptivity and hydraulic conductivity using multiple disc permeameters. *Journal of Soil Science, 40*, 563–568.

Smettem, K. R. J., & Smith, R. E. (2002). Field measurement of infiltration parameters. In R. E. Smith (Ed.), *Infiltration theory for hydrologic application* (Water resources monograph, 15, pp. 135–157). Washington, DC: American Geophysical Union.

Smettem, K. R. J., Parlange, J. Y., Ross, J. P., & Haverkamp, R. (1994). Three-dimensional analysis of infiltration from disc infiltrometer. 1. A capillary-based theory. *Water Resources Research, 30*, 2925–2929.

Smettem, K. R. J., Ross, P. J., Haverkamp, R., & Parlange, J. Y. (1995). Three-dimensional analysis of infiltration from the disk infiltrometer. 3. Parameter estimation using a double-disk tension infiltrometer. *Water Resources Research, 31*(10), 2491–2495.

Smettem, K. R. J., Ross, P. J., Haverkamp, R., Parlange, J. Y., & Gregory, P. J. (1998). Laboratory and field application of a twin disc infiltrometer for measurement of soil hydraulic properties. In P. Dillon & I. Simmer (Eds.), *Shallow groundwater systems* (IAH International contribution to hydrogeology, 18, pp. 41–51). London: Taylor & Francis.

Smiles, D. E., & Knight, J. H. (1976). A note on the use of the Philip infiltration equation. *Australian Journal of Soil Research, 14*, 103–108.

Soilmoisture Equipment Corp. (2008a, June). *2826D20 Tension infiltrometer. Operating instructions*. Santa Barbara: Soilmoisture Equipment Corp.

Soilmoisture Equipment Corp. (2008b, July). *2825K1 Tension infiltrometer. Operating instructions*. Santa Barbara: Soilmoisture Equipment Corp.

Soracco, C. G., Lozano, L. A., Sarli, G. O., Gelati, P. R., & Filgueira, R. R. (2011). Using tension disc infiltrometer to determine infiltration and water-conducting macroporosity and mesoporosity relationships in an agricultural silty loam soil. *Soil Science, 176*(9), 459–463.

Souza, E. S., Antonino, A. C. D., Heck, R. J., Montenegro, S. M. G. L., Lima, J. R. S., Sampaio, E. V. S. B., Angulo-Jaramillo, R., & Vauclin, M. (2014). Effect of crusting on the physical and hydraulic properties of a soil cropped with Castor beans (Ricinus communis L.) in the northeastern region of Brazil. *Soil & Tillage Research, 141*, 55–61.

Sullivan, M., Warwick, J. J., & Tyler, S. W. (1996). Quantifying and delineating spatial variations of surface infiltration in a small watershed. *Journal of Hydrology, 181*, 149–168.

Thony, J. L., Vachaud, G., Clothier, B. E., & Angulo-Jaramillo, R. (1991). Field measurements of the hydraulic properties of soil. *Soil Technology, 4*, 111–123.

Touma, J., Voltz, M., & Albergel, J. (2007). Determining soil saturated hydraulic conductivity and sorptivity from single ring infiltration tests. *European Journal of Soil Science, 58*(1), 229–238.

Turner, N. C., & Parlange, J. Y. (1978). Lateral movement at the periphery of a one-dimensional flow of water. *Soil Science, 118*, 70–77.

UGT GmbH. (2012). *IL-2700 Hood infiltrometer. Manual*. Umwelt-Geräte-Technik GmbH, www.ugt-online.de.

van Genuchten, M. T. (1980). A closed-form equation for predicting the hydraulic conductivity of unsaturated soils. *Soil Science Society of America Journal, 44*, 892–898.

van Genuchten, M. T., & Leij, F. J. (1992). On estimating the hydraulic properties of unsaturated soils. In M. T. van Genuchten, F. J. Leij, & F. J. Lund (Eds.), *Indirect methods for estimating the hydraulic properties of unsaturated soils* (pp. 1–14). Riverside: University of California.

van Genuchten, M. Th., Leij, F. J., & Yates, S. R. (1991). *The RETC code for quantifying the hydraulic functions of unsaturated soils*. Environmental Protection Agency, EPA/600/2-91/065.

Vandervaere, J. P. (2002). 3.5.4.1. Early time observations. In J. H. Dane & G. C. Topp (Co-Eds.), *Methods of soil analysis, Part 4 – Physical methods* (Number 5 in the Soil Science Society of America book series, pp. 889–894). Madison:Soil Science Society of America Inc.

Vandervaere, J. P., Peugeot, C., Vauclin, M., Angulo-Jaramillo, R., & Lebel, T. (1997). Estimating hydraulic conductivity of crusted soils using disc infiltrometers and minitensiometers. *Journal of Hydrology, 188–189*, 209–223.

Vandervaere, J.-P., Vauclin, M., & Elrick, D. E. (2000a). Transient flow from tension infiltrometers: I. The two-parameter equation. *Soil Science Society of America Journal, 64*, 1263–1272.

Vandervaere, J.-P., Vauclin, M., & Elrick, D. E. (2000b). Transient flow from tension infiltrometers: II. Four methods to determine sorptivity and conductivity. *Soil Science Society of America Journal, 64*, 1272–1284.

Vauclin, M., Elrick, D. E., Thony, J. L., Vachaud, G., Revol, P., & Ruelle, P. (1994). Hydraulic conductivity measurements of the spatial variability of a loamy soil. *Soil Technology, 7*, 181–195.

Ventrella, D., Losavio, N., Vonella, A. V., & Leij, F. J. (2005). Estimating hydraulic conductivity of a fine-textured soil using tension infiltrometry. *Geoderma, 124*, 267–277.

Wang, D., Yates, S. R., & Ernst, F. F. (1998a). Determining soil hydraulic properties using tension infiltrometers, time domain reflectometry, and tensiometry. *Soil Science Society of America Journal, 62*, 318–325.

Wang, D., Yates, S. R., Lowery, B., & van Genuchten, M. T. (1998b). Estimating soil hydraulic properties using tension infiltrometers with varying disk diameters. *Soil Science, 163*, 356–361.

Warrick, A. W. (1992). Models for disc infiltrometers. *Water Resources Research, 28*(5), 1319–1327.

Warrick, A. W., & Broadbridge, P. (1992). Sorptivity and macroscopic capillary length relationships. *Water Resources Research, 28*, 427–431.

Watson, K. W., & Luxmoore, R. J. (1986). Estimating macroporosity in a forest watershed by use of a tension infiltrometer. *Soil Science Society of America Journal, 50*, 578–582.

Weir, G. J. (1986). Steady infiltration from large shallow ponds. *Water Resources Research, 22*, 1462–1468.

Weir, G. J. (1987). Steady infiltration from small shallow circular ponds. *Water Resources Research, 23*, 733–736.

White, I., & Perroux, K. M. (1987). Use of sorptivity to determine field soil hydraulic properties. *Soil Science Society of America Journal, 51*, 1093–1101.

White, I., & Sully, M. J. (1987). Macroscopic and microscopic capillary length and time scales from field infiltration. *Water Resources Research, 23*(8), 1514–1522.

White, I., Sully, M. J., & Perroux, K. M. (1992). Measurement of surface-soil hydraulic properties: Disk permeameters, tension infiltrometers, and other techniques. In G. C. Topp, W. D. Reynolds, & R. E. Green (Eds.), *Advances in measurement of soil physical properties: Bringing theory into practice* (SSSA special publication, no. 30, pp. 69–103). Madison: Soil Science Society of America, Inc.

Wooding, R. A. (1968). Steady infiltration from a shallow circular pond. *Water Resources Research, 4*, 1259–1273.

Wu, L., Pan, L., Mitchell, J., & Sanden, B. (1999). Measuring saturated hydraulic conductivity using a generalized solution for single-ring infiltrometers. *Soil Science Society of America Journal, 63*, 788–792.

Xu, X., Kiely, G., & Lewis, G. (2009). Estimation and analysis of soil hydraulic properties through infiltration experiments: Comparison of BEST and DL fitting methods. *Soil Use and Management, 25*, 354–361.

Xu, X., Lewis, C., Liu, W., Albertson, J. D., & Kiely, G. (2012). Analysis of single-ring infiltrometer data for soil hydraulic properties estimation: Comparison of BEST and Wu methods. *Agricultural Water Management, 107*, 34–41.

Yilmaz, D., Lassabatere, L., Angulo-Jaramillo, R., Deneele, D., & Legret, M. (2010). Hydrodynamic characterization of basic oxygen furnace slag through an adapted BEST method. *Vadose Zone Journal, 9*, 1–10.

Zhang, R. (1997a). Determination of soil sorptivity and hydraulic conductivity from the disk infiltrometer. *Soil Science Society of America Journal, 61*, 1024–1030.

Zhang, R. (1997b). Infiltration models for the disk infiltrometer. *Soil Science Society of America Journal, 61*, 1597–1603.

Zhang, Y., Smith, R. E., Butters, G. L., & Cardon, G. E. (1999). Analysis and testing of a concentric-disk tension infiltrometer. *Soil Science Society of America Journal, 63*, 544–553.

Chapter 4
Soils with Specific Features

4.1 Water Repellent Soils

Water infiltration into a water repellent soil is more difficult and will be slower and more variable than into a wettable soil. Related effects of soil water repellency are more runoff and erosion in hilly areas, large variations in soil water content even at short distances, reduced irrigation efficiency, plant available water and hence agricultural crop production, and increased risk of leaching and transport of compounds to groundwater (Ritsema et al. 2008). Infiltration ultimately takes place in water repellent soils, but along specific pathways, commonly named fingers, instead of moving in a uniform wetting front. Moreover, wetting the dry water repellent soil zones between the preferential flow paths is very difficult. Soil water repellency is not geographically or climatically dependent and it can be of consequence anywhere plants are grown or where any source of organic matter (including pollution) may occur. Once an area has become water repellent, its wettability will not necessarily return naturally, even after cool and wet periods (Ritsema et al. 1993). Rather, it is often the case that the severity of the repellency will continue to exist and may in fact increase (Cisar et al. 2000; Kostka 2000).

4.1.1 Causes of Hydrophobicity

The causes of soil water repellency are numerous. Exudates produced by living matter in soil, such as plant roots and microbes, form hydrophobic surface films on soil particles, particularly after drying or heating due, for example, to forest fire, which result in a non-zero contact angle between water and soil (Hallett et al. 2001; Jarvis et al. 2008). Most microbial activity takes place on the surface of the soil aggregates, where the microbial substrates are most available (Hallett and Young 1999). Soil may also contain hydrophobic organic matter and wax from plant

R. Angulo-Jaramillo et al., *Infiltration Measurements for Soil Hydraulic Characterization*,
DOI 10.1007/978-3-319-31788-5_4

leaves. Regular irrigation with treated sewage water can also lead to the development of soil water repellency due to dissolved substances in sewage water (Wallach et al. 2005).

In severe cases, where the coverage of particles by hydrophobic surface films is abundant and the contact angle is higher than 90°, water droplets cannot spontaneously infiltrate the soil (Letey et al. 2000). Soils exhibiting the extreme condition of completely impeded water infiltration are not widespread, leading to the assumption that non-water repellent behavior is the norm (Wallis and Horne 1992; Hallett et al. 2001). In addition to extreme repellency, however, subcritical repellency, i.e. a less extreme manifestation of the same phenomenon, can also occur when the water-solid contact angle is less than 90° but not zero (Tillman et al. 1989). The soil does wet spontaneously when water is applied at zero or at a negative pressure head, but the rate of wetting is reduced by repellency since organic coatings on soil particles increase the contact angle. According to Marshall and Holmes (1979), contact angles greater than zero probably affect the advance of water quite commonly during the wetting of dry soil. The effects are easily observed only in severe cases of water repellency but several authors suggested that most soils exhibit subcritical water repellency (Wallis et al. 1991; Hallett and Young 1999; Hallett et al. 2001).

Soils with coarser grain-size distribution are considered to be more prone to developing soil water repellency due to their small surface area, which easily becomes coated in hydrophobic substances (DeBano 1981). However, water flow and solute transport in clay soils is dominated by structural macropores, which also have a limited contact surface area with the rest of the soil (Jarvis et al. 2008). Furthermore, aggregate surfaces and macropore linings may often be coated with hydrophobic organic material. McGhie and Posner (1980) demonstrated that clay soils could indeed become water repellent and Jarvis et al. (2008) concluded that water repellency in undisturbed clay soils can have significant effects on the occurrence of non-equilibrium water and solute transport in macropores. Therefore, water repellency is a widespread phenomenon that is not confined to sandy soils.

A decrease in water repellency with soil depth can be expected due to the reduced organic matter content at larger depths (Doerr et al. 2000; Vogelmann et al. 2010), but a high repellency can also be detected in the subsurface soil layers because hydrophobic compounds that are not strongly adsorbed on the soil mineral fraction migrate with the draining water to the lower layers of the profile (Vogelmann et al. 2013).

Lower soil water repellency is expected in wet soils compared with dry soils because heating and drying of soil cause wettable (hydrophilic) heads of amphiphilic organic molecules to bond to themselves and soil particles, leaving mainly hydrophobic portions exposed and exacerbating soil water repellency (Dekker and Ritsema 1994; Hunter et al. 2011). Oven-drying of moist samples resulted in much higher levels of hydrophobicity than air-drying in the investigations by Ma'shum and Farmer (1985) and Franco et al. (1995), and Doerr (1998) suggested that, in the experimental determination of water repellency, soil samples should be air-dried rather than oven-dried in order to avoid the possibility of enhancing their degree of hydrophobicity. Repellency can be re-established when wet soil dries below a soil water content threshold value (Ebel and Moody 2013) but a study by Doerr and

Thomas (2000) showed that this is not a common occurrence. According to these authors, complete re-establishment of hydrophobicity after wetting needs a new input and/or redistribution of hydrophobic substances, mainly related to biological activity in or near the root zone during or after the soils dry out. As a matter of fact, the relationship between water repellency and moisture content is not completely understood (Hunter et al. 2011), but it is reasonable to presume that the infiltration data collected with an infiltrometer experiment can vary with the soil water content at the time of the experiment due to its effect on soil water repellency. Consequently, measurements of water repellency can be expected to be useful to properly interpret the infiltration process established by an infiltrometer technique.

Wildfire and vegetation type also play important roles in the occurrence of water repellency (Cerdà and Doerr 2007). Fire induces or enhances repellency, but it also eliminates it when soil heating exceeds a critical threshold. Repellency appears common under rangeland or forestland but it can also occur on agricultural land.

High hydrophobicity levels determine larger and more stable aggregates because wetting is less important, thus reducing energy release and buildup of air pressure in pores (i.e., slaking), and because the cohesive forces between particles act for longer periods (Hallett et al. 2001; Vogelmann et al. 2013). Moreover, accumulation of hydrophobic substances released during burning on the surface of the aggregates acts as a cementing agent, increasing their stability (Mataix-Solera et al. 2011). Hunter et al. (2011) stated that subcritical soil water repellency is beneficial for stabilizing soil structure, preventing dispersion and minimizing erosion.

4.1.2 Hydrophobicity and Water Infiltration

Water repellency complicates the infiltration process, as shown in the examples of Fig. 4.1a (Imeson et al. 1992; Ebel and Moody 2013). Type 1 pattern is the expected pattern from traditional infiltration theory. Type 2 pattern, which is not explained by the traditional theory, has initially low infiltration rates due to initially high soil water repellency but, as soil wets, the infiltration rate increases to a maximum and then declines exponentially with time. Type 3 is like type 2 except that the infiltration rate eventually levels off to a constant value because of macropore flow. Other studies have found other kinds of shape, characterized by only an increase in infiltration rate with time or a decrease at the beginning, followed by an increase before stabilizing at a steady value (Fig. 4.1b; Lassabatere et al. 2012). Other typical examples can be found in Beatty and Smith (2013), where infiltration was investigated in several materials according to the content and the type of organic matter (Fig. 4.1c). These typical shapes reveal hydrophobicity and related cumulative infiltration cannot be treated with usual models, such as that included in the so-called BEST method (Lassabatere et al. 2006). This is one of the causes of failure of traditional methods for soils with high organic matter content.

Anyway, hydrophobicity reduces water infiltration and this effect is more drastic at very low water contents. As a result, Moody and Ebel (2012) showed that a minimum in infiltration rate can occur in what they called hyper-dry conditions (i.e. initial soil water content of less than 0.02 m^3m^{-3}) where traditional theory predicts a maximum. According to Ebel and Moody (2013), if soil water repellency removes the contribution of capillarity from infiltration, then initial infiltration rates would be controlled by gravity, which results in minimal flow due to the very low hydraulic conductivities at such low soil water contents.

Hydrophobicity makes the wetting front advance rapid in random zones, creating a finger wetting pattern (Wang et al. 2000; Letey 2001). For a water repellent soil, Ritsema et al. (1998) stated that fingered flow of water in soils is induced by the instability of the infiltrating wetting front that develops when the pressure gradient for the water phase is opposite (upward) to the direction of flow (downward). Unstable flow can occur for hydrophobic soils, particularly if the wetting front reaches a stratum so water repellent that the pressure head needed to penetrate it is positive (Philip 1975).

The depth of ponded water, h_0, has relatively little effect on infiltration into a wettable soil, but it has pronounced effects on water flow in a water repellent soil (Letey 2001). Water will not infiltrate until h_0 exceeds h_{we}, i.e. the water entry pressure head. Thereafter, the temporal infiltration rate is influenced by the value of h_0. For a stable water repellent condition, i.e. not changing with the time of exposure to water, the h_0/h_{we} ratio controls infiltration rates over time. For example, Feng et al. (2001) found that, for a material having $h_{we} = 8.4$ cm, infiltration rates increased with time when h_0 was 15 or 20 cm. At higher values of h_0, the infiltration rate decreased with time, typical of what would be observed on a wettable soil. Carrillo et al. (2000a) also considered that the depth of the water repellent layer below the soil surface, L, has a comparable effect to that of h_0 on finger development and found that no water penetrated the water repellent layer for values of $(h_0 + L)/h_{we} < 1$, unstable flow developed for values between 1 and 1.5, and a stable wetting front developed for values >1.5. However, temporal infiltration behavior is confounded by the transient nature of repellency, determining a shifting of the contact angle from greater than 90° to less than 90° after some time of contact with

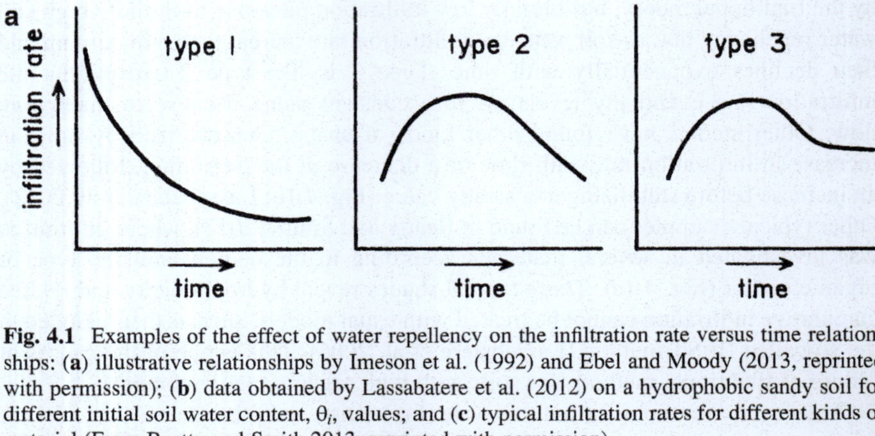

Fig. 4.1 Examples of the effect of water repellency on the infiltration rate versus time relationships: (**a**) illustrative relationships by Imeson et al. (1992) and Ebel and Moody (2013, reprinted with permission); (**b**) data obtained by Lassabatere et al. (2012) on a hydrophobic sandy soil for different initial soil water content, θ_i, values; and (**c**) typical infiltration rates for different kinds of material (From Beatty and Smith 2013, reprinted with permission)

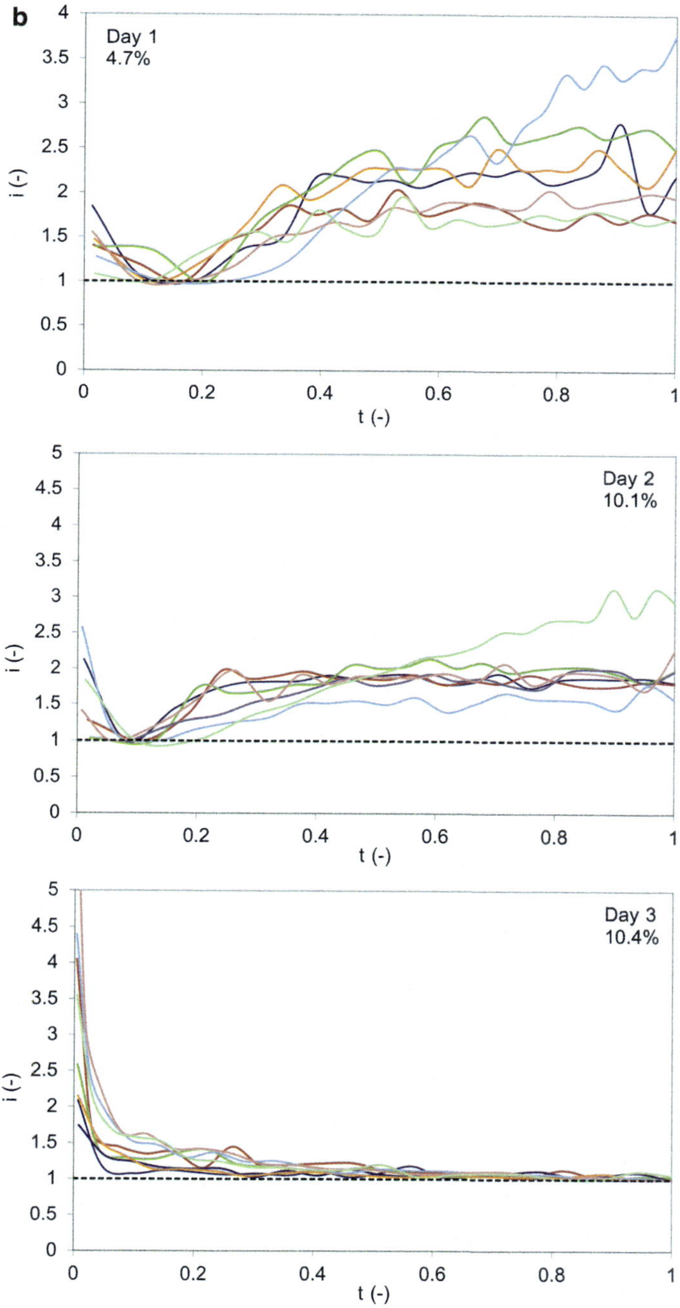

Fig. 4.1 (continued)

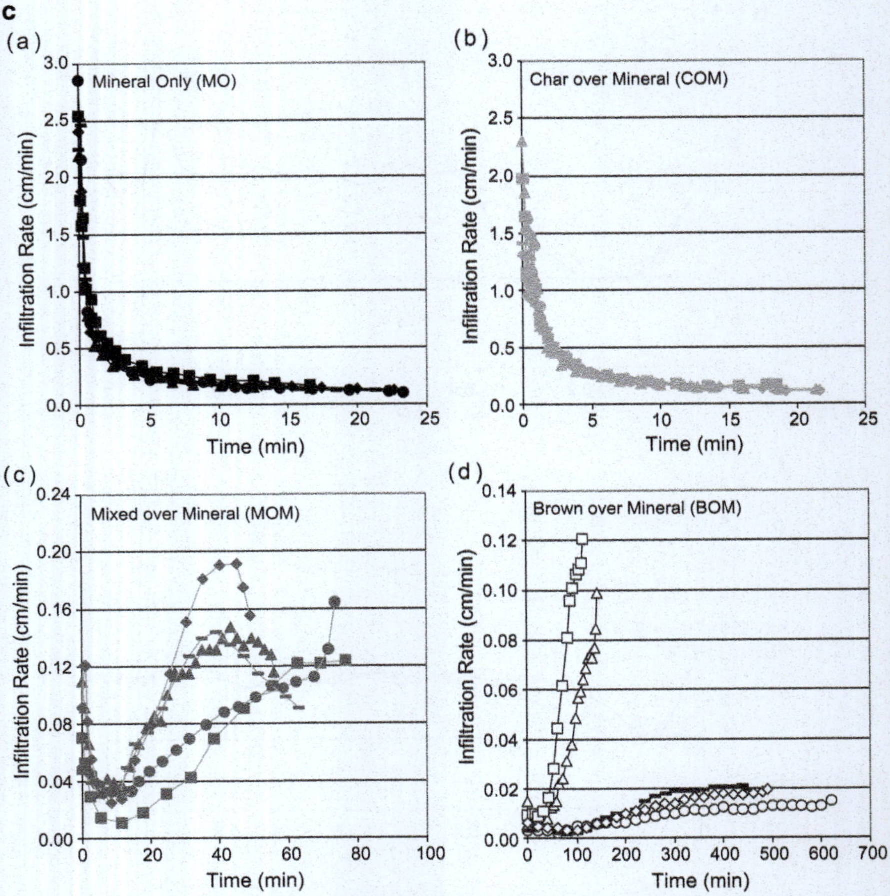

Legend: infiltration rates as a function of time for 20 one-dimensional infiltration columns filled with mineral only (a), char over material (b), mixed over mineral (c) and brown over mineral configuration (d).

Fig. 4.1 (continued)

water. In this case, the water drop penetration time, WDPT, which is the time required to initiate water infiltration in the soil even without a ponded depth (see below) controls unstable flow (Carrillo et al. 2000b).

4.1.3 Hydraulic Characterization of Water Repellent Soils

Reductions in saturated hydraulic conductivity are common in fire-affected soils and one major problem with characterizing infiltration in these water repellent soils is that certain measurement techniques appear to bias results (Ebel and Moody

2013). For example, single-ring infiltrometer measurements with a ponded head of several centimeters may overwhelm soil water repellency and give infiltration rates similar to unburned soils while data collected with a tension infiltrometer at a negative applied pressure head can suggest that fire-affected soils are essentially impermeable (Cerdà 1996; Nyman et al. 2010; Ebel et al. 2012). Moreover, making infiltration measurements in a fire-affected environment where flux is near zero is extremely difficult as negligible errors for large infiltration rates become large errors for near-zero fluxes. Feng et al. (2001) found that the measured hydraulic conductivity of a sand treated to become water repellent was equal to that of the untreated sand when the value of h_0/h_{we} was $\geq$3.1. Below this threshold, the measure of hydraulic conductivity was lower, which was linked to partial water saturation of the sand as the result of hydrophobicity (Letey 2001).

Soil water repellency has a dynamic nature and therefore both the persistence and the degree of the repellency should be determined. Persistence is how long the soil remains repellent in the presence of water whereas the degree of soil water repellency describes how strongly infiltration is inhibited (Hunter et al. 2011). Natural water repellent media are fractionally wettable systems in which some proportion of the particles or particle surfaces are water wettable while the other particles or particle surfaces are water repellent (Beatty and Smith 2013). Even a small percentage of coated soil particles can be sufficient to change a wettable soil into a water repellent medium (Steenhuis et al. 2005). Contact angle dynamics, speaking directly to the persistence of repellency, generates time dependent changes in relative proportions of wettable and non-wettable fractions. The complex interplay between the relative fractions of wettable and non-wettable materials and how those proportions shift over time with the dynamics of contact angles is the fundamental nature of water repellent systems. Combining the concepts of fractional wettability and contact angle dynamics is necessary to understand hydrological processes in these systems (Beatty and Smith 2013, 2014).

A quantitative characterization of soil water repellency can be obtained either in the field or in the laboratory by establishing particular infiltration processes on the soil to be tested (e.g., Letey et al. 2000). The Water Drop Penetration Time test, specifically suited for characterizing persistence, the Molarity of an Ethanol Droplet and the Repellency tests, yielding data on the degree of soil water repellency, are among the most common approaches to characterize water repellency (Letey 2001; Hunter et al. 2011). Alternative approaches have also been developed for characterizing repellency with tension infiltrometer experiments.

The Water Drop Penetration Time (WDPT) test (Letey 1969; Doerr 1998) involves placing a water drop on a soil surface and recording the time taken for its complete penetration. The WDPT can vary from instant penetration to many hours and it increases with the persistence of severe hydrophobicity (contact angle >90°; Wallis and Horne 1992; Jarvis et al. 2008). To be clearer, a water drop that does not initially penetrate indicates that the water-soil contact angle is equal to or greater than 90°. The fact that the water drop eventually penetrates indicates that, with time of exposure to water, the soil surface changes so that the contact angle decreases from more than 90° to less than 90°. Thus, the WDPT test, measuring the time required for the contact angle to change from its original value (>90°) to a value approaching 90°, characterizes the dynamic nature rather than the static state of the repellency (Letey et al. 2000; Letey 2001). To cover all possible WDPTs, the test should be extended until all droplets had infiltrated but this can become very time-consuming and ultimately evaporation of droplets will influence results for very hydrophobic samples.

Doerr (1998) suggested to apply a pre-established number (from 5 to 15) of drops of distilled water at 20 °C on the soil surface of a given sample using a hypodermic syringe. Ten drops (0.05 mL) of distilled water can be used according to Buczko et al. (2002) and Cerdà and Doerr (2007). The penetration time of each drop is recorded and the representative WDPT value is assumed to be the median or the mean of the different time measurements. For laboratory testing of sieved and thus thoroughly mixed samples, a low replicate variability is expected and 3–5 drops can be enough to obtain representative data. A standard protocol regarding the number and volume of applied drops was proposed by Hallin et al. (2013) who suggested that a reliable mean WDPT value (±10 % of the true value) can be obtained with 95 % confidence from 30 drops of 80–200 μL size. Samples with a WDPT in excess of 1 h are covered with lids to prevent drop evaporation, allowing the test to be extended to more than 5 h. However, the duration of the run varies widely with the investigation. For example, recording was stopped after 3600 s by Buczko et al. (2002) and after several minutes in other investigations (Doerr 1998), including that of Lewis et al. (2006) measuring the persistence of the drop on the surface for up to 300 s. Numerous, essentially arbitrary, classifications exist relating WDPT categories to severity levels in hydrophobicity. For example, eleven categories were considered by Doerr (1998), i.e. <5, 5–10, 10–30, 30–60, 60–180, 180–300, 300–600, 600–900, 900–3600, 3600–18,000 and >18,000 s. In the investigations by Buczko et al. (2002) and Cerdà and Doerr (2007), WDPT values of ≤5 s were taken as indicative of no water repellency, 5–60 s as slight, 60–600 s as strong, 600–3600 s as severe and >3600 s as extreme water repellency. Surface water repellency was divided into weak (0–60 s), moderate (61–180 s) and strong (181–300 s) by Lewis et al. (2006). The WDPT test can be applied both in the field and in the laboratory. For laboratory testing by Doerr (1998), soil was air-dried and carefully hand-sieved to remove the skeletal fraction. Approximately 10 g of soil was put into clear circular plastic dishes (50 mm diameter, 10 mm depth) and surfaces were carefully smoothed by hand. Drops were applied from a height of not more than 5 mm to avoid excessive kinetic energy affecting soil-droplet interaction. Buczko et al. (2002) used 100 cm^3 undisturbed soil samples and they discriminated between a potential and an actual water repellency. Potential water repellency was measured on soil samples dried in an oven at 45 °C for 3 days and then equilibrated for approximately 24 h with the ambient temperature. Measurements of actual repellency were made on moist samples, i.e. having the soil water content at the time of sampling. Measurement of the WDPT can be rather subjective when the drop is slowly adsorbed or it is covered in fine dust (Lewis et al. 2006).

The Molarity of an Ethanol Droplet (MED) test, also known as Percentage Alcohol test or Ethanol Percentage test (Watson and Letey 1970; Doerr 1998;

Buczko et al. 2002), is an indirect measure of the surface tension of the soil surface and indicates how strongly a water drop is repelled by a soil at the time of application, i.e. how strongly it will ball up (Doerr 1998). This test is based on the concept that the effective contact angle for ethanol is zero regardless of whether or not the soil is water-repellent (Tillman et al. 1989) and that the higher the concentration of ethanol in a liquid droplet that is adsorbed within a specified time, the higher the degree of water repellency. A droplet with a higher surface tension (lower ethanol concentration) than that of the soil surface will remain on it for some time whereas a droplet with a lower surface tension (higher ethanol concentration) will infiltrate instantly. Therefore, drops with decreasing ethanol concentrations are applied on the soil surface until a drop resists infiltration, allowing the classification of the soil into a surface tension category between two ethanol concentrations. If small concentration increments are used, the surface tension at the time of initial contact can be determined quite accurately. The time that is allowed for the drop to infiltrate varies with the author from 3 to 10 s (King 1981; Crockford et al. 1991; Harper and Gilkes 1994; Doerr 1998; Buczko et al. 2002) but Doerr (1998) suggested that keeping the penetration time short is important to avoid the possible decay of hydrophobicity affecting the results. In the study by this last author, drops (0.05 mL) of solutions with different ethanol concentrations, equal to 0, 3, 5, 8.5, 13, 24 and 36 % by volume, were applied onto a smoothed soil surface from a height of not more than 5 mm using medical droppers. The median of five MED measurements undertaken within an area of approximately 100 cm^2 was taken as being representative of the sample. In the investigation by Buczko et al. (2002), the used ethanol concentrations were 0, 1, 2, 3, 4, 5, 6, 8, 10, 15, 20, 30 and 40 % by volume. The resulting degree of water repellency may be expressed either as a liquid surface tension, as the molarity of the infiltrating ethanol droplet or as the lowest percentage of the solution that infiltrates the soil in less than the chosen penetration time. In fact, the surface tension of the water-ethanol solution, that is non-linearly related with the ethanol percentage, is the physically more fundamental property for characterizing the degree of hydrophobicity of solid surfaces (Watson and Letey 1970; King 1981; Doerr 1998; Letey et al. 2000; Buczko et al. 2002). However, this parameter is more complicated to estimate, explaining the choice of the concentration of ethanol.

A satisfactory correspondence between the results of the WDPT and MED tests was obtained by some authors (e.g., Crockford et al. 1991; Harper and Gilkes 1994) but not in other investigations (Dekker and Ritsema 1994). Doerr (1998) suggested that the correspondence between the two tests should be reasonably good for highly hydrophobic soils and worse for moderately hydrophobic soils. In the investigation by Buczko et al. (2002), the results obtained with the two tests were positively correlated. Yet, a considerable scattering was observed and attributed to differences in the (i) physical meaning of the two tests, i.e. persistence for the WDPT test and degree of water repellency for the MED test, and (ii) applied measurement process and support volumes. In particular, the WDPT test relates to the soil surface with a single water droplet whereas the MED test makes use of several droplets that are applied at neighboring locations on the soil surface. In other terms, small-scale heterogeneity has to be averaged with this latter test. An advantage of the MED test over the WDPT test is its speed so that it is well suited to field investigations where long persistence times make the WDPT technique impossibly laborious (Wallis et al. 1991). Yet, this last test has a more direct hydrological relevance (Doerr 1998). According to Beatty and Smith (2014), the MED test, approximating the surface tension required to initiate infiltration instantly, and the WDPT test, measuring the persistence of water repellency when the soil is exposed to water, are appropriate to capture fundamental information about fractionally wettable and dynamically changing materials through simple and reliable methodologies.

Soils exhibiting subcritical water repellency are difficult to detect with conventional water repellency tests such as the WDPT test because they appear to uptake water readily. Tillman et al. (1989) overcame this problem by developing a sensitive and physically meaningful measurement of water repellency based on sorptivity. By comparing the sorptivity of water against a liquid not influenced by repellency, they were able to define a repellency index that was directly proportional to the reduced infiltration rate (Hallett et al. 2001). As stated before, ethanol readily infiltrates hydrophobic soil because of the solid-liquid contact properties. Its sorptivity therefore provides a measurement of liquid transport in soil that is not influenced by water repellency and is representative of the pore structure. Hallett and Young (1999) combined the miniature, laboratory infiltration device by Leeds-Harrison and Youngs (1997) with the repellency technique developed by Tillman et al. (1989) in order to assess the extent of subcritical water repellency of individual soil aggregates of 10–30 mm in diameter (Hallett et al. 2001; Vogelmann et al. 2013). In particular, water infiltrates into a soil aggregate from a small area (3 or 4 mm in diameter) which produces an expanding wetting bulb that does not reach the boundary of the aggregate during measurement. A sponge tip allowed the establishment of a negative pressure head equal to −2 cm in the investigations by Hallett and Young (1999) and Hallett et al. (2001) in order to reduce macropore flow. Infiltration is measured from the mass loss of water in a reservoir, using a balance accurate to 1 mg. Errors due to evaporative loss during the testing time can be reduced by applying a thin layer of silicone oil to the surface of the water reservoir and by having a hole in the top of the reservoir only slightly larger than the tube used to convey the liquid to the aggregate. The water repellency index, *RI*, is determined from sorptivity measurements conducted for both water and a 95 % ethanol to water solution. Sorptivity, S ($L\ T^{-1/2}$), is determined by the following relationship:

$$S = \sqrt{\frac{Qf}{4br}} \tag{4.1}$$

where Q (L^3T^{-1}) is the steady rate of flow during early-time infiltration (Hallett et al. 2001), f is the fillable porosity, b is a parameter taken as 0.55 (White and Sully 1987) and r (L) is the radius of the infiltrometer tip. The fillable porosity was taken as the air-filled porosity, evaluated using aggregate bulk density measurements obtained with the saran resin approach, by Hallett et al. (2001). For non-repellent soils, the relationship between sorptivity of a 95 % ethanol to water solution, S_e, and sorptivity of pure water, S_w, is:

$$S_w = \left[\frac{(\mu_e/\gamma_e)^{1/2}}{(\mu_w/\gamma_w)^{1/2}}\right] S_e \tag{4.2}$$

where μ_e is the viscosity of 95 % ethanol to water solution (0.0012 N s m^{-2} at 20 °C), γ_e is the surface tension of 95 % ethanol (0.023 N m^{-1} at 20 °C), μ_w is the viscosity of water (0.0010 N s m^{-2} at 20 °C) and γ_w is the surface tension of water (0.073 N m^{-1} at 20 °C). Using these values, Eq. (4.2) reduces to:

$$S_w = 1.95 \cdot S_e \tag{4.3}$$

The repellency index, *RI*, therefore becomes:

$$RI = 1.95\frac{S_e}{S_w} \tag{4.4}$$

According to Tillman et al. (1989), $RI = 1.0$ is expected for a non-repellent, initially dry soil whereas *RI* is infinite for critically repellent soil since sorptivity measured with water will be zero. The presence of resident water alters the sorptivity of both water and ethanol and hence the repellency index. However, the sorptivity of ethanol can never be greater than water if the soil is non-repellent. Therefore, in a non-repellent soil, *RI* cannot be greater than 1.95, which is the ratio of $(\mu/\gamma)^{1/2}$ for the two liquids. The aggregate stays in contact with the tip of the infiltrometer (sponge) for 120–180 s (Hallett et al. 2001; Vogelmann et al. 2013). In the investigation by the former authors, for example, steady-state infiltration occurred after 30 s and remained until testing ended. The *RI* value calculated with Eq. (4.4) was also used by Vogelmann et al. (2013) to estimate the soil-water contact angle with the following relationship (Gryze et al. 2006):

$$\theta = \arccos\left(\frac{1}{RI}\right) \tag{4.5}$$

Hallett et al. (2001) suggested that the RI test should be considered more appropriate than both WDPT and MED tests to determine subcritical water repellency because the latter tests only provide a qualitative measurement that is not sensitive to low levels of subcritical water repellency. In addition, the former test provides a physically meaningful measurement of subcritical water repellency and it is inexpensive and easy to set up, with each test taking more or less 3 min. The *RI* value was found to increase with an increase in soil organic matter content in some investigations (Täumer et al. 2005; Vogelmann et al. 2010, 2013).

Use of tension infiltrometers for determining water repellency in the field is becoming popular. Lewis et al. (2006) and Robichaud et al. (2008) used the mini-disk infiltrometer (MDI) to measure the infiltration rate at a specific tension (i.e. 5 or 10 mm) during the first minute of the infiltration process. Indeed, this rate was found to be inversely related to the WDPT, with correlation coefficients of approximately 0.6–0.7 (Lewis et al. 2006). The choice of the 1-min time interval was arbitrary but it was considered appropriate for detecting water repellent soil conditions and also fast enough to be a useful assessment procedure for teams working in the field. The MDI test results can be grouped into several groups in function of the degree of soil water repellency. Lewis et al. (2006) suggested that a test with the MDI is faster and less subjective than the WDPT test. In addition, Robichaud et al. (2008) specified that the MDI test is particularly suited for post-fire assessment, which requires measurements to be done within few days after fire containment.

Jarvis et al. (2008) used a tension infiltrometer to measure steady-state infiltration using both ethanol and water as infiltrating fluids at supply pressure potentials of -1, -3 and -10 cm and they considered the ratio of ethanol to water flow rates as a measure of the degree of subcritical water repellency. Although the method is normally applied to sorptivity data derived from early-time infiltration behavior, Jarvis et al. (2008) found that, with the large infiltrometers used in their study (diameter 14 cm), they achieved better precision using final infiltration rates, since steady-state was attained rapidly, i.e. within a few minutes at the two larger pressure heads and 1–2 h at the smallest potential. Changes in the degree of water repellency were not observed in the time-frame (<2 h) of the infiltration experiments. The infiltration rates of ethanol were corrected for the difference in viscosity between the two liquids, i.e. they were multiplied by a factor of 1.2, to ensure correct comparisons. Furthermore, the authors considered that the maximum pore size that can conduct ethanol and water at any given supply potential differs due to differences in the surface tension and density of these fluids. Substituting the appropriate values for these physical constants in the capillarity equation showed that, at any given supply potential, the maximum pore diameter conducting ethanol should be smaller than for water by a factor of 2.5. To account for this, ethanol infiltration rates were plotted against a scaled or "effective" supply pressure head, derived by multiplying the actual supply potential by 2.5. The example of Fig. 4.2 shows similar steady-state infiltration rates with both liquids on an arable plot, indicating a lack of water repellency at this site, while, on an undisturbed grassed plot, infiltration rates were larger by a factor of approximately 15 with ethanol than with water, indicating occurrence of water repellent conditions at this site.

The repellency index, *RI*, was determined with Eq. (4.4) on seven sandy soils and a clay soil to assess the impact of tension infiltrometer disk size on the field-measured repellency index (Hunter et al. 2011). In particular, a pressure head of -3 cm was established on the infiltration surface with a standard tension infiltrometer (disk 20 cm in diameter) and with a mini tension infiltrometer (disk 4.5 cm in diameter). Early-time sorptivity was calculated from the first

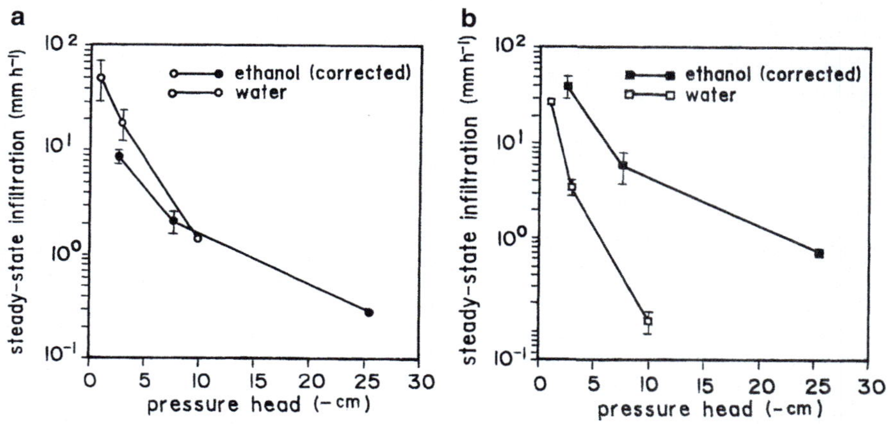

Fig. 4.2 Steady-state infiltration rates of water and ethanol in (**a**) an arable plot and (**b**) a grass plot (From Jarvis et al. 2008, reprinted with permission)

two measurable movements in liquid level by the following relationship (Philip 1969):

$$S = \frac{I}{\sqrt{t}} \tag{4.6}$$

where I (L) is the early-time cumulative infiltration and t (T) is the time. Higher *RI* values and greater variance were associated with the smaller disk size due to the smaller zone of influence. However, differences between the two disk sizes were not statistically significant in most cases and the conclusion by the authors was that the MDI would be well suited for in situ analysis of soil repellency at the site level.

4.2 Crusted Soils

Surface soil seals or crusts have usually a higher bulk density, smaller pores and, consequently, a lower saturated conductivity than the underlying soil. The term seal is used during formation of this particular surface layer whereas crust is used after drying. Determining the hydraulic characteristics of crusted soils is important in hydrological modeling because crusting is a rather common occurrence, especially in arid and semi-arid areas, and the partition between infiltration and runoff at the surface of these soils depends on the hydrodynamic properties of both the crust and the underlying soil even if, generally, the thickness of a surface crust is only of a few millimeters (Assouline and Mualem 2002, 2006; Šimůnek et al. 1998; Vandervaere et al. 1997, 1998). As an example, Fig. 4.3 compares infiltration of a constant intensity rainfall in an undisturbed soil with that occurring when the seal is

Fig. 4.3 Infiltration through a soil with and without a completely formed seal

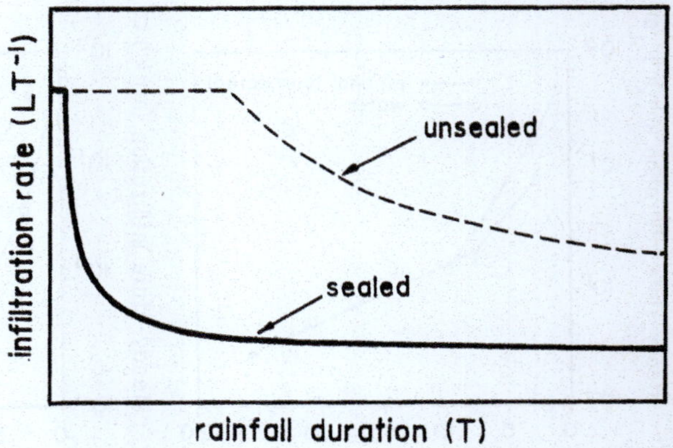

completely formed. Moreover, a surface crust hampers seed germination, acts as a barrier preventing further evaporation and reduces root aeration and water availability, adversely influencing soil physical quality.

4.2.1 Crust Formation

Soil surface sealing may result from different causes such as raindrop impact, fire, biological activity, and mechanical or chemical treatments. Rainfall-induced seals and crusts can be of different type. According to Valentin (1993) two main types of crust are distinguished by their mode of formation: structural crusts and depositional crusts. Structural crusts are defined as a surface layer of the soil, ranging in thickness from a few millimeters to a few centimeters, which is much more compacted than the material beneath. External materials are not involved in the formation of these crusts. Structural crusts can also be formed by physical forces as a result of trampling by livestock or through traffic by agricultural machinery or other vehicles. Depositional crusts form when soil particles, suspended in water, are deposited on the soil surface as the water infiltrates or evaporates. Externally-derived materials are always involved in the development of these crusts. According to Valentin (1993), structural crusts include slaking and sieving crusts. Slaking crusts form when soil contains enough clay (>15–20 %) to entrap and compress air during wetting so that aggregates break down. These crusts are more porous than the surrounding depressions where depositional crusts develop. Sieving crusts can consist of two or three well-sorted layers resulting from a sieving process. In the most developed form, the uppermost layer is composed of loose, coarse sand, the middle layer consists of fine, densely packed gains with vesicular pores, and the lower layer shows a higher content of fine particles with reduced porosity. This lower layer causes the low infiltrability (0–15 mm h^{-1}) of the crust. Depositional crusts are built up with a combination of runoff and "still" crusts. Runoff depositional crusts are characterized by alternate, very thin layers contrasting in texture, and they are formed by sediments deposited in overland flow. Still depositional crusts consist of densely packed and well-sorted particles, the size of which progressively increases with depth. The vertical particle size distribution, with coarser particles at the bottom and finer particles at the top, is the reverse of that observed in the sieving structural crusts. Still depositional crusts form in standing water where the larger particles sink rapidly to form the bottom layer and the finer particles are deposited at the top. Infiltrability of these crusts is low (0–7 mm h^{-1}).

According to Assouline (2011), the thickness of structural crusts formed under high kinetic energy rainfall can vary by more than two orders of magnitude, i.e. from 0.1 to 20 mm. Similarly, the ratio between the saturated hydraulic conductivity of the seal and that of the undisturbed soil underneath presents a huge variability, i.e. from 0.2 to 0.0001. The saturated hydraulic conductivity of the crust is related to the conditions prevailing during seal formation, such as the applied rainfall intensity. The characteristics of the sealed layer vary with depth but, in some conceptual models of the crust layer, a single uniform saturated layer of constant low permeability, 5 mm thick, was assumed (e.g., Hillel and Gardner 1969, 1970). On the other hand, Mualem and Assouline (1989) proposed the model of a non-uniform layer where the bulk density decreases exponentially with depth to that of the undisturbed soil. During rainfall on a bare soil surface, transfer of kinetic energy from the raindrop to the surface and soil wetting occur simultaneously. Consequently, the processes involved in seal formation are destruction of the soil aggregates exposed to the direct impact of the raindrops, compaction, slaking, particle segregation, and pore filling and clogging by wash-in of fine material. Chemical dispersion can also play a significant role. The relative importance of these processes depends upon the initial and boundary conditions that prevail during seal formation. Assouline and Mualem (1997) modeled seal formation in terms of the temporal increase of the bulk density of the soil surface during the rainfall. Rainfall intensity and kinetic energy are decisive in determining the seal properties and the rate of seal formation. Soil properties affecting the cohesive power between the soil particles also play a role in seal formation. These properties include soil mineralogy, texture, structure, aggregate stability, initial bulk density and water content, application of phosphogypsum or polymers to the upper soil layer, slope, organic matter content, soil exchangeable sodium percentage and electrical conductivity of the applied water.

4.2.2 Water Infiltration Experiments for Crust Characterization

The tension infiltrometer (TI) can be used to determine the hydraulic characteristics of crusted soils but particular applicative and data analysis procedures have to be applied. Indeed, in heterogeneous soil profiles presenting a fine layering organization, the classical steady-state water flow analysis based on the Wooding's (1968) equation, which assumes homogeneity and isotropy of the porous medium, is inadequate and may lead to unrealistic results including negative values of soil hydraulic conductivity (Angulo-Jaramillo et al. 2000; Hussen and Warrick 1993; Logsdon and Jaynes 1993).

The field method developed by Vandervaere et al. (1997) to determine the crust hydrodynamic properties near saturation uses a TI and a minitensiometer and is based on the application of transient flow analysis procedures to estimate sorptivity at different pressure heads, h_0 (L), imposed at the soil surface. For each test, and hence for each h_0 value, the matric flux potential is calculated from the corresponding sorptivity value. Finally, the hydraulic conductivity is obtained by differentiating the matric flux potential with respect to h_0. The minitensiometer, placed horizontally at the crust-subsoil interface, is used to detect the arrival of the wetting front below the crust and to select data describing water infiltration regime in the crust solely. It has a porous cup 20-mm-long and 2.2-mm-diam. and is connected to a pressure transducer through a capillary tube with an internal diameter of 1.45 mm (Fig. 4.4). Inserting the porous cup to about mid-distance between the center and the edge of the TI disk is enough for practical application of the method (Vandervaere et al. 1997). Using the infiltration model by Haverkamp et al. (1994), the sorptivity of the crust, S_c (L T$^{-1/2}$), for a given value of h_0 is determined using the cumulative infiltration, I (L), versus time, t (T), data collected

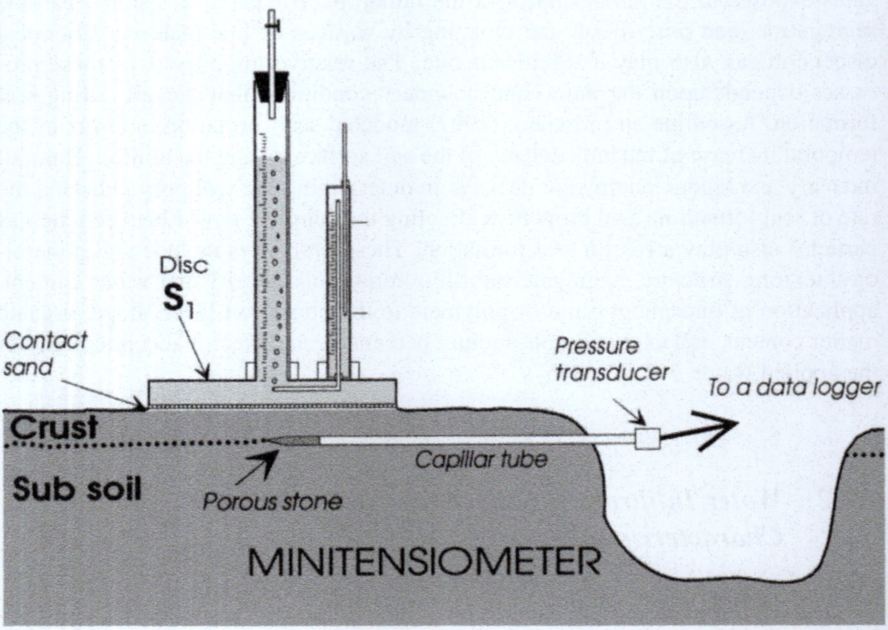

Fig. 4.4 Experimental set-up for measuring hydraulic properties of crusted soils (From Vandervaere et al. 1997 and Angulo-Jaramillo et al. 2000, reprinted with permission)

until the time t_1 (T), that is defined as the time when the wetting front reaches the crust-subsoil interface. During infiltration, this instant is indicated by the response of the minitensiometer to the arrival of the vertical wetting front and it is defined by the maximum dh/dt value, h (L) being the measured soil water pressure by the minitensiometer. Taking into account that a contact sand layer is commonly placed at the soil surface to ensure the hydraulic contact between the disk of the TI and the soil, the so-called differentiated linearization (DL) method (Vandervaere et al. 1997, 2000) is applied to estimate S_c. To recall, this method allows the detection of the early-time influence of the contact sand layer, which is revealed by the first data points falling away from the linear behavior of the data plotted on a $\Delta I/\Delta t^{0.5}$ vs. $t^{0.5}$ plot (see Sect. 3.1.4). The matric flux potential, ϕ_c (L^2T^{-1}), corresponding to the established pressure head value is then calculated:

$$\phi_c = \frac{b\ S_c^2}{\theta_0 - \theta_i} \tag{4.7}$$

where b is a parameter commonly taken equal to 0.55, and θ_0 and θ_i (L^3L^{-3}) are the final and the initial volumetric soil water content, respectively. Sorptivity notoriously depends on both initial and established pressure heads. However, the procedure makes use of the S_c determinations corresponding to different values of h_0 but the same initial soil water pressure head and, hence, the same θ_i value. Therefore, different determinations of S_c have to be carried out at different locations to obtain a set of S_c and ϕ_c data for different h_0 values. The two parameters, i.e. saturated soil hydraulic conductivity, K_s (L T^{-1}), and α (L^{-1}) parameter, of the assumed exponential relationship between soil hydraulic conductivity, K (L T^{-1}), and soil water pressure head, h (L) (Gardner 1958), can be determined by fitting the following linear relationship to the $\ln(\phi_c)$ vs. h_0 data:

$$\ln(\phi_c) = \ln\left(\frac{K_s}{\alpha}\right) + \alpha h_0 \tag{4.8}$$

Equation (4.8), that was derived by combining Eqs. (3.3) and (3.5) (Chap. 3), assumes that variations of α with h_0 are small within the range of potentials covered by the TI, that typically vary between 0 and -150 mm of water, although α cannot be considered constant over the whole range of h (Parlange 1972). The estimated intercept, b_0, of Eq. (4.8) allows simple calculation of K_s:

$$K_s = \alpha\ \exp(b_0) \tag{4.9}$$

Figure 4.5 shows, for an infiltration test carried out with $h_0 = -10$ mm on a structural crust, the measured cumulative infiltration and soil water pressure as a function of time (Fig. 4.5a), the corresponding derivatives with time (Fig. 4.5b), and the sorptivity value obtained by applying the DL method to the infiltration data collected from $t = 0$ to the time $t_1 = 300$ s (maximum dh/dt), that identifies the wetting front arrival to the minitensiometer installation depth (Fig. 4.5c)

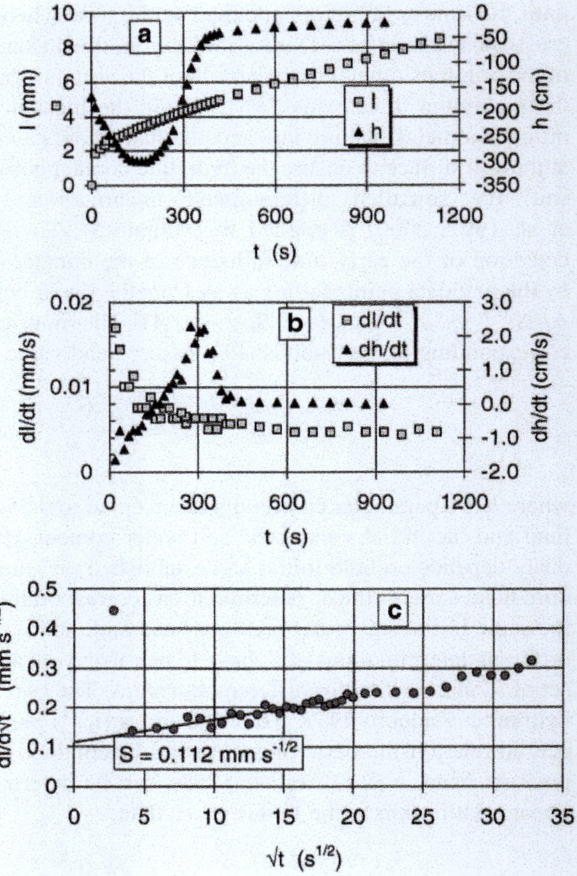

Fig. 4.5 Application of a tension infiltrometer and a minitensiometer to a crusted soil at a pressure head $h_0 = -10$ mm: (**a**) measured cumulative infiltration, I, and pressure head, h, against time, t; (**b**) infiltration flux and rate of water pressure change; and (**c**) estimated sorptivity by the differentiated linearization method (From Vandervaere et al. 1997 and Angulo-Jaramillo et al. 2000, reprinted with permission)

(Vandervaere et al. 1997). The early-time decrease in h is explained by the initial non-equilibrium of the tensiometer with the surrounding soil. For this example, the early influence of the contact sand layer was revealed by the first data point on the $\Delta I/\Delta t^{0.5}$ vs. $t^{0.5}$ plot. Obviously, this point was not considered for the estimation of S_c. Figure 4.6 shows an example of the ϕ_c vs. h_0 relationship obtained by Vandervaere et al. (1997) on a crust layer. According to the authors, the correct linearity which appears despite the important scatter due to the use of squared S_c values validates the approach. Because of the fragility of the crusts, the minitensiometer cannot be installed in dry conditions without the risk of a significant soil disturbance, i.e. detachment, fracture or cracking. Consequently, Vandervaere et al. (1997) drilled a hole into the soil, while injecting a total of about 1 cm^3 of water with a syringe by small increments. Installation of the tensiometer into this hole and the infiltration test were performed 24 h later, when the soil had dried. In any case, the infiltration test should be performed as quickly as possible after installing the tensiometer to avoid losing the hydraulic contact

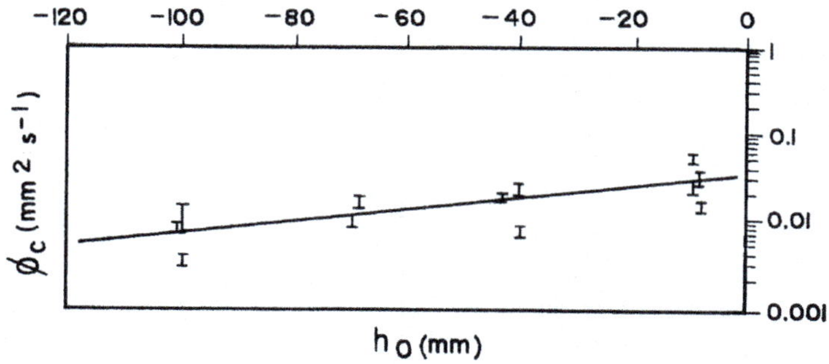

Fig. 4.6 Relationship between the matric flux potential, ϕ_c, and the imposed pressure head, h_0, for a crust layer (From Vandervaere et al. 1997, reprinted with permission)

between the porous cup of the tensiometer and the soil. The need to determine the S_c values for different pressure heads at different locations implies that, due to spatial variability of soil sorptivity, a large number of replicated runs should be carried out for a given h_0 value to obtain reliable estimates. Indeed, the slope of the ϕ_c vs. h_0 relationship must be estimated with enough accuracy to obtain reliable estimates of K_s and α by Eq. (4.8). Because each $\phi_c(h_0)$ data point is obtained at a single sampling point, the slope of ϕ_c vs. h_0 cannot be estimated with reasonable accuracy when the effects of spatial variability dominate those of the imposed pressure heads. Therefore, the transient method should not be applied in the case of a high field heterogeneity, unless a very large number of tests is performed (Vandervaere et al. 1997). Use of an excessively large amount of sand, required for situations with high surface roughness, may result in problematic application of the DL method due to failed linearization of the data on the $\Delta I/\Delta t^{0.5}$ vs. $t^{0.5}$ plot. Vandervaere et al. (1997) recommended not to use the method when crusts have a high surface roughness, because of the need for a thick layer of sand, or when they are thinner than 1 cm, because of the difficulty of placing the minitensiometer at the crust-subsoil interface. However, the method was suggested to be usable for any case of layered soils, since the extra-effort required for installation of the minitensiometers is offset by avoiding the steady-state flow requirement. The experimental/analytical procedure developed by Vandervaere et al. (1997) was found to be usable to predict crust-induced surface runoff at the plot scale by a simple two-layer Green and Ampt (1911) (GA) model that only needs initial and saturated volumetric soil water content and the K_s and α parameters of the Gardner's exponential equation for both the crust and the subsoil (Vandervaere et al. 1998). A point of interest in the investigation by Vandervaere et al. (1998) was that soil hydraulic properties measured at the square meter scale were found to be relevant parameters even for the study of a process involving a scale two orders of magnitude larger. Indeed, a close agreement between measured and calculated runoff was found when accounting for the properties of the soil and the surface crust estimated by their method.

The problem to obtain a complete hydraulic characterization (water retention curve and hydraulic conductivity function) of the two layered crust-subsoil system appears solvable with the method developed by Šimůnek et al. (1998) more or less concurrently with that by Vandervaere et al. (1997). Even the method by Šimůnek et al. (1998) uses the data obtained with a TI experiment but it applies an inverse procedure combining the Levenberg-Marquardt non-linear parameter optimization method with a numerical solution of the axisymmetric variably-saturated flow equation. The objective function is defined in terms of cumulative infiltration curve and final water content measured directly below the disk of the TI at the end of the experiment. Details on the inverse procedure are not reported here since it is fully described elsewhere in this book (see Sect. 3.1.5). The method combines two different multipotential TI experiments that have to be carried out on the two-layered system and on the subsoil, i.e. after removing the surface crust, respectively. At first, the infiltration experiment carried out on the subsoil is analyzed to determine the unsaturated hydraulic functions of the lower layer. Subsequent analysis of the infiltration experiment for the two-layered system with known hydraulic properties of the subsoil provides estimates of the hydraulic properties of the surface crust. The method was applied on a loamy-sand soil, using a 25-cm diameter disk and a thin (approximately 0.3 cm) layer of sand between the disk membrane and the soil surface. Six pressure head values, ranging approximately from −11.5 to −0.1 cm but with small differences between the experiments, were established by Šimůnek et al. (1998) on both the two-layered system and the subsoil. Disturbed soil was collected to determine the initial and final gravimetric water contents. Undisturbed samples were also taken to obtain the dry bulk density needed to convert gravimetric water contents into volumetric values. For the sampled soil, the saturated hydraulic conductivity of the surface crust, equal to less than 0.1 mm h^{-1}, was found to be three orders of magnitude lower than for the subsoil.

A similar approach was used by Yilmaz et al. (2013) to assess the impact of the formation of crust at the surface of basic oxygen furnace (BOF) slag on microstructure and hydraulic properties. Indeed, BOF slags are studied with regards to their recycling in civil engineering (pavement, roads, etc.). Yet, weathering processes, due to contact with water and carbon dioxide, trigger carbonation that may change microstructure, and both mechanical and hydraulic properties. Yilmaz et al. (2013) performed several campaigns of water infiltration experiments over 2 years on an experimental plot made of fresh BOF slag. The water infiltration experiments were conducted using a SW80 infiltrometer (Soil Measurement, Tucson, AZ). The authors applied successively the following pressure heads: −110, −50, −20, −10 and 0 mm. At time zero, the BOF slag was fresh without any crust. With time, a crust formed, as the result of carbonation, i.e. more specifically the dissolution of portlandite and precipitation of calcite. At the pore scale, it was clearly established that calcite covered grains and even clogged pores (Fig. 4.7). The numerical inversion of water infiltration data (Fig. 4.8) provided hydraulic and water retention curves for each sampling date. The data proved that carbonation impacted the saturated hydraulic conductivity but also the water retention curve. The clogging of pores reduced K_s but increased the water retention in the material by capillarity, thus impacting the relative hydraulic conductivity. Geochemical mechanisms and hydraulic properties were clearly proved to be interrelated.

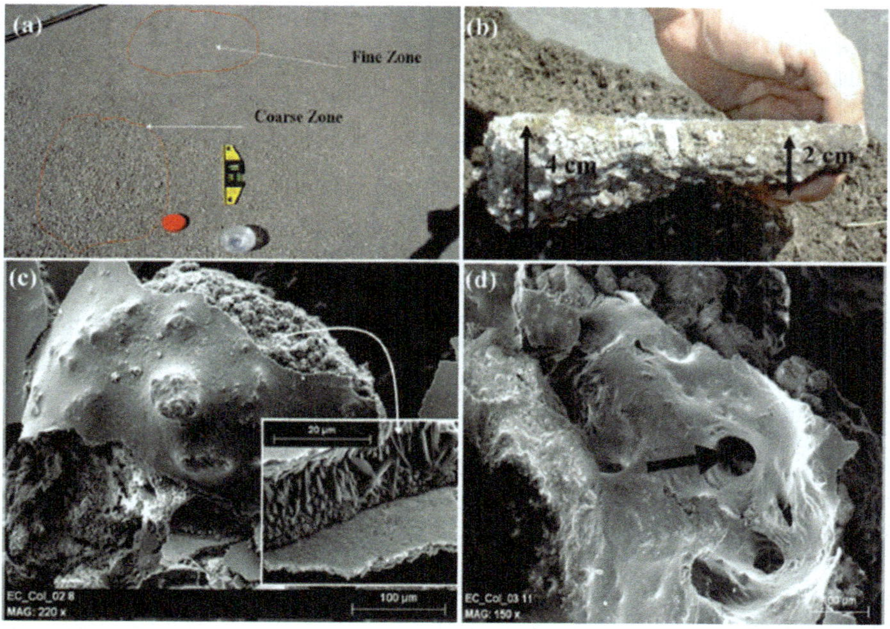

Fig. 4.7 Crust formation in highly reactive BOF (basic oxygen furnace) slag: (**a**) clogging of pores due to carbonation, i.e. the dissolution of portlandite along with the precipitation of calcite with contact with water and carbon dioxide; (**b**) crust layer; and (**c**) and (**d**) scanning electron microscopic images of carbonation and its effect on microstructure at micro-scale (From Yilmaz et al. 2013, reprinted with permission)

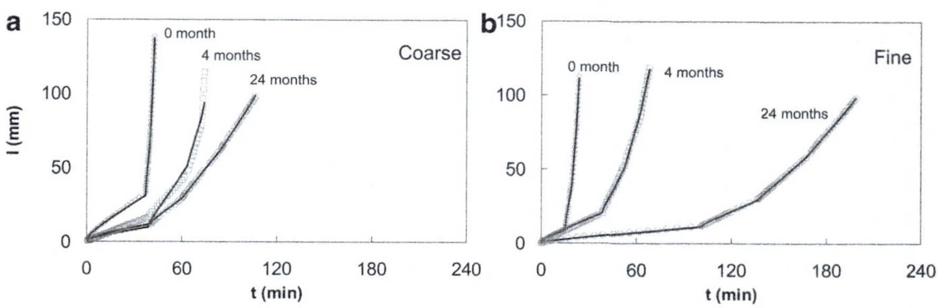

Fig. 4.8 Influence of carbonation and formation of the crust on the cumulative water infiltration, I, versus time, t, relationship. At the beginning, there is no crust. Time allows carbonation thus leading to the formation of a crust that influences water infiltration at the surface of both coarse (**a**) and fine (**b**) zones of the same experimental plot (From Yilmaz et al 2013, reprinted with permission)

Ponding infiltration measurements in combination with either a rainfall simulator or a mini disk infiltrometer (MDI) can also be used to characterize crusted soils in the field. In particular, the method developed by Touma et al. (2011) on the basis of an earlier work by Hillel and Gardner (1969) combines a rain simulation experiment and a single-ring infiltration experiment to characterize a surface crust in terms of its hydraulic resistance, R_c (T):

$$R_c = \frac{L_c}{K_c} \tag{4.10}$$

where L_c (L) is the crust thickness and K_c (L T^{-1}) is the corresponding saturated conductivity. Rainfall simulation yields steady-state infiltration rate through the crust surface. An experimental approach basically similar to the BEST (Beerkan Estimation of Soil Transfer parameters) procedure (see Sect. 3.2) is used to determine the hydraulic properties of the subsoil using ring infiltrometer data collected after removing the crust. In particular, the water retention curve is modeled according to van Genuchten (1980) with the Burdine's (1953) condition whereas the model by Brooks and Corey (1964) is used for the hydraulic conductivity function. Application of the Darcy law to the crust allows us to define the flux, q_c (L T^{-1}), across the surface crust of thickness L_c as follows:

$$q_c = -K_c \frac{h_s - L_c - h_0}{L_c} \tag{4.11}$$

in which h_0 (L) is the water pressure head at the crust surface and h_s (L) is the water pressure head at the crust-subsoil interface. Taking into account that h_0 is 0 for zero ponded conditions, which is the case of a plot where instantaneous runoff occurs since water does not accumulate on the soil surface, and considering that the thickness of the crust in the numerator of Eq. (4.11) is generally negligible as compared with h_s, being of the order of a few mm, Eq. (4.11) becomes:

$$q_c = -K_c \frac{h_s}{L_c} = -\frac{h_s}{R_c} \tag{4.12}$$

For a transient infiltration process, h_s increases with time up to a constant value when steady-state has been reached. The water flux entering the subsoil, q_s (L T^{-1}), can also be calculated by applying the Darcy's law to the subsoil:

$$q_s = -K(h_s)\left(\frac{dh}{dz} - 1\right) \tag{4.13}$$

where $K(h_s)$ is the unsaturated hydraulic conductivity of the subsoil corresponding to h_s and ($dh/dz - 1$) is the hydraulic gradient in the subsoil below the crust-subsoil interface. Due to the continuity of the flux at this interface, q_c has to be equal to q_s. Moreover, when steady-state has been reached, dh/dz in the subsoil at the crust-subsoil interface becomes a negligible quantity and Eq. (4.13) reduces to:

$$q_s = K(h_s) \tag{4.14}$$

Combining Eqs. (4.12) and (4.14), the following equation for the crust hydraulic resistance is obtained:

$$R_c = -\frac{h_s}{K(h_s)} \tag{4.15}$$

When the crust thickness cannot be neglected in the numerator of Eq. (4.11), the crust resistance will be given by:

$$R_c = -\frac{h_s - L_c}{K(h_s)} \tag{4.16}$$

Therefore, estimation of R_c needs determining (i) the water retention curve and the hydraulic conductivity function of the subsoil and (ii) the steady-state infiltration rate through the crusted soil at a null value of the pressure head imposed on the surface of the crust-subsoil system. Under steady-state conditions, $K(h_s)$ is determined by Eq. (4.14). The corresponding volumetric water content at the crust-subsoil interface is obtained from the soil hydraulic conductivity function, and the soil water retention curve is used to determine the associated value of h_s. Finally, crust hydraulic resistance is obtained by Eqs. (4.15) or (4.16). In the investigation by Touma et al. (2011), a rainfall simulator constituted by an oscillating nozzle hanging 4 m above the soil surface and a 1 m^2 plot, equipped with a runoff measuring device, were used. Steady-state infiltration rate, q_s, was computed as the difference between the rainfall intensity and the steady-state runoff rate. A ring having a diameter of 0.15 m and driven about 5 mm into the subsoil, i.e. after removing the crusted layer, was used to establish the infiltration process required by the BEST experimental procedure (Lassabatere et al. 2006). However, the Brutsaert (1977) model was chosen to describe one-dimensional (1D) cumulative infiltration, I_{1D} (L), in substitution of the infiltration model by Haverkamp et al. (1994) that is applied in the original BEST procedure (see Sect. 3.2). Touma et al. (2011) applied their method in a sandy soil and they obtained values of R_c ranging from 1.29 to 6.91 h. The proposed methodology was successfully tested by reproducing numerically the experiments.

A simplification of the method by Touma et al. (2011) was proposed by Alagna et al. (2013) who, under the assumption that the flow in the crust can be considered 1D, used a mini disk infiltrometer (MDI) for measuring steady-state infiltration rates through the surface of the soil crust and the complete BEST procedure (Lassabatere et al. 2006) to determine the hydraulic properties of the subsoil (Fig. 4.9). Compared to the rainfall simulator, the MDI avoids the potential disturbance of the crust, whereas the rainfall simulator may impact the crust surface. In other words, taking into account that soil disturbance is minimal with the MDI, the hypothesis that the measured steady-state infiltration rate is representative of the crust and it does not include changes due to the experiment appears more defensible. Moreover, the MDI samples a small area (i.e., approximately 16 cm^2). Therefore, this device allows assessment of small-scale spatial variability

Fig. 4.9 Experimental procedure for determining the hydraulic resistance of the surface soil crust: (**a**) surface crust; (**b**) mini disk infiltrometer experiment; (**c**) crust removal; (**d**) ring infiltration test (From Alagna et al. 2013)

of hydraulic resistance, determination of crust characteristics at particular points and easy association of the measured infiltration rate to the thickness of the crust. On the other hand, the pressure head established at the base of the disk cannot be higher than approximately −0.5 cm. However, it is always necessary to spread a thin layer of contact sand on the crust to level small irregularities and assure a good hydraulic contact between the porous disk and the soil. Therefore, the imposed pressure head on the infiltration surface is higher than the one established at the base of the device by a quantity that depends on the thickness of the contact material (Reynolds 2006). For the infiltration tests conducted by Alagna et al. (2013), the established pressure head on the crust surface was expected to be close to zero, i.e. higher than −0.5 cm. In addition, considering the fact that the crusted soil has a low porosity and pores are expected to be small, even a slightly negative pressure head on the infiltration surface should not have an appreciable effect on the measured steady-state infiltration rate. The analysis by Alagna et al. (2013) assumes that edge effects are expected to be negligible for the crust, which cannot be perfectly true. This choice was made because the crust is generally thin (i.e. a few mm) and the implicit assumption was that near 1D flow occurred through this layer at steady-state, i.e. that almost all flow divergence occurred in the subsoil. For a clay soil with a crust of 6 mm in thickness, Alagna et al. (2013) reported a mean value of R_c of 1.31 h, or 1.35 h when the crust's thickness was explicitly considered by using

Eq. (4.16), and a saturated hydraulic conductivity of the crust, calculated by Eq. (4.10), of 4.6 mm h^{-1}. The developed method appears simple and suitable to discriminate between different levels of hydraulic resistance of the crust. Another advantage of the method is that both the MDI and the ring infiltration tests can be carried out at exactly the same point of the field, by gently removing the crust a few days after the MDI run. In any case, testing the method on other soils is necessary. Tests should also be carried out to establish if the experimental procedure for measuring the crust infiltration rate has to be improved to better reproduce the assumed theoretical scheme.

More in general, there is a general lack of an extensive field testing of all methodologies described in this section. Due to the hydrological importance of the surface crust layer, future research should improve knowledge on the usability, in practice, of these or other techniques for characterizing crusted soils.

4.3 Low Permeability Soils

In the last two decades, the Canadian school was in the front line in the development of infiltration analysis procedures for determining the saturated hydraulic conductivity, K_{fs}, of low permeability soils (K_{fs} approximately less than 1×10^{-9} – 1×10^{-8} m s^{-1}). These procedures use early-time transient flow into the initially unsaturated porous medium because steady-state flow analysis procedures cannot be applied in practice due to the very long times (days, weeks) required to reach near steady-state conditions and the very small flow rates that can be measured. The initial flow of water from a surface ring into initially unsaturated soil is controlled almost solely by the pressure head and the capillary sorption and it can be considered to be one-dimensional and free from gravitational effects (Odell et al. 1998). In this case, the cumulative infiltration, I (L), vs. time, t (T), relationship and the infiltration rate, i (L T^{-1}) vs. t relationship can be written as (Philip 1957, 1969):

$$I = S_H t^{1/2} \tag{4.17}$$

$$i = \frac{dI}{dt} = \frac{1}{2} S_H \ t^{-1/2} \tag{4.18}$$

where S_H (L $T^{-1/2}$) is the soil sorptivity for the ponded head, H (L), established on the infiltration surface. The flow rate, Q (L^3T^{-1}), is therefore given by:

$$Q = A_0 \ i = \frac{1}{2} A_0 S_H \ t^{-1/2} \tag{4.19}$$

where A_0 (L^2) is the infiltration surface. Using data obtained by a numerical solution of the Richard's equation, Fallow et al. (1994) showed that, with a ponded depth of water $H = 0$ on the surface of an initially unsaturated soil, the Q vs. $t^{-1/2}$ relationship (Fig. 4.10) is linear for approximately 17 h for $K_{fs} = 1 \times 10^{-8}$ m s^{-1} and 70 h when K_{fs} is equal to 1×10^{-9} m s^{-1}. For $H = 1$ m or more, the linear portion of the

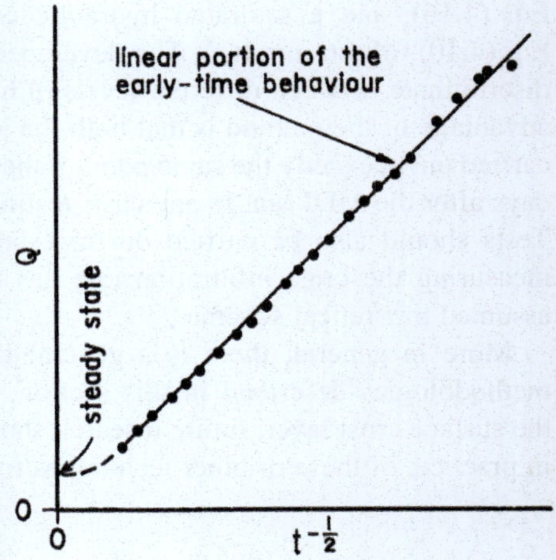

Fig. 4.10 Qualitative representation of the flow rate, Q (L^3T^{-1}), vs. $t^{-1/2}$ relationship, t (T) being the time, for an initially unsaturated, low permeability soil

early-time behavior is shortened by a factor of two or three, still leaving many hours for obtaining valid measurements.

In practice, the experiment can be carried out by establishing a constant (early-time constant-head, ECH, technique) or a falling (early-time falling-head, EFH, technique) head of water on the infiltration surface, or realizing an initial constant-head stage followed by a falling-head phase (sequential early-time constant-head/falling-head, ECFH, technique). A common assumption to the different procedures is that the soil is rigid. This assumption can adversely influence the application of the transient approaches in initially very dry soils with a high content of swelling clay (Odell et al. 1998). Furthermore, these approaches are not even usable in porous media with an initially high water content since the duration of the early stage of infiltration is short in this case. However, the hydraulic conductivity of low permeability wet soils can be determined with the falling-head infiltrometer technique developed by Youngs et al. (1995). In the following, the ECH, EFH and ECFH techniques and the K_{fs} measurement methodology by Youngs et al. (1995) are described. The time domains in an infiltration process are then discussed, taking into account that estimating the expected duration of the early-time phase of the infiltration process makes the choice of the most appropriate methodology to determine K_{fs} easier.

4.3.1 Early-Time Constant-Head (ECH) Technique

During constant-head conditions, S_H is invariant and it can be expressed by the following approximate analytical relationship (White and Sully 1987; Fallow et al. 1994; Elrick et al. 2002):

$$S_H = \left(2\ \Delta\theta\ K_{fs}\ H + S_0^2\right)^{1/2} \tag{4.20}$$

where $\Delta\theta$ is the difference between the field-saturated soil water content, θ_{fs} (L^3L^{-3}), and the initial soil water content, θ_i (L^3L^{-3}), K_{fs} ($L\ T^{-1}$) is the field-saturated soil hydraulic conductivity, H (L) is the constant head of water on the infiltration surface and S_0 ($L\ T^{-1/2}$) is the soil sorptivity for $H=0$, equal to:

$$S_0 = \left(\frac{\Delta\theta\ \phi_m}{b}\right)^{1/2} \tag{4.21}$$

where ϕ_m (L^2T^{-1}) is the field-saturated soil matric flux potential and b is a constant, varying from 1/2 to $\pi/4$, that can be set equal to 0.55 in many cases, with an error of, at maximum, about 10 % in S_H (Elrick et al. 2002). Taking into account that, according to Eq. (2.7), the α^* (L^{-1}) parameter is equal to K_{fs}/ϕ_m, combining Eqs. (4.20) and (4.21) and solving for K_{fs} yields (Fallow et al. 1994; Bagarello et al. 2004):

$$K_{fs} = \frac{S_H^2}{\Delta\theta\left(2H + \frac{1}{b\ \alpha^*}\right)} \tag{4.22}$$

Equation (4.22) can be used to determine K_{fs} from early-time flow data under a constant head (Odell et al. 1998). The procedure, the ECH technique by Fallow et al. (1994), evaluates S_H from the slope of the linear portion of the I vs. $t^{1/2}$ graph, and needs the measurement of $\Delta\theta$ and the estimation of the α^* parameter (see Table 2.1). More recently, Elrick et al. (2002) derived another approximate expression of S_H based on the Green and Ampt (1911) (GA) approach:

$$S_H = \left[2\ K_{fs}\ \Delta\theta\left(H - h_f\right)\right]^{1/2} \tag{4.23}$$

where h_f (L) is the soil water pressure head at the wetting front, being h_f negative. The relationship between h_f and ϕ_m is (Reynolds 2008):

$$\phi_m = -2\ b\ K_{fs}\ h_f \tag{4.24}$$

For a GA soil, having $b=0.5$, Eq. (4.24) becomes (Elrick et al. 2002):

$$\phi_m = -K_{fs}\ h_f \tag{4.25}$$

Comparing Eq. (4.25) with Eq. (2.7) allows us to write:

$$\alpha^* = -\frac{1}{h_f} \tag{4.26}$$

Solving Eq. (4.23) for K_{fs}, considering Eq. (4.26), yields:

$$K_{fs} = \frac{S_H^2}{2\ \Delta\theta\left(H + \frac{1}{\alpha^*}\right)} \tag{4.27}$$

The only difference between Eqs. (4.22) and (4.27) is in the estimation of b.

The ECH technique was developed with the specific objective to determine the saturated hydraulic conductivity of very low permeability soils with an experiment of practically reasonable duration, i.e. not more than a few hours, but it was also applied in other soil types. Fallow et al. (1994) carried out an early-time constant-head experiment driving a 9.6 cm diameter ring to a depth of 5 cm in a soil liner and using a horizontal burette positioned at a pre-established height above the soil surface to maintain a constant depth of water on the infiltration surface. The position of the meniscus was then monitored at given times, allowing to measure K_{fs} values equal, on average, to 6.4×10^{-10} m s^{-1}. Odell et al. (1998) applied the ECH technique on six identically prepared compacted clay soil cores having a height of 50 mm and a diameter of 101 mm. On average, the ECH estimate of K_{fs}, equal to 2.46×10^{-10} m s^{-1}, was 60 % of the value obtained by a modified version of the classical constant-head permeameter apparatus. This difference, that was relatively small, was attributed to a moderate swelling of the cores, resulting in a more porous material at complete saturation, and to the circumstance that the soil properties (i.e. bulk density) of the near-surface depth to which the early-time infiltration reached were not entirely representative of the whole core. Vauclin et al. (1994) used both the ECH analysis procedure and the classical steady-state approach by Reynolds and Elrick (1990) for analyzing 32 infiltration runs carried out in a loam soil with the commercial version of the single-ring pressure infiltrometer. The estimates of K_{fs} obtained with the transient approach, equal on average to 8.23×10^{-5} m s^{-1}, were 2.8 times higher and more variable than the corresponding values obtained with the steady-state approach. These Authors attributed generically the observed differences to the approximations of the transient method. An investigation methodologically similar to that by Vauclin et al. (1994) was carried out by Bagarello and Iovino (1999) with the single-ring pressure infiltrometer by Ciollaro and Lamaddalena (1998) in a sandy-clay soil. In this case, the transient and steady-state approaches showed moderate differences, equal in particular to 20 % with reference to the mean K_{fs} value.

According to Eq. (4.17), the S_H term to be used for the K_{fs} calculation with Eqs. (4.22) or (4.27) has to be obtained by the fit of a straight line passing through the origin of the axes to the experimental ($t^{1/2}$, I) data pairs, in order to satisfy the condition that $I = 0$ at $t = 0$. In practice, Fallow et al. (1994) suggested to estimate S_H by the following relationship:

$$S_H = \frac{I_2 - I_1}{t_2^{1/2} - t_1^{1/2}} \tag{4.28}$$

where I_1 and I_2 (L) are the cumulative infiltration values at the t_1 and t_2 (T) instants, respectively, with $t_2 > t_1$. Alternatively, the same Authors suggested to fit a straight line not forced to pass through the origin of the axes to the experimental ($t^{1/2}$, I) data pairs. This last choice was also made by Odell et al. (1998), as seen in their figure 1. With the single-ring pressure infiltrometer by Ciollaro and Lamaddalena (1998)

(Fig. 2.20b), it is theoretically possible to obtain ($t^{1/2}$, I) data pairs that can be interpolated by Eq. (4.17). The instant $t = 0$ coincides with the raising of the piston, since soil wetting and the starting of the infiltration process occur almost instantaneously. At the beginning of the run, a part of the water volume contained into the reservoir of the device serves to establish the fixed depth of ponding on the infiltration surface. Therefore, the cumulative infiltration at the first measurement time, t_1, is obtained by subtracting the water level drop corresponding to the product between the infiltration surface and the established ponded head from the total water level drop at $t = t_1$. From the second measurement time onwards, the water level drop in the reservoir is entirely transformed in infiltrated water volume. In practice, however, imposing a null intercept of the straight line interpolating the ($t^{1/2}$, I) data can yield less satisfactory results, in terms of representation of the data, than those obtained without this constraint (Bagarello and Iovino 1999).

Example 4.1

A constant head of water, $H = 0.014$ m, was established on a soil core having $\Delta\theta = 0.131$ m^{3}m^{-3}, a diameter of 0.085 m and a height of 0.11 m to determine K_{fs} by the ECH technique. Table 4.1 lists the measured time, t (s), cumulative infiltration, I (m), data pairs during the early phase of infiltration. Linear regression of I against $t^{1/2}$ yields an estimate of $S_H = 1.20 \times 10^{-4}$ m s$^{-1/2}$, with a high coefficient of determination ($R^2 = 0.989$), indicative of the linearity of the relationship described by the experimental data points (Fig. 4.11). Assuming $\alpha^* = 4$ m^{-1} as an appropriate value for the tested fine-textured and poorly structured soil, Eqs. (4.22) and (4.27) yield, respectively:

$$K_{fs} = \frac{\left(1.20 \times 10^{-4}\right)^2}{0.131\ \left(2 \times 0.014 + \frac{1}{0.55 \times 4}\right)} = 2.28 \times 10^{-7}\ \text{m s}^{-1} = 0.82\ \text{mm h}^{-1}$$

$$K_{fs} = \frac{\left(1.20 \times 10^{-4}\right)^2}{2 \times 0.131\left(0.014 + \frac{1}{4}\right)} = 2.08 \times 10^{-7}\ \text{m s}^{-1} = 0.75\ \text{mm h}^{-1}$$

The two estimates of K_{fs} are similar, showing that the approach chosen to analyze the data did not influence appreciably the calculation results. The depth of the soil

Table 4.1 Data for the Example 4.1

Time, t (s)	Cumulative infiltration, I (m)
811	0.0011
1134	0.0017
1368	0.0022
1797	0.0031
2168	0.0034
2348	0.0037
2655	0.0040
3075	0.0044
3600	0.0047

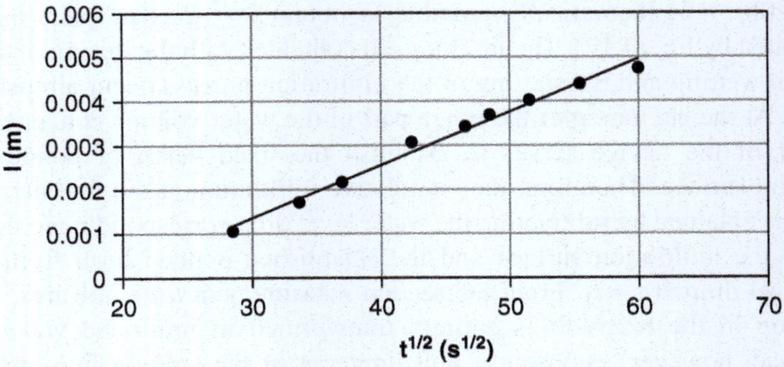

Fig. 4.11 Cumulative infiltration, *I*, vs. the square root of time, *t*, for the data of the Example 4.1

layer explored with the run, d_s (L), can be estimated by the following relationship (Bagarello et al. 2004):

$$V = d_s \ A_0 \ \Delta\theta \tag{4.29}$$

where V (L^3) is the totally infiltrated water volume and A_0 (L^2) is the cross-sectional area of the core. For the considered run, $d_s = 0.036$ m is obtained.

4.3.2 *Early-Time Falling-Head (EFH) Technique*

During early-time falling-head infiltration, S_H is variable and therefore does not exhibit a linear behavior on a graph of I vs. $t^{1/2}$ (Odell et al. 1998). For falling-head conditions, cumulative infiltration is a function of the ponded depth of water on the infiltration surface, H (L), as follows (Elrick et al. 1995):

$$I(t) = \frac{a}{A_0}[H_0 - H(t)] \tag{4.30}$$

where a (L^2) is the cross-sectional area of the falling-head tube, A_0 (L^2) is the cross-sectional area of the infiltration surface and H_0 (L) is the height of the ponded head at $t = 0$ (Fig. 4.12). Considering the falling-head infiltration process as a succession of constant-head conditions, the I vs. t relationship can be written as (Fallow et al. 1994):

$$I(t) = \left[2 \ \Delta\theta \ K_{fs} \ H(t) + \frac{\Delta\theta \ \phi_m}{b}\right]^{1/2} t^{1/2} \tag{4.31}$$

where the impact of the falling head is given by the term $H(t)$. Substitution of Eq. (4.30) into Eq. (4.31) yields (Fallow et al. 1994; Elrick et al. 1995):

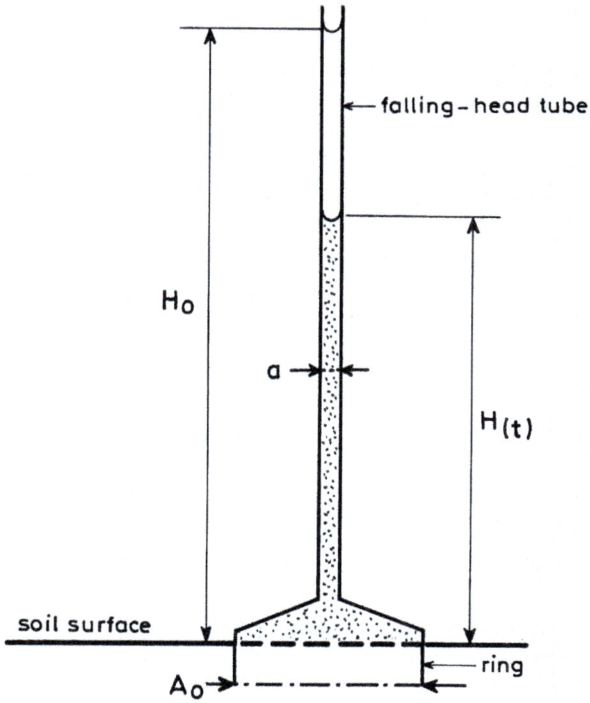

Fig. 4.12 Falling-head infiltrometer: H_0 (L) is the initial ponded head of water, $H(t)$ (L) is the ponded head of water at time t (T), a (L^2) is the cross-sectional area of the falling-head tube and A_0 (L^2) is the infiltration surface area

$$t^{1/2} = \frac{(a/A_0)[H_0 - H(t)]}{\left[2\ \Delta\theta\ K_{fs}\ H(t) + \frac{\Delta\theta\ \phi_m}{b}\right]^{1/2}} \tag{4.32}$$

The EFH technique allows simultaneous estimation of K_{fs} and ϕ_m by a non-linear, least-squares fitting procedure of Eq. (4.32) to the (t, H) data pairs collected during the initial stage of a falling-head infiltration process. Therefore, the EFH technique has the advantage, in comparison with the ECH technique, that both unknowns can be estimated. For slowly permeable soils ($K_{fs} < 1 \times 10^{-9}$ m s^{-1}), H_0 should be set at 1–2 m (Fallow et al. 1994). Initially H falls very quickly, depending on the choice of a and A_0, even in slowly permeable soils. Therefore, readings have to be taken at short time intervals (i.e., $\Delta t = 30$ s) at the beginning of the run and then less frequently as the head falls. For this early-time falling-head solution, the parameters K_{fs} and α^* or ϕ_m are simultaneously identifiable and unique, and the model is stable (Elrick et al. 1995; Odell et al. 1998). The tests for solution identifiability, uniqueness and stability indicated that the $K_{fs} - \phi_m$ combination gives the most reliable best fit parameter values and that the solution can withstand a significant amount of random error in the falling-head data. The H_0 term in Eq. (4.32) is difficult to obtain experimentally with the falling-head procedure due to the time required to fill the infiltrometer and falling-head tube with water.

For this reason, Fallow et al. (1994) suggested to obtain a first guess of H_0 by a linear extrapolation to $t=0$ using the first few readings of H vs. $t^{1/2}$. This estimate of H_0 is then used in the optimization procedure. In other words, H_0 is a third unknown, in addition to K_{fs} and ϕ_m. For a test of the EFH procedure on a compacted clay soil, Fallow et al. (1994) used a core having a cross-sectional area of 8.01×10^{-3} m^2 and a 1.5 m long vertical tube with $a=8.75 \times 10^{-6}$ m^2.

More recently, Elrick et al. (2002) used the GA approach to deduce a new implicit equation in I for falling-head conditions that neglects gravity:

$$t = -\frac{(a/A_0)}{K_{fs}\ \Delta\theta}\left[I(t) + \Delta\theta\ B\ \ln\left(1 - \frac{I(t)}{\Delta\theta\ B}\right)\right] \tag{4.33}$$

where

$$B = \frac{H_0 - h_f}{\Delta\theta}\frac{a}{A_0} \tag{4.34}$$

According to Elrick et al. (2002), Eqs. (4.32) and (4.33) are similar at early times but deviate as time increases. Both expressions are approximations. However, Eq. (4.33) is a true solution under GA falling head conditions, whereas Eq. (4.32) is an approximation based on a succession of constant-head solutions. The GA expression for a falling head that includes gravity was derived by Philip (1992) (Elrick et al. 2002):

$$t = \frac{\Delta\theta}{C\ K_{fs}}\left[\frac{I(t)}{\Delta\theta} - \frac{H_0 - h_f}{C}\ln\left(1 + \frac{C\ I(t)}{\Delta\theta(H_0 - h_f)}\right)\right] \tag{4.35}$$

where

$$C = 1 - \frac{\Delta\theta}{(a/A_0)} \tag{4.36}$$

Equations (4.33) and (4.35) are valid until the falling head drops to zero. Equation (4.35) produces a singularity when the a/A_0 ratio is numerically equal to $\Delta\theta$ since C is equal to 0 in this particular case. In addition, the analysis may become inaccurate as $a/A_0 \rightarrow \Delta\theta$ because this circumstance may generate rounding errors. This problem was tackled by Reynolds (2008) who proposed an alternative relationship for the analysis of the data collected by a falling-head infiltration run.

Elrick et al. (2002) applied Eq. (4.32) with $b=0.5$ and Eqs. (4.33) and (4.35) to the falling-head infiltration data obtained in the laboratory by Fallow et al. (1994), with the aim to establish the dependence of the K_{fs} and α^* estimates on both the applied solution and the run duration. To test this last effect, these authors conducted different fittings by progressively restricting the data to earlier times through sequential elimination of the last data points. The maximum difference between the K_{fs} values obtained with different equations and run durations was equal to 50 %, which is a practically negligible difference in many cases. On the other hand, the α^* values obtained with Eqs. (4.33) and (4.35) were about three times higher than the values calculated by Eq. (4.32). In addition, run duration effects were more noticeable for α^* than K_{fs}. It made little difference if gravity was, or was not, included in the infiltration equation.

Reynolds (2008) developed a new approximate relationship for a simplified determination of K_{fs} with a falling-head experiment:

$$K_{fs} = \frac{(a/A_0)^2}{4\ \Delta\theta\ t} \frac{[H_0 - H(t)]^2 \left[H_0 + H(t) + (b\alpha^*)^{-1}\right]}{\left[H(t) + (2b\alpha^*)^{-1}\right]\left[H_0 + (2b\alpha^*)^{-1}\right]} \tag{4.37}$$

Applying this equation to obtain an estimate of K_{fs} involves assuming an α^* value based on the most appropriate texture-structure category (Table 2.1) and measuring a single (H, t) data pair during the run. Equation (4.37) is usable for $a/A_0 < \Delta\theta$ and $H(t) \geq 0.5\,H_0$. An incorrect selection of the α^* parameter has not a strong impact on the K_{fs} calculations. In particular, Reynolds (2008) showed that an incorrect choice of α^* imparted less than $\pm 10\,\%$ error in the K_{fs} estimates for porous media with capillarity ranging from negligible ($\alpha^* = 100\ \text{m}^{-1}$) to high ($\alpha^* = 4\ \text{m}^{-1}$).

4.3.3 Sequential Early-Time Constant-Head/Falling-Head (ECFH) Technique

To avoid treating H_0 as an unknown, and considering the prolonged temporal validity of Eq. (4.17) in slowly permeable materials, Elrick et al. (1995) suggested to perform the early-time falling-head infiltration run after an initial early-time constant-head phase (ECFH technique). By this approach, the H_0 value at the beginning of the falling-head run is known exactly and does not have to be fitted for as is the case with the falling-head analysis only. In other words, only the variables of specific interest for the soil sample characterization are optimized. The relationship used with the ECFH technique is (Elrick et al. 2002):

$$t = t_c + \frac{(a/A_0)^2 [H_0 - H(t)]^2}{2\ \Delta\theta\ K_{fs}\ \left[H(t) - h_f\right]} \tag{4.38}$$

where t_c (T) is the instant that separates the constant-head and falling-head phases. Elrick et al. (2002) also deduced, with the GA approach, alternative relationships to Eq. (4.38), both neglecting and considering gravity effects.

4.3.4 *Saturated Conductivity of Low Permeability Wet Soils*

In some instances, it can be necessary to determine the saturated hydraulic conductivity of very low permeability soils having an initially high water content. For an almost saturated soil, the intake rate achieves its final constant rate very soon after the start of a test because there is little pore volume left to fill. Therefore, the early stage of the infiltration process is not usable to determine K_{fs}. Youngs et al. (1995) developed an applicative procedure of the single-ring pressure infiltrometer allowing determination of saturated hydraulic conductivity of low permeability, initially very wet soil by a relatively short duration (1–2 h) falling-head run. In the falling-head pressure infiltrometer used by the authors, water is supplied to the soil surface at a head through the sealed top lid from a small capillary tube that also acts as a measuring burette (Fig. 4.13). The ring has a diameter of 98 mm and it could be driven into the soil to a maximum depth of 60 mm. With the lid of the device removed, the ring is pressed vertically into the wet soil to a depth of about 50 mm. The lid is then fitted to make a water-tight seal with the top of the ring. Because of the positive heads used in the tests, the ring is anchored or weighed down to prevent the hydraulic force lifting the ring out of the soil. A vertical Perspex tube of diameter 36 mm and length 1.5 m is fitted with an O-ring seal into the hole provided in the center of the lid. Connection is made via a small connecting tube in the lid to the measuring capillary tube having a diameter of

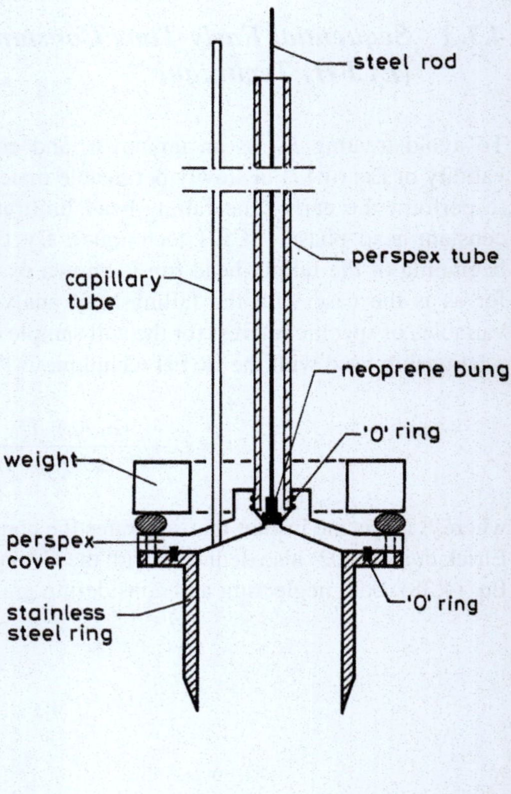

Fig. 4.13 Scheme of the falling-head pressure infiltrometer used by Youngs et al. (1995) to determine the saturated hydraulic conductivity of low permeability wet soils (Reprinted with permission)

0.54 mm. Water is then poured through the large central tube so that the infiltrometer device fills with water and water stands in the capillary tube to the required head of 1 m or more. Infiltration into the soil then takes place. Water supply from the central large tube is stopped with a plug on the end of a long handle when all air bubbles from the system have been expelled, so that water is thereafter supplied only from the capillary tube. The position of the water meniscus in the tube is followed with time during the falling-head infiltration from the infiltrometer. The calculation procedure of K_{fs} developed by Youngs et al. (1995) is based on the analysis by Reynolds and Elrick (1990), that was modified taking into account that the ϕ_m term has a reduced importance in an initially practically saturated soil. In particular, K_{fs} is estimated by the following relationship:

$$\frac{K_{fs} r (t_2 - t_1)}{\pi\ r_t^2 G} = \ln(S_1 H_1) - \ln(S_2 H_2) \tag{4.39}$$

where r (L) is the radius of the ring, r_t (L) is the radius of the vertical measuring tube, G is the shape factor for a steady-state experiment with the single-ring pressure infiltrometer as estimated by Eq. (2.65) (Reynolds and Elrick 1990), H_1 and H_2 (L) are the heads at times t_1 and t_2 (T), respectively, and S_1 and S_2 (L) are the corresponding shape factors given by:

$$S_i = \frac{\pi\ r^2}{H_i} + \frac{r}{G} \tag{4.40}$$

where H_i is the H value at the instant i. Therefore, the shape factor, S_i, varies during the run. Youngs et al. (1995) suggested to plot ln(SH) against time, t. Detecting a linear relationship between these two variables with a slope of $- b_1$ allows to obtain an estimate of K_{fs} by the following relationship:

$$K_{fs} = \frac{b_1 \pi\ r_t^2 G}{r} \tag{4.41}$$

With reference to Fig. 4.14, showing an example of the results obtained with the method, Youngs et al. (1995) recognized that the plot suggested a linear relationship between ln(SH) and t after a short, initial transient stage. In the tests that were carried out, the proposed methodology allowed to determine K_{fs} values varying from 1.3×10^{-10} m s^{-1} to 4.3×10^{-9} m s^{-1}.

4.3.5 *Time Domains for Pressure Infiltrometer Data*

In measuring K_{fs}, an estimate of the expected duration of the early-time transient stage of the infiltration process allows to make a choice on how to conduct the experiment and analyze the data. With reference to a constant-head infiltration process established with the pressure infiltrometer, Groenevelt et al. (1996)

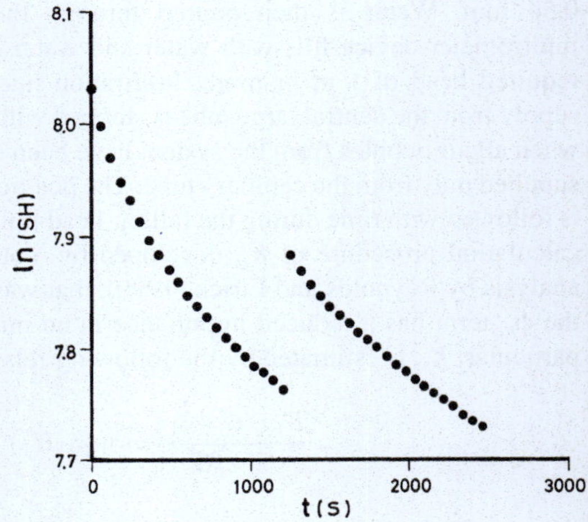

Fig. 4.14 Example of an experimental test showing a plot of ln(*SH*) (S = shape factor in mm; H ponded depth of water in mm) vs. time, t (From Youngs et al. 1995, reprinted with permission)

deduced the following relationship to estimate the critical time, t_{cr} (T), discriminating between the early-time and steady-state flow domains:

$$t_{cr} = \frac{S_H^2}{4\ K_{fs}^2}\left[\frac{H}{\pi\ r\ G} + \frac{1}{2} + \frac{1}{\pi\ r\ G\ \alpha^*}\right]^{-2} \tag{4.42}$$

More in particular, t_{cr} is the time that separates the latest possible early-time flow period from the earliest possible steady-state flow phase. Using Eq. (4.42), Groenevelt et al. (1996) also tested the dependence of t_{cr} on K_{fs}, $\Delta\theta$ and H. The duration of the early-time flow period is equal to zero for an initially saturated soil and it increases with an increase in $\Delta\theta$. Moreover, t_{cr} decreases as K_{fs} and H increase. In other terms, high values of the initial soil water content, the saturated soil hydraulic conductivity and the established ponded depth of water on the infiltration surface indicate a reduced duration of the initial transient phase of the infiltration process. In any case, the t_{cr} values obtained on the basis of Eq. (4.42) cannot be considered as rigidly discriminating between the two temporal domains since there is a time zone around t_{cr} in which neither the early-time analysis nor the steady-state analysis is appropriate, as flow is neither purely dominated by capillary forces nor gravitational forces.

4.3.6 Accounting for Soil Swelling

The early falling head or constant head methods assume that soil is non-deformable, even if these measures are dedicated to fine soils prone to swelling. The application of these methods in such cases may lead to erroneous results for different reasons.

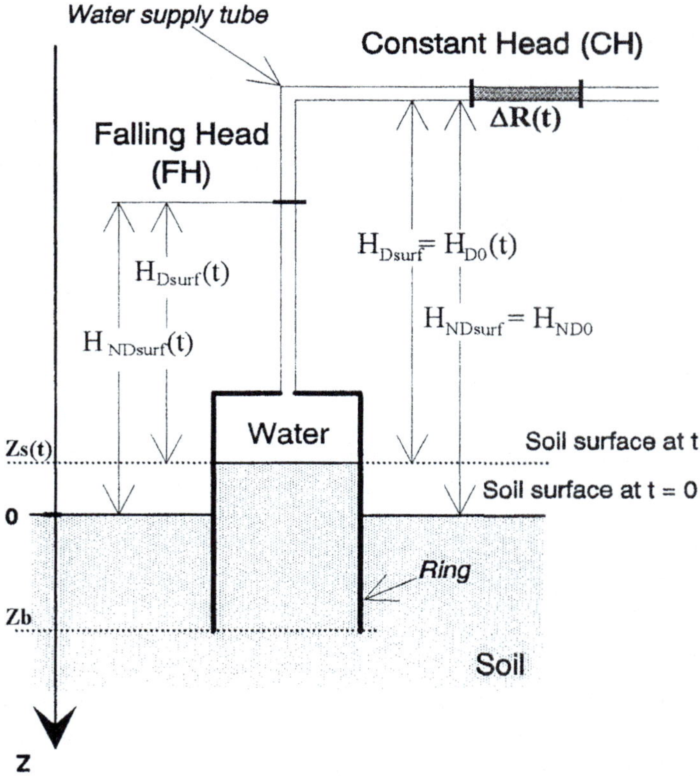

Fig. 4.15 Schematic view of the constant head/falling head pressure infiltrometer for swelling soils (Adapted from Gérard-Marchant et al. 1997, reprinted with permission)

At first, soil swelling may trigger the rise of both soil surface and infiltrometers, which may disturb the contact between the device and the soil below. To alleviate such problems, the use of weights is advised to stabilize the infiltrometer and steady the contact between the device and the soil. Secondly, soil swelling may lead to miss-estimation of the real water flux that infiltrates into the soil. A schematic view of soil rise in infiltrometers is illustrated in Fig. 4.15. In case of soil rise, the amount of water that enters into the soil corresponds to the decrease in total water volume, as revealed by the position of the water meniscus in the water supply tube $\Delta R(t)$, plus the layer of water replaced by the soil in the infiltrometer $\Delta Z_s(t)$. Basically, it can easily be understood that in case of no changes in water level in the water supply tube ($\Delta R = 0$), any rise of the soil of X cm corresponds also to X cm of water infiltrated into the soil. The omission of soil rise may thus lead to under-estimation of the real water cumulative infiltration and thus of saturated hydraulic conductivity. At third, the conceptual models presented above are all based on the hypothesis of non-deformable soils. Consequently, the equations used in such context can no longer be considered for swelling soils. Gérard-Marchant et al. (1997) developed a specific analytical approach to treat data corresponding to water infiltration into swelling soils. They considered the transformation from Eulerian to Lagragian coordinate systems, to account for solid movement and to properly quantify the movement of water relative to soil particles (Gérard-Marchant et al. 1997). They extended the approaches described above to deformable materials and introduced a new concept, i.e. the bulk sorptivity, which characterizes the solid phase movement. They validated their approach using field data obtained for two liners of swelling and slowly permeable materials.

4.4 Macroporous and Bi-modal Soils

In recent decades, field and laboratory studies have demonstrated that preferential flow is more a rule than the exception (Köhne et al. 2009a). Preferential flow can be seen as uneven and rapid movement of water and solutes through restricted zones of the porous media, with slow or even negligible movement of water and solutes elsewhere. Flow pathways develop in restricted zones of the soil (Flury et al. 1994; Alaoui et al. 2011) whereas, at the soil surface, water infiltration may appear usual and gives no evidences of preferential flow. Water infiltrates through a portion of the soil volume, thus reducing the access of pollutants to the soil matrix and their opportunity to adsorb onto soil particles (Lassabatere et al. 2004; Köhne et al. 2009b). The development of preferential flow may be linked to the soil macroporosity (macropores and cracks), soil hydrophobicity or fingered flow due to hydraulic instability. Macroporosity includes cylindrical biopores made by burrowing soil animals and plant roots, planar fissures formed by wetting/drying or freezing/thawing cycles, and irregularly shaped 'packing voids' between denser aggregates in structured topsoils (Jarvis 2007). When water pressure is high enough to activate the macroporosity, flow develops essentially into the macropores, leading to non-equilibrium and preferential flow phenomena in the soil.

Concepts and related models developed for preferential flow are often classified into single porosity with non-equilibrium flow, dual porosity, dual permeability and multi-region models (Šimůnek et al. 2003). In this chapter, the focus is put on dual permeability models. These models implement water flow in and between two pore regions, i.e. matrix with slow flow and a fast-flow region, and they encompass most cases of preferential flow (Šimůnek et al. 2003). The following subjects are successively presented: conceptual models for the description of preferential flow, related analytical and numerical models, methods for detection of preferential flow during water infiltration experiments, and methods to characterize dual permeability behaviors using water infiltration experiments.

4.4.1 Conceptual Models for Preferential Flow

The different models are successively discussed by increasing complexity starting from the simple way to simulate physical non-equilibrium flow developed by Ross and Smettem (2000). These authors consider that water flows through the porous medium following Richards' equation, but water content is not instantaneously in equilibrium with water pressure. Such situation can be encountered when water takes time to reach remote zones (typically zones far from flow pathways) and when equilibration of water contents requires time. Water flow is thus modelled with the following set of equations:

$$\frac{\partial \theta}{\partial t} = \frac{\partial}{\partial z}\left(K(h)\left(\frac{\partial h}{\partial z} - 1\right)\right) \tag{4.43a}$$

$$\frac{\partial \theta}{\partial t} = f(\theta, \theta_e) \tag{4.43b}$$

$$f(\theta, \theta_e) = \frac{(\theta_e - \theta)}{\tau} \tag{4.43c}$$

where θ (L^3L^{-3}) is the volumetric soil water content, *h* (L) is the soil water pressure head, *K* (L T^{-1}) is the soil hydraulic conductivity, *t* (T) is time, *z* (L) is the depth taken positive downward, *f* is a linear function of actual and equilibrium volumetric water contents, θ and θ_e respectively, and τ is an equilibration time constant. One of the main advantages of this model is that it requires only one parameter more than the usual Richards' equation, namely the equilibration time constant. Ross and Smettem (2000) investigated the effect of this parameters onto water infiltration and drainage in a synthetic soil and compared their modeled results with experimental data from Smettem et al. (1994). They could model successfully the downward flow through cores of a structured field soil and concluded that their approach based on the implementation of non-equilibrium flow in a single porosity domain could be an interesting tool for modelling preferential flow.

Dual porosity and dual permeability models are among the most famous approaches for modelling preferential flow in soils. For dual porosity model, water flow is restricted to the fractures or the inter-aggregate pores and macropores, whereas water does not move at all in the matrix (Šimůnek et al. 2003). Richards' equation is considered for the flow of mobile water, with an additional term to account for water exchange between the mobile and stagnant zones. This exchange is modelled by a first-order process, i.e. a linear function of the difference in saturation degree between mobile and stagnant zones. The dual-permeability model considers two single-permeability media separated by a permeable interface. The flow equations for the fast-flow and the matrix regions are ruled by Richards' equation (Gerke and van Genuchten 1993):

$$\frac{\partial \theta_f}{\partial t} = \frac{\partial}{\partial z}\left(K_f \frac{\partial h_f}{\partial z} - K_f\right) - \frac{\Gamma_w}{w} - \Phi_f \tag{4.44a}$$

$$\frac{\partial \theta_m}{\partial t} = \frac{\partial}{\partial z}\left(K_m \frac{\partial h_m}{\partial z} - K_m\right) - \frac{\Gamma_w}{1-w} - \Phi_m \tag{4.44b}$$

where subscripts f and m represents, respectively, the fast-flow and the matrix domains, Γ_w (T^{-1}) stands for water exchange between the matrix and fast-flow regions, w is the fraction of the whole medium occupied by the fast-flow region and Φ (T^{-1}) is a sink/source term. The exchange rate of water between the fast-flow and matrix regions Γ_w is assumed proportional to the difference in pressure heads (Gerke and van Genuchten 1993):

$$\Gamma_w = \alpha_w (h_f - h_m) \tag{4.45}$$

in which α_w (L^{-1} T^{-1}) is a first-order mass transfer coefficient. In opposite to the model developed by Ross and Smettem (2000), the dual permeability model accounts for two concomitant flows in the matrix and the fast flow regions. Yet, this model requires more parameters and a complete characterization of hydraulic conductivity and water retention curves for both the matrix and the fast-flow region along with knowledge of parameter α_w. This parameters is often defined as follows (Gerke and van Genuchten 1993):

$$\alpha_w = \frac{\beta_a}{d^2} K_a \gamma_w \tag{4.46}$$

where β_a and γ_w are scaling factors, d (L) stands for the characteristic length of matrix elements, K_a ($L\,T^{-1}$) stands for the interfacial hydraulic conductivity. Gerke and van Genuchten (1996) investigated several methods for the determination of K_a as a function of the averaged water pressure head at the interface between the fast-flow region and the matrix. They concluded that the best option was to consider K_a as an arithmetic average of hydraulic conductivities obtained for water pressure head in the fast-flow region and in the matrix, namely h_f and h_m:

$$K_a(h) = \frac{1}{2}\left(K_a(h_f) + K_a(h_m)\right) \tag{4.47}$$

The use of Eq. (4.47) implies that the medium is made of geometrically well-defined rectangular or other types of elements (van Genuchten and Dalton 1986). Geometrically based models are conceptually attractive, but their use for field applications may be challenging for several reasons (Šimůnek et al. 2008). Indeed, in structured soils, the shape and size of aggregates and matrix blocks may vary significantly. In addition, some of the parameters of Eq. (4.46) may not be identifiable, in particular β_a and γ_w. Hence, rather than using Eq. (4.46) directly, parameters β_a, d, and γ_w can be lumped into an effective hydraulic conductance K_a^* of the interface between the fast-flow region and the matrix (Šimůnek et al. 2003). It may be noted that when water exchange is quasi-instantaneous, water pressure heads are similar in the matrix and the fast-flow regions. In this case, solving Eqs. (4.44a and 4.44b) is similar to solving Richards' equations with bimodal hydraulic conductivity and water retention functions, as defined by Durner (1994) (Zurmühl and Durner 1996, 1998).

Several alternative models were proposed on the basis of the dual flow approach, but flow equations are modified for the macropore or fast-flow region. Such changes are required by the physics of flow that can no longer be accurately described by Darcy-Buckingham's law and Richards' equations. In such cases, Richards' equation is still considered for flow in the matrix, but alternative laws are considered for the description of flow into the fast-flow region. Some alternatives consider that capillarity is negligible in the fast-flow region because of the large size of macropores. In such a case, flow is mainly driven by gravity and viscous forces due to friction within the fluid itself, and is referred to as viscous flow (Germann 2014). In such case, it can be demonstrated that water flux is related to water content by a power law and that its combination with the mass conservation law leads to the kinematic wave equations (Germann 1985; Germann and Beven 1985):

$$\frac{\partial q}{\partial t} + C\frac{\partial q}{\partial z} + \Gamma_k = 0 \tag{4.48a}$$

$$q = b\theta_f^a \tag{4.48b}$$

$$C = \frac{dq}{d\theta_f} \tag{4.48c}$$

where q (L T^{-1}) is the volumetric flux density, C (L T^{-1}) is the kinematic wave velocity, b (L T^{-1}) is a macropore conductance parameter, a is a kinematic exponent and the term Γ_k (L T^{-2}) represents water exchange from macropores to the matrix. The adequacy of the power law, i.e. Eq. (4.48b), was demonstrated by Germann (2014) for the case of viscous flow into well-defined geometries like cylindrical macropores or planar fractures. For the case of fully saturated cylindrical macropores, Hagen-Poiseuille's law can fully be demonstrated with such an approach (Germann 2014). Its application leads to the quantification of water flux from the number of macropores per unit area of porous medium, N_p, and macropore radius, r_p, as follows (Ahuja et al. 2000):

$$q = \frac{N_p \rho g \pi r_p^4}{8\nu} \tag{4.49}$$

where ρ stands for density of water, g for the gravitational constant, and ν is the water dynamic viscosity. The kinematic wave model proposed by Germann (1985) and Germann and Beven (1985) has the advantages to be more general than the application of Hagen-Poiseuille's law and to encompass non-saturated macropores with complex geometries. This is the reason why the kinematic wave equation has been the subject of several studies and developments (Rousseau et al. 2004; Di Pietro et al. 2003).

The choice of one of the possible approaches mainly depends on the physics of flow. It is thus necessary to apprehend the type of flow in the studied conditions. At this purpose, Coppola et al. (2009) proposed a conceptual classification for preferential flow according to field conditions and types of soil: case (1) preferential flow in real non-capillary macropores; case (2) preferential flow in inter-aggregate pores, which are often capillary pores; case (3) fingering flow due to the instability of the wetting fronts, often encountered at the interface between two materials with a lower unsaturated hydraulic conductivity below, case (4) preferential flow due to spatial irregularities or temporal dynamics in soil wettability. According to these authors, Richards' equation is applicable to the two pore domains (matrix and fast-flow region) in cases 2–4. In the first case, Richards' equation is no longer applicable and the kinematic wave equation or Hagen-Poiseuille's equation should be applied instead. Briefly, dual permeability models based on kinematic wave equation or Hagen-Poiseuille's equation address the case of large cracks or macropores where capillary forces are negligible. Dual permeability models that combine two Richards' equations for the matrix and the fast-flow region should be applied for the cases when macropores and fractures are smaller, leading to non-negligible capillary forces.

To sum up, these approaches do not properly describe complex flow that can be encountered in macropores such as for instance turbulent flow regimes, film or rivulet flow, flow enhancement in the matrix, etc. At first, flow regime is often non-laminar and turbulent in macropores (Jarvis 2007; Logsdon 1995). On second, complex rivulet and film flow may develop below the macropore water-entry pressure (Jarvis 2007; Cey and Rudolph 2009). Basically, films of water, with thickness in the order of several dozens of micrometers, may form and move along macropore walls, leading to turbulent or unstable flow regimes with very high velocities (Tokunaga and Wan 1997; Tokunaga et al. 2000; Or et al. 2000). At third, when water pressure head is increased to full water saturation of macropores, water flow was proved to be drastically enhanced in the matrix surrounding macropores, which contradicts Darcian approaches that predict flow restriction to macropores (Lamy et al. 2009; Lassabatere et al. 2011; Ghodrati et al. 1999). Such flow enhancement in the matrix has a drastic impact on water flow and solute transport and boots the effect of macropores in soils (Ghodrati et al. 1999). In addition to these specific flow patterns into the soil, phenomena at the soil surface may be much more complex than described and impact also the activation of the macropore network. Weiler (2005) found that the activation of macropores within the macropore networks depended on the initiation of flow at surface, this being dependent on soil surface micro-topography. This author also insisted on the necessity to account for such surface initiation that routes water in a small fraction of the whole macropore network and for interactions between close macropores. Finally, Greco (2002) insisted on the need to account for soil swelling during water infiltration and its impact on macropore and crack geometry. These illustrative example show the complexity of preferential flow and the need for additional development of specific conceptual models and tools.

4.4.2 Modelling Water Infiltration into Dual Permeability Soils

Several numerical tools were developed in relation with the conceptual models described above. The following numerical models are cited as illustrative examples, among the most famous models. Dual porosity and dual permeability models utilizing Richards' equation for both the matrix and the fast-flow region were implemented into HYDRUS 2D-3D software (Šimůnek and van Genuchten 2008). Dual permeability model utilizing Hagen-Poiseuille's law for macropore flow was implemented into RZWQM model (Ahuja et al. 2000). In this last model, the Green-Ampt's approach is used to calculate water infiltration in the matrix and water exchange between the macropore and the matrix whereas water redistribution is described by Richards' equation. The use of the Green-Ampt's approach instead of Richards' equation aims at ensuring numerical stability. The combination of Richards' equation for water flow in the matrix with kinematic wave equations for flow into macropores was implemented in MACRO model (Larsbo and Jarvis 2003; Larsbo et al. 2005). Finally, IN3M model (INfiltration-INitiation-INteraction Model) developed by Weiler (2005) and VIMAC model developed by Greco (2002) present illustrative examples of models developed to account for multi-physics like the concomitancy of flow initiation in macropores at surface, soil swelling in activated pores and negative feedback on macropore apertures. Numerical models for preferential flow and solute transport are quite numerous. For more details, the reader could refer to the intensive reviews by Šimůnek et al. (2006) and Köhne et al. (2009a, b).

The boom of computer science and facilities for numerical models may explain the enthusiasm for numerical approaches to the detriment of analytical approaches. Indeed, many analytical tools were developed for prediction of water infiltration into simple permeability systems (Haverkamp et al. 2006) but similar analytical development for the case of water infiltration into dual permeability soils is rare. Lassabatere et al. (2014b) presented an interesting analytical model for the analysis of water infiltration experiments into dual permeability soils. These authors extended the model proposed by Haverkamp et al. (1994) to the case of dual permeability soils. Lassabatere et al. (2014b) applied the model to calculate cumulative infiltration over matrix and fast-flow regions and summed these components, proportionally to ratios of volume occupied by each region, to compute the total cumulative infiltration, I_{2K} (L), from a circular source of radius r (L):

$$I_{2K}(t) = wI_f(t) + (1 - w)I_m(t) \tag{4.50a}$$

$$I_f(t) = I_{f,1D}(t) + \frac{\gamma_f S_f^2}{r\Delta\theta_f} t \tag{4.50b}$$

$$I_m(t) = I_{m,1D}(t) + \frac{\gamma_m S_m^2}{r\Delta\theta_m} t \tag{4.50c}$$

$$\frac{2\Delta K_f^2}{S_f^2} t = \frac{1}{1-\beta_f}\left[\frac{2\Delta K_f}{S_f^2}\left(I_{f,1D} - K_{f,0}t\right) - \ln\left(\frac{\exp\left(\beta_f \frac{2\Delta K_f}{S_f^2}\left(I_{f,1D} - K_{f,0}t\right)\right) + \beta_f - 1}{\beta_f}\right)\right] \tag{4.50d}$$

$$\frac{2\Delta K_m^2}{S_m^2} t = \frac{1}{1-\beta_m}\left[\frac{2\Delta K_m}{S_m^2}\left(I_{m,1D} - K_{m,0}t\right) - \ln\left(\frac{\exp\left(\beta_m \frac{2\Delta K_m}{S_m^2}\left(I_{m,1D} - K_{m,0}t\right)\right) + \beta_m - 1}{\beta_m}\right)\right] \tag{4.50e}$$

where subscripts $2K$, f and m represents, respectively, the dual permeability porous medium, the fast-flow and the matrix domains, I_{3D} (L) is the axisymmetric cumulative infiltration, I_{1D} (L) is the one-dimensional cumulative infiltration, ΔK is the difference between the surface hydraulic conductivity, K_0 [$=K(\theta_0)$], and the initial value of soil hydraulic conductivity, K_i [$=K(\theta_i)$], $\Delta\theta = (\theta_0 - \theta_i)$, S (L $T^{-0.5}$) is soil sorptivity, β and γ are shape parameters and w is a proportionality coefficient corresponding to the fraction of surface occupied by fast-flow domain. To be applied, this model requires some conditions. At first, the model addresses the case of dual permeability media with darcian flows in the two pore domains. Typically, according to Coppola et al. (2009), the macropores should be capillary macropores, and the flow should be laminar with both gravity and capillary forces as the main forces. On second, all requirements must be respected with regards to the validity of the Haverkamp's model for each pore domain (see Sect. 1.4 on infiltration process and equations). In addition, the additivity of cumulative infiltration described by Eq. (4.50a) was stated for the case of separate infiltrations in two pore domains without any interaction.

The adequacy of Eq. (4.50a) for the case of dual permeability soils with non-negligible water exchange was demonstrated by Lassabatere et al. (2014b), using analytically generated data. These authors modelled wetting fronts and cumulative infiltration at soil surface for several values of the interfacial hydraulic conductivity, K_{sa}, which governs water exchange at the interface between the matrix and the fast-flow regions. The results are depicted in Fig. 4.16 for the case of a loamy matrix containing 10 % (v/v) of macropores, 1 mm in diameter (Lassabatere et al. 2014b). In particular, the cumulative infiltration curves are shown in Fig. 4.16a. For zero interfacial hydraulic conductivity, wetting fronts move much faster in the fast flow region with no water exchange between regions

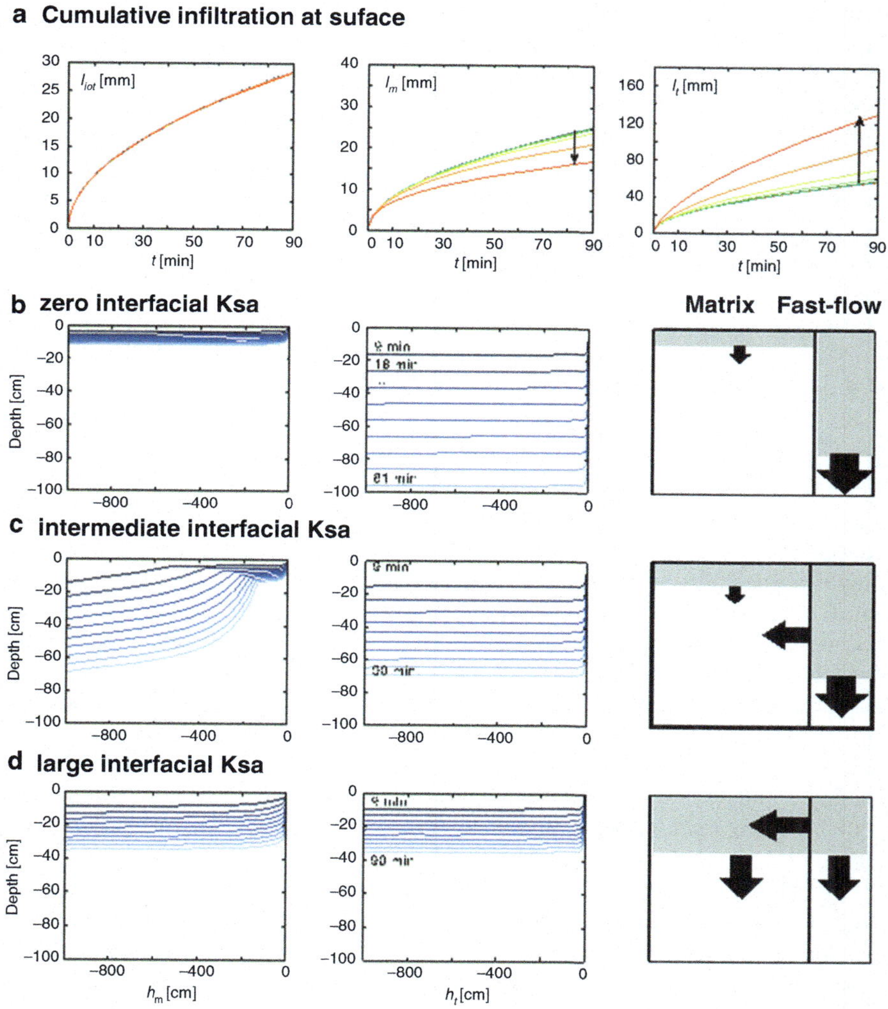

Fig. 4.16 Water infiltration into dual permeability soils as a function of interfacial saturated hydraulic conductivity K_{sa}: cumulative infiltrations at surface (**a**), and water pressure head profiles (**b**–**d**) (Modified from Lassabatere et al. 2014b, reprinted with permission)

(Fig. 4.16b). For intermediate interfacial hydraulic conductivity, wetting fronts move faster in the fast flow region, and a part of water flows from the fast flow region towards the matrix (Fig. 4.16c). For very large values of interfacial hydraulic conductivity, water exchange is instantaneous so that water pressure heads instantaneously equilibrate between the matrix and the fast-flow regions. As a results, wetting fronts move exactly at the same speed (Fig. 4.16d). Clearly, water exchange influences flow pattern in the soil but has no impacts on total cumulative infiltration at surface (Fig. 4.16a). In fact, any rise in K_{sa} increases infiltration over the fast-flow region but decreases infiltration over the matrix (Fig. 4.16a), resulting in no effect on total cumulative infiltration. In conclusion, whatever water exchange and resulting flow pattern in the soil, total cumulative infiltration remains the same at the surface and equals that of the specific case of zero water exchange. Similar results were found by Ma and Shao (2008), who modelled water infiltration at the surface of dual porosity soil with several water exchange constants. Therefore, the proposed model Eq. (4.50a) may be used for any dual permeability soil.

4.4.3 How to Detect Preferential Flow

It seems quite challenging to detect dual permeability behavior with only water infiltration experiment at one value of pressure head at surface. Indeed, several authors have already demonstrated that one sole cumulative infiltration obtained for one pressure head was not adequate to characterize the full set of parameters of single permeability soils (Šimůnek and van Genuchten 1996, 1997). The situation must be even worse for the characterization of dual permeability soils, which requires the estimation of more parameters. Additional data is thus required that may be additional infiltration tests at several values of water pressure heads, additional tests involving the injection of solutes, or direct field observations using dyes. Moreover, the combination of information on both unsaturated and saturated behavior of the soil must be targeted. Indeed, unsaturated behavior mainly results from the behavior of the soil matrix, whereas saturated behavior will be more representative of the hydraulic behavior of soil macropores and cracks.

The Beerkan method along with BEST method is insufficient to detect and characterize dual permeability behavior and may lead to incoherent and unreliable results. The inadequacy of Beerkan method was clearly demonstrated by Lassabatere et al. (2014b), using synthetic numerically generated data. These authors modelled water infiltration experiments using Eqs. (4.50) for several synthetic dual permeability soils and for Beerkan and multi tension water infiltration tests. For Beerkan experiments, they obtained shapes that were completely similar to that obtained for a single permeability soil with a perfect fit onto the single permeability model (Fig. 4.17a, Beerkan STI). In opposite, the fit of the single permeability model onto synthetic data from multiple tension infiltration experiments was much poorer (Fig. 4.17a, MTI). In particular, the analysis of errors showed a strong auto-correlation of errors, with under-estimation of cumulative infiltration for the lowest and highest water pressure heads along with over-estimation of cumulative infiltration at intermediate water pressure heads. Similar results were obtained with real experimental data by Yilmaz et al. (2013)

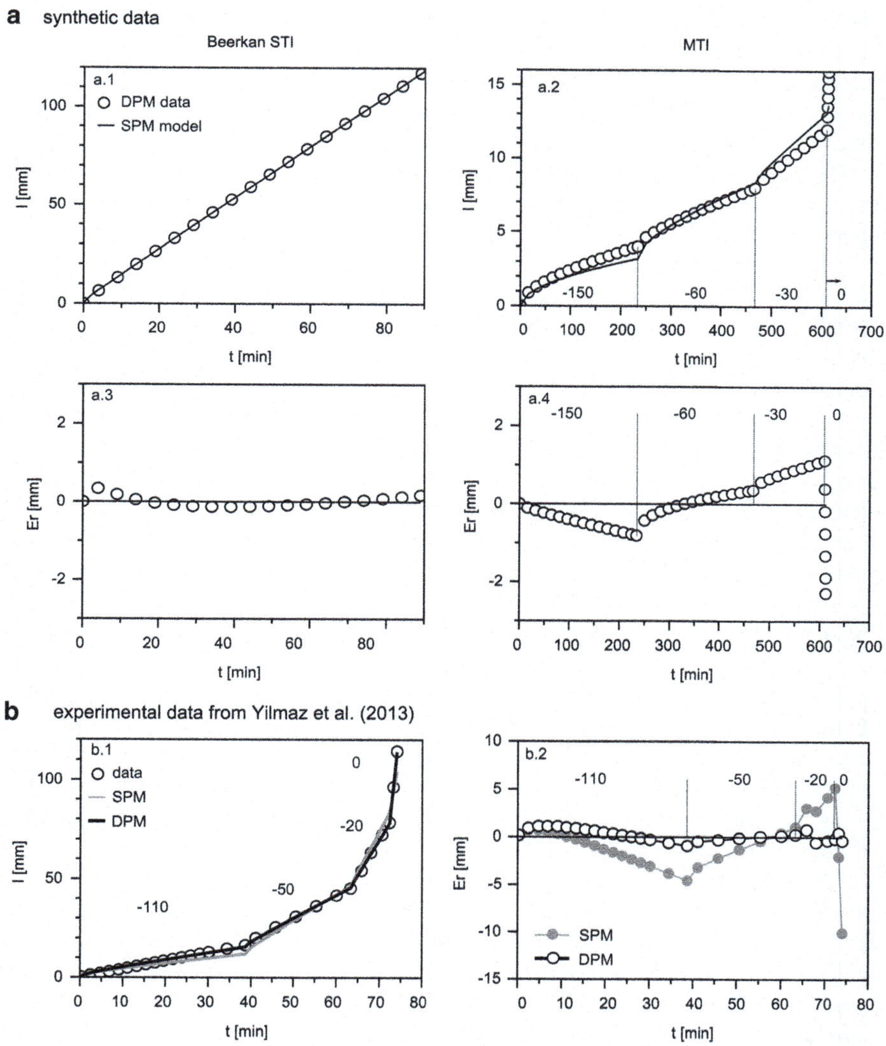

Fig. 4.17 Fit of single permeability model to data obtained for dual permeability behavior, (**a**) synthetic cumulative infiltration obtained for zero water pressure head at surface (*left*) and for a series of water pressure heads (*right*) and (**b**) real experimental dataset for a dual porosity porous medium (Modified from Lassabatere et al. (2014b), the indicated values corresponds to the water pressures heads imposed at surface, reprinted with permission)

(Fig. 4.17b). Dual permeability behavior induces a sharp increase of cumulative infiltration slopes as the water pressure head applied at surface increases, and this slope increase cannot be reproduced properly by single permeability models (Lassabatere et al. 2014b). Based on these results, Lassabatere et al. (2014b) concluded that multi-tension experiments are needed to detect dual permeability soil and proposed the following advices: (i) the values of the applied water pressure head should be chosen so as to activate the matrix and the fast-flow region progressively, (ii) volume of water infiltrated during the application of water pressure heads should be of the same order, meaning much longer times for the lowest water pressure heads.

Dyes are often used to depict flow pathways and to pinpoint preferential flows in the ground (Alaoui et al. 2011; Kodešová et al. 2012). Dyes can also be used to characterize flow pathways under infiltrometers (Cey et al. 2009; Cey and Rudolph 2009; Alberti and Cey 2011; Timlin et al. 1994). Infiltrometers are filled with a solution of dyes and successive water pressure heads are applied at the soil surface. At the end of the experiment, after soil excavation, the soil is photographed at the surface and eventually at several soil sections below the infiltrometer. Several sections can be prepared including vertical sections, usually located along the centerline of infiltrometers, and horizontal soil sections at several depths, as illustrated in Cey and Rudolph (2009). A calibration between soil color and dyes concentrations is recommended, using photographs of the same soils previously stained with increasing known dyes concentrations (Cey and Rudolph 2009). Photographs then undergo a process of image analysis to discriminate between stained and unstained areas and to classify stained areas into several classes of dye concentration (Weiler and Flühler 2004). An illustrative example is proposed in Fig. 4.18, from the study of a west Canadian soil by Cey and Rudolph (2009). The use of dyes helped these authors to detect main wetting fronts under the infiltrometers and also film flow and rivulet flow developing in macropores (white arrows in photographs, Fig. 4.18a). The distribution of dyes in horizontal and vertical sections clearly pinpointed preferential flow below two of the infiltrometers (VK-D3 and VK-D4, Fig. 4.18b–c) with even preferential flow along separate macropores (VK-D4, Fig. 4.18c). These studies show the interest of the use of dyes for the detection of preferential flow.

The use of tracer may also help to map flow pathways under infiltrometers with quantification of the fraction of mobile water, i.e. water active in flow. Infiltrometers are filled with a tracer solution and the soil is sampled for determination of tracer concentration at the end of the experiment (Angulo-Jaramillo et al. 2000; Clothier et al. 1992; Roulier 1999; Roulier et al. 2002). Most of these approaches consider the so-called MIM approach based on the hypothesis that water is fractionated into mobile (M) and immobile (IM) water fractions (Gaudet et al. 1977). Infiltrating water and dissolved solutes move only into the mobile water and diffuse from mobile to immobile water fractions. Yet, it is assumed that diffusion is negligible during the course of the experiment and that tracer concentration remains null in the immobile water fraction. In such a case, the ratio of global tracer concentration to injected concentration corresponds to mobile water fraction (simple dilution). Consequently, the measure of the global concentration C^* along with total water content θ_0, and the knowledge of the injected concentration C_m allows the determination of mobile water content θ_m as follows (Clothier et al. 1992; Angulo-Jaramillo et al. 1996):

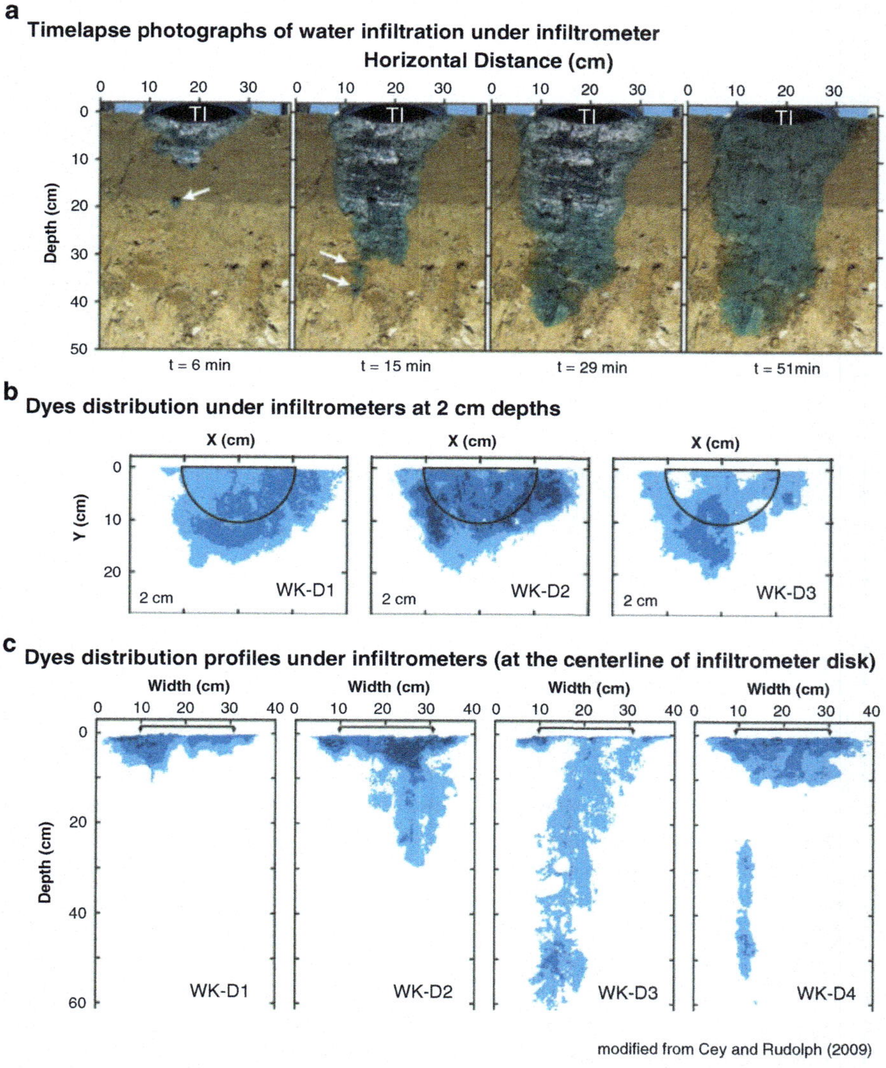

Fig. 4.18 Use of dyes to assess subsurface flow patterns (Cey and Rudolph 2009); photographs of water infiltration under the infiltrometer (**a**), *horizontal sections* below the infiltrometer (**b**), soil *vertical sections* centered along the centerline of the tension infiltrometer disk (**c**); the location of the disk is indicated on the soil *horizontal sections* (**b**) (Reprinted with permission)

$$\theta_m = \theta_0 \frac{C^*}{C_m} \tag{4.51}$$

Angulo-Jaramillo et al. (1996) used this method to get an insight on the spatial variability of flow under the infiltrometer device. These authors found under the disc mobile water fractions below unity with uniform spatial distribution for the highest water pressure heads and more heterogeneous distributions for the two lowest water pressure heads (Fig. 4.19a). They reported an increase in mobile water fraction with the value of applied water pressure head. These authors attributed such an increase to the activation of the macroporous domain (Angulo-Jaramillo et al. 1996). On the same basis, many studies used tracer-filled infiltrometers to characterize mobile water content below the infiltrometer (Clothier et al. 1992, 1995, 1996; Casey et al. 1997, 1998; Jaynes and Shao 1999; Gaudet et al. 1995; Roulier 1999; Roulier et al. 2002). Gaudet et al. (1995) performed an infiltration test on a loamy soil with a solution enriched with ^{18}O as tracer solution (Fig. 4.19b). The concentration of ^{18}O was used to derive mobile water contents using Eq.(4.51). These authors found a strong spatial heterogeneity of ^{18}O concentration and mobile water contents (Fig. 4.19b) and related it to preferential flow under the disk. They analyzed the spatial distribution of these parameters through geostatistical analysis and concluded that the whole variability at short distance was completely held within the disk size, meaning that the disk radius was probably appropriate to encompass flow variability at the soil surface.

4.4.4 *How to Characterize Dual Permeability Soils*

This section presents several methods that use water infiltration experiments for the characterization of dual permeability soils or soils containing macropores. It presents successively the methods that make use only of raw data without any modelling, the methods that use numerical modelling and finally the methods based on analytical approaches similar to BEST method. All the presented methods are extracted from previous studies and show the complexity of the characterization of dual permeability soils together with the need for both extensive data and further developments of strategies for characterizing the hydraulic properties of multi-region or dual permeability soils.

The first method makes use of the methodology developed by Watson and Luxmoore (1986) using only ponded and tension infiltrometers. Ponded infiltrometer experiments must be performed with sufficient water depth at surface to ensure full saturation of the soil and activation of the fast-flow region. In their study, Watson and Luxmoore (1986) used double-ring infiltrometers and derived saturated hydraulic conductivity from steady state falling-head infiltration. Ponded infiltration experiments are then followed by tension infiltration experiments at the same place. The application of increasing tension (i.e. decreasing water pressure head) allows the progressive exclusion of pores with larger diameters. Watson and Luxmoore (1986) suggested to apply tensions of 3, 6 and 15 cm to exclude pore with diameters >1, 0.5 and 0.2 mm. Such proposition is in agreement with the criterion of >1 mm proposed by Luxmoore (1981) to distinguish between macropores and mesopores. The correspondence between tension and pore radius is deduced from the application of Laplace's law. Since the soil has already been wetted during ponded infiltration, water infiltration performed with the tension

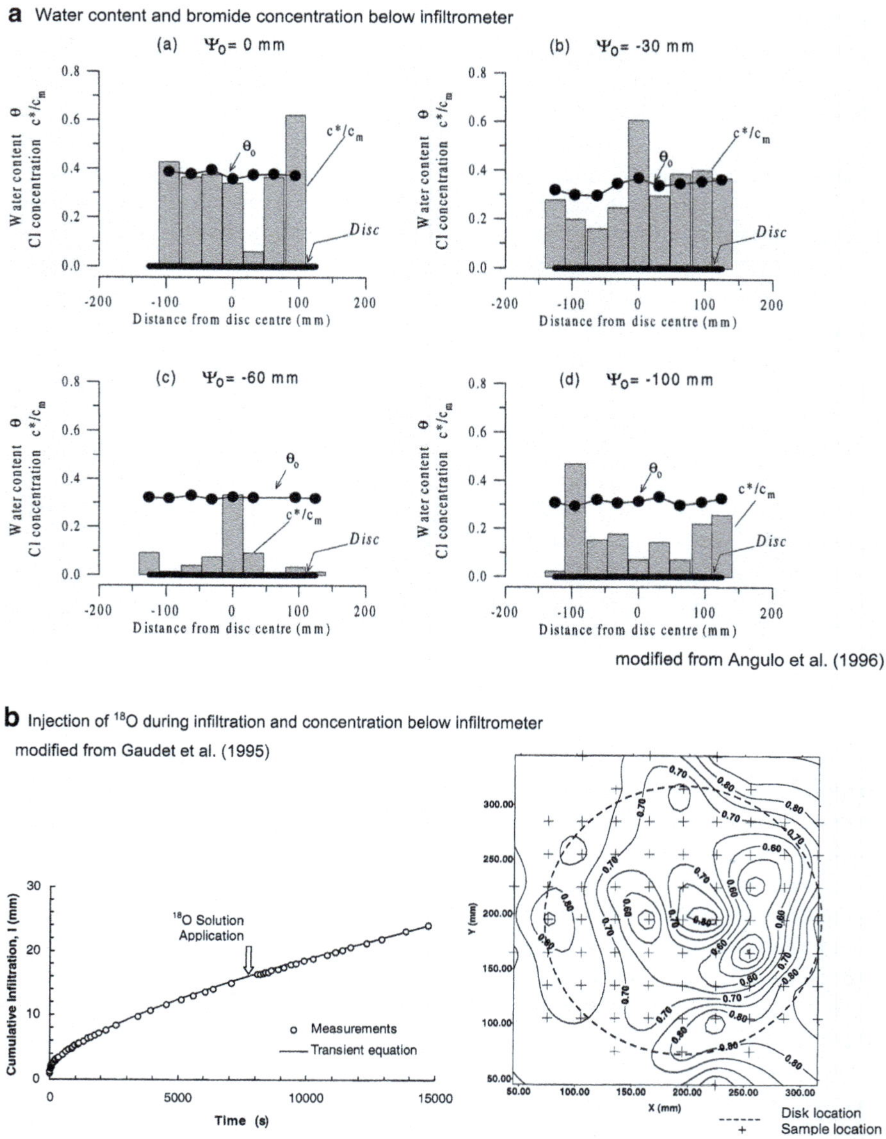

Fig. 4.19 Tracer concentrations at surface and flow pathways under infiltrometers, (**a**) study from Angulo-Jaramillo et al. (1996, reprinted with permission) and (**b**) from Gaudet et al. (1995, reprinted with permission). Water content and tracer relative concentration below infiltrometers, as a function of applied water pressure heads (**a**), cumulative infiltration and spatial distribution of ^{18}O relative concentration below the infiltration disk (**b**)

infiltrometer can be considered as mostly 1D and gravity driven, leading to equality between hydraulic conductivity and steady-state infiltration rates. At the end of the infiltration experiments, a set of hydraulic conductivity values is obtained, corresponding to the applied tensions, i.e. 0, 3, 6 and 15 cm. The difference in hydraulic conductivity between two successive tensions can be used to quantify the contribution of a specific pore domain, i.e. the pores with diameter intervals of >1 mm, 0.5–1 mm, and 0.2–0.5 mm, respectively. For instance, the difference between the values of hydraulic conductivity obtained for ponded condition and for a tension of 3 cm quantifies the contribution of pores with diameter >1 mm. Watson and Luxmoore (1986) suggested to use Poiseuille's law to link the difference in hydraulic conductivity, ΔK_p, to the number of active pores, N_p, and the pore size r_p (L) representative of the pore domain, leading to:

$$N_p = \frac{8\mu\Delta K_p}{\pi\rho g {r_p}^4} \tag{4.52}$$

where μ (M $L^{-1}T^{-1}$) and ρ (M L^{-3}) are the viscosity and the density of water, respectively. Watson and Luxmoore (1986) used the lower boundary of pore size intervals as the pore size representative of the pore domain. The total effective macroporosity can be calculated using:

$$\Theta_p = N_p \pi r_p^2 \tag{4.53}$$

If the realization of four successive water infiltration runs is too time-consuming, two water infiltrations could be performed including one ponded infiltration to activate macroporosity followed by one tension infiltration to activate only the matrix. Timlin et al. (1994) investigated such experiment, considering a tension of 6 cm instead of the value of 3 cm suggested by Watson and Luxmoore (1986). In all cases, the method presented above may appear quite convenient and easy to conduct. The application of water pressure heads by decreasing order aims to avoid capillary driven fluxes and to enhance the 1D character of the infiltration process. Yet, hysteresis may impact water infiltrations and should be accounted for.

With the development of numerical modelling, numerical methods for inverting experimental infiltration data developed as well (Šimůnek and van Genuchten 1996). The two methods presented below are illustrative examples of the use of numerical modelling for the estimation of hydraulic properties of dual permeability soil. The first example, from Kodešová et al. (2010), makes use of the dual permeability model as described by Eqs. (4.44a and 4.44b), which considers two interacting domains superimposed in space: one domain for the matrix and the other for the macropores or the fast-flow region, with water exchange between these two domains. The second example, from Alberti and Cey (2011), illustrates the use of discrete macropore models whereby macropores are explicitly represented within separate subregions of the numerical domain. Dual permeability and discrete macropore models appear to be the two main options when inverting infiltration data for soils containing macropores and/or fast-flow region. According to Alberti and Cey (2011), discrete macropore models require fewer parameters and represent more accurately the physics of preferential flow. But, they require a proper description of macropore geometry and continuity and typically address small scales. Whatever the chosen option, experimental data must combine water infiltration data obtained at least for two water pressure heads, one for the activation of the whole pore domain (i.e. matrix and macropores) and the other for the activation of only the matrix. In practice, Kodešová et al. (2010) combined the tension infiltrometer with the Guelph permeameter, whereas Alberti and Cey (2011) used the same tension infiltrometer device for the application of the two water tensions. Below, the methods used by these authors to determine the whole set of hydraulic parameters are presented.

Kodešová et al. (2010) aimed at characterizing the complete sets of hydraulic parameters of a Haplic Luvisol soil with macropores. They assumed the van Genuchten (1980) model with $m = 1 - 1/n$ for water retention and the Mualem (1976) model for hydraulic conductivity, with the tortuosity parameter l fixed at its default value, i.e. 0.5. For the matrix, the authors estimated the water retention curve parameters and the saturated hydraulic conductivity through inverting the data from tension water infiltration experiments obtained at −2 cm water pressure head. To avoid problems of non-uniqueness of parameter estimate, as already discussed in previous papers (Šimůnek and van Genuchten 1996, 1997), Kodešová et al. (2010) considered additional information from laboratory measures on intact soil cores. Then, the data obtained with the Guelph permeameter were inverted using the dual permeability model to derive the hydraulic parameters of the fast-flow region. The authors considered the following assumptions and additional data. At first, they determined the volume fraction occupied by macropore using image analysis of soil as described by Kodešová et al. (2009) in their micro-morphological characterization of the studied soil. The parameters needed for aggregate geometry (see Eq. (4.46)) were also derived from the same micro-morphological studies and confirmed on the field with the use of dyes (Kodešová et al. 2010). For water retention curve, residual water content was fixed at zero and saturated water content was equaled to the value of porosity, so that both matrix and macropore regions had the same saturated water content. The authors justified such a choice on the basis of previous studies (Gerke and van Genuchten 1996; Gerke et al. 2007). For the other parameters of the water retention curve, several sets of values were fixed a priori for scale parameters α_f and shape parameter n_f. These values were fixed so as to ensure that macropores saturate only for pressure heads close to zero, i.e. a step-like shape for water retention curve, as suggested by Othmer et al. (1991). The value of the interfacial hydraulic conductivity, K_{sa}, could not be estimated directly and several plausible values were tested. The authors used K_{sa} estimates from a previous study dealing with solute injection in soil cores using laboratory column (Kodešová et al. 2009). Finally a set of values was defined for aggregate parameters and for parameters n_f, α_f, and K_{sa}, and the saturated hydraulic conductivity of the macropore domain, K_{sf} was optimized by fitting experimental data from the Guelph permeameter onto the dual permeability model. Such a method looks intensive and shows the complexity of the estimation of the whole set of parameters for dual permeability soils. Despite this, the authors concluded that the applicability of their approach for the assessment of some of the hydraulic parameters of the dual-permeability model (preferably K_{sf} and K_{sa}) is quite conclusive.

For the discrete macropore model, Alberti and Cey (2011) modelled the results from Cey and Rudolph (2009). These data comprise tests exhibiting strong preferential flow (Fig. 4.18, WK-D2-4), and tests without any preferential flow (Fig. 4.18, WK-D1). For numerical modelling, the numerical domain was built to mimic the soil amended with macropores, with most of the domain representing the matrix, and a specific subdomain representing the macropore. This subdomain is vertical, cuts across the whole numerical domain and has a diameter in agreement with field observations (i.e. 0.5 cm). It is positioned below the disk source in agreement with dyes observations. For hydraulic curves, Alberti and Cey (2011) assumed the van Genuchten (1980) model with $m = 1 - 1/n$ for water retention, and the Mualem (1976) model for hydraulic conductivity. The parameters were obtained by fitting the model onto field data, including water cumulative infiltration at surface and the water content at 7 cm below the infiltrometer, representing the depth of dye transport. In addition, they used additional measures in laboratory to constraint estimates and alleviate the problem of non-uniqueness. Basically, water retention curves measured on intact soil cores were used to derive parameters related to the water retention curve of the matrix (Cey and Rudolph 2009). The saturated hydraulic conductivity of the matrix was optimized through fitting the infiltration tests WK-D1 that exhibited no preferential flow. For the macropores, residual and saturated water contents were equaled to those of the matrix. Scale parameter α_f was calculated from macropore radius using Laplace's law. The empirical parameter l_f was taken as 1.0 for a vertically continuous macropore (no tortuosity). Finally, the parameters K_{sf} and n_f had to be optimized through fitting infiltration data for a dataset exhibiting preferential flow, namely WK-D2. The optimized parameters were then used to model the WK-D3 infiltration data that also exhibited strong preferential flow. The adequacy of the model to quantify closely cumulative infiltration, water content and dyes distribution in the soil validates the inversion procedures (Fig. 4.20). On this basis, the authors investigated the influence of macropore density and connectivity on the inversion process and concluded that these were also key parameters. Their study shows the efficiency of tension infiltration data to model film and rivulet flow and to invert part of the hydraulic parameters of the studied soils. Yet, as for the previous study, such inversion procedure requires extensive data. In addition, opposite to the study from (Kodešová et al. 2010) for which Guelph permeameter was used to conduct water infiltration experiments under ponded conditions, all the water infiltration experiments were performed under tension conditions, and full water saturation of the macropores was never attained. This means that the proposed values for hydraulic parameters may be questioned and adapted to the case of full water saturation of the soil.

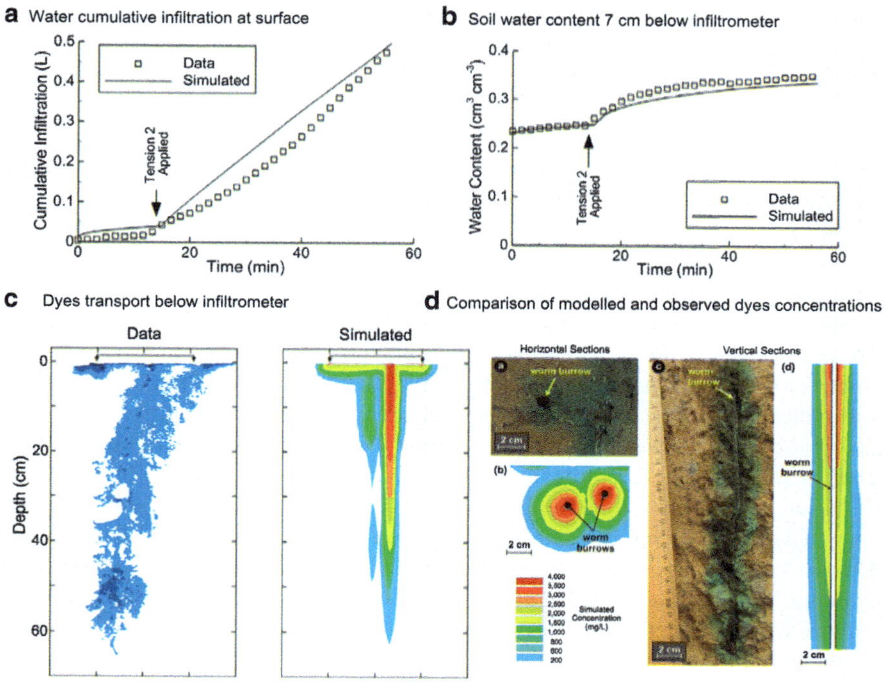

Fig. 4.20 Inverse modeling using discrete macropore models of water infiltration data for a soil containing macropores: accuracy of the model to reproduce cumulative infiltration data (**a**), water content measured 7-cm depth below the infiltrometer (**b**), dyes distribution over a vertical section centered at the centerline of the infiltrometer disk (**c**) and comparison of predicted and observed dyes concentration around macropores (**d**) (From Alberti and Cey 2011, reprinted with permission)

It is clear that numerical methods seem quite convenient but requires data to avoid non-uniqueness of parameter estimates and ill-posed problems. In addition, the two illustrative examples discussed above show that numerical inversion is not yet automatic and requires a lot of assumptions. In both studies, the optimized parameters were adjusted manually without any use of inversion algorithm, not yet available in the current version of HYDRUS 2D (Kodešová et al. 2010; Alberti and Cey 2011). There is a need for simpler methods that could provide robust first estimates, like BEST method for single permeability soils. Some direct and simple analytically based methods were developed, on the basis of the extension of BEST method to dual permeability soils. These are presented below.

Gonzalez-Sosa et al. (2010) used data from mini-disk tension infiltrometer and ponded single ring infiltrometers to characterize a large panel of soils containing macropores. For mini-disk infiltrometer, water tension was fixed at 2 cm to exclude soil macropores from the infiltration process (Gonzalez-Sosa et al. 2010). The matrix hydraulic parameters were derived by running the BEST method on tension infiltrometer data. For the macropores, saturated hydraulic conductivity was deduced from water infiltration experiments under ponded conditions using a falling-head method. The single ring infiltrometer was 40 cm in diameter to encompass spatial heterogeneity and sample enough macropores. Using these parameters, the complete water retention and hydraulic conductivity functions were determined as follows. Water retention curves correspond to that of the matrix for water pressure heads <-20 mm. Total saturated water content was calculated using bulk density and gravimetric water content measurements at the end of the ponded water infiltration experiments. Hydraulic conductivity curves $K(\theta)$ correspond to that of the matrix for water pressure heads below -20 mm. Above, a linear relationship is assumed between the natural logarithm of conductivity versus water content. Finally, water retention and hydraulic conductivity are completely determined for the dual permeability system. Yet, these water retention and hydraulic conductivity curves correspond to that of the whole dual permeability system, with no information on the pore domains, namely the matrix and the fast-flow region. Despite this drawback, the obtained soil hydraulic properties define bimodal water retention and hydraulic functions that can be implemented into Richards' equation to model dual permeability system as suggested by Durner (1994) and Zurmühl and Durner (1998). Note that such approach addresses flows in dual permeability systems with pressure head equilibrium between the matrix and the fast-flow region; which restricts the use of such approach.

Finally, on the basis of previous work, Lassabatere et al. (2014a) proposed an extension of BEST method to dual permeability systems (Fig. 4.21). The method requires the following experimental data: particle size analysis, water infiltration experiments at pressure heads $h = -3$ cm (tension infiltrometer) and $h = 0$ (Beerkan method), and gravimetric water contents at time zero and at the end of the two successive infiltration experiments. Image analysis or field observations can be used to guess the volume fraction occupied by the macropore network, namely w. At first, the measures of water contents performed at -3 cm and at 0 cm water pressure heads, θ_{-3} and θ_0, are used to estimate saturated water content for both the matrix and the fast-flow region. Basically, water content at -3 cm quantifies the contribution of the matrix only, whereas water content at 0 cm quantifies the contribution of the soil pores including the matrix and the fast-flow regions, leading to:

$$\theta_{sm} = \frac{\theta_{-3}}{1 - w} \tag{4.54}$$

$$\theta_{sf} = \frac{\theta_0 - \theta_{-3}}{w} \tag{4.55}$$

where θ_{sm} and θ_{sf} stand for the saturated water contents of the matrix and the fast-flow region, respectively. The particle size distributions are then analyzed using BEST method pedotransfer functions (Lassabatere et al. 2006). In the case of bimodality, particle size distributions are split into two parts, and the first and the second modes are analyzed to derive the shape parameters of the matrix and the fast flow region, respectively (Fig. 4.21a). Then, matrix scale parameters are inferred

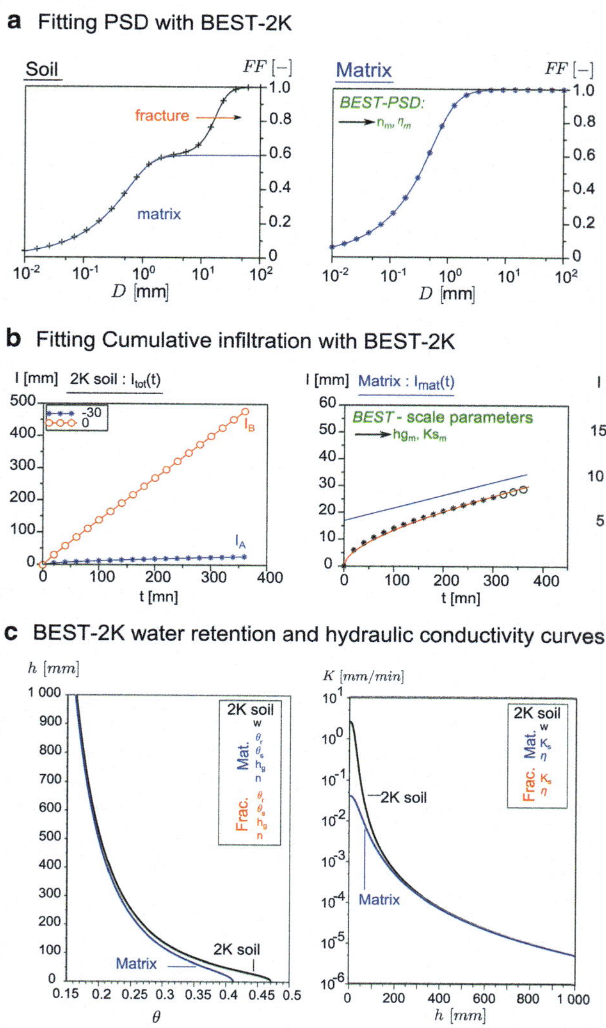

Fig. 4.21 Towards a new method BEST for dual permeability systems: BEST-2K: (**a**) analysis of bimodal particle size distribution and derivation of shape parameters for the matrix and the fast flow region; (**b**) analysis of the cumulative infiltration in the matrix and the macropore region to derive scale parameters, (**c**) estimated water retention and hydraulic conductivity functions (Lassabatere et al. 2014a, reprinted with permission)

from the BEST analysis of water infiltration performed at −3 cm, I_{3cm}. In fact, I_{3cm} quantifies water infiltration over the whole base of the tension infiltrometer, whereas the matrix occupies only a part of it. The cumulative infiltration into the matrix must be corrected as follows:

$$I_m = \frac{I_{-3}}{1 - w} \tag{4.56}$$

At this step, the matrix is completely characterized. The last step consists in the treatment of the infiltration into the fast-flow region. Some simplifications are needed. Firstly, it is considered that cumulative infiltration into the matrix is the same at 0 or −3 cm water pressure head. Indeed, according to Watson and Luxmoore (1986), the value of − 3 cm excludes from flow macropores with diameters >1 mm. The infiltration into the fast-flow region is thus computed from total infiltration subtracted with the infiltration into the matrix, i.e. the cumulative infiltration obtained at −3 cm, leading to:

$$I_f = \frac{(I_0 - I_{-3})}{w} \tag{4.57}$$

Again, cumulative infiltration is divided by the surface occupied by macropores, which explain the division by w. Secondly, the cumulative infiltration into the fracture, I_f, is then analyzed with BEST method to derive the scale parameters of the fast-flow region. Such method can provide the entire set of unsaturated parameters for the matrix and the fast-flow region hydraulic characteristics. It must be stated that such method gives no information on the water exchange between the matrix and the fast-flow region, and in particular no information on the interfacial hydraulic conductivity K_a.

These developments are the very first proposed extensions of BEST method for dual permeability soils. They require further developments and tests on field data. Yet, these methods are quite promising and, along with inverse numerical method address the challenging topic of hydraulic characterization of dual permeability soils. This challenge is a prerequisite for the understanding of preferential water flow and solute transport in the vadose zone.

References

Ahuja, L., Rojas, K., & Hanson, J. (2000). *Root zone water quality model: Modelling management effects on water quality and crop production*. Highlands Ranch, Colorado: Water Resources Publication.

Alagna, V., Bagarello, V., Di Prima, S., Giordano, G., & Iovino, M. (2013). A simple field method to measure the hydrodynamic properties of soil surface crust. *Journal of Agricultural Engineering, XLIV*(s2), e14:74–79.

Alaoui, A., Caduff, U., Gerke, H. H., & Weingartner, R. (2011). A preferential flow effects on infiltration and runoff in grassland and forest soils. *Vadose Zone Journal, 10*, 367–377.

Alberti, D. R., & Cey, E. E. (2011). Evaluation of macropore flow and transport using three-dimensional simulation of tension infiltration experiments. *Vadose Zone Journal, 10*, 603–617.

Angulo-Jaramillo, R., Gaudet, J.-P., Thony, J.-L., & Vauclin, M. (1996). Measurement of hydraulic properties and mobile water content of a field soil. *Soil Science Society of America Journal, 60*, 710–715.

Angulo-Jaramillo, R., Vandervaere, J. P., Roulier, S., Thony, J. L., Gaudet, J. P., & Vauclin, M. (2000). Field measurement of soil surface hydraulic properties by disc and ring infiltrometers. A review and recent developments. *Soil & Tillage Research, 55*, 1–29.

Assouline, S. (2011). Soil surface sealing and crusting. In J. Glinski, J. Horabik, & J. Lipiec (Eds.), *Encyclopedia of agrophysics* (Encyclopedia of earth sciences series 1st ed.). Dordrecht: Springer, XLVI, 1028 p.

Assouline, S., & Mualem, Y. (1997). Modeling the dynamics of seal formation and its effect on infiltration as related to soil and rainfall characteristics. *Water Resources Research, 33*(7), 1527–1536.

Assouline, S., & Mualem, Y. (2002). Infiltration during soil sealing: The effect of areal heterogeneity of soil hydraulic properties. *Water Resources Research, 38*(12), 1286. doi:10.1029/2001WR001168.

Assouline, S., & Mualem, Y. (2006). Runoff from heterogeneous small bare catchments during soil surface sealing. *Water Resources Research, 42*, W12405. doi:10.1029/2005WR004592.

Bagarello, V., & Iovino, M. (1999). Spatial and temporal variability of saturated hydraulic conductivity in a central Sicily soil. In *Proceedings of the international workshop of EurAgEng's field of interest on soil and water "Modelling of Transport Processes in Soils"*, Leuven, Belgium, pp. 249–256.

Bagarello, V., Iovino, M., & Elrick, D. (2004). A simplified falling head technique for rapid determination of field-saturated hydraulic conductivity. *Soil Science Society of America Journal, 68*, 66–73.

Beatty, S. M., & Smith, J. E. (2013). Dynamic soil water repellency and infiltration in post-wildfire soils. *Geoderma, 192*, 160–172.

Beatty, S. M., & Smith, J. E. (2014). Infiltration of water and ethanol solutions in water repellent post wildfire soils. *Journal of Hydrology, 514*, 233–248.

Brooks, R. H., & Corey, C. T. (1964). *Hydraulic properties of porous media* (Hydrology paper 3). Fort Collins: Colorado State University.

Brutsaert, W. (1977). Vertical infiltration in dry soil. *Water Resources Research, 13*, 363–368.

Buczko, U., Bens, O., Fischer, H., & Hüttl, R. F. (2002). Water repellency in sandy luvisol under different forest transformation stages in northeast Germany. *Geoderma, 109*, 1–18.

Burdine, N. T. (1953). Relative permeability calculation from pore size distribution data. *Transactions of the American Institute of Mining and Metallurgical Engineers, 198*, 71–77.

Carrillo, M. L. K., Letey, J., & Yates, S. R. (2000a). Unstable water flow in a layered soil: I. Effects of a stable water-repellent layer. *Soil Science Society of America Journal, 674*, 450–455.

Carrillo, M. L. K., Letey, J., & Yates, S. R. (2000b). Unstable water flow in a layered soil: II. Effects of an unstable water-repellent layer. *Soil Science Society of America Journal, 674*, 456–459.

Casey, F., Horton, R., Logsdon, S., & Jaynes, D. (1997). Immobile water content and mass exchange coefficient of a field soil. *Soil Science Society of America Journal, 61*, 1030–1036.

Casey, F. X. M., Horton, R., Logsdon, S. D., & Jaynes, D. B. (1998). Measurement of field soil hydraulic and solute transport parameters. *Soil Science Society of America Journal, 62*, 1172–1178.

Cerdà, A. (1996). Seasonal variability of infiltration rates under contrasting slope conditions in southeast Spain. *Geoderma, 69*, 217–232.

Cerdà, A., & Doerr, S. H. (2007). Soil wettability, runoff and erodibility of major dry-Mediterranean land use types on calcareous soils. *Hydrological Processes, 21*, 2325–2336.

Cey, E. E., & Rudolph, D. L. (2009). Field study of macropore flow processes using tension infiltration of a dye tracer in partially saturated soils. *Hydrological Processes, 23*, 1768–1779.

Cey, E. E., Rudolph, D. L., & Passmore, J. (2009). Influence of macroporosity on preferential solute and colloid transport in unsaturated field soils. *Journal of Contaminant Hydrology, 107*, 45–57.

Ciollaro, G., & Lamaddalena, N. (1998). Effect of tillage on the hydraulic properties of a vertic soil. *Journal of Agricultural Engineering Research, 71*, 147–155.

Cisar, J. L., Williams, K. E., Vivas, H. E., & Haydu, J. J. (2000). The occurrence and alleviation by surfactants of soil-water repellency on sand-based turfgrass systems. *Journal of Hydrology, 231–232*, 352–358.

Clothier, B. E., Kirkham, M. B., & McLean, J. E. (1992). In situ measurement of the effective transport volume for solute moving through soil. *Soil Science Society of America Journal, 56*, 733–736.

Clothier, B. E., Heng, L., Magesan, G., & Vogeler, I. (1995). The measured mobile-water content of an unsaturated soil as a function of hydraulic regime. *Soil Research, 33*, 397–414.

Clothier, B. E., Magesan, G., Heng, L., & Vogeler, I. (1996). In situ measurement of the solute adsorption isotherm using a disc permeameter. *Water Resources Research, 32*, 771–778.

Coppola, A., Kutílek, M., & Frind, E. O. (2009). Transport in preferential flow domains of the soil porous system: Measurement, interpretation, modelling, and upscaling. *Journal of Contaminant Hydrology, 104*, 1–3.

Crockford, S., Topadilis, S., & Richardson, D. P. (1991). Water repellency in a dry sclerophyll forest – Measurements and processes. *Hydrological Processes, 5*, 405–420.

DeBano, L. F. (1981). *Water repellent soils: A state-of-the-art* (General technical report PSW-46). Berkeley: USDA Forest Service, 21 pp.

Dekker, L. W., & Ritsema, C. J. (1994). How water moves in a water repellent sandy soil: Potential and actual water repellency. *Water Resources Research, 30*, 2507–2517.

Di Pietro, L., Ruy, S., & Capowiez, Y. (2003). Predicting preferential water flow in soils by traveling-dispersive waves. *Journal of Hydrology, 278*, 64–75.

Doerr, S. H. (1998). On standardizing the 'Water Drop penetration Time' and the 'Molarity of an Ethanol Droplet' techniques to classify soil hydrophobicity: A case study using medium textured soils. *Earth Surface Processes and Landforms, 23*, 663–668.

Doerr, S. H., & Thomas, A. D. (2000). The role of soil moisture in controlling water repellency: New evidence from forest soils in Portugal. *Journal of Hydrology, 231–232*, 134–147.

Doerr, S. H., Shakesby, R. A., & Walsh, R. P. D. (2000). Soil water repellency: Its causes, characteristics and hydro-geomorphological significance. *Earth-Science Reviews, 51*, 33–65.

Durner, W. (1994). Hydraulic conductivity estimation for soils with heterogeneous pore structure. *Water Resources Research, 30*, 211–223.

Ebel, B. A., & Moody, J. A. (2013). Rethinking infiltration in wildfire-affected soils. *Hydrological Processes, 27*, 1510–1514.

Ebel, B. A., Moody, J. A., & Martin, D. A. (2012). Hydrological conditions controlling runoff generation immediately after wildfire. *Water Resources Research, 48*, W03529.

Elrick, D. E., Parkin, G. W., Reynolds, W. D., & Fallow, D. J. (1995). Analysis of early-time and steady state single-ring infiltration under falling head conditions. *Water Resources Research, 31*, 1883–1893.

Elrick, D. E., Angulo-Jaramillo, R., Fallow, D. J., Reynolds, W. D., & Parkin, G. W. (2002). Analysis of infiltration under constant head and falling head conditions. In P. A. C. Raats, D. Smiles, & A. W. Warrick (Eds.), *Environmental mechanics: Water, mass and energy transfer in the biosphere* (Geophysical monograph series, 129, pp. 47–53). Washington, DC: American Geophysical Union.

Fallow, D. J., Elrick, D. E., Reynolds, W. D., Baumgartner, N., & Parkin, G. W. (1994). Field measurement of hydraulic conductivity in slowly permeable materials using early-time infiltration measurements in unsaturated media. In D. E. Daniel & S. J. Trautwein (Eds.), *Hydraulic conductivity and waste contaminant transport in soil* (ASTM STP, 1142, pp. 375–389). Philadelphia: American Society for Testing and Materials.

Feng, G. L., Letey, J., & Wu, L. (2001). Water ponding depths affect temporal infiltration rates in a water-repellent sand. *Soil Science Society of America Journal, 65*, 315–320.

Flury, M., Fluhler, H., Jury, W. A., & Leuenberger, J. (1994). Susceptibility of soils to preferential flow of water – A field study. *Water Resources Research, 30*, 1945–1954.

Franco, C. M. M., Tate, M. E., & Oades, J. M. (1995). Studies on non-wetting sands: I. The role of intrinsic particulate organic matter in the development of water-repellency in non-wetting sands. *Australian Journal of Soil Research, 19*, 275–285.

Gardner, W. R. (1958). Some steady state solutions of the unsaturated moisture flow equation with application to evaporation from a water table. *Soil Science, 85*(4), 228–232.

Gaudet, J.-P., Jegat, H., Vachaud, G., & Wierenga, P. J. (1977). Solute transfer, with exchange between mobile and stagnant water, through unsaturated sand. *Soil Science Society of America Journal, 41*, 665–671.

Gaudet, J.-P., Angulo-Jaramillo, R., Thony, J. L., Vauclin, M., Ladouche, B., Bariac, T., Huon, S., Ambroise, B., & Auzet, A. V. (1995). Mesures in situ de la fraction immobile de l'eau du sol avec de l'eau enrichie en 18O dans un infilltromètre à succion contrôlée. In *Proc. Conf. Isot. Water Resour. Manag.*, pp. 28–35. ASEA-SM-336/130P, Vienna.

Gérard-Marchant, P., Angulo-Jaramillo, R., Haverkamp, R., Vauclin, M., Groenevelt, P., & Elrick, D. E. (1997). Estimating the hydraulic conductivity of slowly permeable and swelling materials from single-ring experiments. *Water Resources Research, 33*(6), 1375–1382.

Gerke, H. H., & van Genuchten, M. T. (1993). A dual-porosity model for simulating the preferential movement of water and solutes in structured porous-media. *Water Resources Research, 29*, 305–319.

Gerke, H. H., & van Genuchten, M. T. (1996). Macroscopic representation of structural geometry for simulating water and solute movement in dual-porosity media. *Advances in Water Resources, 19*, 343–357.

Gerke, H. H., Dusek, J., Vogel, T., & Köhne, J. M. (2007). Two-dimensional dual-permeability analyses of a bromide tracer experiment on a tile-drained field. *Vadose Zone Journal, 6*, 651–667.

Germann, P. (1985). Kinematic wave approach to infiltration and drainage into and from soil macropores. *Transactions of the ASAE-American Society of Agricultural Engineers (USA), 28*, 745–749.

Germann, P. F. (2014). *Preferential flow: Stokes approach to infiltration and drainage*. Bern: University of Bern, Institute of Geography.

Germann, P. F., & Beven, K. (1985). Kinematic wave approximation to infiltration into soils with sorbing macropores. *Water Resources Research, 21*, 990–996.

Ghodrati, M., Chendorain, M., & Chang, Y. J. (1999). Characterization of macropore flow mechanisms in soil by means of a split macropore column. *Soil Science Society of America Journal, 63*, 1093–1101.

Gonzalez-Sosa, E., Braud, I., Dehotin, J., Lassabatere, L., Angulo-Jaramillo, R., Lagouy, M., Branger, R., Jacqueminet, C., Kermadi, S., & Michel, K. (2010). Impact of land use on the hydraulic properties of the top soil in a French catchment. *Hydrological Processes, 24*, 2382–2399.

Greco, R. (2002). Preferential flow in macroporous swelling soil with internal catchment: Model development and applications. *Journal of Hydrology, 269*, 150–168.

Green, W. H., & Ampt, G. A. (1911). Studies in soil physics. I. The flow of air and water through soils. *Journal of Agricultural Science, 4*, 1–24.

Groenevelt, P. H., Odell, B. P., & Elrick, D. E. (1996). Time domains for early-time and steady-state pressure infiltrometer data. *Soil Science Society of America Journal, 60*, 1713–1717.

Gryze, S., Jassogne, L., Bossuyt, H., Six, J., & Merckx, R. (2006). Water repellence and soil aggregate dynamics in a loamy grassland soil as affected by texture. *European Journal of Soil Science, 57*, 235–246.

Hallett, P. D., & Young, I. M. (1999). Changes to water repellence of soil aggregates caused by substrate-induced microbial activity. *European Journal of Soil Science, 50*, 35–40.

Hallett, P. D., Baumgartl, T., & Young, I. M. (2001). Subcritical water repellency of aggregates from a range of soil management practices. *Soil Science Society of America Journal, 65*, 184–190.

Hallin, I., Douglas, P., Doerr, S. H., & Bryant, R. (2013). The role of drop volume and number on soil water repellency determination. *Soil Science Society of America Journal, 77*(5), 1732–1743.

Harper, R. J., & Gilkes, R. J. (1994). Soil attributes related to water-repellency and the utility of soil survey for predicting its occurrence. *Australian Journal of Soil Research, 32*, 1109–1124.

Haverkamp, R., Ross, P. J., Smettem, K. R. J., & Parlange, J. Y. (1994). Three-dimensional analysis of infiltration from the disc infiltrometer. 2. Physically based infiltration equation. *Water Resources Research, 30*, 2931–2935.

Haverkamp, R., Debionne, D., Viallet, P., Angulo-Jaramillo, R., & de Condappa, D. (2006). Soil properties and moisture movement in the unsaturated zone. In J. W. Delleur (Ed.), *Handbook of groundwater engineering* (pp. 1–59). Boca Raton: CRC Press.

Hillel, D., & Gardner, W. R. (1969). Steady infiltration into crust-topped profiles. *Soil Science, 108*, 137–142.

Hillel, D., & Gardner, W. R. (1970). Transient infiltration into crust topped profiles. *Soil Science, 109*, 69–76.

Hunter, A. E., Chau, H. W., & Si, B. C. (2011). Impact of tension infiltrometer disc size on measured soil water repellency index. *Canadian Journal of Soil Science, 91*, 77–81.

Hussen, A. A., & Warrick, A. W. (1993). Alternative analyses of hydraulic data from disc tension infiltrometers. *Water Resources Research, 29*(12), 4103–4108.

Imeson, A. C., Verstraten, J. M., van Mulligen, E. J., & Sevink, J. (1992). The effects of fire and water repellency on infiltration and runoff under Mediterranean type forest. *Catena, 19*, 345–361.

Jarvis, N. (2007). A review of non-equilibrium water flow and solute transport in soil macropores: Principles, controlling factors and consequences for water quality. *European Journal of Soil Science, 58*, 523–546.

Jarvis, N., Etana, A., & Stagnitti, F. (2008). Water repellency, near-saturated infiltration and preferential solute transport in a macroporous clay soil. *Geoderma, 143*, 223–230.

Jaynes, D., & Shao, M. (1999). Evaluation of a simple technique for estimating two-domain transport parameters. *Soil Science, 164*, 82–91.

King, P. M. (1981). Comparison of methods for measuring severity of water repellence of sandy soils and assessment of some factors that affect its measurement. *Australian Journal of Soil Research, 19*, 275–285.

Kodešová, R., Vignozzi, N., Rohošková, M., Hájková, T., Kočárek, M., Pagliai, M., Kozák, J., & Šimůnek, J. (2009). Impact of varying soil structure on transport processes in different diagnostic horizons of three soil types. *Journal of Contaminant Hydrology, 104*, 107–125.

Kodešová, R., Šimůnek, J., Nikodem, A., & Jirku, V. (2010). Estimation of the dual-permeability model parameters using tension disk infiltrometer and Guelph permeameter. *Vadose Zone Journal, 9*, 213–225.

Kodešová, R., Němeček, K., Kodeš, V., & Žigová, A. (2012). Using dye tracer for visualization of preferential flow at macro-and microscales. *Vadose Zone Journal, 11*. doi:10.2/36/vzj2011.0088, 10 pp.

Köhne, J. M., Köhne, S., & Šimůnek, J. (2009a). A review of model applications for structured soils: a) Water flow and tracer transport. *Journal of Contaminant Hydrology, 104*, 4–35.

Köhne, J. M., Köhne, S., & Šimůnek, J. (2009b). A review of model applications for structured soils: b) Pesticide transport. *Journal of Contaminant Hydrology, 104*, 36–60.

Kostka, S. J. (2000). Amelioration of water repellency in highly managed soils and the enhancement of turfgrass performance through the systematic application of surfactants. *Journal of Hydrology, 231–232*, 359–368.

Lamy, E., Lassabatere, L., Béchet, B., & Andrieu, H. (2009). Modeling the influence of an artificial macropore in sandy columns on flow and solute transfer. *Journal of Hydrology, 376*, 392–402.

Larsbo, M., & Jarvis, N. (2003). *MACRO 5.0: A model of water flow and solute transport in macroporous soil: Technical description*. Uppsala: Department of Soil Sciences, Swedish University of Agricultural Sciences.

Larsbo, M., Roulier, S., Stenemo, F., Kasteel, R., & Jarvis, N. (2005). An improved dual-permeability model of water flow and solute transport in the vadose zone. *Vadose Zone Journal, 4*, 398–406.

Lassabatere, L., Winiarski, T., & Galvez-Cloutier, R. (2004). Retention of three heavy metals (Zn, Pb, and Cd) in a calcareous soil controlled by the modification of flow with geotextiles. *Environmental Science & Technology, 38*, 4215–4221.

Lassabatere, L., Angulo-Jaramillo, R., Soria Ugalde, J. M., Cuenca, R., Braud, I., & Haverkamp, R. (2006). Beerkan estimation of soil transfer parameters through infiltration experiments – BEST. *Soil Science Society of America Journal, 70*, 521–532.

Lassabatere, L., Lenoir, T., Peyneau, P., Lamy, E., Bechet, B., Angulo-Jaramillo, R., Simionovici, A., & Manceau, A. (2011). Experimental lighting of water velocity around macropores in a porous media. In *AGU Fall Meet. Abstr.*, p. 1194.

Lassabatere, L., Loizeau, S., Angulo-Jaramillo, R., Winiarski, T., Rossier, Y., Delolme, C., Gaudet, J. P. (2012). *Influence of the initial soil water content on Beerkan water infiltration experiments*. EGU General Assembly 2012, 22–27 April, Vienna, Austria, p. 2278.

Lassabatere, L., Angulo-Jaramillo, R., Yilmaz, D., & Peyrard, X. (2014a). Hydraulic characterization of dual-permeability unsaturated soils using tension disc infiltration experiments: BEST-2K method. In *EGU Gen. Assem. Conf. Abstr.*, p. 1848.

Lassabatere, L., Yilmaz, D., Peyrard, X., Peyneau, P. E., Lenoir, T., Šimůnek, J., & Angulo-Jaramillo, R. (2014b). New analytical model for cumulative infiltration into dual-permeability soils. *Vadose Zone Journal, 13*. doi:10.2136/vzj2013.10.0181, 15 pp.

Leeds-Harrison, P. B., & Youngs, E. G. (1997). Estimating the hydraulic conductivity of aggregates conditioned by different tillage treatments from sorption measurements. *Soil & Tillage Research, 41*, 141–147.

Letey, J. (1969). Measurement of contact angle, water drop penetration time, and critical surface tension. In *Proceedings of the symposium on water-repellent soils*, 6–10 May 1968, University of California, Riverside, pp. 43–47.

Letey, J. (2001). Causes and consequences of fire-induced soil water repellency. *Hydrological Processes, 15*, 2867–2875.

Letey, J., Carrillo, M. L. K., & Pang, X. P. (2000). Approaches to characterize the degree of water repellency. *Journal of Hydrology, 231–232*, 61–65.

Lewis, S. A., Wu, J. Q., & Robichaud, P. R. (2006). Assessing burn severity and comparing soil water repellency, Hayman Fire, Colorado. *Hydrological Processes, 20*, 1–16.

Logsdon, S. D. (1995). Flow mechanisms through continuous and buried macropores. *Soil Science, 160*, 237–242.

Logsdon, S. D., & Jaynes, D. B. (1993). Methodology for determining hydraulic conductivity with tension infiltrometers. *Soil Science Society of America Journal, 57*, 1426–1431.

Luxmoore, R. J. (1981). Micro-, meso-, and macroporosity of soil. *Soil Science Society of America Journal, 45*, 671–672.

Ma, D., & Shao, M. (2008). Simulating infiltration into stony soils with a dual-porosity model. *European Journal of Soil Science, 59*, 950–959.

Ma'shum, M., & Farmer, V. C. (1985). Origin and assessment of water repellency of a sandy South Australian soil. *Australian Journal of Soil Research, 23*, 623–626.

Marshall, T. J., & Holmes, J. W. (1979). *Soil physics*. Cambridge: Cambridge University Press.

Mataix-Solera, J., Cerdà, A., Arcenegui, V., Jordán, A., & Zavala, L. M. (2011). Fire effects on soil aggregation: A review. *Earth-Science Reviews, 109*, 44–60.

McGhie, D. A., & Posner, A. M. (1980). Water repellence of a heavy textured Western Australian surface soil. *Australian Journal of Soil Research, 18*, 309–323.

Moody, J. A., & Ebel, B. A. (2012). Hyper-dry conditions provide new insights into the cause of extreme floods after wildfire. *Catena, 93*, 58–63.

Mualem, Y. (1976). A new model for predicting the hydraulic conductivity of unsaturated porous media. *Water Resources Research, 12*, 513–522.

Mualem, Y., & Assouline, S. (1989). Modeling soil seal as a non-homogeneous layer. *Water Resources Research, 25*(10), 2101–2108.

Nyman, P., Sheridan, G., & Lane, P. N. J. (2010). Synergistic effects of water repellency and macropore flow on the hydraulic conductivity of a burned forest soil, south-east Australia. *Hydrological Processes, 24*, 2871–2887.

Odell, B. P., Groenevelt, P. H., & Elrick, D. E. (1998). Rapid determination of hydraulic conductivity in clay liners by early-time analysis. *Soil Science Society of America Journal, 62*, 56–62.

Or, D., Tuller, M., & others. (2000). Flow in unsaturated fractured porous media: Hydraulic conductivity of rough surfaces. *Water Resources Research, 36*, pp. 1165–1177.

Othmer, H., Diekkrüger, B., & Kutilek, M. (1991). Bimodal porosity and unsaturated hydraulic conductivity. *Soil Science, 152*, 139–150.

Parlange, J. Y. (1972). Theory of water movement in soils: 4. Two and three dimensional steady infiltration. *Soil Science, 113*, 96–101.

Philip, J. R. (1957). The theory of infiltration. 4. Sorptivity and algebraic infiltration equations. *Soil Science, 84*, 257–264.

Philip, J. R. (1969). Theory of infiltration. *Advances in Hydroscience, 5*, 215–305.

Philip, J. R. (1975). Stability analysis of infiltration. *Soil Science Society American Proceedings, 39*, 1042–1049.

Philip, J. R. (1992). Falling head ponded infiltration. *Water Resources Research, 28*(8), 2147–2148.

Reynolds, W. D. (2006). Tension infiltrometer measurements: Implications of pressure head offset due to contact sand. *Vadose Zone Journal, 5*, 1287–1292.

Reynolds, W. D. (2008). Alternative unsaturated flow analyses for the falling-head ring infiltrometer. *Vadose Zone Journal, 7*, 131–135.

Reynolds, W. D., & Elrick, D. E. (1990). Ponded infiltration from a single ring: I. Analysis of steady flow. *Soil Science Society of America Journal, 54*, 1233–1241.

Ritsema, C. J., Dekker, L. W., Hendrickx, J. M. H., & Hamminga, W. (1993). Preferential flow mechanism in a water repellent sandy soil. *Water Resources Research, 29*, 2183–2193.

Ritsema, C. J., Dekker, L. W., Nieber, J. L., & Steenhuis, T. S. (1998). Modeling and field evidence of finger formation and finger recurrence in a water repellent sandy soil. *Water Resources Research, 34*(4), 555–567.

Ritsema, C. J., Dekker, L. W., Oostindie, K., Moore, D., & Leinauer, B. (2008). Soil water repellency and critical soil water content. In S. Logsdon, D. Clay, D. Moore, & T. Tsegaye (Eds.), *Soil science: Step-by-step field analysis* (pp. 97–112). Madison: Soil Science Society of America.

Robichaud, P. R., Lewis, S. A., & Ashmun, L. E. (2008). *New procedure for sampling infiltration to assess post-fire soil water repellency* (Research note RMRS-RN-33). Fort Collins: USDA Forest Service, 14 pp.

Ross, P. J., & Smettem, K. R. J. (2000). A simple treatment of physical nonequilibrium water flow in soils. *Soil Science Society of America Journal, 64*, 1926–1930.

Roulier, S. (1999). *Caractérisation hydro-dispersive in situ de sols non saturés par infiltration d'eau et de soluté: cas de sols structures et de sols hétérogènes*. Université Joseph-Fourier-Grenoble I.

Roulier, S., Angulo-Jaramillo, R., Bresson, L. M., Auzet, A. V., Gaudet, J. P., & Bariac, T. (2002). Water transfer and mobile water content measurement in a cultivated crusted soil. *Soil Science, 167*, 201–210.

Rousseau, M., Ruy, S., Di Pietro, L., & Angulo-Jaramillo, R. (2004). Unsaturated hydraulic conductivity of structured soils from a kinematic wave approach. *Journal of Hydraulic Research, 42*, 83–91.

Šimůnek, J. (2006). Models of water flow and solute transport in the unsaturated zone. *Encyclopedia of Hydrological Sciences, 6*, 78.

Šimůnek, J., & van Genuchten, M. T. (1996). Estimating unsaturated soil hydraulic properties from tension disc infiltrometer data by numerical inversion. *Water Resources Research, 32*, 2683–2696.

Šimůnek, J., & van Genuchten, M. T. (1997). Estimating unsaturated soil hydraulic properties from multiple tension disc infiltrometer data. *Soil Science, 162*, 383–398.

Šimůnek, J., & van Genuchten, M. T. (2008). Modeling nonequilibrium flow and transport processes using HYDRUS. *Vadose Zone Journal, 7*, 782–797.

Šimůnek, J., Angulo-Jaramillo, R., Schaap, M. G., Vandervaere, J. P., & van Genuchten, M. T. (1998). Using an inverse method to estimate the hydraulic properties of crusted soils from tension disc infiltrometer data. *Geoderma, 86*, 61–81.

Šimůnek, J., Jarvis, N. J., van Genuchten, M. T., & Gärdenäs, A. (2003). Review and comparison of models for describing non-equilibrium and preferential flow and transport in the vadose zone. *Journal of Hydrology, 272*, 14–35.

Šimůnek, J., van Genuchten, M. T., & Sejna, M. (2008). Development and applications of the HYDRUS and STANMOD software packages and related codes. *Vadose Zone Journal, 7*, 587–600.

Smettem, K., Kirkby, C., & Chittleborough, D. (1994). Hydrologic response of undisturbed soil cores to simulated rainfall. *Soil Research, 32*, 1175–1187.

Steenhuis, T. S., Hunt, A. G., Parlange, J.-Y., & Ewing, R. P. (2005). Assessment of the application of percolation theory to a water-repellent soil. *Australian Journal of Soil Research, 43*, 357–360.

Täumer, K., Stoffregem, H., & Wessolek, G. (2005). Determination of repellency distribution using soil organic matter and water content. *Geoderma, 125*, 107–115.

Tillman, R. W., Scotter, D. R., Wallis, M. G., & Clothier, B. E. (1989). Water-repellency and its measurement by using intrinsic sorptivity. *Australian Journal of Soil Research, 27*, 637–644.

Timlin, D. J., Ahuja, L. R., & Ankeny, M. D. (1994). Comparison of 3 field methods to characterize apparent macropore conductivity. *Soil Science Society of America Journal, 58*, 278–284.

Tokunaga, T. K., & Wan, J. (1997). Water film flow along fracture surfaces of porous rock. *Water Resources Research, 33*, 1287–1295.

Tokunaga, T. K., Wan, J., & Sutton, S. R. (2000). Transient film flow on rough fracture surfaces. *Water Resources Research, 36*, 1737–1746.

Touma, J., Raclot, D., Al-Ali, Y., Zante, P., Hamrouni, H., & Dridi, B. (2011). In situ determination of the soil surface crust resistance. *Journal of Hydrology, 403*, 253–260.

Valentin, C. (1993). Chapter 9: Soil crusting and sealing in West Africa and possible approaches to improved management. In *Soil tillage in Africa: Needs and challenges* (FAO soils bulletin 69). Rome: Food and Agriculture Organization of the United Nations. ISBN ISBN 92-5-103442-7.

van Genuchten, M. T. (1980). A closed-form equation for predicting the hydraulic conductivity of unsaturated soil. *Soil Science Society of America Journal, 44*, 892–898.

van Genuchten, M. T., & Dalton, F. N. (1986). Models for simulating salt movement in aggregated field soils. *Geoderma, 38*, 165–183.

Vandervaere, J. P., Peugeot, C., Vauclin, M., Angulo-Jaramillo, R., & Lebel, T. (1997). Estimating hydraulic conductivity of crusted soils using disc infiltrometers and minitensiometers. *Journal of Hydrology, 188–189*, 209–223.

Vandervaere, J. P., Vauclin, M., Haverkamp, R., Peugeot, C., Thony, J. L., & Gilfedder, M. (1998). Prediction of crust-induced surface runoff with disc infiltrometer data. *Soil Science, 163*, 9–21.

Vandervaere, J. P., Vauclin, M., & Elrick, D. E. (2000). Transient flow from tension infiltrometers: I. The two-parameter equation. *Soil Science Society of America Journal, 64*, 1263–1272.

Vauclin, M., Elrick, D. E., Thony, J. L., Vachaud, G., Revol, P., & Ruelle, P. (1994). Hydraulic conductivity measurements of the spatial variability of a loamy soil. *Soil Technology, 7*, 181–195.

Vogelmann, E. S., Reichert, J. M., Reinert, D. J., Mentges, M. I., Vieira, D. A., de Barros, C. A. P., & Fasinmirin, J. T. (2010). Water repellency in soils of humid subtropical climate of Rio Grande do Sul, Brazil. *Soil and Tillage Research, 110*, 126–133.

Vogelmann, E. S., Reichert, J. M., Prevedello, J., Awe, G. O., & Mataix-Solera, J. (2013). Can occurrence of soil hydrophobicity promote the increase of aggregates stability? *Catena, 110*, 24–31.

Wallach, R., Ben-Arie, O., & Graber, E. R. (2005). Soil water repellency induced by long-term irrigation with treated sewage effluent. *Journal of Environmental Quality, 34*, 1910–1920.

Wallis, M. G., & Horne, D. J. (1992). Soil water repellency. *Advances in Soil Science, 20*, 91–146.

Wallis, M. G., Scotter, D. R., & Horne, D. J. (1991). An evaluation of the intrinsic sorptivity water repellency index on a range of New Zealand soils. *Australian Journal of Soil Research, 29*, 353–362.

Wang, Z., Wu, Q. J., Wu, L., Ritsema, C. J., Dekker, L. W., & Feyen, J. (2000). Effects of soil water repellency on infiltration rate and flow instability. *Journal of Hydrology, 231–232*, 265–276.

Watson, C. L., & Letey, J. (1970). Indices for characterizing soil water-repellency based upon contact angle-surface tension relationships. *Proceedings of the Soil Science Society of America, 34*, 841–844.

Watson, K. W., & Luxmoore, R. J. (1986). Estimating macroporosity in a forest watershed by use of a tension infiltrometer. *Soil Science Society of America Journal, 50*, 578–582.

Weiler, M. (2005). An infiltration model based on flow variability in macropores: Development, sensitivity analysis and applications. *Journal of Hydrology, 310*, 294–315.

Weiler, M., & Flühler, H. (2004). Inferring flow types from dye patterns in macroporous soils. *Geoderma, 120*, 137–153.

White, I., & Sully, M. J. (1987). Macroscopic and microscopic capillary length and time scales from field infiltration. *Water Resources Research, 23*(8), 1514–1522.

Wooding, R. A. (1968). Steady infiltration from a shallow circular pond. *Water Resources Research, 4*, 1259–1273.

Yilmaz, D., Lassabatere, L., Deneele, D., Angulo-Jaramillo, R., & Legret, M. (2013). Influence of carbonation on the microstructure and the hydraulic properties of a Basic Oxygen Furnace slag as a potential alternative construction material. *Vadose Zone Journal, 12*(2), 1–15.

Youngs, E. G., Leeds-Harrison, P. B., & Elrick, D. E. (1995). The hydraulic conductivity of low permeability wet soils used as landfill lining and capping material: Analysis of pressure infiltrometer measurements. *Soil Technology, 8*, 153–160.

Zurmühl, T., & Durner, W. (1996). Modeling transient water and solute transport in a biporous soil. *Water Resources Research, 32*, 819–829.

Zurmühl, T., & Durner, W. (1998). Determination of parameters for bimodal hydraulic functions by inverse modeling. *Soil Science Society of America Journal, 62*, 874–880.

Chapter 5
Appendix – Additional Measurements

5.1 Soil Bulk Density

The soil bulk density, ρ_b (M L^{-3}), is the ratio between the mass, M_s (M), of the oven-dried soil and the corresponding bulk soil volume, V_t (L^3), at some specified soil water content, usually that at sampling:

$$\rho_b = \frac{M_s}{V_t} \tag{5.1}$$

Destructive methods are commonly applied to determine ρ_b although non-destructive methods were also developed (Culley 1993; Grossman and Reinsch 2002). The *core method* and the *excavation method* are among the most common methods for determining ρ_b destructively. In both cases, the soil surface to be sampled has preliminarily to be exposed by removing the extraneous material such as litter or, for a sampling at some depth of the soil profile, the uppermost soil layer.

For the core method, an ideally undisturbed soil sample of known bulk volume is collected in the field. At this aim, a thin-walled, stainless steel cylinder with a pre-established internal volume and a sharp bottom edge is inserted into the soil. The diameter of the cylinder is chosen in function of the type of soil and conditions so as to ensure the representativeness of the sample (see below). The cutting at the bottom edge should be carried out on the outer side of the cylinder not to alter the sampled soil volume when the sampler is driven into the soil. According to Raper and Erbach (1985), samplers should be pushed into loose soil and hammered into more compact soils. In particular, the cylinder can be inserted into the soil by placing a wooden tablet on the upper edge of the sampler and gently hammering on the top of the tablet with a rubber mallet. As the embedding of the cylinder proceeds, the soil surrounding its outer side should be removed to reduce frictional resistances. In soils exhibiting soil-metal adhesion, application of mineral oil to the

R. Angulo-Jaramillo et al., *Infiltration Measurements for Soil Hydraulic Characterization*,
DOI 10.1007/978-3-319-31788-5_5

cylinder may be beneficial to reduce soil disturbance (Culley 1993). When the cylinder has been driven into the soil almost completely, hammering on the tablet may determine compaction at the surface of the soil volume to be collected. In the terminal stage of the cylinder's insertion, it is therefore advisable to interpose a short cylinder (e.g. 20-mm-high), having the same diameter as the cylinder used for sampling, between the sampler and the tablet. The presence of this short cylinder allows a complete insertion of the sampler into the soil without any compaction at the surface layer due to hammering on the top of the exposed soil surface. The insertion direction of the sampler must remain constant. For a vertical sampling, this condition can easily be checked by repeatedly positioning a level on the upper edge of the sampler along two orthogonal directions. If the cylinder tilts during insertion, no attempt should be made to straighten it because the stress induced by straightening can trigger an alteration of the soil structural characteristics, possibly leading to an unreliable bulk density value. Rather, it is preferable to interrupt the sampling and to start again the procedure at a new location. Ring insertion is a particularly delicate phase of the ρ_b determination with the core method in clay-rich soils when the water content is either high (compaction-prone conditions) or low (shattering-prone conditions). According to Costantini (1995), the utility of any core sampler will be limited in very dry soils where either core penetration is difficult and/or a lack of soil consistency makes extraction of the sampler together with the sample difficult. After completing insertion of the cylinder, the undisturbed soil core has to carefully be removed for transport to the laboratory. The pit surrounding the sampler is enlarged and the soil is cut a few centimeters under the bottom of the sampler with a spatula or a small trowel. Therefore, the soil extracted from the pit is partly contained in the sampler and in part protrudes above its upper edge, if the sampler has been driven a few millimeters below the exposed soil surface, and particularly from the bottom side of the sampler. The collected sample is examined for visible signs of soil alteration and the ends of acceptable cores are trimmed flush with the end of the cylinder. To prevent slippage of the soil outside of the sampler, the ends of the core are covered by plastic caps that are fixed with adhesive tape.

In the laboratory, the adhesive tape and the caps are removed and the core, placed on a porcelain capsule or another appropriate weighing tin, is placed in an oven set at 105 °C, in general for 24–48 h. The oven-drying duration can also be longer, depending on both the soil textural characteristics and the size of the soil sample. According to Culley (1993), for example, cores of about 350 cm^3 usually require about 72 h of drying in ovens equipped with circulating fans, but smaller cores require less time. The soil sample is then cooled in a desiccator. The mass of the oven-dried soil, M_s, is determined by weighing with a balance sensitive to 0.01 g. The core is weighed before tare, including the stainless steel sampler and the weighing tin. Therefore, the M_s value to be used in Eq. (5.1) is obtained by subtracting this tare from the total measured mass.

In some instances, such as an intensive sampling of remote areas, transporting a large number of samplers can be impractical. A single cylinder can be re-used many times if, after collection, the soil in the sampler is pushed out into a small plastic bag that is then labeled, closed and transported to the laboratory for oven-drying. In this case, the measured oven-dried mass of the disturbed soil is easily referred to a known bulk volume that is the internal volume of the sampler.

Soil bulk density can also be determined when the core does not completely fill the cylinder but, in this case, there is the additional need to measure the core volume not occupied by the soil (Culley 1993). In the laboratory, a graduated cylinder of volume V_1 is filled with glass beads (260 μm). Using this cylinder, glass beads are poured onto the soil and they are leveled to the top of the cylinder with a spatula. The glass beads in excess are returned to the graduated cylinder and the volume of glass beads remained in this cylinder, V_2, is recorded. The bulk volume of the soil sample is given by the internal volume of the sampler minus the volume of beads used to fill the core, equal to $V_1 - V_2$. Obviously, only the mass of the soil has to be considered in the calculation of ρ_b with Eq. (5.1).

Core size can be expected to influence the determination of ρ_b. In particular, higher ρ_b estimates with smaller cores can be due to the occurrence of soil compaction phenomena when the cylinder is hammered into the soil whereas smaller ρ_b values may depend on incomplete filling of the cylinder and/or shattering during penetration (Costantini 1995; Page-Dumroese et al. 1999). According to Blake and Hartge (1986), cores of 48 mm in diameter can be used in soils with few coarse fragments. In gravelly to stony soils, use of small diameter cores may be hindered by large stones resulting in systematic under-representation of this fraction and lower than actual bulk density values (Flint and Childs 1984; Page-Dumroese et al. 1999). The effect of the sampler size on the ρ_b estimates was tested by Costantini (1995) on different soils using samplers 10-mm-high and with different internal diameters, i.e. 34.8, 48.3, 59.8 and 91.2 mm. A small but significant reduction in the estimated ρ_b was associated with increased sampler size but the accuracy of bulk density assessment was not improved by increasing sampler diameter beyond 59.8 mm. Another study showed that small (diameter = 50 mm, volume = 98.2 cm^3) and large (85 mm, 652.6 cm^3) samplers yielded similar results in terms of both mean (1.28 and 1.25 g cm^{-3}, respectively) and variability (coefficients of variation of 5.8 % and 5.0 %, respectively) of the estimated bulk density data of a clay soil (Fig. 5.1). Another comparison between small (48 mm) and large (100 mm) diameter cores was carried out by Page-Dumroese et al. (1999). The

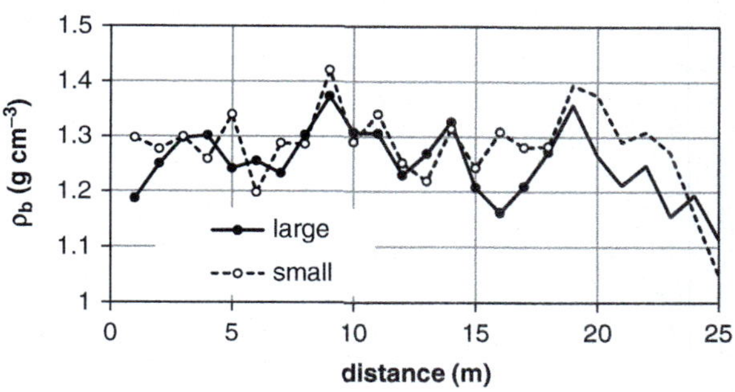

Fig. 5.1 Surface soil bulk density, ρ_b, values obtained along a transect established on a clay soil with both small (98 cm^3) and large (653 cm^3) samplers (unpublished data)

estimates of ρ_b given by the small cores were more variable and equal to or higher than the values obtained with the large cores, depending on the depth of sampling. This last result, i.e. smaller volumes yielding higher ρ_b values, was attributed to incomplete filling or soil loss from the larger sampler and to difficulties in measuring soil core length. According to Grossman and Reinsch (2002), the cores should be at least 75–100 mm in diameter, the height should not exceed the diameter and the cylinder wall should be 0.5–0.6 mm. Similar diameters were suggested by McIntyre and Loveday (1974), who also suggested that a diameter of 50 mm might be sufficient for purposes not requiring high accuracy.

With the excavation method, a hole is excavated to the desired depth and width, all soil material is removed and collected for mass determination after oven-drying in the laboratory, and the volume of the hole is measured. The volume of the sampled soil can easily be adapted to the size of coarse fragments in the soil. Filling the hole with water or expanding polyurethane foam are the most common methods to determine the hole volume (Muller and Hamilton 1992; Page-Dumroese et al. 1999). In the former case, a plastic bag or film has to be used as a soil liner to hold the water, and protruding stones and roots must be removed from the soil hole surface to avoid puncturing the plastic. Water is then poured in the hole until it is completely filled up. Equation (5.1) can easily be applied to determine ρ_b because the poured water volume coincides with the bulk soil volume. The volume of the hole can also be determined by forming a cast using expanding polyurethane foam, commercially available in small pressurized cans. The foam is best applied in a circular fashion beginning at the deepest recess of the hole and continuing until the hole is filled. A slight excess of foam is applied to the hole and a cardboard plate with a weight is placed across the surface. The plate ensures that continued expansion during the curing process fills all irregularities that may exist at the surface of the hole. Following curing, the cast is excavated and attached soil particles are removed by washing with water. The top of the cast is then trimmed flush with the top of the hole. The volume of the cast, which represents the bulk soil volume, is simply determined by volume displacement of water in a glass container of an appropriate size. In a test by Muller and Hamilton (1992), three repeated measures on each of 20 casts showed that this volume determination was in error by less than 1 %. The cured cast consists of closed cells, and water adsorption during volume determination is minor, i.e. less than 1 % of the measured volume (Muller and Hamilton 1992). However, to minimize water adsorption, Brye et al. (2004) suggested to seal the cut surface by spraying it with a thick coat of water-resistant, clear gloss urethane, commonly used as an indoor-outdoor wood finisher, and to allow it to dry for at least 24 h. Holes of approximately 1000 and 1300 cm^3 were dug by Muller and Hamilton (1992) and Brye et al. (2004), respectively. The polyurethane foam offers several advantages over the water bag (Muller and Hamilton 1992; Page-Dumroese et al. 1999). Large stones and protruding roots can remain in the holes, and cans of polyurethane foam are easier to carry than large volumes of water. Moreover, unlike polyurethane foam, water needs horizontal working surfaces. Disadvantages include higher costs of the procedure and the fact that the foam must cure for 8–24 h before it can be removed from the hole, which may require returning to the site the following day.

The excavation method with polyurethane foam was also applied to vertically exposed alluvium with high coarse-fragment content (Brye et al. 2004). In this case, a horizontal cavity was excavated and the foam was sprayed into the cavity starting with the rear and moving toward the front of the cavity until it was completely filled with foam. The front of the cavity was usually left open to atmosphere but sometimes the semi-liquid foam tended to pour out of the cavity. To avoid this occurrence, a flat piece of Plexiglass was braced over the front of the cavity to keep the foam in the cavity. The foam then just expanded against and around the Plexiglass. After foam curing, a marker was used to trace the perimeter of the front of the cavity onto the foam mold. The entire foam mold was excavated from the surrounding alluvium and the excess foam that expanded and cured beyond the front of the cavity was cut away along the marker trace using a hacksaw blade. According to Brye et al. (2004), it is better to excavate the cavity with a subtle downward tilt, such that the cavity opening is at a slightly higher elevation than the back of the cavity, to facilitate complete expansion of the foam into the volume excavated and minimize the chance of semi-liquid foam spilling out of the cavity under the force of gravity, leaving part of the cavity unfilled with foam. However, during this process, too much downward slope may make complete collection of the excavated material difficult.

Sampling with a core sampler may produce higher bulk density values than the excavation method because some compaction may occur during core insertion but core sampling may also give lower bulk density values than large hole excavation because the cylinder does not sample rock fragments larger than the diameter of the cylinder (Lichter and Costello 1994; Muller and Hamilton 1992; Page-Dumroese et al. 1999). The excavation method should be preferred over the core method for soils high in large rock fragments, for fragile horizons and for thin horizons (Grossman and Reinsch 2002). In these cases, soil excavation was suggested to be more practical and accurate than core sampling because of larger sample sizes and reduction of sample variability (Page-Dumroese et al. 1999).

A technique for measuring the bulk density of thin layers (10 mm) of soil is the so-called wax casting technique, that was developed by Frasier and Keiser (1993) with specific reference to loose, non-uniform soils with low moisture levels. The technique involves the removal of the soil in shallow layers. As each layer is removed, the hole is filled to the original soil level of the layer with a molten paraffin wax. After the wax has solidified, the wax casting is removed. The volume of the wax casting is determined by dividing its mass by the density of the paraffin wax, that can easily be determined by filling a known volume container with molten wax and waiting for cooling. The soil removed from the hole is dried and weighed. The temperature of the wax should be as close to the solidifying temperature as possible. If the wax is too hot, pouring may disturb the soil particles resulting in incorporation of soil into the wax casting. Also, cooler wax will not flow into the pores of the surrounding soil, which can occur if the wax temperature is too hot. Frasier and Keiser (1993) suggested to use wax with a melting point of not less than

53 °C. The reason was that, on summer days with high soil temperatures, low melting point waxes are slow to solidify and the casting are soft and may deform when being removed from the soil or transported. After the wax has solidified, the casting is removed from the hole. Soil particles clinging to the wax casting are rinsed off with cool water and brushed away. Wax shrinks as it cools which leaves a depression in the top surface. Molten wax is poured into the depression in sufficient quantity to "bead" above the surface. After the wax has cooled, the excess wax above the sides of the original sample is removed using a sharp knife. The sampling procedure can be repeated as necessary for deeper layers in the soil profile.

Regardless of the applied method, a ρ_b determination is obviously referred to a soil sample composed by particles differing in size, that may also include gravel (diameter > 2 mm). For certain applications, the bulk density of the fine soil fraction, ρ_{bf} (M L^{-3}), defined as those particles of less than 2 mm in diameter, is of interest (Culley 1993). In this case, after determining M_s and V_t, the oven dried soil is sieved through a 2 mm sieve. The material retained on the sieve is washed and dried. The mass of the coarse fragments, M_{sc} (M), is then obtained by weighing and the corresponding volume, V_{sc} (L^3), can be determined by measuring the displacement of water in a graduated cylinder when the fragments are added. The bulk density of the fine soil fraction is then obtained by the following relationship:

$$\rho_{bf} = \frac{M_s - M_{sc}}{V_t - V_{sc}} \tag{5.2}$$

Page-Dumroese et al. (1999) applied the following relationship by Andraski (1991) to determine ρ_{bf} using the total bulk density value, ρ_b, including both fine (<2 mm) and coarse particles:

$$\rho_{bf} = \rho_b \frac{1 - g_r}{1 - v_r} \tag{5.3}$$

where g_r is the gravimetric rock fragment content, obtained by dividing the mass of rock fragments by total sample mass, and v_r is the volumetric rock fragment content that was determined with the following relationship:

$$v_r = \rho_b \frac{g_r}{\rho_{br}} \tag{5.4}$$

where ρ_{br} (M L^{-3}) is the rock fragment density, that was assumed to be 2.65 g cm^{-3} by Page-Dumroese et al. (1999). In this case, mineral density must be properly determined using for instance a mercury or gas pycnometer (Klute and Dirksen 1986).

5.2 Soil Water Content

Soil water content can be measured with either direct or indirect methods (Topp and Ferré 2002a). Direct methods involve removal of water from the porous medium and the direct measurement of the amount of water removed. Indirect methods are based on the measurement of physical or chemical properties of the soil that depend on its water content, such as dielectric constant, electrical conductivity, heat capacity or H content (Santini 1997a). The thermo-gravimetric method is the most frequently applied direct method, since it is simple and allows precise measurements of water content (Topp and Ferré 2002b). This method is based on the determination, through weighing, of the amount of water lost by a soil sample as a result of drying in an oven. The impossibility to repeat twice the measurement at the same site, due to the destructive nature of the sampling, and the long duration of the drying process are the main disadvantages of the method. The indirect methods do not cause, in general, an appreciable disturbance to the sampled site and they make use of sensors that can be installed permanently into the soil to monitor its water content. However, a high precision of the measurement needs the calibration, for the soil of interest, of the relationship between the soil water content and the measured soil physical-chemical property, which can imply a considerable effort. Over the past 35 years, the TDR (Time Domain Reflectometry) method has become a widely applied method, also because it allows automatic collection of data. The TDR method is now commonly applied for monitoring agro-environmental processes and it has become a standard method of water content measurement, second only to the thermo-gravimetric method (Ferré and Topp 2002).

With the thermo-gravimetric method, the collected soil sample is placed immediately into an airtight container for the transport to the laboratory. A balance, accurate to 0.01 g, is used to weigh the moist sample that is then dried in an oven set at 105 °C until a practically steady weight is attained. Forced air circulating ovens will dry samples more rapidly, but passive convective ovens are also suitable, if adequately vented (Topp and Ferré 2002b). The sample removed from the oven is placed in a desiccator until it has cooled to ambient room temperature. Then, the sample is weighed again for determining the mass of dry soil. The water content on a mass basis, U (M M^{-1}), is given by:

$$U = \frac{(M_u + M_t) - (M_s + M_t)}{M_s} \tag{5.5}$$

where M_u (M) is the mass of the moist soil, M_s (M) is the mass of the oven-dried soil and M_t (M) is the tare.

For hydrological applications, the soil water content is commonly expressed on a volumetric basis and it is denoted by the symbol θ. In particular, θ (L^{3}L^{-3}) is the ratio between the volume of water in the sample, V_w (L^3), and the bulk soil volume, V_t (L^3). The volumetric soil water content is related to the gravimetric water content through the bulk density of the soil, ρ_b (M L^{-3}), and the density of water, ρ_w (M L^{-3}), equal to 1 Mg m^{-3} under standard conditions of pressure and temperature:

$$\theta = \frac{V_w}{V_t} = \frac{\rho_b}{\rho_w} U \tag{5.6}$$

Due to soil spatial variability, determining both ρ_b and U on the same soil sample can be advisable. At this aim, an undisturbed soil sample of known bulk volume has to be collected in the field. In view of its simplicity, the thermo-gravimetric method is considered practically free from error and for this reason it is used to calibrate indirect methods. However, the duration of the oven-drying process can influence the quality of the measurement. As a matter of fact, some soils, and particularly those with a high content of colloidal substances, can continue to lose water for several days when they are maintained at a constant temperature of 105 °C (Gardner 1986). Conventionally, the sample is considered dry when its weight does not vary by more than 0.1 % in one (Santini 1997a) or six (Topp and Ferré 2002b) hours. Usually, 24–48 h is considered to be an adequate drying duration for most mineral soils. For soil samples having more than 5 % organic matter, drying at 105 °C may result in mass losses arising from oxidation and volatilization of organic compounds. Notwithstanding this, Topp and Ferré (2002b) suggested to dry all samples at 105 °C because there is no absolute temperature at which water can be removed without determining some loss of organic substances.

In soils with stones or gravel (particle diameter > 2 mm), the volumetric water content of the fine soil fraction (<2 mm), θ_f, is given by (Topp and Ferré 2002b):

$$\theta_f = \frac{\theta}{1 - v_r} \tag{5.7}$$

where θ is the volumetric water content of the bulk soil sample and $v_r = V_{sc}/V_t$ is the volumetric stone and gravel content. If the stones and gravel are porous, the water held in the stone fraction has also to be considered (Gardner et al. 1991).

The TDR method is based on the determination of the propagation velocity, v (L T^{-1}), of an electromagnetic signal through a probe, commonly of the parallel rod type (Fig. 5.2), inserted into the soil (Ferré and Topp 2002). This velocity depends on the dielectric characteristics of the porous medium according to the following relationship:

$$v = \frac{c}{\sqrt{\varepsilon_r}} \tag{5.8}$$

where c (L T^{-1}) is the speed at which light propagates in vacuum ($c = 3 \times 10^8$ m s^{-1}) and ε_r (–) is the relative dielectric permittivity, sometimes called the dielectric constant, equal to the ratio between the permittivity, ε, of the considered medium and that of the vacuum, ε_0 ($\varepsilon_r = \varepsilon/\varepsilon_0$, with $\varepsilon_0 = 8.85 \times 10^{-12}$ F m^{-1}). The soil is a multi-phase system since it is constituted by a matrix of solid particles separated by

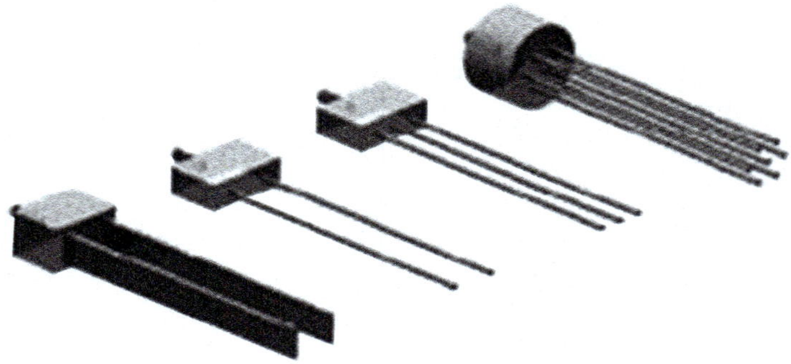

Fig. 5.2 Types of TDR probes

spaces filled with fluids (generally air and water) in varying proportions. The dielectric permittivity of this porous medium is some average of the dielectric permittivities of the individual components. However, the dipolar nature of the water molecules provides to bulk water unique electric properties, since water has a value of the relative dielectric permittivity ($\varepsilon_r \approx 80$ at 20 °C) appreciably higher than that of both air ($\varepsilon_r = 1$) and mineral soil components ($\varepsilon_r = 3 - 7$). The very large dielectric permittivity of water relative to that of air and dry soil results in the permittivity of a wet soil being dominated by the volumetric water content. More precisely, the bulk dielectric permittivity of a soil will be a function of the volumetric water content, with only a slight dependence on the volume fraction of solids. The value of the relative dielectric permittivity of the multi-phase porous medium, ε_b, measured by the TDR is named apparent relative permittivity or apparent dielectric constant since it also includes the effects of energy dissipation. The assumption that $\varepsilon_b \approx \varepsilon_r$ allows the use of the measured value of ε_b for estimating the soil water content.

As described in more details by Ferré and Topp (2002), a sequence of fast rise-time (<200 ps) electromagnetic square wave pulses is transmitted by a pulse generator onto the transmission line leading to and into the soil. The impedance changes along the transmission line reflect signal back to the source instrument, where the sampling receiver measures the sum of the transmitted and reflected voltages. The record of magnitude, either as voltage or reflection coefficient, as a function of time is displayed as a TDR trace or waveform. For water content determination, the two-way travel time of the TDR signal in the soil is determined by the difference between the time of arrival of the signal reflected from the end of the probe, t_2 (T) (Fig. 5.3), and the time of arrival of the signal reflected from the beginning of the probe, t_1 (T). In other words, the time difference $\Delta t = t_2 - t_1$ is a measure of the two-way travel time over the length of the rods. This calculation only involves measurements made on the horizontal time axis on the waveform. The intersection of the tangential lines on either side of the identifying signal

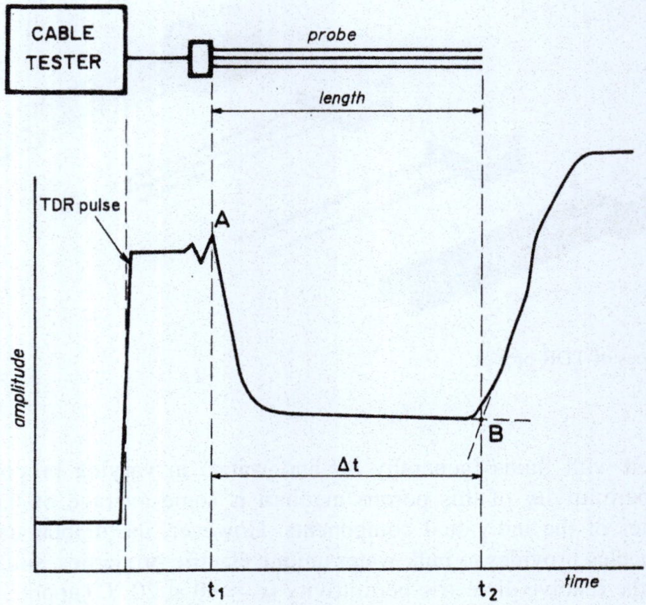

Fig. 5.3 Schematic view of the TDR probe and the corresponding signal

reflection is the most precise indication of the desired times (Ferré and Topp 2002). The propagation velocity of the signal is given by:

$$v = \frac{2L}{\Delta t} \tag{5.9}$$

where L (L) is the length of the probe. Substituting Eq. (5.9) into Eq. (5.8) allows the determination of the apparent relative permittivity by the following relationship:

$$\varepsilon_b = \left(\frac{c\Delta t}{2L}\right)^2 \tag{5.10}$$

The measurements of ε_b can be related with the volumetric soil water content, θ (L^3 L^{-3}), by either empirically derived calibration equations or conceptual models. Topp et al. (1980) carried out one of the most well-known applications of the empirical approach. These authors, using mineral agricultural soils of texture varying from sandy-loam to clay, showed that the determination of ε_b is practically independent of soil temperature (10–36 °C) and bulk density (1.14–1.44 Mg m^{-3}), hysteresis of the water retention curve, and salt content. The proposed calibration curve by Topp et al. (1980) is:

$$\theta = -5.3 \times 10^{-2} + 2.92 \times 10^{-2}\varepsilon_b - 5.5 \times 10^{-4}\varepsilon_b^2 + 4.3 \times 10^{-6}\varepsilon_b^3 \quad (5.11)$$

The broad applicability of Eq. (5.11) has been confirmed by different investigations carried out in soils of different texture (Topp et al. 1982; Nadler et al. 1991; Dasberg and Hopmans 1992). However, other investigations have shown that, with this relationship, θ is overestimated in organic soils (Roth et al. 1990; Herkelrath et al. 1991) and underestimated in soils with a high clay content (Dirksen and Dasberg 1993). Alternatives to empirically derived calibrations are often based on dielectric mixing models (Roth et al. 1990; Dirksen and Dasberg 1993; Ferré and Topp 2002). According to these last authors, a "square root" mixing model is equivalent to a linear relationship between $\varepsilon_b^{0.5}$ and the TDR travel time, Δt, in Eq. (5.10), which leads to a linear calibration for water content:

$$\theta = a + b\ \varepsilon_b^{0.5} \quad (5.12)$$

where the a and b parameters can be estimated by a simple, two-point calibration for a given medium. Furthermore, the slopes of the linear relationships found for all non-clayey soils are very similar (Ferré and Topp 2002). Therefore, changes in water content can be determined using any of the linear calibrations. The difference in the intercepts among the linear relationships requires calibration at a single, low water content to determine the absolute water content. Table 5.1 lists some empirical or semi-empirical calibration curves for different soils. In general, it is recommended to develop specific calibration curves for soils with a low bulk density, particular mineralogical properties (e.g., Fe-rich soils), high clay and/or organic matter content and, in any case, when the soil water content has to be determined with great precision.

In general, the probe for a TDR measurement has to be suitable to minimize the attenuations of the signal while limiting soil disturbance during insertion into the soil. A complete contact between the rods and the porous medium is required as the void spaces in direct contact with the probe represent preferential pathways for water and solute transport and they can affect the measured value of ε_b (Baker and Lascano 1989). For these reasons, the two-rod and three-rod probes are the most frequently used probes (Fig. 5.2). The length of the probe can influence the precision of the measurement. Probes of less than 0.1 m in length can induce large errors in the soil water content data, especially in dry soil conditions, due to the uncertainties in the determination of Δt (Heimovaara 1993). The maximum length of the probe is conditioned by the attenuation of the signal that can make detection of the second reflection uncertain, especially in soils with high clay and/or salt content. A criterion for the choice of the maximum length of the probe is that the reflected voltage should be at least 10 % of the transmitted voltage (Dalton and van Genuchten 1986). In non-saline soils, this criterion is fulfilled with probes approximately 0.3 m long. In practice, appropriate probe lengths for most field applications of the TDR method vary from 0.15 to 0.3 m. Spacing rods further apart increases the ease with which they can be inserted into the soil but increases the

Table 5.1 Calibration curves for the estimation of the volumetric soil water content as a function of the apparent relative permittivity, ε_b

References	Calibration curve	Note
Topp et al. (1980)	$\theta = (-530 + 292\ \varepsilon_b - 5.5\ \varepsilon_b^2 + 0.043\ \varepsilon_b^3) \times 10^{-4}$	Mineral soils (clay = 9–66 %), organic soils, vermiculite, glass beads
Herkelrath et al. (1991)	$\theta = 0.1273\sqrt{\varepsilon_b} - 0.051$	Loam soils with gravel
Nadler et al. (1991)	$\theta = (-725 + 367\ \varepsilon_b - 12.3\ \varepsilon_b^2 + 0.15\ \varepsilon_b^3) \times 10^{-4}$	Sandy-loam soils
Dasberg and Hopmans (1992)	$\theta = (-751 + 424\ \varepsilon_b - 18.5\ \varepsilon_b^2 + 0.38\ \varepsilon_b^3) \times 10^{-4}$	Sandy-loam soils
Dasberg and Hopmans (1992)	$\theta = (-1096 + 581\ \varepsilon_b - 22.7\ \varepsilon_b^2 + 0.32\ \varepsilon_b^3) \times 10^{-4}$	Clay soils
Jacobsen and Schjønning (1993)	$\theta = (-701 + 347\ \varepsilon_b - 11.6\ \varepsilon_b^2 + 0.18\ \varepsilon_b^3) \times 10^{-4}$	Soils from sand to sandy-clay-loam (clay = 2.5–23 %)
Hook and Livingston (1996)	$\theta = 0.1193\sqrt{\varepsilon_b} - 0.1849$	Semi-empirical relationship valid for non-clay soils
Regalado et al. (2003)	$\theta = (-400 + 430\ \varepsilon_b - 10\ \varepsilon_b^2 + 0.09\ \varepsilon_b^3) \times 10^{-4}$	Volcanic soils, sandy-loam, bulk density = 0.65 Mg m^{-3}, porosity = 76.7 %

magnitude of the resistance measured across the rods that is detrimental when soil electrical conductivity is measured in addition to soil water content. Moreover, it leads to more energy storage closer to the surface of the rods where, due to compaction effects, the soil might be less representative than undisturbed soil (Robinson et al. 2003). Large rods narrowly spaced optimize electromagnetic energy distribution between the rods, but result in large disturbed areas (Ferré and Topp 2002). Knight (1992) and Knight et al. (1994) suggested that the rod diameter should not be less than 1/10 of rod spacing to reduce concentration of electromagnetic energy close to the rods and hence to explore larger soil volumes. Ferré and Topp (2002) found that 6 mm rods spaced at 50 mm worked well in a variety of studies in tilled and untilled agricultural soils. For some laboratory investigations on more homogeneous porous media, the rod size can be reduced although the diameter should be ten times larger than the mean size of the soil particles (Santini 1997a).

The shape of the soil volume sampled by the electromagnetic signal is more complex in more heterogeneous soils. However, the measurement volume tends to be more confined with the three-rod probes than the two-rod ones. Topp and Davis (1985) suggested that the sampled soil volume by a two-rod probe corresponds to a cylinder with diameter and height equal to 1.4 times the rod spacing and the rod length, respectively. In a laboratory investigation, Baker and Lascano (1989) established that the influence zone has an elliptic or rectangular shape. In particular, a probe with a rod spacing of 50 mm samples, in the cross-sectional direction, a surface of 20×65 mm^2, that allows accurate measurements also in the surface soil layers. According to Knight et al. (1994), 95 % of the energy in an electromagnetic wave is confined within a diameter $\approx$ 1.5 times the rod separation distance. Rods of the probe should be installed in parallel but minor deviations from parallel alignment do not lead to significant errors unless the rods come into contact with each other (Noborio 2001; Ferré and Topp 2002).

Use of vertical TDR probes is generally simple and appropriate for extemporaneous measurements in which the probe is removed after the reading. If vertical probes are installed permanently, cracks determining preferential ways to either wetting or drying of the subsurface soil layers can develop, especially in clay rich soils. Horizontally installed probes provide a more precise profile of water content. With parallel rods installed from the soil surface 45° off the vertical, the resulting water content profile is a single vertical profile, it is less affected by lateral variability, and each depth increment provides, for equal magnitude, lateral and vertical integration (Ferré and Topp 2002).

TDR measurements are easily automated using a computer or data-logger, and analysis of waveforms is commonly completed during measurement, or may be analyzed later if waveforms are saved. Automated measurement of θ requires only a few seconds for each probe. As many as eight or even 16 probes may be attached to a single multiplexer, and several multiplexers may be connected in series to provide a large array of spatially distributed measurements, although the use of multiplexers introduces additional signal deterioration. Practical distances from the probe to the TDR unit are typically limited to 20–30 m due to signal deterioration with cable length. However, even longer cable lengths may provide reliable readings where soil salinity and clay content are low. The possibility to obtain high resolution time series measurements at multiple locations (e.g. depths) using automated and multiplexed TDR is particularly useful for both research and practical purposes.

5.3 Soil Water Matric Potential

Soil water matric potential can be measured in the subsurface soil by tensiometers. In essence, the device consists of a porous cup, a water reservoir and a measurement gauge (Fig. 5.4). Energy equilibrium between the tensiometer and the surrounding soil is achieved through water movement across the porous cup (Young and Sisson 2002). When the matric potential in the soil is lower than the water potential inside the tensiometer, water will move into the surrounding soil through the pores of the cup. Conversely, when the soil water matric potential is higher than the water inside

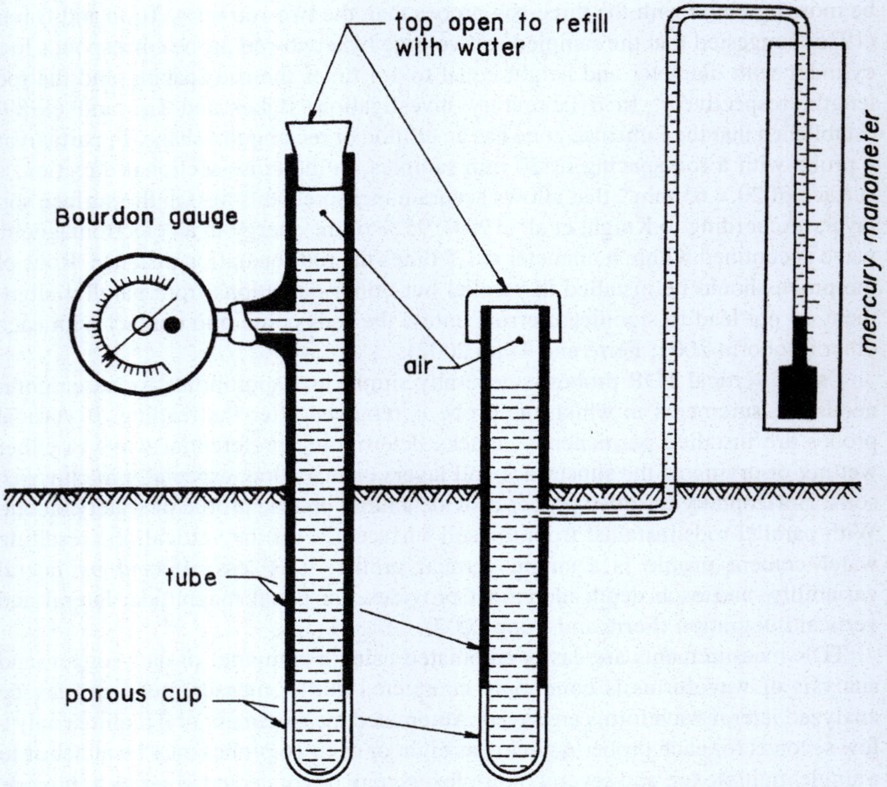

Fig. 5.4 Schematic view of the constituent parts of a tensiometer

the tensiometer, water will flow into the tensiometer through the porous cup. Water movement will continue until the difference in potential between the water inside the tensiometer and the soil water is zero, at which point static equilibrium is achieved. The negative pressure in the device is measured by mercury-filled manometers, Bourdon gauges or pressure transducers. A widely used class of pressure transducers measures the water pressure relative to atmospheric pressure, transforming the difference to an electrical response that can be caused by the deformation of a silico crystal onto which a piezoresistive circuit is fused (Young and Sisson 2002). As the deformation increases, resistance on the circuit changes, and thus also the voltage measurement. Matric potential values greater than −1 atm (approximately, −10 m of water) are measurable, due to the theoretically admissible negative pressure for water. In practice, the lowest measurable potential is of approximately −8 m because, for more negative pressures, water vapor will spontaneously come out of solution, causing water to "boil" (Young and Sisson 2002). At this point, the vacuum in the tensiometer body reduces drastically and the readings are no longer representative of the soil water matric potential. The limited range of matric potential measurable by the tensiometer does not reduce the interest for the device because this range accounts for more than 50–75 % of the soil water available for plants in many agricultural soils, i.e. it actually encompasses the greater part of the soil-wetness range (Hillel 1998).

The porous cup has to guarantee hydraulic continuity between the soil water and the water in the device, allowing free passage of water but excluding passage of soil gases, and it must have enough mechanical strength to withstand installation under a wide range of field conditions. Ceramic is the most commonly used material for tensiometer cups though stainless steel, teflon and plastic materials can also be used in particular situations such as in the presence of solutes that can alter the physical-chemical characteristics of the ceramic cups or when the tensiometer is also used to extract the soil solution. The characteristics of the porous material control both the conductance, k (L^2T^{-1}), and the air-entry potential of the cup. Larger pores typically lead to higher cup conductance and hence to shorter gauge response times. However, they also determine higher air-entry potentials, i.e. less negative pressure heads, which implies a reduced range of measurement. The reason is that hydraulic continuity within the cup is maintained until the largest continuous pores of the porous material remain saturated. When the soil water matric potential falls below the air-entry potential, the cup desaturates and hydraulic continuity is interrupted. Therefore, the characteristics of the chosen cup represent a compromise between response time of the device and measurable range. In any case, the air-entry potential has to be lower (more negative) than the lowest soil water matric potential to be measured by the tensiometer. The most commonly used tensiometer cups for field applications have an outer diameter of 2.2 cm (42.3 cm^2 surface area). Larger diameter cups, with larger surface areas, can lead to smaller measurement variability between field locations but this advantage is offset by the disadvantage of a lower measurement resolution.

The cup is connected to a rigid, normally plastic, pipe functioning as water reservoir. The connection assures air- and water-tight seals both mechanically and chemically stable for long periods of time. An equal diameter of the cup and the tube simplifies installation of the device and favors a good contact between the tensiometer and the soil. The manometer used to measure negative pressure in the device is collocated at the top of the tube (Fig. 5.4). The height difference between the porous cup and the manometer has to be considered to calculate matric potential. With reference to the scheme of Fig. 5.5 and taking into account that the hydraulic potential, H (L), at equilibrium has to be constant along the vertical, it can be written $H = h_1 + z_1 = h_2 + z_2$, where z_1 and z_2 (L) are the heights of the cup and the manometer, respectively, relative to a reference plane, h_1 (L) is the soil water matric potential close to the cup, here expressed in terms of energy per unit weight of water and more properly named matric head (Dane and Hopmans 2002a), and h_2 (L) is the negative pressure reading at the manometer. Therefore, h_1 is equal to $h_2 + (z_2 - z_1)$ ($L = z_2 - z_1$ in Fig. 5.5). For example, if the reading at the manometer is -3 m, the matric potential (head) close to the cup of a 1-m long vertical tensiometer, i.e. at the depth of 1 m, is $h = -3 + 1 = -2$ m. An implication of these

Fig. 5.5 Scheme for the correction of the reading at the manometer

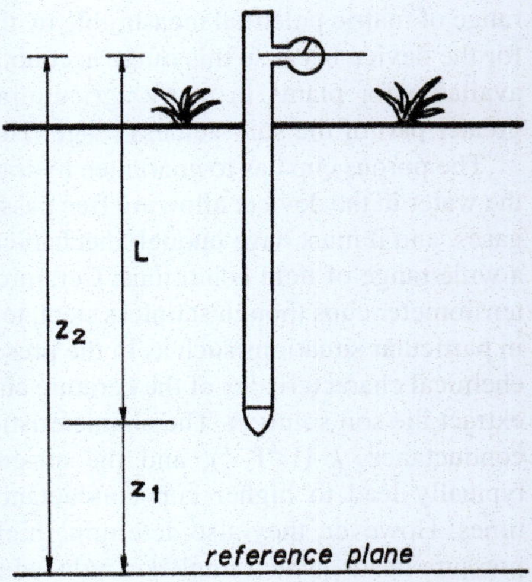

calculations is that longer tensiometer tubes reduce the possible measurement range.

Mercury-filled manometers allow to manually obtain reasonably precise measurements in many situations (±0.01 m of water column height) but their use in the field can be cumbersome due to the presence of external connecting pipes. Bourdon gauges provide readings with a precision no better than ±0.1 m. Therefore, they can be used when a high precision is not required. Pressure transducers allow precise measurements and they can be used to automatically monitor matric potential. Moreover, some transducers are equipped with a thermocouple to automatically offset the effects of the ambient temperature on the experimental readings. In the laboratory, water- or mercury-filled manometers or pressure transducers are commonly employed.

The sensitivity of the tensiometer, S_T (L^{-2}), is:

$$S_T = \frac{dh}{dW} \tag{5.13}$$

where dW (L^3) is the change in the volume of water in the tensiometer system necessary to obtain a change in gauge reading equal to dh (L) (Santini 1997b). Pressure measuring devices, such as manometers and Bourdon gauges, requiring a large displacement of liquid before registering a pressure change, have a low sensitivity. Devices equipped with pressure transducers have a higher sensitivity, since they require a smaller change in W to indicate a given change in h. If soil properties are not a limiting factor, the tensiometer response time, t_T (T), that is the time necessary for the pressure inside the tensiometer to equilibrate with the soil water matric head at the external surface of the porous cup, is given by (Klute and Gardner 1962):

$$t_T = \frac{1}{S_T\ k} \tag{5.14}$$

For given characteristics of the porous cup, tensiometers with pressure transducers have a shorter response time than devices with Bourdon gauges or mercury-filled manometers. This peculiarity makes pressure transducers particularly suited to monitor soil water matric potentials during transient stages of infiltration determining rapid variations in the soil water energy status. If soil is a limiting factor, the response time depends on both the characteristics of the tensiometer cup and the soil hydraulic conductivity. In general, however, response times are adequate to detect changes in matric potential occurring in the soil.

Tensiometers are generally installed in vertical boreholes excavated using a hand auger (Young and Sisson 2002). The diameter of the drill bit should be slightly smaller than the diameter of the tensiometer cup to ensure a snug fit at the bottom of the borehole. In addition, it is advisable to pour a slurry at the bottom of the borehole for porous cup contact. The slurry should be realized using sieved soil and much water. The same slurry has to be used to backfill the annular space between the tensiometer and the walls of the borehole. The tensiometer is filled in the laboratory, immersing the cup in de-aerated water for 24 h. De-aerated water is also used to complete the process in the field.

5.4 Soil Water Retention Curve

The experimental determination of the relationship between the volumetric soil water content, θ, and the matric potential, ψ_{pm}, or the matric head, h, is generally carried out in the laboratory with the tensiometric method for the highest (i.e., less negative) values of the potential and with the pressiometric method for the lowest potentials. Regardless of the method used, the experiment consists of equilibrating a soil sample to a given value of the matric potential and then determining the corresponding value of θ. Therefore, the experimental determination of selected points of the soil water retention curve implies that a series of h values have to be applied on the same soil sample. The $\theta(h)$ relationship is hysteretic and therefore it depends on the sequence (increasing or decreasing values) of the applied potentials. Taking into account that the draining, or drying, retention curve has particular interest for many agronomic purposes, the soil sample is commonly subjected to a monotonic decreasing sequence of h values. For the matric head values close or relatively close to saturation, soil water retention is largely influenced by the so-called structural porosity, i.e. the pore space between the micro-aggregates (Dexter et al. 2008). In this case, undisturbed soil samples should be used since they maintain the same pore arrangement occurring in the field. For low matric heads, soil samples prepared by packing dried and sieved soil (particle diameter < 2 mm) into small diameter cylinders can be used. The reason is that, for high suctions, water retention is mainly dependent on matrix porosity, i.e. the pore space within micro-aggregates and between individual soil mineral particles (Dexter et al. 2008).

The tensiometric method (Dane and Hopmans 2002b) is based on the functioning principle of the field tensiometer. The soil sample is placed on the porous plate of a glass filter funnel in which a given negative pressure value is established. Due to the pressure difference, water drains out of the soil sample until the water matric potential of the soil on the porous plate equilibrates with that of the water close to the porous body. A variant of the filter funnel is the suction table, i.e. a container filled with a porous material having particles of given size (sand, kaolin), allowing to process up to 30 samples simultaneously. In any case, the used material (ceramic septum, layer of inert material) must have an air entry value lower than the lowest applied pressure head. This condition ensures that the largest continuous pores of the porous material remain saturated, which guarantees hydraulic continuity of the system. The negative pressure within the tensiometric system can be realized by a vacuum pump or, for negative pressure heads up to 1–2 m, by an outflow end adjustable in height. In the so-called hanging water column apparatus (Burke et al. 1986), the porous plate of the filter funnel has an air entry potential of −2 m and the funnel is connected to a graduated burette of 50 cm^3 that can be moved vertically along a measuring tape having the reference level (zero value) at the surface of the porous plate. The soil sample, initially saturated on the porous plate, is subjected to a monotonically decreasing sequence of matric head values by lowering the burette along the measuring tape and waiting for the achievement of equilibrium conditions. Due to drainage, the water level into the burette raises. Therefore, the position of the burette has to be repeatedly adjusted to maintain the suction applied to the sample practically constant. When equilibrium conditions have been reached for the lowest pressure head of the sequence, the soil sample is extracted from the funnel and its volumetric water content is determined by the thermo-gravimetric method. The soil water content values for the higher values of the established pressure heads are determined by progressively summing the drained water volumes. The equilibration times depend on the soil type, the height of the sample and the established pressure head. For undisturbed soil samples 0.05-m-high, each step of the sequence implies a run duration varying from a few hours for pressure heads of nearly −0.1 m to a few days for $h < -1$ m. The measurable pressure head values with the tensiometric method are theoretically those greater than −1 atm (nearly −10 m of water) due to the theoretically admissible negative pressure for water. In practice, the lowest measurable potential is of approximately −8 m because, for more negative pressures, water vapor will spontaneously come out of solution, causing hydraulic discontinuities.

The pressiometric method (Dane and Hopmans 2002c) consists of placing a sample of saturated soil into a pressure chamber with a saturated pressure plate or membrane at its base. Within the chamber, a given pressure, higher than the atmospheric one, is established by injecting gas (compressed air or nitrogen). An in-house air compressor, or an air or nitrogen tank, can be used as pressure supply source. The pressure difference between the internal and the external environments pushes the water contained into the pores of the soil sample to move outwards through the porous septum. The draining process stops when the matric potential of the water in the soil sample is equal, in absolute value, to the gas pressure or the applied pneumatic potential, ψ_{pa}. As a matter of fact, at equilibrium, hydraulic potential of the soil-porous septum system has to be constant. In particular, on the external surface of the porous septum the total potential is null because $\psi_{pm} = 0$ (the septum is saturated and the excess water drains) and $\psi_{pa} = 0$ (the atmospheric pressure exists). Therefore, neglecting the gravitational component (the thickness of the soil-porous septum system is of a few cm at most), hydraulic potential has to be null at each point of the sample which means $\psi_{pm} = -\psi_{pa}$. The pressiometric method is largely used to determine the points of the water retention curve in the matric potential range of main interest for water supply to plants ($0.03 \leq \psi_{pm} \leq 1.5$ MPa; 1 MPa $\approx$ 100 m). The lowest limit for ψ_{pm} obtainable with the pressiometric method essentially depends on the construction characteristics of the chamber and the gas application system, that have to assure adequate conditions of safety for the personnel. The characteristics of the porous septum have to be chosen in relation to the applied pneumatic potential. In particular, the air entry potential has to be higher, in absolute value, than ψ_{pa} to be sure that the continuous largest pores remain saturated during the run. Otherwise, the porous septum desaturates which breaks hydraulic continuity. However, smaller pores have a lower conductance and this circumstance implies longer equilibration times. Generally, ceramic porous plates are usable for pressures not exceeding 2 MPa.

5.5 Water Stability of Soil Aggregates

Different mechanisms of aggregate breakdown by water can occur, i.e. slaking, differential swelling and subsequent shrinkage, mechanical breakdown and physical-chemical dispersion (Le Bissonnais 1996). Slaking is due to the compression of air entrapped inside the aggregate during wetting and it can occur when dry aggregates are rapidly wetted. This breakdown mechanism mainly produces micro-aggregates and it decreases as the initial soil water content increases due to the reduction of both the volume of entrapped air during wetting and the gradients of matric potential. An increasing clay content implies less opportunities for slaking and larger fragments resulting from aggregate breakdown although Ben-Hur and Lado (2008) concluded that more clay, increasing aggregate strength, also increases the slaking forces. Differential swelling and shrinkage during wetting and drying, respectively, of clay soils result in a micro-cracking of the aggregates that produces both macro- and micro-aggregates. Breakdown by differential swelling increases with increasing clay content. Mechanical breakdown, mainly due to raindrop impact in natural field situations, has a dominant role on wet soils because the aggregates are weaker when the soil is wetter. The fragments resulting from raindrop detachment are generally small, being either elementary particles or small micro-aggregates (<100 μm). Physical-chemical dispersion occurs due to the reduction of the attractive forces between colloidal particles while wetting, and the resulting fragments are constituted by elementary particles. Monovalent cations cause dispersion, and the exchangeable sodium percentage of the soil, the electrolyte concentration of both the soil solution and the applied water and mechanical disturbance from raindrop impact or shaking are factors influencing this breakdown phenomenon. The initial soil water content has a large influence on aggregate breakdown since slaking is reduced when aggregates are wet but cohesion is also reduced (Le Bissonnais 1996). According to Cerdà (1998), aggregate stability to water can be expected to be higher under initially moist conditions as compared with initially air-dry conditions, with lower differences between wet and dry states for stable soils. Soil swelling is essentially a reversible process whereas clay dispersion is irreversible (Shabtai et al. 2014). Organic matter can play a significant role as well by creating a bound between mineral particles and increasing bulk cohesion. Organic matter can constitute cores for the formation of aggregates or linkage between particles (Badin et al. 2009). In addition to its positive effect on aggregate cohesion, organic matter may lower the wettability, thus preventing water from entering and thus weakening aggregates (Müller and Deurer 2011).

The link between soil structure and infiltration is well known (e.g., Wilson and Luxmoore 1988) and, in principle, all aggregate breakdown mechanisms can occur when infiltration is measured for soil hydraulic characterization. Slaking is a possible occurrence when an initially dry soil is suddenly wetted by establishing a ponding depth of water on the infiltration surface. Wetting a soil volume from a limited source area may locally induce swelling phenomena, possibly affecting the monitored infiltration process. Differential shrinkage occurring after the infiltration run alters the soil particle arrangement within the sampled soil volume and this circumstance can influence a measurement carried out at the same sampling point days or weeks later. Mechanical breakdown is another possible occurrence when a positive head of water is abruptly established on the infiltration surface. Dispersion can occur if distilled water is used for the run. The consequences of breakdown by differential swelling on infiltration are expected to be less severe than those of slaking or dispersion due to the larger size of the resulting fragments (Le Bissonnais 1996). In addition, reduction of soil hydraulic conductivity may need the association of dispersion with slaking because dispersed clay particles are mobile, easily illuviated and not able to clog large conducting pores (Abu-Sharar et al. 1987).

Different methods can be used to assess water stability of soil aggregates. In the following, some accepted methods by the scientific community are described in short, also taking into account the need to experimentally discriminate between different breakdown mechanisms.

The *water drop test*, developed by McCalla (1944), consists of counting the number of water drop impacts of known force required to break down a soil aggregate to a certain state of disruption. According to Imeson and Vis (1984), pretreatment involves gently sieving the 4–4.8 mm fraction of soil aggregates from a bulk sample and moistening these at pF = 1, i.e. at a tension of 10 cm of H_2O, for 24 h with distilled water. Drops of 0.1 g in weight (5.8 mm diameter) are allowed to fall 1 m on aggregates placed on a 2.8 mm metal sieve. The time interval between drops is generally of 0.8–1 s (Cerdà 1998; Jimba and Lowery 2010). Two different application procedures of the method have been suggested (Imeson and Vis 1984). The first procedure, denoted with the acronym CND, involves Counting the Number of water Drop impacts required to disrupt the aggregate sufficiently for it to pass through the 2.8 mm sieve. Energy is transmitted to the aggregate in a series of equal value pulses. The aggregate can fail to respond to the first 40 or 50 impacts, because the strength of the bonds between the elements that compose the aggregate forms a threshold which is not overcome by the discrete impacts of the individual drops. For this reason, the CND procedure is best suited for unstable soils, that are considered to be those in which aggregate disruption occurs prior to 20–30 impacts. The second procedure (TDI) consists of subjecting aggregates 4–4.8 mm in diameter to Ten water Drop Impacts. The weight of the aggregates retained on the 2.8 mm sieve is determined, but the weight of the aggregates of other sizes can also be determined if, for example, a comparison with other experimental methods has to be established. In any case, a particular fraction is expressed as a percentage of the air dry weight of the tested aggregates. The TDI procedure was suggested to reduce a disadvantage of the CND approach, related to the fact that aggregates break down in different ways under drop impact. In particular, these ways include: (i) gradually lost fragments (micro-aggregates) until the aggregate is forced through the 2.8 mm diameter mesh; (ii) breakdown into 2–4 fragments of more or less equal size; (iii) rapidly forcing soft aggregates in a semi-liquefied state through the sieve; (iv) loss of fine dispersed material from the aggregate surfaces; and (v) rapid breakdown into many micro-aggregates. Therefore, the fact that an aggregate is retained on a 2.8 mm sieve does not imply that the drop impacts have not detached material from it. By weighing the material retained on the sieve, this limitation is reduced. The test should be replicated on 10–20 aggregates to obtain representative data for a soil sample (Imeson and Vis 1984; Cerdà 2000; Jimba and Lowery 2010). Jimba and Lowery (2010) proposed an automation of the CND procedure using digital observations of aggregate breakdown and electronic recording of the water drop count.

The *single sieve stability test*, originally developed by Kemper and Koch (1966) and later improved by Kemper and Rosenau (1986), is considered a standard test to assess water stable aggregation (Nimmo and Perkins 2002). According to Amezketa et al. (1996), four grams of 1–2 mm diam. air dry aggregates are placed on a 7 cm diam. sieve with 0.25 mm openings and vapor wetted to saturation. The sieves are placed in a modified Yoder apparatus (Díaz-Zorita et al. 2002; Nimmo and Perkins 2002) and they are raised and lowered through a 1 cm vertical distance at 36 cycles per minute for 5 min in a 500 mL beaker of distilled water. Three min of sieving through a 1.3 cm vertical distance were suggested by Kemper and Rosenau (1986). The material remaining on the sieve is oven dried at 105 °C and weighed to obtain the mass of stable aggregates, *SA*. After weighing, this material is dispersed by sonication in 60 mL of distilled water three times for 10 min each, and wet sieved for 5 min. The fraction remaining on the 0.25 mm sieve is oven dried and weighed to obtain the mass of > 0.25 mm sand, *SM*. The percentage of water stable aggregates, *WSA (%)*, is calculated by:

$$WSA = \frac{SA - SM}{original\ \ soil\ \ mass - SM} 100 \tag{5.15}$$

The *Modified Emerson Water Dispersion* (MEWD) *test* (Emerson 1967; Cerdà 1998, 2000) involves immersing 10 aggregates (4–4.8 mm) in 40 mL of distilled water and determining the degree of aggregate dispersion at time intervals of 5, 120 and 1440 min by using the following scale: (0) no dispersion; aggregate completely entire; (1) dispersion of some particles; milkiness close to the aggregate; (2) aggregate partly dispersed or divided into different smaller aggregates; (3) considerable dispersion; most of the aggregates have been dispersed and the milkiness is very large; (4) total dispersion; the aggregate does not exist. The MEWD test has a lower energy than other tests such as the water drop and the ultrasonic dispersion tests (see below), and it appears to be more or less subjective. However, this test permitted to discriminate between different land uses since agricultural soils were found to be less stable than scrubland soils (Cerdà 1998) and vegetation covered soil was found to be more stable than bare soil (Cerdà 2000).

The *Ultrasonic Dispersion* test is based on the principle that transmission of vibrating sound waves in a soil suspension produces microscopic bubbles inducing breakdown of the soil structural units (Díaz-Zorita et al. 2002). Soil aggregates are initially broken by ultrasounds into micro-aggregates before dispersion takes place (Edwards and Bremner 1967). Measuring the amount of energy absorbed by the material being tested before occurrence of disruption into micro-aggregates or dispersion was suggested to allow estimation of aggregate stability by North (1976) and Koenigs (1978). Ultrasonic procedures can be expected to be too aggressive for very weak aggregates but the so called UD30 procedure, using a Sanfier 1312 cell destructor (Branson Sonic Power Company, Danbury, CT) with a power output set at 30 W, was recommended by Imeson and Vis (1984) for very stable soils. Fifty moistened (pF = 1) 4–4.8 mm diam. aggregates are placed in 100 mL of water in a 37 mm diameter polythene test tube and then are subjected to the probe output for 20 s. The amount of material > 2.8 mm following this treatment is weighed, but other fractions can also be determined. A slightly different methodology was applied by Cerdà (1998, 2000). Ten aggregates (4–4.8 mm) moist at pF = 1 were immersed in 40 mL of distilled water in a cylindrical container (4 cm depth), and then were subjected to the probe output for 5 or 10 s with the probe tip placed 10 mm under the water surface. The power applied varied between 30 and 115 W. After the treatment, the remaining aggregates (>2.8 mm) and the aggregate fragments (<2.8 mm) were weighed and used as indices of aggregate stability. The reason why different power values were applied was that stable aggregates may have a threshold of disruption related to the power level of the test and not to the amount of energy supplied at levels below the threshold of disruption (Imeson and Vis 1984).

With more recent experimental approaches, more specific information on the aggregate breakdown mechanisms can be obtained or the stability of both macro- and micro-aggregates is considered.

The *multiple sieve method* (Le Bissonnais 1996; Amezketa et al. 1996) consists of determining the particle size distribution, PSD, and the mean weight diameter, *MWD*, of the stable fractions remaining after three treatments with different levels of energy applied to the aggregates. In the fast wetting treatment, 4 g of air dry 1–2 mm diam. aggregates are placed in a 0.25 mm diam. sieve and gently immersed for 10 min in a 250 mL beaker filled with 100 mL of deionized water. The sieve is then transferred to the modified Yoder apparatus, where disaggregation is accomplished by sieving in 95 % ethanol. In the slow wetting procedure, an additional 4 g of 1–2 mm aggregates are wetted to saturation in a vapor wetting chamber. Then, they are transferred to the modified Yoder apparatus and disaggregation is again measured in ethanol. For the third treatment, an additional 4 g of air dry 1–2 mm aggregates are gently immersed for 10 min in a 250 mL beaker filled with 50 mL of ethanol. The ethanol is carefully removed with a pipette, the aggregates are carefully transferred to a 250 mL Erlenmeyer flask filled with 50 mL of deionized water, and the level is adjusted to 200 mL. The flask is corked and agitated end over end 20 times, and left 30 min to allow coarse particles to settle. Excess water and suspended material is then carefully removed by pipette. The remaining soil-water mixture is transferred to a 0.25 mm sieve and disaggregation is accomplished by sieving in ethanol. For any of the three treatments, disaggregation consists of mechanically moving the 0.25 mm sieve immersed in ethanol up and down 20 times over a distance of 1 cm. The > 0.25 mm fraction is collected from the 0.25 mm sieve, oven dried and dry sieved for 1 min on a column of four 6.5 cm diam. sieves with hole openings of 2, 1, 0.5 and 0.25 mm using a standard mechanical sieve shaker. The mass percentage of each fraction is calculated from the dry mass remaining on each sieve. From these values, the PSD and the *MWD*, which is the sum of the fraction of soil left on each sieve multiplied by the mean inter-sieve size, is calculated. The *MWD* for each treatment serves as a measure of aggregate stability. Aggregates of 3–5 mm diam., instead of 1–2 mm, were used by Le Bissonnais (1996). The method yields MWD_{fast}, MWD_{slow}, and MWD_{stir} for the three treatments. In particular, the objective of the third treatment is to determine the stability of the aggregates with respect to mechanical breakdown. Ethanol is used for pre-wetting the sample because it allows to remove air from the aggregate without causing slaking. Therefore, pre-wetting with ethanol permits to test the wet mechanical cohesion of aggregates independently of slaking. Comparison of *MWD* among the three treatments allows to identify the mechanisms responsible for the loss of stability and the forces that bind particles together. Fast wetting can emphasize slaking but it should give a better separation of values than slow wetting for soils containing much organic carbon. Moreover, fast wetting represents a good way to compare the behavior of a large range of soils on rapid wetting (e.g., heavy rain storms in the summer) whereas slow wetting corresponds to a field condition of wetting under gentle rain. This last treatment is less destructive than fast wetting and may allow a better discrimination between unstable soils. Data collected with the different treatments can be combined between them. For example, the stability index, SI, can be calculated (Amezketa et al. 1996):

$$SI = \frac{MWD_{fast}}{MWD_{slow}} \tag{5.16}$$

With this index, soil samples can be compared on a relative scale of zero to one. A value of one indicates maximum stability and occurs when aggregates subjected to a fast wetting treatment exhibit no structural changes relative to aggregates subjected to the slow wetting treatment. A stability value of zero indicates a complete breakdown of aggregates.

Boix-Fayos et al. (2001) suggested that the stability of both macro- and micro-aggregates has to be determined taking into account that the degradation of macro-aggregates (>250 μm in diameter) creates aggregates of about 20–250 μm in diameter that are known to be considerably more stable than the larger ones (Oades and Waters 1991). According to Boix-Fayos et al. (2001), the aggregate stability of macro-aggregates (4–4.8 mm) has to be determined for both air dried (pF = 6.1) and prewetted (pF = 1) aggregates using the water drop test and counting the number of drops required to destroy an aggregate up to a maximum of 200 drops (Imeson and Vis 1984). The median value of drops for aggregate disruption is used as the stability indicator. The percentage of water stable micro-aggregates in the < 0.105 mm fraction is determined by two runs using a Microscan II Quantachrome Particle Analyser. The soil sample is dry sieved at the fraction < 0.105 mm for approximately 3 min. Initially, the sample is introduced in distilled water without applying any further energy and a first PSD is obtained with the Microscan. Then, the same sample is dispersed with sodium pyrophosphate 0.1 M (1 mL) and ultrasound at an energy level of 1800 J for 1 min with the objective of breaking down existing bonds between the aggregated particles. From this treatment, a second size distribution of primary particles is obtained. The differences between the two PSDs are used to determine the water stable micro-aggregates existing in the sample.

Finally, a set of stability tests was recently proposed by Ben-Hur et al. (2009) (see also Shabtai et al. 2014) to determine the sensitivity of soils to slaking, swelling and dispersion. The *slaking test*, showing similarities with some of the treatments included in the multiple sieve method (Amezketa et al. 1996), makes use of both fast wetting and slow wetting under vacuum. For fast wetting, 5 g of oven-dried (40 °C) aggregates 2–4 mm are immersed in a beaker containing 50 mL of deionized water. After 10 min, during which the aggregates remain at rest, the water is carefully removed under suction using a pipette. The soil fragments are transferred to a 50-μm sieve which had previously been immersed in ethanol, and they are gently moved up and down five times in ethanol to separate fragments < 50 μm from the larger fragments. The > 50 μm fraction is oven dried and then gently sieved by hand through a column of sieves of mesh sizes 2, 1, 0.5, 0.25 and 0.1 mm. The weight of each fraction is measured and that of the < 50 μm fraction is calculated as the difference between the initial weight and the sum of the weights of the other six fractions. The aggregate stability of the soil sample is expressed in terms of the mean weight diameter, *MWD* (MWD_f for the fast wetting treatment). For slow wetting, 5 g of oven dried (40 °C) aggregates 2–4 mm are placed on a cotton cloth, the edges of which are immersed in deionized water, inside a desiccator. The aggregates are wetted slowly under vacuum in the desiccator for 24 h. The wet aggregates are then transferred to a 50 μm sieve, which had previously been immersed in ethanol, and they are then sieved as described for the fast wetting test. The *MWD* value of the aggregates after slow wetting (MWD_s) is then determined. The slaking value, *SLV*, is calculated by the following relationship:

$$SLV = \frac{MWD_s}{MWD_f} \tag{5.17}$$

Performing the test on smaller aggregates (i.e. < 1 mm) is expected to yield unreliable results. For the *swelling test*, 20 oven dried aggregates of 2–4 mm are placed in a Petri dish and scanned using a flat-bed scanner for determining the image area of each aggregate. The aggregates are considered to be spherical and their volumes are calculated from the radius of a circle with an area equal to the area of the scanned aggregates. After scanning, the aggregates are wetted slowly with deionized water under vacuum in a desiccator and then they are scanned again. The swelling value, *SWV*, is calculated by the following relationship:

$$SWV = \frac{\sum_{i=1}^{n} (I_{wi} - I_{di})/I_{di}}{n} \tag{5.18}$$

where n is the number of aggregates, and I_{wi} and I_{di} are the calculated volumes of the aggregate i after and before wetting, respectively. Also in this case, performing the test on smaller aggregates (i.e. < 1 mm) is expected to yield unreliable results. For the *dispersion test*, 2 g of soil are suspended in 0.07 L of deionized water in a 0.1 L centrifuge tube. The tubes containing the suspension are shaken on a reciprocal shaker for 30 min at 20 rpm and then are immediately centrifuged and the concentration of dispersed clay in the turbid supernatant is determined by a spectrophotometer. The dispersion value, *DV*, is then determined by the following relationship:

$$DV = \frac{M_d}{M_t} \tag{5.19}$$

where M_d and M_t are the mass of the dispersed clay in the turbid supernatant and the total clay content in the soil sample.

References

Abu-Sharar, T. M., Bingham, F. T., & Rhoades, J. D. (1987). Reduction in hydraulic conductivity in relation to clay dispersion and disaggregation. *Soil Science Society of America Journal, 51*, 342–346.

Amezketa, E., Singer, M. J., & Le Bissonnais, Y. (1996). Testing a new procedure for measuring water-stable aggregation. *Soil Science Society of America Journal, 60*, 888–894.

Andraski, B. J. (1991). Balloon and core sampling for determining bulk density of alluvial desert soil. *Soil Science Society of America Journal, 55*, 1188–1190.

Badin, A.-L., Méderel, G., Béchet, B., Borschneck, D., & Delolme, C. (2009). Study of the aggregation of the surface layer of technosols from stormwater infiltration basins using grain size analyses with laser diffractometry. *Geoderma, 153*(1–2), 163–171.

Baker, J. M., & Lascano, R. J. (1989). The spatial sensitivity of time-domain reflectometry. *Soil Science, 147*(5), 378–384.

Ben-Hur, M., & Lado, M. (2008). Effect of soil wetting conditions on seal formation, runoff, and soil loss in arid and semiarid soils – A review. *Australian Journal of Soil Research, 46*, 191–202.

Ben-Hur, M., Yolcu, G., Uysal, H., Lado, M., & Paz, A. (2009). Soil structure changes: Aggregate size and soil texture effects on hydraulic conductivity under different saline and sodic conditions. *Australian Journal of Soil Research, 47*, 688–696.

Blake, G. R., & Hartge, K. H. (1986). Bulk density. In A. Klute (Ed.), *Methods of soil analysis, Part 1* (Agronomy 9, pp. 363–375). Madison: ASA.

Boix-Fayos, C., Calvo-Cases, A., Imeson, A. C., & Soriano-Soto, M. D. (2001). Influence of soil properties on the aggregation of some Mediterranean soils and the use of aggregate size and stability as land degradation indicators. *Catena, 44*, 47–67.

Brye, K. R., Morris, T. L., Miller, D. M., Formica, S. J., & Van Eps, M. A. (2004). Estimating bulk density in vertically exposed stoney alluvium using a modified excavation method. *Journal of Environmental Quality, 33*, 1937–1942.

Burke, W., Gabriels, D., & Bouma, J. (1986). *Soil structure assessment*. Balkema: Rotterdam.

Cerdà, A. (1998). Soil aggregate stability under different Mediterranean vegetation types. *Catena, 32*, 73–86.

Cerdà, A. (2000). Aggregate stability against water forces under different climates on agriculture land and scrubland in southern Bolivia. *Soil & Tillage Research, 57*, 159–166.

Costantini, A. (1995). Soil sampling bulk density in the coastal lowlands of south-east Queensland. *Australian Journal of Soil Research, 33*, 11–18.

Culley, J. L. B. (1993). Chapter 50: Density and compressibility. In M. R. Carter (Ed.), *Soil sampling and methods of analysis, Canadian society of soil science* (pp. 529–539). Boca Raton: Lewis Publishers.

Dalton, F. N., & van Genuchten, M. T. (1986). The time-domain reflectometry method for measuring soil water content and salinity. *Geoderma, 38*, 237–250.

Dane, J. H., & Hopmans, J. W. (2002a). 3.3.2.1 Introduction. In J. H. Dane & G. C.Topp (Co-Eds.), *Methods of soil analysis, Part 4, Physical methods* (Number 5 in the Soil Science Society of America book series, pp. 675–680). Madison: Soil Science Society of America, Inc.

Dane, J. H., & Hopmans, J. W. (2002b). 3.3.2.2 Hanging water column. In J. H. Dane & G. C. Topp (Co-Eds.), *Methods of soil analysis, Part 4, Physical methods* (Number 5 in the Soil Science Society of America book series, pp. 680–688). Madison: Soil Science Society of America, Inc.

Dane, J. H., & Hopmans, J. W. (2002c). 3.3.2.4 Pressure plate extractor. In J. H. Dane & G. C. Topp (Co-Eds.), *Methods of Soil Analysis, Part 4, Physical Methods* (Number 5 in the Soil Science Society of America book series, pp. 688–692). Madison: Soil Science Society of America, Inc.

Dasberg, S., & Hopmans, J. W. (1992). Time domain reflectometry calibration for uniformly and nonuniformly wetted sandy and clayey loam soils. *Soil Science Society of America Journal, 56*, 1341–1345.

Dexter, A. R., Czyż, E. A., Richard, G., & Reszkowska, A. (2008). A user-friendly water retention function that takes account of the textural and structural pore spaces in soil. *Geoderma, 143*, 243–253.

Díaz-Zorita, M., Perfect, E., & Grove, J. H. (2002). Disruptive methods for assessing soil structure. *Soil & Tillage Research, 64*, 3–22.

Dirksen, C., & Dasberg, S. (1993). Improved calibration of time domain reflectometry soil water content measurement. *Soil Science Society of America Journal, 57*, 660–667.

Edwards, A. P., & Bremner, J. M. (1967). Dispersion of soil particles by sonic vibration. *Journal of Soil Science, 18*, 47–63.

Emerson, W. W. (1967). A classification of soil aggregates based on their coherence in water. *Australian Journal of Soil Research, 5*, 47–57.

Ferré, P. A., & Topp, G. C. (2002). 3.1.3.4 Time domain reflectometry. In J. H. Dane & G. C. Topp (Co-Eds.), *Methods of soil analysis, Part 4, Physical methods* (Number 5 in the Soil Science Society of America book series, pp. 434–446). Madison: Soil Science Society of America, Inc.

Flint, A. L., & Childs, S. (1984). Development and calibration of an irregular hole bulk density sampler. *Soil Science Society of America Journal, 48*, 374–378.

Frasier, G. W., & Keiser, J. (1993). Thin layer measurement of soil bulk density. *Journal of Range Management, 46*, 91–93.

Gardner, W. H. (1986). Water content. In A. Klute (Ed.), *Methods of soil analysis, Part 1* (2nd ed., pp. 493–544). Madison: ASA and SSSA.

Gardner, C. M. K., Bell, J. P., Cooper, J. D., Dean, T. J., Hodnett, M. G., & Gardner, N. (1991). Soil water content. In R. A. Smith & C. E. Mullings (Eds.), *Soil analysis – Physical methods* (pp. 1–73). New York: Marcel Dekker.

Grossman, R. B., & Reinsch, T. G. (2002). 2.1 Bulk density and linear extensibility. In J. H. Dane & G. C. Topp (Co-Eds.), *Methods of soil analysis, Part 4, Physical methods* (Number 5 in the Soil Science Society of America book series, pp. 201–228). Madison: Soil Science Society of America, Inc.

Heimovaara, T. G. (1993). Design of triple-wire time domain reflectometry probes in practice and theory. *Soil Science Society of America Journal, 57*, 1410–1417.

Herkelrath, W. N., Hamburg, S. P., & Murphy, F. (1991). Automatic, real-time monitoring of soil moisture in a remote field area with time domain reflectometry. *Water Resources Research, 27* (5), 857–864.

Hillel, D. (1998). *Environmental soil physics*. San Diego: Academic, 771 pp.

Hook, W. R., & Livingston, N. J. (1996). Errors in converting time domain reflectometry measurements of propagation velocity to estimates of soil water content. *Soil Science Society of America Journal, 60*, 35–41.

Imeson, A. C., & Vis, M. (1984). Assessing soil aggregate stability by water-drop impact and ultrasonic dispersion. *Geoderma, 34*, 185–200.

Jacobsen, O. H., & Schjønning, P. (1993). A laboratory calibration of time domain reflectometry for soil water measurement including effects of bulk density and texture. *Journal of Hydrology, 151*, 147–157.

Jimba, S. C., & Lowery, B. (2010). Automation of the water-drop method for soil aggregate stability analysis. *Soil Science Society of America Journal, 74*, 38–41.

Kemper, W. D., & Koch, E. J. (1966). *Aggregate stability of soils from western United States and Canada* (USDA-ARS tech. bull. 1355). Washington, DC: U.S. Gov. Print. Office.

Kemper, W. D., & Rosenau, R. C. (1986). Aggregate stability and size distribution. In A. Klute (Ed.), *Methods of soil analysis, Part 1. Physical and mineralogical methods* (Agron. monogr. 9 2nd ed., pp. 425–442). Madison: ASA and SSSA.

Klute, A., & Dirksen, C. (1986). Hydraulic conductivity and diffusivity: Laboratory methods. In A. Klute (Ed.), *Methods of soil analysis, Part 1. Physical and mineralogical methods* (Agron. monogr. 9 2nd ed., pp. 687–734). Madison: ASA and SSSA.

Klute, A., & Gardner, W. R. (1962). Tensiometer response time. *Soil Science, 93*, 204–207.

Knight, J. H. (1992). Sensitivity of time domain reflectometry measurements to lateral variations in soil water content. *Water Resources Research, 28*, 2345–2352.

Knight, J. H., White, I., & Zegelin, S. J. (1994). Sampling volume of TDR probes used for water content monitoring. In K. M. O'Connor et al. (Eds.), *Time domain reflectometry in environmental, infrastructure and mining applications* (Spec. publ. SP, 19–94, pp. 93–104). Evanston: U.S. Bureau of Mines, Minneapolis and Northwestern University.

Koenigs, F. F. R. (1978). Comments on the paper by P.F. North (1976): "Towards an absolute measurement of soil structural stability using ultrasound". *Journal of Soil Science, 29*, 117–119.

Le Bissonnais, Y. (1996). Aggregate stability and assessment of soil crustability and erodibility: I. Theory and methodology. *European Journal of Soil Science, 47*, 425–437.

Lichter, J. M., & Costello, L. R. (1994). An evaluation of volume excavation and core sampling techniques for measuring soil bulk density. *Journal of Arboriculture, 20*, 160–164.

McCalla, T. M. (1944). Water-drop method of determining stability of soil structure. *Soil Science, 58*, 209–214.

McIntyre, D. S., & Loveday, J. (1974). Bulk density. In J. Loveday (Ed.), *Methods for analysis of irrigated soils* (Comm. Bureau of Soils, Tech. comm. no. 54). Farnham Royal, Buckinghamshire: Commonwealth Agricultural Bureaux.

Müller, K., & Deurer, M. (2011). Review of the remediation strategies for soil water repellency. *Agriculture, Ecosystems & Environment, 144*, 208–221.

Muller, R. N., & Hamilton, M. E. (1992). A simple, effective method for determining the bulk density of stony soils. *Communications in Soil Science and Plant Analysis, 23*, 313–319.

Nadler, A., Dasberg, S., & Lapid, I. (1991). Time domain reflectometry measurements of water content and electrical conductivity of layered soil columns. *Soil Science Society of America Journal, 55*, 938–943.

Nimmo, J. R., & Perkins, K. S. (2002). 2.6 Aggregate stability and size distribution. In J. H. Dane & G. C. Topp (Co-Eds.), *Methods of soil analysis, Part 4, Physical methods* (Number 5 in the Soil Science Society of America book series, pp. 317–328). Madison: Soil Science Society of America, Inc.

Noborio, K. (2001). Measurement of soil water content and electrical conductivity by time domain reflectometry: A review. *Computers and Electronics in Agriculture, 31*, 213–237.

North, P. F. (1976). Towards an absolute measurement of soil structural stability using ultrasound. *Journal of Soil Science, 27*, 451–459.

Oades, J. M., & Waters, A. G. (1991). Aggregate hierarchy in soils. *Australian Journal of Soil Research, 29*, 815–828.

Page-Dumroese, D. S., Jurgensen, M. F., Brown, R. E., & Mroz, G. D. (1999). Comparison of methods for determining bulk densities of rocky forest soils. *Soil Science Society of America Journal, 63*, 379–383.

Raper, R. L., & Erbach, D. C. (1985). *Accurate bulk density measurement using a core sampler*. Am. Soc. Agr. Eng. Paper No. 85–1542.

Regalado, C. M., Munoz-Carpena, R., Socorro, A. R., & Hernandez Moreno, J. M. (2003). Time domain reflectometry models as a tool to understand the dielectric response of volcanic soils. *Geoderma, 117*, 313–330.

Robinson, D. A., Jones, S. B., Wraith, J. M., Or, D., & Friedman, S. P. (2003). A review of advances in dielectric and electrical conductivity measurements in soils using time domain reflectometry. *Vadose Zone Journal, 2*, 444–475.

Roth, K., Schulin, R., Fluhler, H., & Attinger, W. (1990). Calibration of time domain reflectometry for water content measurement using a composite dielectric approach. *Water Resources Research, 26*(10), 2267–2273.

Santini, A. (1997a). Contenuto d'acqua del suolo. In M. Pagliai (Ed.), *Metodi di Analisi Fisica del Suolo, Parte VIII, Ministero per le Politiche Agricole, Società Italiana di Scienza del Suolo* (pp. 1–35). Milano: Franco Angeli (in Italian).

Santini, A. (1997b). Potenziale dell'acqua nel suolo. In M. Pagliai (Ed.), *Metodi di Analisi Fisica del Suolo, Parte VIII, Ministero per le Politiche Agricole, Società Italiana di Scienza del Suolo* (pp. 37–53). Milano: Franco Angeli (in Italian).

Shabtai, I. A., Shenker, M., Edeto, W. L., Warburg, A., & Ben-Hur, M. (2014). Effects of land use on structure and hydraulic properties of Vertisols containing a sodic horizon in northern Ethiopia. *Soil & Tillage Research, 136*, 19–27.

Topp, G. C., & Davis, J. L. (1985). Measurements of soil water content using Time Domain Reflectometry (TDR): A field evaluation. *Soil Science Society of America Journal, 49*, 19–24.

Topp, G. C., & Ferré, P. A. (2002a). The soil solution phase – 3.1 Water content – 3.1.1 General information. In J. H. Dane & G. C. Topp (Co-Eds.), *Methods of soil analysis, Part 4, Physical methods* (Number 5 in the Soil Science Society of America book series, pp. 417–419). Madison: Soil Science Society of America, Inc.

Topp, G. C., & Ferré, P. A. (2002b). 3.1.3 Methods for measurement of soil water content – 3.1.3.1 Thermogravimetric using convective oven-drying. In J. H. Dane & G. C. Topp (Co-Eds.), *Methods of soil analysis, Part 4, Physical methods* (Number 5 in the Soil Science Society of America book series, pp. 422–424). Madison: Soil Science Society of America, Inc.

Topp, G. C., Davis, J. L., & Annan, A. P. (1980). Electromagnetic determination of soil water content: Measurements in coaxial transmission lines. *Water Resources Research, 16*(3), 574–582.

Topp, G. C., Davis, J. L., & Annan, A. P. (1982). Electromagnetic determination of soil water content using TDR: II. Evaluation of installation and configuration of parallel transmission lines. *Soil Science Society of America Journal, 46*, 678–684.

Wilson, G. V., & Luxmoore, R. J. (1988). Infiltration, macroporosity, and mesoporosity distributions on two forested watersheds. *Soil Science Society of America Journal, 52*, 329–335.

Young, M. H., & Sisson, J. B. (2002). Tensiometry. In J. H. Dane & G. C. Topp (Co-Eds.), *Methods of soil analysis, Part 4, Physical methods* (Number 5 in the Soil Science Society of America book series, pp. 575–608). Madison: Soil Science Society of America, Inc.

Index

R. Angulo-Jaramillo et al., *Infiltration Measurements for Soil Hydraulic Characterization*,
DOI 10.1007/978-3-319-31788-5

The manufacturer's authorised representative in the EU is Springer Nature Customer Service Centre GmbH, Europaplatz 3, 69115 Heidelberg, Germany. If you have any concerns regarding our products, please contact ProductSafety@springernature.com

Printed and bound by CPI Group (UK) Ltd, Croydon, CR0 4YY
01/10/2026
02225860-0013